The Ethics of Everyday Medicine

The Ethics of Everyday Medicine

Explorations of Justice

Erwin B. Montgomery Jr.

ACADEMIC PRESS

An imprint of Elsevier

Academic Press
125 London Wall, London EC2Y 5AS, United Kingdom
525 B Street, Suite 1650, San Diego, CA 92101, United States
50 Hampshire Street, 5th Floor, Cambridge, MA 02139, United States
The Boulevard, Langford Lane, Kidlington, Oxford OX5 1GB, United Kingdom

Notices
Knowledge and best practice in this field are constantly changing. As new research and experience
broaden our understanding, changes in research methods, professional practices, or medical treatment
may become necessary.

Practitioners and researchers must always rely on their own experience and knowledge in evaluating
and using any information, methods, compounds, or experiments described herein. In using such
information or methods they should be mindful of their own safety and the safety of others, including
parties for whom they have a professional responsibility.

To the fullest extent of the law, neither the Publisher nor the authors, contributors, or editors, assume
any liability for any injury and/or damage to persons or property as a matter of products liability,
negligence or otherwise, or from any use or operation of any methods, products, instructions, or ideas
contained in the material herein.

Library of Congress Cataloging-in-Publication Data
A catalog record for this book is available from the Library of Congress

British Library Cataloguing-in-Publication Data
A catalogue record for this book is available from the British Library

ISBN: 978-0-12-822829-6

For information on all Academic Press publications
visit our website at https://www.elsevier.com/books-and-journals

Acquisitions Editor: Wolff, Andre Gerhard
Editorial Project Manager: Fernandez, Billie Jean
Production Project Manager: Raviraj, Selvaraj
Cover designer: Bilbow, Christian J.

Typeset by SPi Global, India

Working together
to grow libraries in
developing countries

www.elsevier.com • www.bookaid.org

To Lyn Turkstra whose love is warmth and light,

to Arthur "Buddy" White because of whom I can call myself a philosopher,

and to Herman Turkstra, barrister and solicitor extraordinaire, who answered my incessant questions.

Table of contents

Preface   xv

Introduction   1
  To be argued…   1
  Everyday Medicine Is Everywhere and Every Time   2
  Justice   3
  Ethics, Morality, Principles, Theories, Laws, Case Precedents,
     and Virtues—Definitions Proposed Here   6
  Example of Everyday Ethics at the Most Fundamental Level—Diagnosis   7
  Cases   9
  Ethical Induction   12
  Contributions of Philosophy   14
  The Practice of Everyday Medicine Is Far from Simple, but Is
     Demanding   17
  Procedural Knowledge and Ethics   18
  Procedural Statistical Significance and Clinical Meaningfulness   20
  What Constitutes the Ethics of Everyday Medicine?   21
  Beneath the Radar   24
  Corporate Personhood as a Stakeholder as well as a Shareholder   26
  Dissociation between Healthcare Expected and Healthcare Delivered   27
  Justice as Institutional   28
  Best of All Possible Worlds   29
  Asymmetry in Power and Consequence   30
  The Structure of This Text   32

Part 1

Case 1—A speech language pathologist was asked to evaluate a hospitalized patient with Parkinson's disease who was complaining of choking and coughing each time she ate or drank. Following the evaluation, the speech language pathologist recommended to the attending physician that the patient undergoes a video fluoroscopic examination of swallowing (VFES), otherwise known as a modified barium swallow. The attending physician refused. When asked, the physician said it was not necessary. The attending physician turned, walked away, and refused further discussion.

Case 2—A physician requested a video fluoroscopic evaluation of swallowing (VFES) for his patient with Parkinson's disease who was experiencing difficulty with swallowing. A VFES involves videotaping an X-ray of the mouth and throat as the patient swallows radio-opaque solids and liquids of different consistencies. In this manner, the pathway of the food and liquids can be followed as the patient swallows. The physician was instructed to call a central outpatient coordination service to arrange for the outpatient study. The service informed the physician that they were unable to provide a VFES but would provide a "bedside" clinical examination. A clinical or "bedside" examination involves a speech-language pathologist observing and listening to the patient as the patient swallows liquids, looking for evidence of choking or gurgling or residue in the mouth following an attempt to swallow.

Case 3—A 63-year-old male was seen in consultation requested by a family physician at which time the diagnosis of early and mild Parkinson's disease was confirmed. The consultant recommended the patient start selegiline in order to slow the progression of the disease. The consultant provided detailed information regarding the evidence to support slowing of disease progression, dosing information, a list of potential adverse effects, and medications that are contraindicated when combined with selegiline. The patient had no conditions or treatments that would contraindicate the use of selegiline and there was no reasonable expectation that the situation would change in the near future. There were no financial reasons to prevent the patient from receiving the medication. The consultant offered to help at any time and volunteered his cell phone number. The family physician refused to prescribe selegiline stating that was up to the consultant.

Case 4—A 72-year-old man followed by a movement disorders neurologist specializing in Parkinson's disease was admitted to a hospital because of fever and increased confusion. He was found to have a urinary tract infection and was treated with antibiotics. A general neurologist saw the patient in consultation while the patient was in the hospital. The consulting neurologist, without discussions with the treating movement disorders neurologist, markedly reduced the patient's dose of carbidopa/levodopa used to treat the patient's Parkinson's disease. The patient was discharged to a nursing facility. The movement disorders neurologist later was called by the patient's wife because the patient was unable to move, essentially confined to bed, and unable to care or feed himself. He was unable to converse with the facility staff. The movement disorders neurologist visited the patient in the nursing facility and restored the previous dose of carbidopa/levodopa. Within a week, the patient returned to his previous abilities to care for himself, get out of bed, and interact with the nursing staff, family, and friends.

Case 5—A neurologist does not prescribe dopamine agonists to patients with Parkinson's disease but rather prescribes levodopa, although there is concern about greater long-term risks with early and sole use of levodopa. The neurologist does not think that the risk of long-term complications associated with levodopa outweighs the additional cost for the dopamine agonists. The neurologist does not mention the possibility of treatment with dopamine agonists to the patients.

Case 6—A 78-year-old female with a long history of multiple medical problems complained of excessive shortness of breath (dyspnea) after modest exertion and fatigue.

The patient e-mailed her physician regarding the recommendation of an angiogram (an X-ray procedure where a "dye" is injected into the heart and shows on the X-ray). The patient was concerned that the angiogram might be abnormal but may not be the cause of the shortness of breath and fatigue. The cardiologist stated that he was very busy and suggested e-mailing specific questions. The cardiologist went on to write that his responsibility was to review the cardiovascular issues, which he did, implying that he was not responsible for other noncardiac medical issues, as this was the standard of the discipline. He went on to say that it was the patient's decision regarding proceeding with the angiogram and just to let him know.

Case 7—A nurse asked a physician to stop sending copies of the physician's evaluation and recommendations to patients because the patients would call and ask questions.

Case 8—The patient is a 67-year-old female with a 15-year history of idiopathic Parkinson's disease. Her disease is currently manifested by motor fluctuations where she fluctuates from having severe slowness of movement and tremor to periods where she has severe involuntary uncontrollable movements. She has tried a wide range of treatment regimens, including all reasonable commercially available medications. The patient heard about the possibility of deep brain stimulation and inquired of her neurologist whether she would be a good candidate for it. The patient's physician dismissed the patient, saying "you are not bad enough."

Case 9—A 47-year-old female was seen with complaints of neck pain and a twisting of her head on her neck. The symptoms began 4 years previously. Each time she sought medical care, she was told that it was psychological. Other than the pain and twisting in her neck, she did not have other complaints and her past medical history was unremarkable. To be sure, the pain and twisting were stressful and an embarrassment and occasionally she would admit to being depressed. However, any episodes of feeling depressed lasted only a day or two. Her health otherwise had been excellent. Her only medication was a selective serotonin reuptake inhibitor antidepressant, which she took only after persistent entreaties by her past physician. After seeing her latest neurologist, she was diagnosed as having idiopathic cervical dystonia and her symptoms were resolved with intramuscular injections of botulinum toxin.

Case 10—A 32-year-old female was admitted following a motor vehicle accident. She asked for a consultation by a neurologist as she was complaining of weakness in her right upper extremity since the accident. There were no sensory complaints except for headache. The examination was remarkable for "give way" weakness in the right upper extremity, meaning the patient would generate an inconsistent muscle force, such as involved in pushing away the examiner's hand when asked to make a sustained maximum force. When the patient was asked to make a maximal tight handgrip with the wrist flexed (bent down), she failed to dorsiflex her wrist (bend upward), suggesting that the patient was not actually making a maximal effort. The conclusion was that the weakness likely was a conversion symptom or psychogenic and not evidence of physical injury to the nervous system. The neurologist suggested that keeping the patient in the hospital overnight as is routinely done in patients who may have sustained a serious trauma. In the presence of medical students and residents, the attending physician responded that he was going to "kick" the patient out. No arrangements for follow-up care were made.

Case 11—A 72-year-old male had advanced Parkinson's disease, but was otherwise in good health. The patient responded to medications, such as levodopa. However, the response was complicated by the wearing-off effects where the symptoms, slowness and tremor, appeared before the next dose of medications was due and uncontrollable involuntary movements (dyskinesia) would appear approximately 1–2 h after taking a dose of levodopa. Both the wearing-off effects and the dyskinesias reduced the patient's quality of life greatly. The patient's neurologist said the patient was ineligible for deep brain stimulation because he was over 70 years old. The neurologist held that age 70 was the cutoff from publications he had read.

Case 12—A patient presented to the emergency room with suicidal thoughts. Assessing that the patient was at risk of suicide, the physician admitted the patient to the psychiatry service. The patient's admission was voluntary. The following day, the physician was informed by the patient's insurer that the insurer would not cover the cost of hospitalization. The physician asked on what basis was the coverage denied? The insurer stated that the information and methods were proprietary and would not share them. However, the insurer suggested that the decision was based on an artificial intelligence (AI) algorithm. The insurer specifically stated that the insurer was not directing the physician in the practice of medicine but only that the issue for the insurer centered on the insurer's contractual obligations. The patient, fearful of incurring significant medical financial costs, asked to be discharged. As the physician continued to be concerned about the risk for suicide, the physician and the hospital required the patient to sign a waiver indicating that the physician advised continued hospitalization. Such waivers are termed "discharge against medical advice." The patient left the hospital after signing the waiver.

Case 13—A 10-year-old boy was brought to see a neurologist for consideration of deep brain stimulation (DBS) for treatment of the child's dystonia. After completing the assessment, the neurologist began to explain the nature of DBS and the potential benefits and risks to the child and the parents. The parents interrupted the neurologist. One parent then took the child out of the examination room, while the remaining parent stated that he did not want the child to hear about the discomfort associated with the surgeries and the possible risks. The neurologist explained that she was very experienced in explaining medical and surgical procedures to children and that the hospital had a comprehensive program to prepare children for such surgeries. The program had been very successful in coaching children through such procedures. Further, for optimal outcomes, the child should be awake during some parts of the procedure in order to assist with proper placement of the DBS electrodes. The neurologist also stated to the parent that it is considered standard of care to obtain the child's assessment prior to surgery in addition to the parent's consent. Nevertheless, the parents insisted that these issues not be discussed with the child and that the child remains under anesthesia during the entire procedure.

Case 14—The patient is a 68-year-old-male member of the Mohawk Nation of Canada, who was diagnosed with Parkinson's disease by a nonindigenous physician. When asked by the physician how he was doing regarding his symptoms and disabilities from Parkinson's disease, the patient would respond after a minute with it was "OK." The physician pressed the patient to give more details. The patient responded after a delay that he was "OK." When the physician asked the patient to rate the severity of his Parkinson's disease, with 10 being almost normal and 0 being very disabled, the patient said "5." It was said in such a manner that the physician, from his perspective, suspected the accuracy or veracity of the answer. Exasperated, the physician recommended having the patient spend an entire day in the clinic where his symptoms and disabilities could be rated on an hourly basis. The physician's hope was that a series of objective evaluations by the clinic nurse would provide sufficient information about the patient's response to medications for the physician to recommend an improved medication regimen. When the physician asked the patient if he could spend the day in the clinic, the patient said nothing in reply. The physician asked the patient several more times, and the patient said that it sounded like a good idea and he would consider it. The physician then turned to the patient's wife and asked her whether she thought her husband was depressed or had a problem with apathy. The wife denied either was the case.

Other cases

Part 2

Chapter 1 – Justice   **297**
    To be argued...   297
    Justice Is Paramount   298
    Justice Is Primary—Biologically   299
      Prior considerations   *300*
      Psychology versus neurophysiology   *303*
      Neurophysiological representations of just and unjust behaviors   *305*
      Inborn mechanisms generate behaviors to which a sense of
        justice can be attributed   *307*
      Neurophysiology of common morality   *311*
      Challenges exist in the variety and apparent inconsistency
        of behaviors to which justice is attributed   *313*
    Injustice and Biology   314
    Justice Is Primary—Ethically   317
    Ethical Principles, Case Precedents, and Virtues as Induction from
      Microlevel Justice; Moral Theories from Macrolevel Justice   319
    Operationalization of Justice   322

The Changing Nature of Disease Burden and the Effects of
   "Social Hazard" on Morality and Justice   322
Neural Dynamics and Implications for Ethical Principles, Moral
   Theories, Laws, Case Precedents, and Virtues   324
Not Starting from Random: Evolution—Biological and Social   329
Ethics as an Exercise in Chaos and Complexity   329
Initial Conditions and Principlism and Anti-Principlism   336
Neurophysiology and Moral Culpability   337
Conflation of Neuropharmacology and Genetics with
   Neurophysiology—A Categorical Error   339

Chapter 2 – Ethical Systems   **347**
To be argued…   347
Variety as Diversity or Variation   348
Transcendent Principles, Theories, Case Precedents, and Virtues
   and the *A Priori* Problem   351
Principlism and Anti-Principlism—Two Sides of the Same Coin   356
Reproducibility as Justice   359
The Nature of Abstraction   364
Epistemic Risk   366
Ethics and Consciousness   367
Chaos and Complexity   370
Examples of Anti-Principlist Ethical Theories   371
   Phenomenological Ethics   *373*
   Narrative Ethics   *378*
   Feminist Ethics   *381*
   Care Ethics   *383*
   Virtue Ethics   *383*

**Chapter 3 – Reflective Equilibrium and Ethical Induction from
the Perspective of Logic and Statistics**   **389**
To be argued…   389
Moral and Ethical Objective Truth from Ethical Induction   390
The Necessity of Ethical Induction   393
Ontological Issues with Induction   394
Epistemological Issues with Induction   395
Meno's Paradox and the *A Priori* Problem of Induction   397
Discovery versus Judgment   398
Logic and Philosophy   399
   Evolution of inherent probability and statistical thinking and logic   *400*
   The centrality of certainty   *401*
   The case for philosophy   *402*
   Deductive syllogistic logic   *403*
The Fallacy of Four Terms   409
Implications of Qualitative Research for Ethics   410

Grounded theory   *413*
Mixed-methods research   *416*
Ergodicity   417
Continuous versus Dichotomous Character of Principles,
    Theories, Laws, Case Precedents, and Virtues   418
Coherence and Correlation   419
The Experts, Judges, and Officials as Those Who Make the Consensus   422
Chaos, Complexity, Nonlinear Dynamics, and Far-from-Equilibrium
    Steady States   426
Artificial Intelligence and Ethical Induction   432
Application—Judging   433
Reflection   435

Chapter 4 – The Patient as Parenthetical   **439**
To be argued…   439
Current Status   439
Patient-Centric Ethics   440
Qualifying Patient-Centeredness   444
Physician-Centric Ethics   445
Rise of Business- and Government-Centric Ethics   450
Transactional Medicine: Physician-Centric versus Patient-Centric Ethics   455
Scientist/Academic-Centric Ethics (Scientific Paternalism)   456
Consumer-Centric Ethics   459
Romanticism and Misplaced Virtue   459
Asymmetries in the Patient–Physician/Healthcare Relationship   461
Feminist Bioethics   463
Toward a More Patient-Centric Ethics of Everyday Medicine   464
    External regulation to establish patient centricity   *465*
    Internal regulation to establish patient centricity   *468*
Ethics and Ethos   469
The Best of All Possible Worlds   471
    Societal level   *471*
    Individual level   *473*

Chapter 5 – A Workable Framework   **479**
To be argued...   479
For the Patient, Justice Cannot Wait   479
The Centrality of Hypothesis   480
Intellectual Virtues   482
    Open-mindedness   *484*
    Carefulness   *488*
    Imagination   *492*
    Discipline   *493*
    Perseverance   *493*
    Embrace the iconoclast   *494*

Procedure to Actualize the Intellectual Virtues   495
    Check your gut at the door   495
    Who's got skin in the game?   495
    Who is doing what to whom?   495
    What's at stake?   496
    What is the narrative?   496
    Pick a tool to decide   497
Calling the Question   497

Postscript: COVID-19 pandemic and Black Lives Matter   **501**
Index   **507**

Preface

A reasonable question is asked: By virtue of what credentials and qualifications do I ask the reader to engage this book? Over 40 years in academic medicine and in the practice of caring for patients allows me a certain perspective likely not common to a reader younger. Perhaps more importantly, my career spanned a time of dramatic and tumultuous changes in patient care. I dare say my professors retiring at the time of my graduation from the medical school in 1976 would scarcely recognize the practice of medicine today. More importantly, most physicians and healthcare professionals today would scarcely recognize the practice of medicine at the time of my professors. There have been incredible changes in drugs, biologics, medical devices, surgical procedures, and laboratory and radiologic diagnostics, but my professors of medicine would have assimilated those changes easily. What they likely would not understand and not accept is the change in their roles as physicians and healthcare professionals.

As an intern in internal medicine, health maintenance organizations (HMOs) consolidated the health needs of a patient within a group practice with essentially fixed, prepaid care. The prepayments supported the prevention of disease and maintenance of health prospectively, a rarity otherwise. The alternative system was one of reacting to crises, essentially putting out fires instead of preventing them. HMOs of that time would not be recognized today. Rather, HMOs have become largely middle-level managers who, by controlling physicians' and healthcare professionals' access to patients, in effect control the physicians and healthcare professionals. One response has been for physicians and healthcare professionals to create their own middle-level managers, but it is not clear that these organizations restore the previous relationships between the patient and the physician or healthcare professional. Just as troubling, these changes were promoted by the US government. The jarring change in the relationship of physicians to patients through the medium of HMOs (often referred to as managed care or preferred provider networks) was echoed in a plea I heard from medical students I taught as an assistant professor. They asked their professors to stop complaining about HMOs and teach the students how to live with them. Given the vast increase in the complexity of everyday medicine by intrusion of commercial and governmental interests, the ecology in which ethical decisions are made has increased greatly in complexity, as evidenced by the cases presented in Part 1. Acknowledgment of this complex ecology is what distinguishes this book from the many other works.

As a medical student, intern, and resident, I worked in charitable hospitals, typically run by local governments, and on charity wards of private hospitals. At least from the perspective of a young and perhaps naïve physician, there was a remarkable sense of commitment to the patient. In part, perhaps due to the fact that the patient on these wards was our patient in contrast to working on private wards where we were often treated more like "scut monkeys," chasing down laboratory and radiology reports and implementing the directions of the private attending physicians. To be sure, as students, interns, and residents on the ward service, we were supervised closely by the more senior residents and attending physicians, typically academic physicians. Nonetheless, there was a great source of pride in being the one responsible for the patient. It also bred a sense of comradery among the fellow interns and residents. That same sense of comradery that existed then is described beautifully in "God's Hotel: A Doctor, a Hospital, and a Pilgrimage to the Heart of Medicine" by Victoria Sweet (2012). When commoditization of Medicaid for the care of medically indigent attracted commercial interests and charity hospitals and wards have largely disappeared. No one wishes for a return of charity hospitals and wards, but something special was lost.

My career spanned a remarkable refutation of perhaps what physicians and healthcare professionals pride themselves as being uniquely able to offer, their *raison d'être*—clinical judgment. Initially, evidence-based medicine recognized multiple levels of evidence to support clinical judgments, including case reports, case series, consensus among experts, and expert opinions, as well as randomized controlled trials. Now, evidence-based medicine essentially has become exclusively synonymous with randomized controlled trials, the implementation of which in everyday medicine obviates independent clinical judgment. No reasonable person blinded to their own self-interest would oppose evidence gleaned from randomized controlled trials. Yet, there is a sense of solipsism in current evidence-based medicine that presumes statistical significance is synonymous with clinical meaningfulness such that the concern for meaningfulness seems to have evaporated (Montgomery, 2019a, 2019b; Montgomery and Turkstra, 2003).

The solipsism that has accompanied this change in evidence-based medicine has had a corrosive effect. For example, a group of learned physicians and scientists in 2019 wrote "Unilateral Ventralis intermedius thalamic DBS [Deep Brain Stimulation] … were considered possibly useful" (Ferreira et al., 2019) after decades of analysis and clinical experience, although perhaps not of the type of randomized controlled trials expected? Which one of those experts would hesitate, with all other things being equal, recommending thalamic deep brain stimulation in patients proven refractory to pharmacological therapy and able to tolerate surgery? Likely none, especially if it were for someone the physician had feelings for. Unsophisticated physicians and unscrupulous insurers could use such comments by professional task forces to deny the patient access to deep brain stimulation. Who would be ethically responsible for this? In many ways, the noble attempt to render clinical practice more robust by the original conception of evidence-based medicine has often become a tool of sophistry; at least I believe a good case for that indictment can be made (Montgomery, 2019a).

Late in my career I briefly practiced medicine in Ontario, Canada, allowing me to view a different form of healthcare delivery. Contrasts with the United States' system were striking both positively and negatively. Perhaps most striking, I never encountered a Canadian patient who worried about his personal financial bankruptcy due to medical expenses or had to choose between healthcare and paying for food or rent. Further, the Ontario system is very Egalitarian, everyone is treated the same; frequently treated well, other times not. The Libertarian system in the United States secures the best care for the financially secure; for the financially insecure, perhaps not. These are observations, not value judgments.

Each system has their unique benefits and liabilities. In some ways, each is a different response to the attempt to be just in the face of moderate scarcity of resources (Rawls, 1999). It is not clear, or at least settled, that neither the United States nor Canada (or any other country) has enough resources, or a the least the will, to guarantee the best healthcare to all persons. The question then is how are the resources to be distributed? Distributive justice has a profound effect on the practice of medicine and the ethics associated with that care.

If the cases presented in Part 1 are representative, only the most empathically obtuse would not be moved to concern for patients as well as for physicians and healthcare professionals. I came to my training with a "gut" ethic that medicine was a noble calling, admittedly naïve. At the start of my medical experiences, physicians were limited in the care they could provide by a lack of knowledge. There was the belief that it was the responsibility of the physician to put the patient first and foremost even if the effect was limited. By the time I finished training, physicians knew and could offer more but were limited by what society, through governmental and private insurance, was willing to do. For a while, the old loyalty to the patient held, but now seems to be only a distant echo. How can a physician and healthcare professional do their upmost when society won't?

In an interesting study, not likely permissible by today's standards, James Brady (1958) subjected pairs of nonhuman primates to repeated electrical shocks. One of the pairs could sometimes block the electrical shocks given to both monkeys. The rate of gastric ulcers was higher in the nonhuman primate that had partial control of the pain (the executive monkey) compared to the passive nonhuman primate. The analogy to medicine is that in the pre-1970s, physicians were largely autonomous, perhaps with greater authority than accountability. One might suggest that since the 1970s, physicians have retained the same responsibilities and self-reproach for failing (the potential for electrical shocks in the aforementioned nonhuman primate studies) with less authority to meet the responsibilities (the executive monkey). It would not be surprising that physicians and healthcare professionals would seek to be more "passive." Clearly, this has implications for patient care and is an ethical issue.

A natural question would be why did all these things described happen? Also, it is natural to avoid asking why, as asking such a question suggests some obligation to change things. Doing so could put the questioner in the position of the "executive monkey" described earlier. Each person must make his or her own choice.

I was extremely fortunate to receive education and training where asking "why?" was central and pervasive, that being philosophy. During my fellowship and assistant professorship, I took a great many graduate school courses in philosophy as a non-matriculating student at Washington University in St. Louis and later at McMaster University in Hamilton, Ontario, Canada. I became a visiting scholar in the Department of Philosophy at Thiel College where I taught courses in the ethics of everyday medicine in collaboration with Professor Arthur "Buddy" White. I also had the great privilege of working with the ethicist, Dr. Joseph J. Fins, on a planning grant for clinical trials in deep brain stimulation in minimally conscious states.

My interest in philosophy is epistemology (theory of knowledge, what it is and how it is gained) and logic (fundamentally about methodologies to achieve certainty in knowledge). I used my philosophical knowledge, at least implicitly, as well as my medical training in my research in physiology in the basal ganglia-thalamic-cortical system of the brain and how this physiology is altered in Parkinson's disease. Later, my knowledge and training in epistemology and logic were applied directly and explicitly to wider considerations of medical reasoning and education and in biomedical research (Montgomery, 2019a, 2019b). Ethics frequently was a topic within these works (Montgomery, 2015, 2019a; Montgomery and Turkstra, 2003). I now turn to the ethics of everyday medicine.

My education and training in philosophy have led to specific modes of thought and argumentation perhaps uncommon in other disciplines. Further, I could take advantage of thousands of years of philosophical discussions uncovering recurring themes, paradoxes, conundrums, arguments, and counterarguments. As in any field of knowledge, shorthand descriptors become placeholders for much more extensive discussions. To those not experienced in formal philosophy, these shorthand descriptors can be seen as jargon, and reasonably so. For example, a philosopher might describe a position or claim as "trivial." But this is not a value judgment that the position or claim is unimportant. Rather, the philosopher typically means that the position or claim, while true, is self-evident and unenlightening, such as "all unmarried men are bachelors," which is true by definition but itself does not lead to new knowledge other than creating labels. I may make references to shorthand descriptors, although always with an effort to explain them when initially encountered. I ask some forbearance for the occasional unintended slip.

Over my many years of clinical medicine, I have learned that the practice of medicine, fundamentally, is an exercise in ethics. How medical science is brought to bear on the lives of humans and how the benefits of medical science are administered, are fundamentally questions of ethics—ethics as defined as the manner by which humans treat each other. This has been true even before the ancient Greeks, such as Hippocrates, and it is just as true today. The fundamental ethical nature of medicine perhaps is brought to attention vividly when in stark relief with impending death, such as assisted suicide, withdrawal of life support, or withholding life-saving medicine. But hopefully it will be demonstrated that ethical questions also are the substance of everyday medicine as will be discussed in the Introduction.

Ethics is fundamental to nearly every medical decision, not only as a matter of necessarily having to interact with others, but also based on epistemology and logic. Consider diagnosis utilizing a laboratory test. Use of the test in any individual patient will be associated with a probability of being a true positive, true negative, false positive, or false negative. As a true-positive patient will receive benefits that accrue with subsequent treatment and a true-negative patient will be spared the risk of adverse effects, then the ethical issue is balancing the risks and benefits to the patient who is a true positive and the costs of doing the test. However, the patient may test as a false positive and with subsequent treatment is exposed to adverse effects, considered in its widest connotation, without benefit, thus violating the obligation to nonmaleficence. Similarly, the patient who tests as a false negative will be denied the potential benefits of treatment, thus violating the obligation to beneficence. Further, society likely will absorb the expense with no return on the investment in the case of a false positive while absorbing the expense associated with an untreated disease in the case of a false negative.

How does one weigh the costs, in the fullest connotation of the word, associated with false positives and false negatives? It is not as though 2 units of benefit equal 2 units of maleficence such that a ratio of benefit to risk greater than 1 would argue for the test. To be sure, there have been attempts to quantize benefit and risk, but the large majority do so by translation to financial costs per degree of benefit obtained. The ethical question though is not avoided, as the ethical question becomes which option is too expensive? Also, there are psychological, sociological, legal, political, financial, and moral costs. For example, the risk of dementia due to vitamin B_{12} deficiency is very rare, but nearly every physician tests for the deficiency in patients presenting with dementia. The false-positive rate would be exceedingly large, yet the test is done because the cost of "condemning" the patient to dementia that could be improved with simple vitamin B_{12} replacement is considered ethically unacceptable. This conundrum is inherent in any and every medical decision even if the stakes are minimal, and even if such decisions are made out of habit, protocol, or deliberation.

The fundamental ethical nature of every medical decision is rarely appreciated and therefore engenders little attention or discussion. Ethics too often is limited to those issues that make the headlines. Often, the headlines, particularly in the context of imminent death, are made stark by the introduction of arguments such as religion. Not to denigrate or challenge the place or authenticity of religion in an individual's life by any means, pluralistic modern liberal democracy would quarantine religion from the public forum on ethics. History, such as the Thirty Years' War and the English Revolution, demonstrates that the politicization of religion often is fraught with strife. Theologians may be exemplars of virtue, but what principles can the theologian offer in ethics other than received knowledge (from the Word of God), unless the theologian eschews religiosity in his or her creation of ethical systems? The critical question then becomes by what methods does one construct ethical systems if one does not want to default to religion or dictates of an absolute sovereign? These issues are taken up in Part 2 of this book.

Ethics enters more explicitly when the notion of the quality of life enters medical decision-making. Yet, notions of quality of life calculations in everyday medicine are fundamentally ethical decisions (Leplège and Hunt, 1997). Notions of quality of life gained academic attention as a means to monitor clinical trials in cancer where prior emphasis on survival seemed less relevant (Leplège and Hunt, 1997). Subsequently, greater values in return for expenses are demanded increasingly by insurers to justify healthcare expenditures (Chee et al., 2016), although only relatively small percentages consider patient satisfaction (Damberg et al., 2014). Still, the ethics of the quality of life is buffeted by the ever-changing notions of obligation, duty, and charity of one human to another. Many insurers do not cover the cost of medications for erectile dysfunction in a man robbed of potency by a disease such as diabetes mellitus (Polinski and Kesselheim, 2011) to the degree they would support breast reconstruction surgery for a woman following a mastectomy [Centers for Medicare and Medicaid Services.

Women's Health Care Rights Act (WHCRA). https://www.cms.gov/CCIIO/Programs-and-Initiatives/Other-Insurance-Protections/whcra_factsheet.html; Stavrou et al., 2009]. Again, this is not a value judgment but an observation that, nonetheless, invites consideration as to how this discrimination can be justified—made just.

As will be explored in the Introduction, many physicians, healthcare professionals, and administrators of healthcare systems would greet concerns about the ethics of everyday medicine with a bit of a shrug. It just seems that the practice of everyday medicine in advanced countries such as the United States and Canada goes on generally unruffled by injustices committed in the clinic and at the bedside. Large percentages of patients are totally satisfied with their healthcare perhaps reacting to injustices in healthcare. But the animosity seems directed at insurers—private or governmental—physicians and healthcare professionals being relatively immune, justly or unjustly.

Following an Introduction, Part 1 provides a series of cases from the "everyday experience" of this author selected because of ethical concerns. At first, the cases may not seem like a big deal and just the cost of doing business in the extensive and complex business that constitutes healthcare. The ethical concerns will be explored in the analyses that follow each case. That these cases would not seem unusual is precisely the point. They are there in front of the noses of nearly every physician and healthcare professional. But, as George Orwell wrote, "To see what is in front of one's nose needs a constant struggle" (http://orwell.ru/library/articles/nose/english/e_nose). It is only when the questionable unethical actions are seen that they can be corrected and prevented.

The thesis that ethics are central to the everyday practice of medicine, yet the depth and breadth of the ethical implications are unrecognized, means that no precise and reproducible set of guidelines can be offered, yet. Rather, at this point only questions can be presented. In the past, most considerations of medical or clinical ethics have been narrow in their perspective. The cases presented in Part 1 expand greatly the ecology in which ethical issues in everyday medicine are embedded. The purpose of this text is to make the risk of injustice apparent as the first step toward increasing justice. Means to do so are the theme of Part 2.

This book advances a thesis that one reason there is less awareness of ethics is that the practice of medicine is largely procedural; just doing, as opposed to deliberate calculations that can be declared (verbalized). There is a logic embedded in the actual procedures and one generally does not have to "talk oneself through it" as would be evidence of declarative knowledge. In part, this is because much of medical practice largely is one of pattern matching and following "illness scripts," unfortunately (Montgomery, 2019a). Indeed, some ethicists point to the practice of ethics as pattern recognition. As the ethics are embedded in the procedures of medical practice, it is not surprising that they would escape attention that would be possible by declarative, deliberative, and conscious analyses.

The cases presented are very specific and particular. However, it would not take much effort for any physician, healthcare professional, or interested or concerned person to recognize the same issues in any number, indeed a large number, of examples of everyday medicine in virtually every discipline of medicine. The hope of this author is that while the cases are neurological in nature (the author being a neurologist), every internist, family physician, surgeon, obstetrician, gynecologist, psychiatrist, nurse, physician assistant, therapist, administrator, clerk, technician, patient, friend, family member, and caregiver, among many others, will feel a resonance with their own experience.

Part 2 begins an expositional analysis of the nature of ethics as it relates to the practice of medicine. The term "exposition" is not to be confused with "theoretical," which often invites a connotation as being "academic," so much metaphysical speculation, and thus irrelevant to the concerns of those involved in everyday medicine. Exposition is used here to include ideas that go beyond a sheer enumeration of particular instances or cases, where going beyond enumeration is an epistemological necessity, thus abstractions such as generalizations, principles, theories, laws, case precedents, and virtues. Describing the process of abstraction as theorizing would be a disservice. We want to ensure reproducibility of our ethical decisions so that different persons at different times in similar situations and circumstances are treated the same, although no two situations and circumstances are exactly the same. Inconsistent ethical decisions likely would be viewed as unjust. Yet, what is it that is common to all cases and particulars that will provide for consistency, even if what is common must be abstracted? This is the question that drives the work in Part 2.

Discovering what is common to a set of particulars is the process of induction, in this case ethical induction. Others describe a similar process of intuition. However, the notions of induction and intuition are problematic, whether it is in science or in ethics. Indeed, the problematic nature of induction has been a subject of intense interest to philosophers since the ancient Greeks, and over time recurring themes have been recognized that have implications for the development of ethical systems. Nevertheless, a considerable amount of intellectual work has been done on induction and intuition and that critical understanding will be important to the development of rigorous ethical systems.

One way to organize behaviors into sets of just and unjust actions is based on the underlying mechanisms that generate such behavior. One can explore such mechanisms in terms of psychology, that is, a psychological predisposition to behave in a certain way. Further, psychological constructs can be invented to provide some explanation. The problem is that such psychological constructs could be entirely self-consistent and explain many phenomena but not reflect the physical mechanisms that actually drive behavior. Yet, enough is known that it can be said that the fundamental foundations of any human behavior relate to the electrical states of neurons embedded in vast and complex networks that comprise the nervous system.

My interest in neurophysiology involved discovering neuronal mechanisms underlying behaviors at the level of changes in the voltage of neurons recorded through electrodes in the brain as nonhuman primates performed movements. My work consisted of attempting to create specific correspondences between neuronal activities and the behaviors. There is no *a priori* reason in principle that such a correspondence between neuronal activities and just and unjust behaviors cannot be achieved, unless one believes that ethical behaviors are the result of some immaterial agent, such as a soul. Note that this comment is not meant to denigrate the notion of the soul or religion, only that in a modern liberal democracy containing a great many different views on religion, it is problematic to base any ethical system on a religious basis. Pursuing an analysis of ethics in a physical sense from which considerations of religion have been bracketed out at least avoids the problems of invoking religion.

There are a great many reasons to doubt that a complete explication of just and unjust behaviors, a neurophysiological ethics, can ever be fully achieved. However, this cannot be achieved in other domains of science as well, but this does not denigrate the value in attempting to do so. There has never been a complete explicit empirical demonstration of the principle of inertia—an object in motion will continue to move in a straight line forever unless acted upon by an outside force—as no one has ever experienced an object moving forever [for those interested in how the notion of inertia is justified by the Process Metaphor, see Montgomery, 2019a]. Yet the principle of inertia is powerful in understanding physics.

Attempts are underway to understand the notion of justice in neurophysiological terms that, at the very least, may avoid some of the self-referential circularity inherent in many psychological explanations. Using insights from neurophysiology, particularly in the context of Chaos and Complexity, may help in understanding the vagaries encountered when attempting any wholly consistent and complete explication of what are just and unjust behaviors. An attempt to begin this analysis is the topic of Part 2.

As always, my gratitude to Melissa Revell for translating this author's dyslexic English. I particularly thank Dr. Arthur "Buddy" White, professor and chair of philosophy, for the inspiration to formalize and develop these notions of the ethics of everyday medicine by his kind invitation to teach at the Thiel College and for the many enlightening and joyous conversations we've had.

Erwin B. Montgomery Jr.

References

Brady, J.V., 1958. Ulcers in "executive" monkeys. Sci. Am. 199 (3), 95–104.

Chee, T.T., Ryan, A.M., Wasfy, J.H., Borden, W.B., 2016. Current state of value-based purchasing programs. Circulation 133 (22), 2197–2205. https://doi.org/10.1161/CIRCULATIONAHA.115.010268.

Damberg, C.L., Sorbero, M.E., Lovejoy, S.L., Martsolf, G.R., Raaen, L., Mandel, D., 2014. Measuring success in health care value-based purchasing programs: findings from an environmental scan, literature review, and expert panel discussions. RAND Health Q. 4 (3), 9.

Ferreira, J.J., Mestre, T.A., Lyons, K.E., Benito-León, J., Tan, E.K., Abbruzzese, G., Hallett, M., Haubenberger, D., Elble, R., Deuschl, G., On behalf of MDS Task Force on Tremor and the MDS Evidence Based Medicine Committee, 2019. MDS evidence-based review of treatments for essential tremor. Mov. Disord. https://doi.org/10.1002/mds.27700.

Leplège, A., Hunt, S., 1997. The problem of quality of life in medicine. JAMA 278 (1), 47–50.

Montgomery Jr., E.B., 2015. Twenty Things to Know About Deep Brain Stimulation. Oxford University Press, Oxford.

Montgomery Jr., E.B., 2019a. Medical Reasoning: The Nature and Use of Medical Knowledge. Oxford University Press, Oxford.

Montgomery Jr., E.B., 2019b. Reproducibility in Biomedical Research: Epistemological and Statistical Problems. Academic Press, New York.

Montgomery Jr., E.B., Turkstra, L.S., 2003. Evidence based medicine: let's be reasonable. J. Med. Speech Lang. Pathol. 11 (2), ix–xii.

Polinski, J.M., Kesselheim, A.S., 2011. Where cost, medical necessity, and morality meet: should US government insurance programs pay for erectile dysfunction drugs? Clin. Pharmacol. Ther. 89 (1), 17–19. https://doi.org/10.1038/clpt.2010.179.

Rawls, J., 1999. A Theory of Justice, revised ed. Belknap Press of Harvard University Press.

Stavrou, D., Weissman, O., Polyniki, A., Papageorgiou, N., Haik, J., Farber, N., Winkler, E., 2009. Quality of life after breast cancer surgery with or without reconstruction. Eplasty 9, e18.

Sweet, V., 2012. God's Hotel: A Doctor, a Hospital, and a Pilgrimage to the Heart of Medicine. Riverhead Books.

Further reading

Jonsen, A.R., 2003. The Birth of Bioethics. Oxford University Press, Oxford.

Introduction

To be argued...

Although perhaps scarcely appreciated, virtually every decision made in medicine every day is fundamentally an ethical decision. Indeed, in the very large majority of the instances, the ethical dimensions are not realized; even the outlines of the ethical issues are not seen. Yet, the implications can be profound for patient care. Ignorance of the ethics of everyday medicine is not like the land labeled *terra incognita* (unknown land) on an ancient map. It is more like the world that exists behind one's head—it is not seen, it is not of concern (except rarely), it is just not there. At least the *terra incognita* usually has some boundaries, in terms of landmarks, that can serve as a stepping-off point for further exploration. The purpose of this book is to embark on an exploration, not to point out destinations or routes to get there. Too little is yet known.

It is hoped that this Introduction will demonstrate that there is a vast world of ethics "behind one's head" by demonstrating that much of what is considered ethics in everyday medicine today is limited and narrow. Thus, the possibility exists that current ethical understanding does not meet the needs of patients, family members, caregivers, physicians, healthcare professionals, and society in general. Current ethics of everyday medicine are narrow because only a relatively small spotlight has been shown on it and then it often only is the glaring light of conflicts that make the headlines. Many excellent books with exemplar cases have been published, but most do not take into account the ecological complexities surrounding ethical decision-making, whether explicitly or implicitly. Ideally, a map to the ethics, distilled from the cases, should match the complexity of what it is trying to represent. The cases presented in Part 1 demonstrate the complexities.

Part 2 serves to examine the tools with which the explorations can proceed. In a sense, a reliable compass needs to be fashioned. There are ontological issues (the nature of what the compass is to point to) and epistemological issues (how it is that the compass comes to point to what is to be discovered). The argument is raised that perhaps the wrong things were looked (ontology) for in the wrong way (epistemology). Yet, if true, it is critical to understand what went wrong because very dedicated, intelligent, talented, and skilled people may have gone a bit in the wrong direction.

Part 2 looks at the efforts from perhaps different perspectives. The first is ontological, what is the nature of ethics. It will be argued that ethics derives from an affective (emotional) neurophysiological response to observations of a particular situation or circumstance. This is very different from psychology that has been used to explore ethics. Thus, findings related to neurophysiology will have important implications to notions of ethics.

The second perspective is epistemological, that is, the methods used to gain knowledge in ethics. Ethics will be pursued as an empirically based knowledge domain by virtue that ethical systems are based on the will of those subjects in a pluralistic modern liberal democracy, where religious dogma or dictates of absolute sovereigns do not prevail. The primary method is induction from experience, for example, the opinions of reasonable persons blinded to their own self-interest extracted from a review of a series of experiences. This is little different than a scientist abstracting generalized principles from observations of experiments. The argument is that the logic and statistics that drive science can be brought to bear on ethics to render ethics more robust, thereby helping to ensure justice.

Ethics, like any other empirically driven knowledge domain, is confronted with having to make sense of the potentially infinite variety of experiences. The fact necessitates an ontological choice. Is the variety diversity where every experience is unique and de novo or is variety variation on an economical set of principles, theories, laws, case precedents, or virtues? If the former, how does one avoid the Solipsism of the Present Moment where no help can be borrowed from other experiences, past or present. If the latter, what humans know are the particular experiences and thus how is it possible to abstract any notion that would match, in some way, every variety of experiences? The epistemic choice in response to the fundamental ontological problem shapes empirically derived knowledge, including ethics. Critically, attention to the ontological problem and the epistemic choice could lead to better ethics and more justice in the everyday practice of medicine.

Everyday medicine is everywhere and every time

As will be demonstrated, the impact of the ethics of the everyday practice of medicine is enormous, far greater than the number of lives whose concerns make the headlines. In fact, issues in the ethics of everyday medicine are so extensive that they are largely unseen, which itself produces an ethical conundrum. As Erich Fromm wrote in *The Sane Society* in 1955 (1955), "The fact that millions of people share the same vices [author – taken as ethical lapses in everyday medicine] does not make these vices virtues...." Indeed, the ethical lapses are not appreciated because they would only be detectable if the ethics of everyday medicine were recognized and understood. The ethics of everyday medicine are so far "under the radar" that some ethicists question whether there is such a thing as the ethics of everyday medicine. Those who address the issue typically consider them "minor league" or "ethics-lite" versions of the kinds of ethical concerns that generate headlines, such as passive or active euthanasia and abortion. In many ways, these "headline" ethics are relatively easier to deal with as they generally are predetermined by the perspectives of received political (dictates of law) and religious (word of God) moral theories that often obstruct any deeper and open ethical discussion that would be appropriate, indeed demanded, in a pluralistic modern liberal democracy. This is not to say that issues of passive or active euthanasia and abortion are never encountered in the practice of medicine, but not on an everyday basis outside of hospice care, palliative medicine, and obstetrics, respectively. Further, these issues typically are readily recognizable. These issues are not addressed in this examination of the ethics of everyday medicine.

Most physicians and healthcare professionals recognize that access to care is an ethical issue, but, as will be demonstrated in the cases of Part 1, many implications go unrecognized, and those are the subjects of this effort. In these discussions, physicians and healthcare professionals largely hold themselves as spectators on the sideline, believing themselves bound by contract or law, even if they believe it is possible to act in a more ethical or moral way. For example, a physician or healthcare professional waving deductibles and co-pays for the poor can be accused of fraud. Further, some insurers limit the extent of charitable care that can be given to the medically indigent. The situation is exacerbated by the fact that physicians increasingly are employees and thus bound by their institutional policies. Yet, it is not clear that physicians and healthcare professionals do not bear any responsibility for the limitations or for the advancement of patients' interests.

The marked changes in the authority of physicians and healthcare professionals in today's healthcare delivery systems risk ethical nihilism. Many current texts avoid, by omission or commission, discussion of the regulatory issues, whether legal or contractual. Yet, there can be little doubt about the impact of regulatory issues on the everyday practice of medicine, as will be exemplified by the many cases presented in Part 1 and in Chapter 4. Thus, most discussions of ethics in the absence of consideration of regulatory issues risk being artificial and of limited benefit at best and counterproductive at worst.

The ethics of everyday medicine go much deeper than access to resources. As will be seen, every diagnosis and every recommendation of treatment inherently involve ethics. This will be amply displayed in the cases presented in Part 1. A great many of them do not fundamentally depend on limited access or prioritization of resources. Rather, many of them depend on the decisions that physicians and healthcare professionals do have in their power to make every day. Other examples will be discussed later.

Justice

Justice is fundamental to ethics, as will be explored in detail in Chapter 1. One perspective to be perused here is that justice in the form of laws and court decisions is a type of ethics, when ethics is defined as how humans treat each other. Laws, both statutory and case or common law, are one form of ethics. Laws and court findings are a model by which to view ethics in a larger domain, such as the ethics of every day medicine. The notion of law is taken in its broadest connotation, for example, to include contracts.

Frequently, ethical disputes arising in the everyday practice of medicine are settled based on contracts, implicit or explicit. Paramount is the contractual relationship in insurances that bind patients, physicians, and healthcare professionals that are enforceable by law. There is the social contract where society, through its proxies of governmental representatives, establishes implicit contracts with physicians and healthcare professionals. Society grants physicians and healthcare professionals licenses, a form of monopoly, as licensure is not available simply to anyone. In a real sense, licensure has commercial value and becomes a good that is exchanged with physicians and healthcare professionals. In return, society requires physicians and healthcare professionals to practice competently and to provide care to everyone and anyone who present for care in an emergency regardless of the patient's ability to pay. Society subsidizes the education of physicians and healthcare professionals and, in exchange, physicians and healthcare professionals are contracted with patients to operate under fully informed consent of the patient or the patient's representative. There are a great many examples; only some of them surface in the cases presented in Part 1.

Further, it is a fact of life that any medical decision made any and every day in the practice of medicine ultimately is subject to legal review, even if the vast majority of decisions never come under legal review. It would be naïve to think that physicians and healthcare professionals operate oblivious to the expectations legally held by patients and society. But physicians and healthcare professionals are held under review by any number of persons or institutions that have suasive powers by their teachers, mentors, supervisors, colleagues, expectations of their professions, and even themselves. It would seem naïve to ignore these influences in formulating an ethics of everyday medicine.

This book, consequently, draws heavily on texts on justice such as John Rawls' "A Theory of Justice" (Rawls, 1999), which is unlike most other texts on the ethics of everyday medicine. Rawls' discussion proceeds from a consideration of where laws come from. Historically, sources of justification have been the work of God in theocracies and the dictates of absolute sovereigns in dictatorships. But even in those there has been some appeal to the good of the citizens, whether it is the saving of one's soul to receive a reward in the afterlife or to ensure civil order and the presumed security it provides in this life. Serving the good of the citizens in these contexts is taken as justice. It is likely that the ethics of everyday medicine would be quite different in a theocracy or a dictatorship.

If the word of God and the dictates of an absolute sovereign are disallowed, where then will justice come from? Perhaps starting with Thomas Hobbes in his "Leviathan: or The Matter, Forme, & Power of a Common-Wealth Ecclesiasticall and Civil" (Hobbes, 1651), governance by a social contract has become the norm, at least in pluralistic modern liberal democracies of the West. Hobbes' pessimistic social contract was offset by the more optimistic social contract of Jean-Jacques Rousseau's "The Social Contract; or, Principles of Political Rights" (Rousseau, 1762). The power of theologians emanated from the pulpit and the divine right of kings was symbolic and perhaps perfunctory. The question becomes what social contract will affect the ethics of everyday medicine in a pluralistic modern liberal democracy?

The notion of pluralism relates to the fact that generations of immigrations have rendered most societies very heterogeneous with respect to ethical and moral values, in particular. In other words, most societies are necessarily pluralistic. The implication is that if governance is by the will of the governed and the governed are heterogenous, then any governance must be respectful of the heterogeneity, hence pluralistic, or it must seek to homogenize those whose interests are protected by the governance, risking resistance from the "others" who may be disenfranchised. Yet, the successes of asymmetric warfare suggest it would be unwise to discount even the weakest "others."

The notion of liberal is complex with differing interpretations, connotations, and uses. Liberal likely relates to some notion of liberty. While there can be wide debates on the dimensions of liberty, particularly the actions it allows, such as owning property, perhaps the most fundamental is the freedom of thought and conscience. This is not to say that thought and conscience cannot be manipulated—indeed it happens frequently. To be sure, throughout history, efforts have been made to limit the freedom of thought and conscience, but these nearly all have failed, often violently. Indeed, the 30 Years' War (1618–1648) brought about religious and political pluralism as a consequence of the devastation caused on the continent of Europe and in England. For good or bad, the current state of neurophysiology, one's thoughts and conscience seem most difficult to constrain compared to, for example, bodily freedom. Further, it is likely that most reasonable persons blinded to their self-interest would agree and support freedom of thought and conscience even if they disagree with the freedom of speech, gun ownership, and private property, among many others.

If the mechanisms of governance are to be derived from the will of those governed, how is that will determined? Given the plurality, at least of thoughts and conscience, how can mechanisms be constructed? It does not seem effective to have a single mechanism for each individual. Rather, some economical set of mechanisms is necessary but this presumes that the variety of thoughts and consciences are variations on an economical set of themes. Presumably, any set of mechanisms must be abstracted, perhaps by induction, from the particulars of all the individuals. This conundrum comes back to the fundamental ontological problem and the necessary epistemic choice. As the number of individuals increases, how is the will of the governed determined? One response is to have groups of individuals represented by a single person, a republic rather than a democracy, although the two terms are often conflated. But a republic injects the problem of sampling from the population, as will be discussed in detail in Chapter 3.

Rawls speaks of democracy and liberal socialism that seem to differ based on the ownership of private property or the distribution of goods, more generally. Yet, what is in common is the notion of liberalism, one that seems to default from the ideal operations of a democracy, with private ownership, and from a socialism that seems to enforce the liberal ideals derivative of the operations of democracy, exempting private ownership. In any event, it would seem that the phrase "modern liberal democracy," while not mentioned in Rawls' "A Theory of Justice," would seem consistent, even if possibly a redundancy in terms. The key purpose here is that much of the discussion in Rawls' "A Theory of Justice" finds import in the considerations of the ethics of everyday medicine. As will be discussed, this is particularly true of Reflective Equilibrium as a means to define laws or ethical principles, moral theories, and precedents (Chapter 3). Also important is the notion of law as a consensus formed by persons blinded to their self-interest by a "veil of ignorance." In many ways, this is analogous to the "reasonable persons" ethicists such as Beauchamp and Childress (2012) argue as the source of ethical principles and moral theories.

A modern liberal democracy presumably is the one that all reasonable persons blinded to their self-interest would choose to live in. Pluralism derives from the fundamental notion of freedom of thought and conscience, and as there may be many different "thoughts" or "consciences," respecting the multitude becomes pluralism.

Another critical concern is the notion of moderate scarcity (Rawls, 1999) that leads to accommodations and compromises in the distribution of goods; in this case, what is good for patients in the everyday practice of medicine. This manifests itself in the prioritization and rationing (in its broadest connotation) of healthcare delivery. If there were no scarcity, then there would be little or no conflicts. If there is a severity of resources, it is likely that no accommodation or compromise likely would be achieved and it becomes every person for himself or herself. As can be readily appreciated from the cases presented in Part 1, the issue of moderate scarcity in healthcare delivery is a dominant factor.

Ethics, morality, principles, theories, laws, case precedents, and virtues—Definitions proposed here

There are a great many different definitions and uses of the terms ethics and morality. They have in common a notion of how humans conduct themselves. Given that humans do not live in absolute isolation, then how one conducts himself or herself also entails how one interacts with other things, such as other human beings. As amply demonstrated in this book and elsewhere, understanding how humans conduct themselves and interact with others is immensely complex and yet immensely important. How one comes to understand how humans conduct themselves and interact with others is central to ethics, morality, and law.

Very often the terms "ethics" and "morality" seem conflated or used interchangeably, perhaps leading to confusion. This text makes a distinction between ethics and morality. Nevertheless, the term ethics or ethical will also subsume the totality of relevant human interactions of which ethical mechanisms and morals will be components.

There are a number of dimensions along which ethics and morality, as they are used here, are distinguishable. In this book, ethics is taken as first descriptive of how individual humans interact in a way to create justice (fairness) or injustice (unfairness), particularly at the ethical microlevel. (The notion of micro- and macrolevel ethics is discussed in detail in Chapter 1.) For example, the microlevel includes the dynamics between individuals. Typically, this involves pairs of individuals with conflicting or different interests; this pairwise relation is called dyadic. As will be seen, morality here is taken as how humans interact on a large collective macroscale where terms such as "right" or "wrong" often are appended.

Another dimension that differentiates ethics and morality is that ethics, as developed in this text, is a continuum. For example, the obligation to ethical principles or virtues defines the dyadic relationship and the obligations are continuous variables capable of an infinite set of values, for example, the degree of obligation. In contrast, justice demands an action, where even not taking an action is a kind of action.

The physician will use treatment A fully in the case of diagnosis B even if the physician is not confident in the diagnosis. The physician will not use partial treatment in proportion to the physician's confidence in the diagnosis. For determining any ethical situation or circumstance, a decision is either just or unjust. To be sure, the consequence can be apportioned but typically only after a finding of justice or injustice. The necessity of coming to a dichotomous variable, either true or false, just or unjust, right or wrong, is necessary and what distinguishes obligations from ethical principles or virtues. The dichotomous variable is the output of the moral machinery.

In a pluralistic modern democracy, the value of the dichotomous variable that results from moral machinery must be reproducible when it is applied to multiple cases of the relevant ethical question. Any moral machinery that produces different outcomes given token cases of the same ethical type would be taken *prima facie* as unjust. Thus, morality must operate on the macrolevel. Reproducibility of moral determinations is discussed in Chapter 1 and in detail in Chapter 3. As will be discussed frequently, moral theories are used to weigh the obligations at the microlevel to come to a moral outcome.

The term "ethical mechanisms" was used so as to avoid exact specification, until now. As will be seen throughout the text, there may be different ethical mechanisms. In the Principlist ethical system, ethical mechanisms may be defined as ethical principles, for example, beneficence (to good), nonmaleficence (do not to harm), and autonomy (respect). To be sure, these are not exhaustive and indeed new ones may be necessary. The appeal of principles is that they are often used as though they were premises in a deductive argument that is supposed to provide confidence in the outcomes of their use (Beauchamp and Childress, 2012). As discussed elsewhere, the typical use is not truly deductive but may misleadingly appear to be so. Another set of ethical mechanisms includes virtues, which are characteristics of persons acting ethically, and when operational in physicians and healthcare professionals, patients will receive justice. Virtues are the substrate for virtue ethics, which is taken as an Anti-Principlist ethical system. Other ethical systems have mechanisms analogous to those described here as ethical mechanisms (see Chapter 2).

Example of everyday ethics at the most fundamental level—Diagnosis

Diagnosis is perhaps the most fundamental act in medicine. Indeed, the privilege to diagnose has been guarded jealously by physicians for thousands of years (Montgomery, 2019a). Note that diagnosis is not just of the disease state, as the same inherent processes in diagnosing the disease also operate in predicting the future course (prognosis) and what treatments to use. Thus, if there is an ethical dimension in diagnosis, then medicine is fundamentally ethical at its core and in virtually every act, whether the ethical dimension is apparent or not, appreciated or unappreciated. As will be seen, the factors considered in medical diagnosis are rarely dichotomous variables, yet a dichotomous outcome is necessary as a treatment is necessary. Conversion from continuous inputs to the dichotomous output is problematic and, as will be seen, determined fundamentally by the consequences that can only be interpreted in ethical terms. Further, the physical examination/laboratory test/radiological tests rarely are pathognomonic (a positive test meaning one and only one thing) and thus inherently uncertain. Most often, how to deal with the consequences entailed by the uncertainty, the immediate necessity to act, is not a scientific or statistical question—it is an ethical question.

Essentially, despite the number of proposed approaches, medical diagnosis is fundamentally uncertain, even in the face of evidence-based medicine (Montgomery, 2019a). It is uncertain in the diagnosis, uncertain in the treatment effects, and uncertain in the consequences for the individual patient. Consider the situation in patients with Parkinson's syndrome whose disability warrants treatment (how this is decided is another ethical issue) and deep brain stimulation is being considered. Parkinson's syndrome is a group of disorders with similar manifestations. Of patients encountered with Parkinson's syndrome, approximately 75% will have what is called idiopathic Parkinson's disease, otherwise denoted as typical Parkinson's disease. This means that 25% of presenting patients have other conditions, collectively denoted as atypical parkinsonism. Accept for the sake of discussion, clinical criteria used to distinguish typical Parkinson's disease from atypical parkinsonism have a specificity and sensitivity of approximately 80% (Montgomery, 2015). This means that out of 100 patients presenting with Parkinson's syndrome in need of treatment, 75 will have typical Parkinson's disease, of which 60 will be diagnosed as having typical Parkinson's disease and receive treatment, meaning that 15 patients with typical Parkinson's disease will not be diagnosed and not receive treatment, perhaps a violation of the obligation to beneficence. Similarly, of the 25 patients with atypical parkinsonism, 20 patients will be diagnosed with atypical parkinsonism and not be treated, as treatments typically are ineffective for atypical parkinsonism, but they will not be exposed to adverse effects, consistent with an obligation to nonmaleficence. However, five patients with atypical parkinsonism, who have little probability of benefiting, will be exposed to adverse effects; perhaps a violation of the obligation to nonmaleficence. Note the fact that 15 patients denied treatment who would benefit and the 5 patients exposed to adverse effects with little hope of benefit is inescapable. The unavoidable ethical question is whether providing beneficence to 60 patients diagnosed with typical Parkinson's disease and providing nonmaleficence to the 20 patients with atypical parkinsonism is justifiable given that 20 patients will not be treated appropriately? Perhaps more important is the question "who gets to decide" (see Case 8)?

Consider screening for patients with a cognitive decline for vitamin B_{12} deficiency. The incidence of dementia from a vitamin B_{12} deficiency is exceedingly rare, yet any patient presenting with dementia has an assessment of his or her vitamin B_{12} status despite the direct costs and the follow-up costs of any "positive" test whether true or false. Why? It would be deemed a tragedy if a single patient continued to have dementia that could have been reversed with a vitamin B_{12} replacement. Yet, whether it truly is a tragedy very much depends on the informing moral theory, which determines the ethics of either testing or not testing for a vitamin B_{12} deficiency. Yet, it is highly unlikely that a physician or healthcare professional actively contemplates the ethical basis for assessing the vitamin B_{12} status of a patient with dementia, they just do it as a matter of procedure.

The example just described here and in the cases in Part 1 do not nearly exhaust the issues of unrecognized problems in the ethics of everyday medicine. The purpose of these examples is to demonstrate that ethical issues are involved in the most fundamental tasks of everyday medicine. As fundamental, they are ubiquitous.

Cases

Perusing the cases in Part 1 demonstrates that they are not typical of cases discussed in many studies on clinical or practical ethics (Lo, 2019). Typical of those studies are ethical problems such as when patients or their representatives demand care that physicians or healthcare professionals deem futile, for example, the demand by cancer patients for the US Food and Drug Administration (FDA) to release laetrile, ultimately rejected in 1980 by the US Supreme Court (*United States v. Rutherford*, 544 U.S. 442, 99 S.Ct. 2470, 61 L.Ed.2d 68). Conversely, patient refusal of treatment is another typical topic. Other typical cases are derivative of physician and healthcare conduct, such as lying to insurers to obtain healthcare coverage or having sexual relations with patients.

The types of cases typically addressed in texts on medical or clinical ethics are those already recognized as problematic by physicians and healthcare professionals. These cases arise in salience proportional to the ethical discomfort that they engender in the physician or healthcare professional and perhaps only parenthetically in the patient (Lo, 2019). The authors of other texts are right to address these typical type cases, as their goal is to help relieve the ethical discomfort of physicians and healthcare professionals, as well as the patients caught up in those problems. The goal of this effort is to address what might be unseen ethical problems, such as the world behind one's head, that are ubiquitous in the practice of everyday medicine, especially as they generally do not cause ethical discomfort in physicians or healthcare professionals. Also, such ethical discomfort likely will not be experienced by patients as well, although the patients may experience discomfort in other ways, such as adverse effects on their health.

Many of the cases may seem just the natural course of doing business in everyday medicine. However, on closer examination the ethical implications and the fault lines that define the ethical implications become clear. Although drawn from the author's over 40 years of professional experience as a neurologist, most physicians and healthcare professionals, on a moment of reflection, will find that many of their own experiences resonate with the cases presented. As such, they demand attention from the ethically conscientious and virtuous physician or healthcare professional. Assuming this author is correct in that the cases presented do entail serious ethical issues, the larger question is why don't these types of cases engender ethical discomfort—an apparent prerequisite for them to be recognized? This concern will be explored in detail later (Chapter 4).

The cases presented in Part 1 are also atypical as they do not resolve the issues raised—they do not provide a "practical" answer nor can the analyses of the cases. These cases invoke ethical principles, such as beneficence, nonmaleficence, and autonomy, just as a great many other cases and their discussions have in typical texts. The primary reason why no "practical" resolutions are offered is that the far-ranging complexities of the ethical ecology of the cases have yet to be considered sufficiently. The dyadic ethical relationship between the patient and the physician or healthcare professional is insufficient, as described earlier. It may well be that entirely new ethical principles, moral theories, laws, case precedents, and virtues need to be discovered. Making those discoveries may require new and more robust methodologies (see Chapter 3).

As will be seen, nearly all the cases presented in Part 1 are complex as they involve discussions of law, governmental and insurer policies, public opinion and expectations, and historical trajectories that propelled these issues in addition to formal ethical systems. To this author's knowledge, this greatly expanded context is very different from what is typically considered. Indeed, some clinical ethicists purposefully exclude discussions of law. For example, Bernard Lo (2019) makes a clear distinction between ethics and law. Lo recognizes that law does represent a consensus in how humans should treat each other, which is the very definition of ethics. As such, law can be considered a subset of ethics; however, Lo draws perhaps too fine a distinction. For example, Lo appears to equate law with criminality, and thus right and wrong, in a moral sense, whereas in ethics, as Lo appears to believe, there is no clear notion of criminal right and wrong, particularly any dichotomous notion of right and wrong. However, in most respects, right and wrong are dichotomous because some binary decision is necessary in order for consequences to be implemented. Otherwise, without consequences, ethical considerations become an empty exercise. Lo seems to not consider civil law, which is law nonetheless, and criminality is not an issue. In reality, civil law is the appropriate metaphor as in many aspects it parallels what a bioethics committee might do.

Lo goes on to say that law does not adjudicate many ethical problems. But "does not" is not the same as "cannot" or "will not." Rather, most courts have been reluctant to be drawn into complex and difficult medical ethical problems, preferring to leave it to the experts. This often results in "out of court" settlements but being "out of court" does not mean "outside the bounds of law." Nevertheless, as demonstrated repeatedly, courts of law have stepped in where settling out of court has failed and even successful out of court settlements are subject to judicial review.

This book makes considerable use of laws and court findings to inform the ethics of everyday medicine. It is not a book on the law or the interpretation and application of law to any particular case and the author is not an attorney and, therefore, offers no legal opinion as to matters of law. Rather, the author views laws and court cases as metaphors by which to attempt to understand the ethics of everyday medicine. Nevertheless, the ultimate goal of ethics is justice in the interactions among humans. Adjudication of justice means nothing if it cannot be enforced, an established legal principle since the Roman *ubi jus, ibi remedium* (where there is a legal right, there is also a legal remedy). What would it mean to say where there is an ethical right but there is no remedy?

Perhaps another reason that considerations of law have been excluded in clinical ethics may relate to some ontological difference; a difference in the nature, between ethical and legal principles. The notion of the Rule of Law may suggest that the laws are immutable and thus trump any ethics. However, this is not true as noted in Chapter 1. Citizens, through their representatives of citizens, make and change the laws.

Ethics, however, does not seem so immutable, but rather tentative. However, every law is subject to revision and thus also is tentative. The major difference appears only to be the formality of the process of change. Derivation of ethical principles and moral theories in a pluralistic modern liberal democracy also comes from the consensus of those involved. Beauchamp and Childress (2012) write of ethical principles being the inevitable consensus of reasonable persons, just as John Rawls speaks of justice (by extension laws) as a result of consensus among persons blinded to their self-interest in what Rawls refers to as the veil of ignorance (Rawls, 1999).

In most discussions of ethics, the procedure to achieve consensus is relatively unspecified, certainly not like the procedures used to establish statutory law. But in many countries, many legal decisions are not based solely on statutory law but rather invoke case law where particular settled cases become precedents or exemplars by which subsequent cases are compared. In case law, a consensus, of a fashion, is often achieved and is the case precedent. However, the precedent may change over time. As such, precedents become a form of "running consensus building" with each subsequent case considered. The manner by which it changes can have interesting implications, perhaps ultimately counterproductive as discussed in Chapter 3.

The cases also address the ethical issues of medical errors and poor medical practice, unlike most texts. Yet, medical errors and poor medical practice are unethical in the sense that they violate the obligations to the ethical principles of beneficence, nonmaleficence, and autonomy or ethical virtues. In 1999, the Institute of Medicine reported that medical errors caused between 44,000 and 98,000 deaths every year in American hospitals, and medical errors are the third leading cause of death in the United States (Kohn et al., 2000). It is likely that not much has changed. A survey of medical records found only 23% of patients have their hypertension under control, approximately 20% of elderly patients are prescribed inappropriate and potentially harmful medications, only 40% of hospital visits and 60% of office visits include smoking cessation counseling, only 67% of adults have had their cholesterol checked within the past 2 years, and only 69% of patients who sustain heart attacks are prescribed beta blockers. The boy from the "Emperor's New Clothes" might yell out, "everyday medical care in the United States is naked!" Even if one wanted to say that these alarming problems are because of bad doctors and healthcare professionals practicing poor everyday medical care, the ethical and moral question is still how can the healthcare system countenance such widespread and dangerous poor practices? Where are the medical and professional schools, postgraduate training programs, professional societies, and governing agencies?

Note that this is not to say that medical errors and poor practice are unjust. It may be that the expenditure of resources to eliminate medical errors and poor medical practice is too much. Indeed, it may be that making those expenditures may divert resources away from other priorities. Thus, a discussion of medical errors and poor medical practice in the ethics of everyday medicine is important, as will be seen in many of the cases presented in Part 1.

A detailed analysis follows from each case. The analyses emphasize a Principlist system, for the most part. In particular, the cases focus on the ethical principles of beneficence, nonmaleficence, and autonomy and moral theories that include Egalitarianism, Deontology, Libertarianism, and Utilitarianism. To be sure, there are many other ethical systems (see Chapter 2) and it is hoped that new and more effective systems will be developed (see Chapter 3). These principles and theories in the Principlist tradition were chosen because they are used most frequently in discussions of medical and clinical ethics. The emphasis on Principlism is not to deny the importance of Anti-Principlist approaches such as feminist, phenomenological, virtue, and care ethics. In fact, as discussed in Part 2, the distinction between Principlism and Anti-Principlism will be demonstrated as artifactual. Both represent the two different sides of the same epistemological coin. Construction of ethical systems is an iterative process moving between Principlism and Anti-Principlism, just as any empirical-based system of knowledge iterates between induction from particulars to generalities and then testing of generalities against particulars.

Ethical induction

In many ways, confidence in medical ethics depends greatly on the means to establish a consensus from which ethical principles, moral theories, laws, case precedents, and virtues are derived. This book offers an approach to increase the rigor by which consensus can be constructed that has not typically been considered, at least from this author's knowledge (Chapter 3). Consensus building can be considered an exercise in induction, which is the extraction (induction) of generalities, often called principles, theories, laws, case precedents, and virtues, from observations of particulars—ethical experiences. An example of induction, the judgments or actions of reasonable persons blinded to their self-interest, in a set of similar situations and circumstances, are the materials from which a consensus is induced to form a principle, theory, law, case precedent, or virtue. Similarly, the judgments of persons blinded to their self-interest, such as legislators (ideally) who vote and the proposition adopted by the majority, constitute a consensus that becomes law.

Induction, however, is a difficult and slippery concept long recognized by philosophers such as John Stuart Mill (1843). The actual genesis of induction, how a generalization is created from a set of particular, is difficult (see Chapter 3); indeed, some attempted to sidestep any rigorous explication of induction by calling the process intuition. Some philosophers have attempted to make intuition more rigorous (Osbeck and Held, 2014). One cannot just pick the case experiences on which to intuit or induct a consensus that will become the basis for an ethical system. The risk is that the selection criteria can predetermine the consensus, leading to poor generalizability and reproducibility for all persons subsequently subject to the ethical system.

The question is how to make the induction robust. John Stuart Mill (1843) provided methods for testing the robustness, not the origin, of the inductions, with many based on logic involving the Principle of the Excluded Middle. His method of concomitant variations is the basis for correlational statistical analyses adopted in quantitative research, such as science, and likewise applies to ethics (Chapter 3).

Some ethicists have explicitly argued that logic and scientific methods are not applicable or useful to the work of ethicists attempting to create ethical systems. These ethicists argue that the subject matter is not quantifiable. However, this may not be true. For those instances where it might be true, there are rigorous research methodologies based on the statistical concepts if not methods (Chapter 3). Further, conceptually, statistics are derivative of logic, particularly the conundrum poised by the necessary use of logical fallacies (Montgomery, 2019b). Developing ethical systems likewise necessitates the judicious use of logical fallacies. By no means does this position hold that ethics are reducible to traditional statistical analyses. Indeed, as discussed in Chapter 3, statistical analysis cannot determine the content of ethics. Rather, statistics is an internal measure used to determine the consistency of conclusion (content) developed in ethical systems; statistics does not determine the content, for example, ethical principles, moral theories, laws, case precedents, or virtues. However, the internal consistency demonstrated by statistics is an important means to establish certainty and confidence in the ethics that result. Further, the limitations of traditional statistical analyses in the setting of Chaos and Complexity particularly germane to ethics are discussed.

The critical question in ethics, particularly the ethics of everyday medicine, is how to assure reproducibility in ethics, which is addressed in detail in Chapter 3? The critical ontological conundrum is that it is extremely unlikely that any two or more ethical experiences are exactly the same so how is one to assess reproducibility? If they were identical, then reproducibility would not be a problem but rarely, if ever, is this the case. The fundamental epistemological (methodological) approach is to find a generality, such as a principle, theory, law, case precedent, or virtue, that covers every relevant ethical experience, even if not identical to any particular ethical experience. This is not easily done.

What humans are presented with are particular instances—Dr. John Doe is in conflict with patient Mary Smith over a specific treatment. The problem is that another instance can occur where Dr. Sally Jones is in disagreement with patient Robert Brown regarding the same specific treatment. Yet, it is unlikely that Drs. Doe and Jones are exactly alike and that patients Smith and Brown are exactly alike. Indeed, there is the potential for an infinite number of such particular instances and the central epistemic problem is how to deal with this variety. One approach is to argue that variety is variations on a set of principles, theories, laws, and precedents. Each particular instance is considered as just some combination of the principles, theories, and laws. Alternatively, the variety of experiences are considered as diversity where each experience is unique and de novo. In this case, there are no ethical principles, moral theories, laws, case precedents, or virtues. Ethics is not impossible in the context of variety as diversity, it may simply be transactional.

The root of Principlism, which underlies the majority of texts on clinical or medical ethics, is that there are such things as ethical principles, such as autonomy, beneficence, and nonmaleficence. Further, they apply to all particular cases and provide the means for resolving disputes. Interestingly, the Principlist approach is fundamentally similar to Rationalist/Allopathic medicine in that both presume the variety of experience as variations on an economical set of mechanism. As Rationalist/Allopathic medicine is the dominant form (although not exclusive) of medicine today (Montgomery, 2019a), perhaps it is not surprising that medical or clinical ethics would be dominated by Principlism. The alternative and competing school of medicine, the Empirics, is analogous to the Anti-Principlists in many ways.

One approach to the reproducibility of ethical treatments in the absence of ethical principles and moral theories is to argue the notion of virtuous physicians and healthcare professionals who, by their character, will ensure ethical treatment. As that virtuous character is of a particular and specific type, reproducibility of that character type will enforce the reproducible ethical treatment of patients. This is the basis for virtue and care ethics that are increasingly studied and applied in professional schools (see Chapter 2).

Contributions of philosophy

Virtue ethics is not new, as both Aristotle and Immanuel Kant made similar arguments. Indeed, Kant argued that a virtuous person is far more likely to reproducibly produce actions that are just than operating from principle, with virtue being an aspect of practical reasoning that both Aristotle and Kant held as different from pure reasoning. As discussed later, Aristotle, and particularly Kant, set much of the framework for virtue ethics, important in modern ethics of everyday medicine. It would seem then that today's physician, healthcare professional, ethicist, and everyone involved in the actual conduct of medicine might benefit from what Aristotle, Kant, and other philosophers had to say. Indeed, the contributions of philosophers in understanding the ethics of everyday medicine will be illustrated in much of this text. Thomas L. Beauchamp, one of the authors of the highly influential and foundational text "Principles of Biomedical Ethics" (2012), is a philosopher at Georgetown University and a David Hume scholar (Hume was an important philosopher of the 18th century). The importance of philosophy to ethics has been well discussed by Kwame Anthony Appiah in "Experiments in Ethics" (2008). Many people in the past identified themselves as philosophers—sociologists, philosophers—psychologists, philosophers—historians, or natural philosophers, only for later generations to drop the philosopher to become ethicists and scientists, although continuing the same efforts. Most early ethicists received their advanced degrees from departments of philosophy, and only later did independent degrees conferring departments and institutes of ethics arise. Consequently, references to philosophical concepts will be made frequently in this book, but hopefully in a manner demonstrating their relevance to the nonphilosopher.

It is interesting that the relevance of philosophy to the ethics of everyday medicine is not appreciated, when clearly many philosophers have addressed the same concerns that confront physicians, healthcare professionals, and everyone involved in the conduct of medicine. Philosophers or those trained in philosophy continue to make important contributions to medical and biomedical ethics, as reviewed by Albert R. Jonsen in "The Birth of Bioethics" (1998).

Perhaps there may be a notion of medical and clinical ethics as uniquely modern, resulting from a medicalization of ethics, particularly in the face of rapidly advancing modern scientific medicine. No need for philosophizing and hence no need for philosophers. There may be the presumption that modern medical science will supersede the need to consider ethics; indeed, this was one source of opposition to the first code of ethics of the American Medical Association in 1847 (Montgomery, 2019a). Thus, ethics should be left to the physicians practicing scientific medicine without the need of philosophers. Similar sentiments were found in opposition to the Belmont Report (1978). Good biomedical research ethics are just what good biomedical researchers do (clearly an argument for virtue ethics). Analogously, good ethics in everyday medicine are what good physicians, healthcare professionals, and all those involved in the conduct of medicine do. Subsequent history clearly demonstrates that such hubris is ill-founded in both cases.

Another reason for underappreciated philosophy, or at least a philosophical attitude, lies in the reduction of everyday medicine to procedures that would seem to obviate a need for explicit declarative conscious deliberation, hence philosophy. This issue is addressed later. Further, ethical considerations became proceduralized as an exercise in the application of ethical principles, first in the ethics of human subjects' research and then extended to the practice of medicine. Indeed, it was in the Belmont Report that the ethical principles of beneficence, autonomy, and justice, which figure predominantly in medical or clinical ethics, were codified (the principle of nonmaleficence was subsumed under the principle of beneficence). The Belmont Report (1978) states:

> Three principles, or general prescriptive judgments, that are relevant to research involving human subjects are identified in this statement. Other principles may also be relevant. These three are comprehensive, however, and are stated at a level of generalization that should assist scientists, subjects, reviewers and interested citizens to understand the ethical issues inherent in research involving human subjects. These principles cannot always be applied so as to resolve beyond dispute particular ethical problems. *The objective is to provide an analytical framework that will guide the resolution of ethical problems* [italics added] arising from research involving human subjects.

The authors were careful to note the difference between research and medical practice, with the latter including innovative practice, suggesting a dichotomy between the ethics of research and medical practice. The basis for that distinction was who benefited, the patient directly—hence medical practice—versus society—hence research. Presumably, the benefits of medical practice are presumed while the benefits of research participation cannot be. Thus, the benefit-to-risk ratio, better conceptualized as the Principle of Double Effect initially attributed to Thomas Aquinas in his "Summa Theologica" written in the mid-13th century, is more problematic in research, thus requiring additional protections. Further, many scientists base their careers on human subject research, thus placing them at an immediate conflict of interest.

Interestingly, schema for the implementation of human subject research includes (1) informed consent, (2) compensation, (3) voluntariness, and (4) assessment of risks and benefits, particularly a systematic assessment of risks and benefits

(Belmont Report, 1978). With the exception of compensation, the schema offered is considered standard of care in medical practice, even if only perfunctory, as demonstrated in several of the cases in this book. In some ways, the ethics of human experimentation have been conflated with modern medical or clinical ethics (Rothman, 1992), perhaps accomplishing *fait accompli* so that all that remains to do is to apply. Future controversies in medical and clinical ethics would seem now only a matter of responding to new technologies, such as genetic engineering, in terms of the established ethical framework and that predominantly within the context of Principlism. Indeed, many texts proceed from just this presumption and would suggest that the philosophical work is over. However, there is sufficient experience to argue that such is not the case. Just as applications of ethics to advancing technologies are necessary, so too is continued reexamination of the nature of ethics employed necessary. This book attempts this in Part 2.

Another interesting development in bioethics that affects, directly or indirectly, the ethics of everyday medicine has been institutionalization of the ethical review of human subjects' research, as well as standing bioethics committees in clinical service providers to help resolve disputes. This has led to development of the professional bioethicist (Evans, 2012). Bioethical professionals train in philosophy, law, sociology, and theology, for example, but self-identify as bioethicists, eschewing identification with philosophy, law, sociology, and theology. Many centers for bioethics were constructed without any reference to philosophy, law, sociology, and theology. Thus, philosophy and the others receded from the limelight, although not their contributions or fundamental significance.

Virtue ethics, as discussed previously, would seem to resonate with professional codes of conduct that featured prominently in Aristotle and Kant. As such, virtue would seem to figure in the ethics of everyday medicine and, indeed, has been codified in medical codes of ethics (see Chapter 4). Lo (2019) appears to discount professional codes of conduct as meaningful to ethical conduct in everyday medicine. Lo lists many good reasons for doubting professional codes of conduct as implemented. Indeed, many of these points are demonstrated in the historical development of ethics as described in Chapter 4. Perhaps Lo's position overgeneralizes to deny any possible benefit professional codes of conduct could have. Perhaps this may be a situation of "throwing the baby out with the bathwater."

To be sure, Lo does argue for training in professionalism, but writes "Professionalism training assumes that all positions share a common view on what maxims should guide physician behaviors and what actions are appropriate in a specific clinical situation; that is, there is a shared understanding of what is right to do." Lo then discounts, to some degree, professionalism training by arguing that professionalism is a vague term, as though vague terms cannot be useful. Love and beauty are vague terms but seem to be very forceful in driving certain behaviors. Lo also indicates that physicians bring with them their own cultural, secular, and religious values into their profession. But is not quarantining those personal idiosyncratic values necessary in a pluralistic modern liberal democracy? Indeed, that is the point of John Rawls' "Veil of Ignorance" necessary to provide means to justice. One would find physicians or healthcare professionals overtly imposing their religious values on a patient as unjust. But is it less unjust if the imposition was hidden? It might be too much to ask patients, physicians, healthcare professionals, insurers, or government agents to give up their own cultural, secular, and religious values; nonetheless, these can be quarantined, hopefully sufficiently, to allow for justice. Interestingly, judges are expected to quarantine their values and defer to the law and precedence that are contrary to their values, at least most of the time. Specific methods for such quarantining of personal values are addressed in Chapters 3 and 5.

The interesting and operative words in the quote from Lo are "knows what is the right thing to do." The question left unanswered is how does the professional know "what is the right thing to do?" Indeed, a major theme of this book is how does one go about knowing what is the right thing to do and how to have certainty in knowing what is the right thing to do? Lo's use of the term "maxims" appears to support a Principlist ethics, although there is a considerable reason to be skeptical or, at the very least, mitigate conclusions from a Principlist perspective, perhaps in line with narrative, phenomenological, virtue, care, and feminist ethics (discussed in Chapter 2).

The practice of everyday medicine is far from simple, but is demanding

The driving force in this text is to see justice done. Consider the virtue of righteous indignation. Aristotle wrote

> The deficiency, whether it is a sort of "inirascibility" or whatever it is, is blamed. For those who are not angry at the things they should be angry at are thought to be fools, and so are those who are not angry in the right way, at the right time, or with the right persons; for such a man is thought not to feel things nor to be pained by them, and, since *he does not get angry, he is thought unlikely to defend himself; and to endure being insulted and put up with insult to one's friends is slavish* [italics added]
> (Aristotle, Nicomachean Ethics, Book IV Section 5., Translated by W.D. Ross, http://classics.mit.edu/Aristotle/nicomachaen.html)

Thomas Aquinas wrote

> [H]e that is angry without cause, shall be in danger; but he that is angry with cause, shall not be in danger: for without anger, teaching will be useless, judgments unstable, crimes unchecked, and concludes saying that "to be angry is therefore not always an evil."
>
> Aquinas (1485)

Jesus expelled the moneylenders out of the temple (Gospel of Matthew 21), perhaps an example of righteous indignation.

Aristotle noted that anger is positive as it is self-protective, and perhaps for those charged with the care of others when done in the "right way, at the right time, or with the right persons." Note that the "right" way, time, or persons is precisely the domain of justice. If one considers a form of maleficence (harm), it still may be justice following from Thomas Aquinas' Principle of Double Effect (see Case 8).

In this text, justice is seen as an outcome. The great challenge is dealing with the complexity and diversity of ethical experiences. The ecology in which the ethics of everyday medicine is embedded is far more complex and multidimensional than typically encountered in "headline" ethics, which tend to be dyadic. Even a cursory deliberative consideration of who are the shareholders and stakeholders involved in the everyday practice of medicine demonstrates a very wide array to include patients, physicians, healthcare professionals, insurers, governmental regulatory agencies, legislators, governors, presidents, educators, and society. Even the division into shareholders and stakeholders creates a simplifying dichotomy that likely is false when the full complexity of the ethical ecology is considered.

Procedural knowledge and ethics

The notion of procedure here is a behavior, where the course of the behavior is not generated directly by any conscious deliberative declaration. The success of the behavior argues for a procedural knowledge that is different in important ways from deliberative declarative knowledge, the kind of knowledge exemplified in a conscious description. For example, one can drive home from work and later not remember anything about the drive, such as where the person turned right or left. Clearly, there is knowledge because otherwise the behavior would have been random. This is termed procedural knowledge. Interestingly, this phenomenon was termed "reflexive" introspective consciousness by David Armstrong (1978), which would strike many as an oxymoron, but this does illustrate the problem of understanding behaviors that are just or unjust (see Chapter 1). To be sure, the driver could tell you, declare if asked, where each turn is and whether it is right or left. This would be evidence of declarative knowledge but it likely would be post hoc reconstruction, not evidence of the type of knowledge actually used.

As noted earlier, most ethical knowledge is procedural and not subject to deliberative, declarative conscious analyses. Note that procedural knowledge in ethics does not obviate a principled ethics of the type usually associated with declarative knowledge. Indeed, careful, deliberative, declarative conscious considerations operate on the intrinsic neurophysiological mechanisms that generate the procedural behavior. As the procedures are done all the time, they are "in front of one's nose," but as George Orwell wrote, "To see what is in front of one's nose needs a constant struggle" (George Orwell, In Front of Your Nose, Narrative Essays, Harvill Secker). The latter constant struggle may very much require a deliberative, declarative, conscious effort.

There are a number of reasons why the ethics of everyday medicine may be procedural, that is, ethical or unethical behavior is part of the procedures made every day in medicine. Only afterward are reasons posited in the declarative sense and then, only if asked. The classic mantra of clinical teaching is "see one, do one, teach one" procedures, like memes that are handed around without necessarily any conscious awareness of the transmission. The distinction emphasized here is the difference between procedural knowledge, as evidenced in nonrandom behaviors, and declarative knowledge that can be described verbally, suggesting a conscious and self-conscious deliberative awareness. Acting based on procedural knowledge is not necessarily a bad situation. Indeed, Immanuel Kant recognized the importance of ethical instincts, inherently procedural, as the best way to ensure ethical behavior. Kant wrote

> In the physical constitution of an organized being, that is, a being adapted suitably to the purposes of life, we assume it as a fundamental principle that no organ for any purpose will be found but what is also the fittest and best adapted for that purpose [author – Kant is suggesting a type of evolution nearly a century before Darwin's "On the Origin of Species" in 1859 and anticipating modern ethicists use of a teleological evolution to explain ethical or moral behavior such as Joshua Greene, 2014]. Now in a being which has reason and a will [author – taken here as deliberative declarative knowledge], if the proper object of nature were its conservation, its welfare, in a word, its happiness, then nature would have hit upon a very bad arrangement in selecting the reason of the creature to carry out this purpose. For all the actions which the creature has to perform with a view to this purpose, and the whole rule of its conduct, would be far more surely prescribed to it by instinct [author – taken here as procedural knowledge], and that end would have been attained thereby much more certainly than it ever can be by reason. Should reason have been communicated to this favored creature over and above, it must only have served it to contemplate the happy constitution of its nature, to admire it, to congratulate itself thereon [author – taken here as examples of post-hoc declarative commentary on the consequence of procedural knowledge], and to feel thankful for it to the beneficent cause, but not that it should subject its desires to that weak and delusive guidance, and meddle bunglingly with the purpose of nature. In a word, nature would have taken care that reason should not break forth into practical exercise [author – taken as procedural knowledge], nor have the presumption, with its weak insight, to think out for itself the plan of happiness, and of the means of attaining it. Nature would not only have taken on herself the choice of the ends, but also of the means, and with wise foresight would have entrusted both to instinct.
>
> Kant (2001)

To summarize in a cruder sense, when confronted with a social interaction with others that could significantly affect one's well-being (and ability to procreate), it is better to just react (appropriately) than to have to think about it. Further, thinking about things in a deliberative way may lead to a "paralysis" of initiative because of increasing uncertainty with increasing analyses in complex situations and circumstances.

As Kant suggested, reasoned (declarative) knowledge is a commentary on the actions driven by procedural knowledge just as subsequent modern ethicists argue (Hauser, 2006; Appiah, 2008; Greene, 2014). This theme is taken up in detail in Chapters 1 and 2. What may be different from previous approaches is the concept that the procedures that represent procedural knowledge of justice are neurophysiological, ultimately the dynamics of changes in the voltages in neurons (see Chapter 1).

The analogy to the neurophysiology of ethics is that post hoc rationalizations to different principles, theories, laws, case precedents, and virtues do not mean that the different principles, theories, laws, case precedents, and virtues reflect a fundamentally different ontology (ethical realities). In other words, the different principles, theories, laws, case precedents, and virtues need not be incommensurable. Adoption of one principle, theory, law, or precedent need not obviate the others even as they appear contradictory. Reconciliation between what appears to be incommensurables may be successful and thereby productive, which is explored in Chapter 2.

The value of ethical instincts, as they determine the ethics of the procedures or actions, presupposes that what good physicians and healthcare professionals do defines the sound ethics of everyday medicine as it follows from the ethical instincts in the good physician or healthcare professional. Indeed, fostering the expression of these instincts is fundamental in virtue ethics (Chapter 2). However, one should be cautious not to think that the ethical instincts are created by the virtue ethics.

Some have argued that further advances in medical knowledge and practice would create better medical practices and procedures in the hands of virtuous physicians, which will naturally lead to better ethics, thus obviating any need for a declared code of ethics. This position was explicit in a number of scientists and academically inclined physicians opposed to the first code of ethics of the American Medical Association in 1847 (Montgomery, 2019a). Similarly, the same objection was raised by biomedical scientists against the Belmont Report (Office of the Secretary, U.S. Department of Health, Education and Welfare, 1979) that established institutional review boards (IRBs) to regulate human subject experimentation. Good biomedical research ethics are just what good biomedical scientists do. Yet, ethically questionable human research continued subsequent to the Belmont Report (Lurie and Wolfe, 1997; Stokstad, 2004; Branca, 2005; Resnik and Wing, 2007).

The ascendency of evidence-based medicine, as instantiated in randomized controlled trials, could be a metaphor for a type of procedural ethics inherent in the everyday practice of medicine. For example, if a randomized controlled trial demonstrates that *treatment with a ground up kitchen sink* results in a statistically significant greater improvement over a placebo or alternative therapies for *disease B*, then one would have little recourse but to use a *ground up kitchen sink*. One need not, even if one could, explain declaratively how a *ground up kitchen sink* works; the onus is to use a *ground up kitchen sink*, even though one could not offer any further justification other than the procedure of the randomized controlled trials. One would hope that there is some pathophysiological explanation, expressible declaratively, for why a *ground up kitchen sink* should work, but this need not be the case. Presumably, actions taken by administration of a *ground up kitchen sink* do address the pathophysiology in a process independent of what can be expressed declaratively. Unfortunately, the significant concerns about the lack of reproducibility in biomedical research provide little reassurance that this always will be the case (Montgomery, 2019b).

Procedural statistical significance and clinical meaningfulness

Randomized controlled trials can only indicate statistical significance, which is only an internal measure. This epistemic device does not necessarily translate to clinical meaningfulness (Montgomery and Turkstra, 2003; Montgomery, 2019a). Clearly, the clinical meaningfulness of the study is what is important to the practice of medicine, not the statistical significance, and ethics is central to clinical meaningfulness. The lack of reproducibility in biomedical research demonstrates that statistical significance can exist even when clinical meaningfulness is absent. Thus, it is not clear that evidence-based medicine will supplement the need for an independent ethics of everyday medicine. Indeed, evidence-based medicine risks are becoming dogma such that any ethical concerns are buried.

The usefulness of any result from randomized controlled trials requires some threshold applied to the statistical measure, such as at what point does the mean change in *disease B* with *treatment A* become statistically significant and another, and perhaps different, threshold at which the result becomes clinically (and ethically) significant. Indeed, the threshold for statistical significance should be derivative of the clinical meaningfulness (Montgomery, 2019b). For example, consider *treatment A* for *disease B*. *Treatment A* has no adverse effects and costs virtually nothing. *Disease B* is horrible, debilitating, and painful. However, the randomized controlled trial demonstrates that the improvement was associated with a level of significance where $p < 0.1$ rather than the typical $p < 0.05$. The $p < 0.1$ significance means that there was a 10% chance that the benefit of *treatment A* was a fluke, whereas $p < 0.05$ means that there was only a 5% chance of a fluke. But who would not take *treatment A* if they had *disease B* even if there was a 10% chance of a fluke? In this case, $p < 0.1$ would also be clinically meaningful; perhaps $p < 0.4999$ would be clinically meaningful.

The question becomes what is meant by clinically meaningful? Ultimately, this is an ethical question related not simply to the quality of life or the quality of death for the patient. Rather, the ethical questions relate to the full range of consequences that go well beyond the patient. What are the resources expended to provide the treatment? What are the resources expended if the treatment is not provided? How does the expenditure of resources directly on the patient affect other priorities? What is the moral obligation to expending the resources? What is the precedent set for future similarly situated patients? Thus, the application of results from randomized controlled trials is not a statistical issue or even strictly a scientific issue, it is an ethical issue encountered daily in everyday medicine (Montgomery and Turkstra, 2003).

What constitutes the ethics of everyday medicine?

The ethics of everyday medicine would seem to be intuitive and procedural rather than declarative. But as primarily procedural, subsequent declarative explication is likely problematic. Yet, discussions require declarations. In an excellent literature review, Zizzo et al. (2016) described the vagueness of the term, everyday medicine, and sought to clarify the terrain from the perspective of what was published. While an admirable and necessary effort, it risks a certain ascertainment problem in that those who did write on the subject may have been motivated in a manner that is not characteristic of the vast majority of practitioners in everyday medicine. Further, the notion of everyday medicine labors in its conflation with other forms of ethics, such as biomedical ethics and human studies ethics.

Bioethics is very often discussed in the context of human experimentation, which is not directly related to the situation of everyday medicine, except perhaps related to the issue of the off-label use of approved drugs, biologics, and devices, such as those approved by the FDA for other indications. Interestingly, the Belmont Report, which established institutional review boards governing human experimentation, expressly stated that the review of standard medical care was not to be the purview of IRBs (Taylor, 2010). Bioethics in the research setting typically involves trials of diagnostic or treatment methods whose beneficence or risk of maleficence to the subject is unknown and the obligations to subject's autonomy reflect the unknown. In large part, the research ethics is judged by the potential beneficence, maleficence, and autonomy of society where the subject benefits to the degree they are a member of society. The distinction from the ethics of everyday medicine can be seen in the context of innovative therapies that are being used every day (somewhere) in medicine. The question becomes when do innovative therapies represent a reasonable practice of medicine rather than research? The Belmont Report specifically excluded innovative therapies, holding these were not research. The fineness of that distinction has not been lost, with some institutions establishing committees to review proposed innovative therapies separate from institutional review boards.

Interestingly, the US FDA has complicated the matter of the distinction between research and innovative care by requiring IRB approvals for some approved devices, for example, that were allowed marketing under a humanitarian device exemption. Typically, these devices were FDA approved but not through the typical randomized controlled trial paradigm. Often, sufficient case experience warrants approval as safe and effective. The rationale for requiring IRB participation in what should be considered standard medical care is that the informed consent on the part of the patient or the patient's legal representative is difficult precisely because of the lack of randomized controlled trials. Being experts in informed consent, the IRBs may have been thought better to ensure informed consent than physicians and healthcare professionals. The result is a further conflation between human research ethics and the ethics of everyday medicine, with the power to enlighten and to confuse.

In looking for dimensions by which to characterize the ethics of everyday medicine, Zizzo et al. (2016) used the following description of everyday medicine in contrast to other forms of ethics. Everyday medicine was not "headline" material as in news accounts; not acute, not involving significant loss of life or limb; and, perhaps most importantly, overlooked and unrecognized. However, this dimension does not allow for a ready categorization as this dimension likely is a continuum. Such a continuum only allows ready identification of the extremes with the risk that the extremes come to be the definitions of archetypes. Zizzo et al. (2016) go on to define four core features of everyday ethics:

> First, everyday ethics encompasses real-life issues; everyday ethics is not hypothetical, it includes events that occur often and affect the many.

> Second, everyday ethics is situated in common interactions between people; issues may be especially tied to relational and contextual factors, but also systemic and organizational factors.

Third, everyday ethics varies depending on the agent or stakeholder, including clinicians (nurses, physicians, et cetera), patients, and their relatives and caregivers. Notably, while everyday ethics affects more than just clinicians, it is often associated with professionalism.

Fourth, everyday ethics is often captured in a folk taxonomy, or nonexpert ethics language, and its ethical dimension may not always be apparent to stakeholders.

It is difficult to see how these four features cleanly separate everyday ethics from any other form. For example, it is very difficult to see how any form of ethics would not be concerned about real-life issues. Indeed, analogical approaches to ethics, the Anti-Principlists, distinguish themselves from Principlist ethics on the interest of the real-life issues of the individuals involved while, perhaps falsely, describing the Principlist as aloof. However, even Principlists shy away from hypotheticals (Beauchamp and Childress, 2012).

As discussed in Chapter 2, the Principlist derives ethics from an empirical observation of behaviors from a group, collection, or sequence of experiences of the individual or particular, from which ethical principles, moral theories, laws, case precedents, and virtues are induced. The induced ethical principles, moral theories, laws, case precedents, and virtues then fall back on the individual cases and issues to be adjudicated. To hold the Principlist otherwise would be like arguing that a biological scientist never considers an exact real-life biological issue. Similarly, the analogical Anti-Principlist ethicist cannot avoid using some notion of generalized ethical principles, moral theories, laws, case precedents, and virtues.

On the third count that "everyday ethics varies depending on the agent or stakeholder, including clinicians (nurses, physicians, et cetera), patients, and their relatives and caregivers. Notably, while everyday ethics affects more than just clinicians, it is often associated with professionalism," again it is not clear that this constitutes a real distinction (as discussed earlier). It would seem that the ethics of other than everyday medicine are more patient-centric. However, historically, most ethics in everyday medicine are not actually patient-centric but rather physician-, business-, and government-centric (see Chapter 4).

As will be argued, there are not many different ethical mechanisms in the ethics of everyday medicine compared to the ethics of other situations or circumstances. Rather, it is the obligation implied by these mechanisms that differs. For example, when contrasting medical ethics in the routinely accepted practice of medicine with those of biomedical research, both hold to the same ethical mechanisms. However, in routine medical practice, the object of benefit or good is primarily and directly the immediate patient. In biomedical research, the object of benefit or good is society, as benefit or good to the immediate patient cannot be assured. In both settings, an obligation to patient respect is obtained; however, implementation of the obligation to patient respect, in the sense of informed consent, differs. In routine medical care, informed consent often is implicit based on a common understanding—the patient seeking the help of a physician or healthcare professional understands that certain actions will be taken, such as being examined. In the case of biomedical research, the same principle of respect operates, but the obligation is explicitly codified and not left to be implicit between the patient and the researcher because governmental intervention was necessary by human research scandals.

Fundamentally, the difference between the ethics of everyday medicine and other ethics, such as dramatic or "headline" ethics, is a matter of degree in terms of the obligations to the ethical mechanisms, be they ethical principles, moral theories, laws, case precedents, or virtues. The range and potency of the obligations to the ethical mechanisms in the interests of all relevant shareholders and stakeholders differ in degree between the ethics of everyday medicine and the others. The ethics of everyday medicine differs in degree between the ethics affected by procedure versus conscious deliberate declarative consideration directly necessitated by explicit informed consent. Yet, quantitative differences can be so great as to result in qualitative differences. The addition of another single grain of sand can cause a progressively increasing size of a sand pile to collapse. Perhaps the addition of single grains of sand before the collapse is analogous to the ethics of everyday medicine, and only when the sand pile collapses does the ethics become the dramatic and "headline" ethics. The response to either situation may be different, even if the underlying ethical mechanisms are the same. As such, the ethics of everyday medicine deserve their time and are worthy of effort on the part of all shareholders and stakeholders—to be supported by the dedicated scholarship.

Beneath the radar

There is the intuition that somehow the ethics of everyday medicine are plainly different. For "headline" ethics, it may be only a matter of degrees on the dimensions of notoriety, being made the foil for other interests (political or religious, for example), severity of consequences, or the disruptiveness. A premise from which this text operates is that everyday medicine is at one extreme of the dimensions with headline ethics at the other extreme. The operations of the ethics of everyday medicine are not notorious, being far more often taken for granted. As these ethics are taken as "standard operating procedures," there appears to be a general tacit acquiescence to them and, therefore, little use as a foil for other interests (political or religious, for example). Certainly, on appearance, they are not particularly disruptive. The severity of the consequence is not a concern, as the consequence, however severe for those affected, it is taken as the natural history of the human condition.

One should not conclude by what might appear as a timidity of ethical issues of everyday medicine that the vast majority of physicians and healthcare professionals operate in an ethical vacuum and that someone has to step in to fill the void. This is not likely the case. It would not be difficult to overlook something that is inherent in the procedures of everyday medicine as they do not generate apparent conflict, or only rarely. One need only ask any physician or healthcare professional whether they think ethical practice is important and whether they practice ethical medicine in their everyday actions—the answer would be unanimously yes, most likely. Ask how often in their practice are they concerned about their own ethical behavior and the answer likely is extremely rarely.

In a study of 285 patients totaling 562 office visits, ethical problems were identified in 84 (30%) patients and in 119 (21%) office visits (Connelly and DalleMura, 1988). On a general inpatient medical service, the incidence of ethical problems identified by residents occurred in 7 of 179 cases (3.9%), yet when a bioethicist engaged in discussions, the incidence increased to 16 of 92 cases (17%). This discrepancy suggests a clear difference in the perception of ethical issues between medical residents and ethicists. However, it would be a mistake to think that medical residents are naïve to ethics. More likely, their notion of ethics is different from those of ethicists. The fact that medical residents could come to a position closer to ethicists is encouraging. Yet, a study in 2006 showed that 61% of biomedical ethics committees of hospitals reported less than 6 cases per year.

In a survey of hospital-based outpatient physicians in their use of ethics committees, factors related to the use or nonuse of ethics committees were explored (Orlowski et al., 2006). Nonusers had more hours of education related to ethics than users, suggesting that the nonusing physicians felt themselves capable of dealing with ethical problems, although there could be other explanations. Users as well as nonusers did not think that ethicists were uniquely qualified to resolve ethical problems. This was most clear in the opinion of surgeons who felt that members of the ethics committee could not grasp the issues involved in surgical cases. This could be considered evidence that the ethics of everyday medicine are not absent or overlooked. Rather, they are embedded in practice and are a form of procedural, if not declarative, knowledge. As the ethicist is not necessarily an expert in the procedures, it may seem that ethicists are in no position to give advice.

In an interesting survey of physicians selected randomly from the American Medical Association Master List of Physicians and Medical Students for Mailing Purposes utilizing a computer-assisted telephone interview, 344 subjects responded at a response rate of 64% (Hurst et al., 2005). Of the respondents, 90% could recall a recent event that rose to the level of an ethical concern. Certainly, this suggests that there is not an insensitivity to ethics. It is interesting that the recollection of an ethical concern was most common among critical care specialists, then oncologists, with general internists the least. The themes of issues in the recollected ethical problems were pretty much as might be expected. These included incurable or dying patient, 65%; disagreement, 54%; situations including others besides physician and patient, 79%; doing what is best for the patient, 92%; doing what is best for the family, 9%; respect for the patient's self-determination, 47%; and patient advocacy, 38%. Likely, these concerns were recognized by the physicians and healthcare professionals. Their ethical discomfort raised the cases to a level sufficient for deliberative review. This is not likely in the cases of ethical concerns considered in Part 1 of this text, which very frequently escape notice. The study did not, nor could it, delve into the exact actions taken associated with these themes and whether the incident, which was the subject of concern, was resolved appropriately.

More telling are the conclusions reached, albeit qualitative, but they do constitute expert opinions and thus are not of insignificant credibility. The authors wrote "The avoidance of conflict, between any parties, emerged as a goal in its own right. It often seemed to take priority over other goals… Looking for assistance and avoiding conflict both contributed to protecting, or attempting to protect, the integrity, or wholeness, of respondents' conscience and reputation, and the integrity of the group of individuals participating in the decision." From this description it would appear that accommodation rather than justice was the primary goal, unless one is willing to argue that accommodation is synonymous with justice. A particular response quoted in the paper was "Um ... first [decision] was mine, to bring issues up with the family. Second [was] to leave [the] patient out of [the] decision-making process. [The] third decision was to involve [the] family ... [and] to allow [the] family's expressed wishes to supersede those of the patient ... [the] other decision is leaving nurses and other people out of the process" (Hurst et al., 2005). Perhaps a questionable ethical response, at least it does not appear to be particularly patient-centric but rather heavily physician-centric (see Chapter 4).

The descriptions of the responses quoted in the paper all held the patient as the focus of the discussion, and the issues seem to be how to implement what was best for the patient in the context of other persons and situations involved. Yet, the descriptions of the patients are brief and dispassionate, perhaps less than they should be in a patient-centric ethic (Chapter 4). One could infer that the patient was a foil, a stakeholder but not a shareholder, with others acting as proxies for the patient's interests, at best, or acting paternalistically, at worst. To be sure, it is not clear what "patient-centric" ethics might be. Indeed, it is hoped that the cases presented in Part 1 and the approaches described in Part 2 may lead to a clearer notion of patient-centric ethics in everyday medicine. One of the purposes of this book is to raise awareness of the ethical issues inherent in the everyday practice of medicine even if they would not be recognizable generally. Laws are not invalidated just because violators are not caught. At the very least, the review of cases presented in Part 1 may offer new insights into what is occurring today and, perhaps, if there is the inclination, how things may change.

Corporate personhood as a stakeholder as well as a shareholder

A central theme of this text is the complex ecology surrounding ethical problems. The role of entities other than the patient, physician, and healthcare professionals, such as insurers and regulators, was considered in terms of the interest of society as it relates to the patient's healthcare. Governments act by proxy for the citizens and justification for that empowerment was that governments are expected to act on the best interests of the citizens so governed. The government gained shareholder status by proxy and also has stakeholder status because the government depends on the assent of the citizens. Insurers are empowered by the contracts entered into, for example, with the patient, physician, healthcare professional, and governmental regulators. The discussions did not presume deontological rights for the governments or insurers.

The situation has changed by the establishment of corporations as persons, in the extent that corporations are engaged in the provision of healthcare, such as providing insurance. Recent legal decisions appear to provide corporations on par with individuals. Thus, whatever deontological rights are granted to individuals, they also are granted to corporations (with the notable exception of casting a vote, although not prevented from influencing the vote of others). Corporate personhood allows corporations to hold assets and enter into contracts as they are natural groups. The result is now the personal interests of corporations, particularly those privately held, competing for the interests of patients. Any relevant ethical system for the everyday practice of medicine cannot ignore the potential impact of corporate personhood, for example, the US Supreme Court case of *Burwell v. Hobby Lobby*, 573 U.S. (2014), where the Hobby Lobby corporation has the right to not pay for birth control agents for its employees. This is discussed further in Chapter 4.

Dissociation between healthcare expected and healthcare delivered

Ethical problems can manifest when there is a distinction between what the patient expects and what is delivered (Cases 2, 3, and 6). Such difference is not necessarily unjust. For example, from a Contractualist perspective, both sides acknowledge the difference and come to some mutual acceptance of the difference. Lack of transparency may not be unjust. For example, it may be that citizens recognize that they, as a collective, cannot come to agreement on an important issue. However, the collective of citizens may agree to let someone else decide; further, the decision stands even in the face of opposition by any citizen or group of citizens, as in a Ulysses pact. Further, they may recognize that transparency in the decider's decision-making would cause maleficence, harm to the citizen or body politic.

The critical point is that an establishment of acceptable lack of transparency results from the transparent establishment of the Ulysses pact that could be revoked. As seen in many of the cases presented in Part 1, there were significant differences between what a patient expected, or should have expected, and that is expected to be provided by the physician, healthcare professional, healthcare delivery system, insurer, or governmental agency. Actions taken in the face of such differences by acts of hard or soft paternalism (overriding a patient's choice or not offering a choice, respectively) or failure to obtain full informed consent generally require a lack of transparency. The critical question is who is justifying the lack of transparency? If the lack of transparency originates in the decision of the physician, healthcare professional, healthcare delivery system, or insurer, there would be significant challenges to just actions. Even a justifiable lack of transparency on the part of the government first requires full transparency in the transfer of the right to not providing transparency.

In the everyday practice of medicine there are numerous opportunities for discrepancies between healthcare expected and healthcare delivered. For example, the patient may be operating from a Deontological moral theory that the patient is entitled to a particular service. The physician or healthcare professional may respect that deontological right but the actions are determined by a Libertarian moral theory affected by contractual obligations to the insurer or governmental policy. The physician or healthcare professional simply could make the discrepancy clear, thereby forcing the conflict to be between the patient and the insurer or governmental agency. Yet, such actions are fraught and thus rarely conducted, as discussed in Case 1.

Another example is that patients may hold they have a right to healthcare, whereas the government holds that the patient only has a right to purchase healthcare, often through insurance. The healthcare provided through such purchases is limited by contract. Rarely is this discrepancy ever discussed in a fully transparent manner. A classic example of the lack of transparency is in "gag" laws prohibiting physicians and healthcare professionals from discussing anything a patient might benefit from if it is not provided by the insurer. Note that such "gagging" may not be explicit, indeed it rarely is. However, this does not mean that "gagging" does not take place enforced by indirect punishments. The risk is that some physicians and healthcare professionals will "gag" themselves.

Justice as institutional

In writing about justice and institutions, John Rawls makes the distinction between the ideal of justice and the social and political practices that are meant to facilitate achievement of that ideal (Rawls, 1999). He notes that the ideal held may be correct and therefore just, but the social and political practices are unjust. However, it is hard to envision situations where the ideal is unjust yet the social and political practices are just.

Rawls' discussion of justice in the context of institutions is applicable to the relationships among physicians, healthcare professionals, and patients (including legal representatives for patients). Indeed, these relationships are strongly shaped by governmental regulations, policies and bylaws of clinics and hospitals, contractual relations with insurers, expectations of peers, and traditions established during the course of training and inculturation into the professions. Gone are the days of solo practitioners; in 2011, 35.3% of physicians were employees of organizations with 50 or more physicians (Welch et al., 2013). The trend suggests that even more physicians are in large organizations. Indeed, moves to establish accountability of care and capitation reimbursement schemes only reinforce the notion of a collective rather than individual independent contractors.

There are situations or circumstances where institutional justice is necessary. The premise is that justice follows from the consensus of reasonable persons blinded to their own self-interest. Yet, like Diogenes searching for the honest man, it just may not be possible for everyone or anyone to be reasonable or blinded to their own self-interests. Nonetheless, some consensus so as to fashion ethical principles, moral theories, laws, case precedents, or virtues is necessary. Means are needed to accomplish such consensus in the face of no guarantee of reasonableness or blindness to one's self-interest. However, the consensus is derived from the collective by the consensus being institutionalized and taken out of the hands of the individual who could not possibly be expected to be reasonable or blinded to self-interest to the degree necessary, perhaps on the basis of a Ulysses pact as described earlier. However, this is not to say that the institutional policies are unreproachable, and indeed require constant surveillance. However, their advantage is all the more reason that institutional policies should be considered in ethical decision-making in everyday medicine. These issues are explored in more detail in Chapter 3.

Best of all possible worlds

It may be that the current status of the ethics of everyday medicine is not the "best" but the "best of all possible worlds." Interestingly, the notion "best of all possible worlds" was an attempt by the philosopher Gottfried Wilhelm Leibniz (1710) to reconcile evil in a world created by an omnipotent loving god. The implication for ethics is that the current system may not be the "best" but it is the best of all possible systems and any change likely would be for the worse. Clearly, there is tension between striving for the "best" at the risk of ending up with what is less than the best of all possible worlds. At the worst, acceptance of the "best of all possible worlds" may bred fatalism or, at the very least, an inertia that leads to complacency (likely a bit of both in medical ethics).

The question in assessing the ethical consequence often becomes one of culpability of those participating in the delivery of poor care, which often is seen as depending on their intentions, *mens rea*. It is possible, although extremely unlikely, that a physician or healthcare professional intentionally commits poor medical practice. Rather, as the practice of medicine very often operates from procedural knowledge, for example, in pattern recognition or what are called illness scripts (Montgomery, 2019a), it is likely the poor practice may not even be recognized in a conscious, declarative, and deliberative sense.

Often, the issue is whether physicians, healthcare professionals, or other responsible persons should have recognized the actual or potential poor medical practice. However, this question is misleading as it presupposes a declarative knowledge, which may not be appropriate, as the practice of medicine very often operates from procedural knowledge as described earlier. In that case, the culpability may apply to those who instilled the procedural knowledge, as mentors to the physicians and healthcare professionals or in guidelines or decision trees offered by experts. Further, culpability may lie with those who supervise the physician or healthcare professional. For example, in US health law, hospitals are charged with the accreditation, credentialing, and supervision of the physicians and healthcare professionals who operate within the institution (Furrow, 2017, pp. 434–436).

In this context, it may be instructive to examine the use of notion of Continuing Medical Education (CME) as a means of ensuring best care. A critical view suggests a very vague answer to whether CME is effective (Marinopoulos et al., 2007). Most published literature on CME focuses on pedagogy. A controlled trial of CME regarding acute care of patients with myocardial infarction demonstrated statistically significant increases in objective measures of good care following CME programs. The average score improved from 48.5% to 60% ($P < 0.001$) (White et al., 1985). To be sure, there was a statistically significant increase but was this ethically meaningful, as 40% still did not meet the objective measures even after the CME? From a Utilitarian moral theory, perhaps an expenditure of resources to increase appropriate care to 100% would reduce resources needed elsewhere in a situation or circumstance of moderate scarcity. From an Egalitarian perspective, the situation may not be unethical if every person has the same risk of being among the 40% not receiving optimal care. But then, optimal care becomes a lottery, which may be ethical if everyone is exposed to the same chance. However, there is reason to believe that optimal care may turn on socioeconomical status or geographic location and it is not clear that most reasonable persons blinded to their self-interest would find this just.

In the author's experience, compounding the problem is that CME is most often provided as an institutional function such as grand rounds. The issue becomes one of supporting CME efforts, where financial support by pharmaceutical and medical device manufacturers is discouraged. Yet, in one survey, only 42% of physicians were willing to increase registration fees to reduce dependence on commercial support (Tabas et al., 2011) and the fees paid did not significantly approach the actual cost of the CME. Further, often the topics are not directly relevant to identifying and correcting poor medical practices and certainly not in addressing problems in the ethics of everyday medicine.

The question then becomes who is responsible for any ethical problems that result from poor medical practices? It would not be an overstatement to say that the issues are complex. The first approach will be to pursue the issue in some of the cases in part I of this text. The cases set the stage and it is hoped that the subsequent analyses will expand and enlighten the issues. The cases are presented first to demonstrate their placement in procedural knowledge. In other words, these cases seem ordinary, and thus taken for granted as standard procedures, even though analyses suggest that such standard procedures are fraught ethically. The analyses are decidedly deliberative and declarative in nature in Part 2 of this text. This is to demonstrate that declarative and deliberative knowledge, such as principles, moral theories, and approaches that can be articulated in a conscious fashion, can affect behavior by the effort of relating the declarative knowledge to the procedures, thereby modifying the procedural knowledge.

Asymmetry in power and consequence

The asymmetry of power in affecting healthcare creates ethical risks by its very nature. Patients literally have to deal with fears that the physician or healthcare professional does not. This fear was described beautifully by John Donne in 1624 in his aphorism given here:

> I observe the Phisician, with the same diligence, as hee the disease; I see hee fears, and I feare with him, I overrun him in his feare, and I go the faster, because he makes his path slow; I feare the more, because he disguises his fear, and I see it with more sharpnesse, because hee would not have me see it...

> He [Phisician] knowes that his feare shall not disorder the practise, and exercise of his Art, but he knows that my fear may disorder the effect, and working of his practise.

Franz Ingelfinger, a physician and esteemed editor of the *New England Journal of Medicine*, the preeminent medical journal, wrote of his own experience when diagnosed with cancer. He wrote

> I received from physician friends throughout the country a barrage of well-intentioned but contradictory advice. As a result, not only I, but my wife, my son and daughter-in-law (both doctors), and other family members became increasingly confused and emotionally distraught. Finally…one wise physician friend said, "What you need is a doctor." *He was telling me to forget the information I already had and the information I was receiving from many quarters, and to seek instead a person who would…tell me what to do, who would in a paternalistic manner assume responsibility for my care. When that excellent advice was followed, my family and I sensed immediate and immense relief* [italics added].
>
> Ingelfinger (1980)

It would seem that even the most expert are not immune to fear.

The longing to have the weight of fear removed from one's shoulders is powerful, perhaps so powerful that any other obligation to ethics may be trumped. Whether or not one can waive one's autonomy without jeopardizing the autonomy of others is problematic. For example, there is the notion of Ulysses contracts or pacts from Homer's "Odyssey" where Ulysses had himself tied to the mast and instructed his crew not to obey any of his orders so he could listen to the beautiful songs of the sirens as the sirens lured the crew to the ship's destruction (the crew had wax placed in their ears). The problem is when should the crew start listening to Ulysses, when Ulysses asked them to remove the wax? Ulysses pacts have been used in psychiatry, for example (Walker, 2012). When a patient requests suspension of his or her right to control actions to be taken, out of fear or exhaustion, how will the physician or healthcare professional know when to return it (Davis, 2008)? While understandably human, perhaps physicians and healthcare professionals need to be cautious in taking such paternalism as beneficence, even if implored by the patient.

It is tempting to think that the Ulysses pact is of concern only in episodic psychiatric disorders. When the patient is relatively well and capable of making good judgments, the patient enters into a Ulysses pact to be executed when the patient is not well—the question of deciding when the patient is not well notwithstanding. The question is whether a more subtle Ulysses pact exists in the practice of everyday medicine when the patient or the patient's relative says to the physician or healthcare professional "do what you think is best" and the physician or healthcare professional proceeds to do so without further discussion. At what point does the physician or healthcare professional "reconsult" to see if the directive "do what you think is best" continues in force? It would seem highly unlikely, but what if "do what you think is best" actually was taken as "do not reconsult?" This problem is compounded by the possibility of the physician or healthcare professional in doing what they think as best precludes adequate discussion to achieve a valid informed consent, whether the consent is implicit or explicit.

It is difficult to identify any profession with less accountability relative to their power to do good and to do harm than that enjoyed by physicians and healthcare professionals (Montgomery, 2019a). Even US Supreme Court justices with their lifetime appointments are subject to public scrutiny and possible impeachment, although only one justice in the history of the United States has been impeached but was not removed from the bench by the Senate. Further, the legislative and executive branches could alter the law that Supreme Court justices would be forced to follow. Although physicians can be sued for malpractice and have their license revoked, this is at best a minor constraint (Montgomery, 2019a). In the end it would seem that it is up to the individual physician or healthcare professional to demand of herself or himself what justice would demand.

It would be naïve to think that this asymmetry of power and consequences would not affect the ethical status of everyday medicine, as discussed in Cases 6–8. As such, asymmetries are inherent in gender roles and status, where feminist ethics have clarified many of the consequences of such asymmetries. Their insights are important in consideration of the ethics of everyday medicine (Chapter 2), but perhaps in ways that may seem counterintuitive.

A potential remedy to the asymmetry of power is to recognize where the power comes from as is discussed in Cases 2, 5, and 8 and in Chapter 2. In one sense, the power stems from the knowledge of the physician or healthcare provider. It is the demonstration of that knowledge that leads to governmental agencies providing a type of monopolistic privilege, that is, a license to practice. It is that knowledge that allows healthcare delivery systems to appoint and privilege physicians and healthcare professionals. Yet, there is an "equalizer" and that is informed consent. The process of informed consent is to translate and transfer sufficient knowledge to the patient or the patient's legal guardian transferring power to the patient. Thus, asymmetry of knowledge is rendered moot by the process of full informed consent. To be sure, there are other asymmetries, such as the power to prescribe controlled substances and referral to tests and treatments. Yet, even in this, the obligations to the patient by physicians and healthcare professionals impose responsibilities that, at the very least, qualify the independence of the physician or healthcare professional.

The structure of this text

Following this introduction, the text is divided into two parts. Part 1 consists of a number of case studies derived from common everyday experiences in the clinic and at the bedside that raise ethical questions. At first glance, many of the cases would not seem as ethical conflicts, as most would deem the cases not out of the ordinary and generally acceptable. That is precisely the point, ethical problems exist in the ordinary practice of everyday medicine even if not recognized as such.

The hope is that physicians and healthcare professionals will recognize many cases and situations in their own experience that resonate with the cases presented. Each case is then followed by an analysis in a declarative and deliberative fashion to be brought to bear on the ethics presupposed by the procedural knowledge that drives the behaviors. An additional set of cases are presented but without detailed discussion. The intent was that these cases would serve as a catalyst for the reader's own explorations. Part 2 is an epistemic and ontological exposition of the ethics of everyday medicine. The hope will be that after the analyses, a rational framework can be offered from which to begin to understand the cases of everyday medicine.

The cases are drawn primarily from the author's experience in over 40 years of practicing academic medicine. The author specialized in the diagnosis and treatment of movement disorders, such as Parkinson's disease, which is a subspecialty in neurology. However, the reasonable presumption is that the ethical problems encountered would resonate readily with clinicians of many stripes.

For each case study, a specific and particular case is provided. The reader is urged to come to his or her own response to the case before reading the analysis that follows. The reason for doing so is that readers do not come to ethical problems as a *tabula rasa* or blank slate. Even 15-month-old infants appear to have a sense of fairness (Schmidt and Sommerville, 2011). In teaching medical ethics to undergraduate freshman college students, the author has never seen a case where a student did not have a prior held opinion. Experience demonstrates that prior ethical notions had to be uprooted before new and sounder ethical reasoning could take hold, but this is no different from education in general, for example, uprooting Aristotelean physics that are intuitive to students in order to replace it with Newtonian physics. Note that such uprooting is not likely to eliminate the intuitive use of Aristotelian physics, but perhaps to quarantine such intuitive predispositions. The same can be said of ethical understanding and this topic is explored in Chapter 5.

References

Appiah, K.A., 2008. Experiments in Ethics. Harvard University Press, Cambridge, MA.

Aquinas, T., 1485. Summa Theologica.

Armstrong, D., 1978. Nominalism and Realism. Universals and Scientific Realism. vol. 1. Cambridge University Press.

Beauchamp, T.L., Childress, J.F., 2012. Principles of Biomedical Ethics, seventh ed. Oxford University Press, Oxford.

Belmont Report: Ethical Principles and Guidelines for the Protection of Human Subjects of Research, National Commission for the Protection of Human Subjects of Biomedical and Behavioral Research, Department of Health, Education and Welfare. 1978. United States Government Printing Office, Washington, DC.

Branca, M.A., 2005. Gene therapy: cursed or inching towards credibility? Nat. Biotechnol. 23 (5), 519–521.

Connelly, J.E., DalleMura, S., 1988. Ethical problems in the medical office. JAMA 260 (6), 812–815.

Davis, J., 2008. How to justify enforcing a Ulysses contract when Ulysses is competent to refuse. Kennedy Inst. Ethics J. 18 (1), 87–106.

Donne, J., 1624, Meditations VI. In Devotions Upon Emergent Occasions, https://www.gutenberg.org/files/23772/23772-h/23772-h.htm

Evans, J.H., 2012. The History and Future of Bioethics: A Sociological View. Oxford University Press, Oxford.

Fromm, E., 1955, The Sane Society, Open Road Integrated Media, New York

Furrow, B.R., 2017. Medical malpartice liability: of modest expansions and tightening standards. In: Cohen, I.G., Hoffman, A.K., Sage, W.M. (Eds.), The Oxford Handbook of U.S. Health Law. Oxford University Press, Oxford.

Greene, J., 2014. Moral Tribes: Emotion, Reason, and the Gap Between Us and Them. Penguin.

Hauser, M., 2006. Moral Minds: The Nature of Right and Wrong. Harper Perennial.

Hobbes, T., 1651. Leviathan: Or the Matter, Forme & Power of a Common-Wealth Ecclesiasticall and Civil.

Hurst, S.A., Hull, S.C., DuVal, G., Danis, M., 2005. How physicians face ethical difficulties: a qualitative analysis. J. Med. Ethics 31 (1), 7–14.

Ingelfinger, F.J., 1980. Arrogance. N. Engl. J. Med. 303 (26), 1507–1511.

Jonsen, A.R., 1998. The Birth of Bioethics. Oxford University Press, Oxford. (Chapter 3).

Kant, I., 2001. Fundamental principles of the metaphysics of morals. In: Wood, A.W. (Ed.), Basic Writings of Kant. The Modern Library, New York, pp. 153.

Kohn, L.T., Corrigan, J.M., Donaldson, M.S. (Eds.), 2000. Institute of Medicine (US) Committee on Quality of Health Care in America; To Err Is Human: Building a Safer Health System. National Academies Press, Washington, DC.

Leibniz, G.W., 1710. Essais de théodicée sur la bonté de Dieu, la liberté de l'homme et l'origine du mal (essays of theodicy on the goodness of God, the freedom of man and the origin of evil).

Lo, B., 2019. Resolving Ethical Dilemmas: A Guide for Clinicians, sixth ed. Wolters Kluwer.

Lurie, P., Wolfe, S.M., 1997. Unethical trials of interventions to reduce perinatal transmission of the human immunodeficiency virus in developing countries. N. Engl. J. Med. 337 (12), 853–856.

Marinopoulos, S.S., Dorman, T., Ratanawongsa, N., Wilson, L.M., Ashar, B.H., Magaziner, J.L., Miller, R.G., et al., 2007. Effectiveness of continuing medical education. Evid. Rep. Technol. Assess. (Full Rep.) (149) 1–69.

Mill, J.S., 1843. A System of Logic.

Montgomery Jr., E.B., 2015. Twenty Things to Know About Deep Brain Stimulation. Oxford University Press, Oxford.

Montgomery Jr., E.B., 2019a. Medical Reasoning: The Nature and Use of Medical Knowledge. Oxford University Press, Oxford.

Montgomery Jr., E.B., 2019b. Reproducibility in Biomedical Research: Epistemological and Statistical Problems. Academic Press, New York.

Montgomery Jr., E.B., Turkstra, L.S., 2003. Evidence-based practice: let's be reasonable. J. Med. Speech Lang. Pathol. 11 (2), ix–xii.

Office of the Secretary, U.S. Department of Health, Education and Welfare, 1979. Protection of human subjects; notice of report for public comment (PDF). Fed. Regist. 44 (76), 23191–23197.

Orlowski, J.P., Hein, S., Christensen, J.A., Meinke, R., Sincich, T., 2006. Why doctors use or do not use ethics consultation. J. Med. Ethics 32 (9), 499–502.

Osbeck, L.M., Held, B.S. (Eds.), 2014. Rational Intuition: Philosophical Roots, Scientific Investigations. Cambridge University Press.

Rawls, J., 1999. A Theory of Justice, revised ed. Belknap Press of Harvard University Press.

Resnik, D.B., Wing, S., 2007. Lessons learned from the Children's Environmental Exposure Research Study. Am. J. Public Health 97 (3), 414–418.

Rothman, D.J., 1992. Strangers by the bedside: a history of how law and bioethics transformed medical decision-making. J. Am. Hist. 79 (1), 346–347.

Rousseau, J.-J., 1762. The Social Contract; or, Principles Of Political Rights.

Schmidt, M.F., Sommerville, J.A., 2011. Fairness expectations and altruistic sharing in 15-month-old human infants. PLoS One 6 (10), e23223. https://doi.org/10.1371/journal. pone.0023223.

Stokstad, E., 2004. Children's Exposure Study Put Hold. (November 10). Available from: https://www.sciencemag.org/news/2004/11/childrens-exposure-study-put-hold.

Tabas, J.A., Boscardin, C., Jacobsen, D.M., Steinman, M.A., Volberding, P.A., Baron, R.B., 2011. Clinician attitudes about commercial support of continuing medical education: results of a detailed survey. Arch. Intern. Med. 171 (9), 840–846. https://doi.org/10.1001/ archinternmed.2011.179.

Taylor, P.L., 2010. Overseeing innovative therapy without mistaking it for research: a function-based model based on old truths, new capacities, and lessons from stem cells. J. Law Med. Ethics 38 (2), 286–302. (Summer) https://doi.org/10.1111/j.1748-720X.2010.00489.x.

Walker, T., 2012. Ulysses contracts in medicine. Law Philos. 31 (1), 77–98.

Welch, W.P., Cuellar, A.E., Stearns, S.C., Bindman, A.B., 2013. Proportion of physicians in large group practices continued to grow in 2009-11. Health Aff. (Millwood) 32 (9), 1659–1666. https://doi.org/10.1377/hlthaff.2012.1256.

White, C.W., Albanese, M.A., Brown, D.D., Caplan, R.M., 1985. The effectiveness of continuing medical education in changing the behavior of physicians caring for patients with acute myocardial infarction. A controlled randomized trial. Ann. Intern. Med. 102 (5), 686–692.

Zizzo, N., Bell, E., Racine, E., 2016. What is everyday ethics? A review and a proposal for an integrative concept. J. Clin. Ethics 27 (2), 117–128. (Summer).

Part One

Cases and explorations

Case 1

Case 1—A speech language pathologist was asked to evaluate a hospitalized patient with Parkinson's disease who was complaining of choking and coughing each time she ate or drank. Following the evaluation, the speech language pathologist recommended to the attending physician that the patient undergoes a video fluoroscopic examination of swallowing (VFES), otherwise known as a modified barium swallow. The attending physician refused. When asked, the physician said it was not necessary. The attending physician turned, walked away, and refused further discussion.[a]

Medical facts of the matter

As a medical fact, the speech language pathologist was correct. Pneumonia is the single leading cause of death in patients with Parkinson's disease—most likely from aspiration from dysphagia (problem with swallowing) (Akbar et al., 2015; D'Amelio et al., 2006; Suttrup and Warnecke, 2016; Wirth et al., 2016). Dysphagia affects more than 80% of patients with Parkinson's disease. It presents with oropharyngeal dysphagia (a type that involves the upper food passages, resulting in swallowing impairments) over the course of disease progression (Suttrup and Warnecke, 2016; Wirth et al., 2016). If the risk is recognized and the nature of the altered swallowing mechanisms determined, treatments can be instituted to reduce the risk of aspiration pneumonia and death. Treatments include modification of the diet, how the patient swallows, and, if necessary, finding an alternative means of nutrition that does not involve ingesting food or liquids. The treatments differ depending on the exact nature of the swallowing disorder.

The central problem in this case is how to recognize if the patient is at risk and, if so, how to reduce the risk. Apparently, the physician made an assessment of the risk being sufficiently low, as otherwise a refusal to conduct tests would have been arbitrary and capricious. Perhaps the physician made the decision because he never heard the patient manifest dysphagia, for example, by coughing or choking. However, many patients at risk for aspiration, as evidenced by subsequently developing pneumonia, never have a history of choking, coughing, or similar symptoms—silent aspirators. In a study of patients without symptoms of dysphagia, such as coughing or choking, over 50% demonstrate abnormal swallowing on objective testing, such as direct visualization of the swallowing act such as the VFES as recommended by the speech language pathologist (Suttrup and Warnecke, 2016; Wirth et al., 2016).

A number of alternative assessments exist for the risk of dysphagia. These include a clinical or bedside examination during which the patient is given foods and liquids of

[a] The case presented is hypothetical, although modeled on experience. No judgment is made relative to the correctness, appropriateness, or value of any actions described on the part of any of the participants. The purpose here is to describe and explore the variety of ethical considerations and consequences that depend on the perspective of different relevant ethical systems.

The Ethics of Everyday Medicine. https://doi.org/10.1016/B978-0-12-822829-6.00001-1

different viscosities and consistencies and the speech language pathologist listens for sounds of choking or coughing. In addition, the speech language pathologist looks into the mouth and throat for any residual foods or liquids (Hassan and Aboloyoun, 2014; O'Horo et al., 2015). From systematic reviews, it was concluded that the majority of bedside examinations reviewed "lack the sensitivity to be used as a screening test for dysphagia across all patient populations examined" and that "no bedside screening protocol has been shown to provide adequate predictive value for presence of aspiration" (O'Horo et al., 2015). Yet, the clinical or bedside examination is widely used for the diagnosis and treatment of dysphagia (Hassan and Aboloyoun, 2014).

Another problem with a clinical or bedside examination is its lack of specificity for the actual nature of the dysphagia. The specificity for the question of risk is very high. In other words, a patient who tests positive on a bedside or clinical examination, that is, demonstrates findings indicative of aspiration, has a high probability of actually having aspiration. However, even if the clinical or bedside examination was "positive," demonstrating a risk for aspiration, one cannot know the mechanisms and, therefore, cannot know what treatment to recommend. Thus, a patient with a positive clinical or bedside examination will have to be referred for objective testing but only after incurring a delay (Steele et al., 2007), often lengthy (see Case 2). Also, the sensitivity of the test is poor. Even those who "pass" or test negative on the clinical or bedside evaluation still have a high risk of aspiration and, consequently, should undergo objective testing. Thus, a referral for clinical or bedside testing only delays what needs to be done (see Case 2).

A VFES is an objective test with high specificity and high sensitivity even in patients who are silent aspirators (Martin-Harris and Jones, 2008; Miles and Allen, 2015). An alternative objective test is a fiberoptic endoscopic examination of swallowing (FEES), where a fiberoptic cable is inserted through the nose and lies just above the throat internally. The speech language pathologist is able to visualize the parts of the throat through the fiberoptic cable. Comparisons between VFES and FEES are complicated with some authors, suggesting an equivalence in diagnostic utility (Giraldo-Cadavid et al., 2017). However, in this author's opinion, often there is a confusion of terms, for example, claims to accuracy based on specificity, sensitivity, or some combination. The FEES looks for material administered by mouth remaining in the throat following a swallow. However, the FEES cannot visualize the actual swallow because the soft palate elevates and the pharynx constricts, obstructing the view. Also, FEES cannot provide the extent and range of information that is obtained by a VFES (Karkos et al., 2009).

The issue in this case is not whether the VFES or the FEES be performed. Rather, the question is whether any objective test beyond the clinical or bedside evaluation should be done. The speech language pathologist's opinion was that the VFES should have been performed, while the physician did not. The science that is known would support the position of the speech language pathologist, and one could argue that failure to perform the VFES was unjust.

Ethics from the "gut"

When the case was presented to a Master's graduate degree class of speech language pathology students, there were several different initial responses. Perhaps the most common was to default to the physician's decision. Note that this was not the same as deferring to the physician based on some presumed greater knowledge or expertise on the part of the physician. Rather, the response was in the vein that the physician is the "Captain of the Ship" and has the ultimate authority. (The origin of the concept of Captain of the Ship is interesting and derived from the problematic nature of assessing medical liability as discussed in Chapter 1. Captain of the Ship was not a statement of authority but rather responsibility. The latter has morphed into a concept of authority.) In this case, it is likely that the speech language pathologist knew considerably more than the physician about the diagnosis of swallowing, which is different than a diagnosis of the cause (in this case Parkinson's disease), and treatments that can be offered by the speech language pathologist. A good physician would take the expertise of the speech language pathologist into consideration, even if ultimately not pursuing the recommendations.

The question is what is the origin of this default deference to the physician? It is not due to a presumed greater knowledge on the part of the physician. In part, such deference may be institutionalized through history, such as a physician-centric ethic (see Chapter 4), and in the very structure of the practice of everyday medicine. Perhaps it is the historical institutionalization that renders the default to the physician as a "gut reaction," meaning requiring little declarative deliberative consideration. It is just "procedural" that the physician has defaulted to. It would be hard to deny that some part is the historical male-gendered physician dominance in medical decision-making. Feminist ethics shines a valuable light on this history. Most of these future speech language pathologists would not have pursued the matter further.

A smaller proportion of graduate students would have sought out the physician for further discussion in an attempt to convince the physician of the necessity of a VFES. When it was pointed out that the physician refused further discussion, most of these students then defaulted to the physician's decision and would not pursue the matter further, as described previously. A small number of students suggested discussing the situation with the patient and the patient's family; thinking that, at the very least, the patient and family should know what the correct course should have been. Perhaps there was the unspoken hope that the patient and the family would lobby the physician to perform the VFES. A rare student sought support from policies of their professional organization supporting the role of speech language pathologists in administrating a VFES.

The first observation is that every student had an opinion, the ethical soundness of that opinion notwithstanding. This should not be surprising, as most humans, even children as young as 2 years have a sense of justice (see Introduction and Chapter 1) (Li et al., 2016; Surian and Franchin, 2017). Whether that sense of justice is innate or learned rapidly is an interesting subject that will be pursed in Chapter 1. The question of whether such reactive initial ethical opinions (gut reactions) on the part of the students help or hinder subsequent ethical judgments is an open question. Note that this is not to deny the role of such reactions, indeed such reactions likely are the neurophysiological first response and the necessary precursor for all ethics. However, philosophers and ethicists have long held that such a reaction is primary but emphasized the ability and hence necessity to mitigate or channel those initial responses through learning and deliberative experiences. In this case, it is possible that the initial "gut" reaction may be counterproductive, leading to injustice (see Chapter 5).

Complex ethical ecology

At first, it would appear that the ethical challenge involves two parties, the physician and the speech language pathologist. The patient at this juncture is the object of the challenge instead of a means of or participant in resolving the challenge. It can be stipulated that both the physician and the speech language pathologist want good for the patient, which, in the Principlist approach, can be defined as an obligation to the ethical principles of beneficence and nonmaleficence for the patient. Alternatively, the intent to do good can be seen as a virtue, in virtue and care ethics. The issue is how can justice, good for the patient, result when the perspectives of both the physician and the speech language pathologist are different? Also, at this point, achieving justice or producing injustice would seem to center over the issue of obligation to the autonomy due each of the physician and speech language pathologist, the obligation to autonomy of each relative to each other. Certainly, a few speech language pathologists sought to involve the patient and the patient's family, thereby invoking an obligation to the autonomy to the patient; however, this is very problematic for many reasons. Yet, the obligation to autonomy of the patient would seem central to any pluralistic modern liberal democracy. In other words, a patient-centric ethic. Yet, historically, this is rarely the case (Chapter 4).

What would happen if the speech language pathologist, acting as an advocate, told the patient that she needed a VFES in order to effectively treat her swallowing problem, which, if left untreated, increases the risk of pneumonia and death? When the patient subsequently asks the physician about what the speech language pathologist said, the physician disagrees. When the speech language pathologist and the physician disagree, who is the patient to believe? Can the patient believe any physician or healthcare professional now or in the future? Clearly, the patient is unlikely to have the knowledge to adjudicate the difference on her own. What is the patient's friend going to think when the patient tells him of her predicament? Indeed, it can be argued that all of society has a stake in the resolution of differences between the physician and the speech language pathologist. All involved are benefited when the integrity and respect for everyone involved are protected and there is confidence that the challenge was resolved fairly, whatever be the nature of the resolution. Indeed, the necessity to protect the image of the physician or the healthcare professional as a person with integrity by keeping interprofessional controversies and disagreements from the public has been a key component of professional ethics since the ancient Greeks (see Chapter 4). Thus, the speech language pathologist who successfully turns her patient into an advocate for the speech language pathologists may win the battle but perhaps lose the war.

Other than a physical brawl between the physician and the speech language pathologist, how to resolve the challenging notions of autonomy between the physician and the speech language pathologist? No clear or easy solution from within the case presented appears to offer itself. Typically, in these situations, it is helpful to look for metaphors by which to "borrow" insight, resolution, and validation. These metaphors act as case precedents that affect the ethical decision-making process. This is true whether the metaphor is in the form of a deductive-like logical argument, as would be the case in a Principlist approach, or if the metaphor was an appeal to virtue as in virtue ethics, which can be taken as an Anti-Principlist approach (see Chapter 2). Indeed, the notion of precedent and *stare decisis* (see Introduction) itself is a metaphor of the type "the solution of the case today is to the solution of the case in the past previously held just." However, as discussed in Chapter 3, metaphors risk the logical Fallacy of Four Terms and thus the metaphor may be at a great risk to be invalid unless precautions are taken.

A ready metaphor seized upon by most of the students was the Captain of the Ship metaphor. Originally, the Captain of the Ship doctrine was a means to recover damages from a surgeon based on maleficence on the part of someone else in the operating room of a charity hospital through no act of commission or omission by the surgeon in order to preserve the charity hospital (Murphy, 2001). Originally, the Captain of the Ship was a telishment (Rawls, 1999), a microlevel injustice to the physician to serve macrolevel justice to the society of preserving charitable hospitals. In some ways, the original telishment has morphed into a notion of vicarious liability. However, it also has been used to shield those in the operating room whose following of instructions from the surgeon led to malpractice. In a case where the anesthesiologist did what he was told by the surgeon and the action resulted in injury to the patient, the court rejected the anesthesiologist's argument that the surgeon was liable, not the anesthesiologist. The court argued that the anesthesiologist should have known better and, importantly, that the entire surgery is not under the complete control of the surgeon. The anesthesiologist holds knowledge that the surgeon could not be expected to have reasonably.

To the degree that the speech language pathologist has knowledge and skills that the physician reasonably cannot be expected to have, the speech language pathologist may have an independent responsibility that cannot be delegated to the physician. Certainly, the speech language pathologist is in a position of providing beneficence by providing appropriate care that first requires an accurate understanding of the relevant pathophysiological mechanisms. No one would suggest otherwise. But the question becomes is there an obligation to beneficence and to whom is that obligation accorded? Is the obligation to the physician, who, in ordering a speech language pathologist evaluation, expects a report of the speech language pathologist's best professional opinion to be at the service of the physician? Providing the best professional opinion does not require the physician to act in any manner suggested by the opinion. In other words, is the speech language pathologist performing a service to the physician, which the physician is free to accept or not, or to someone else?

To be sure, the physician takes on the liability from acting or not acting on the findings and recommendations of the speech language pathologist. However, it is unlikely that prosecution of that liability would come at the hands of the speech language pathologist. It is more likely to come from the patient or the patient's surrogate, the patient's attorney, for example, in a medical malpractice complaint. Yet, it is highly unlikely that most speech language pathologists hold themselves in strictly contractual terms as might be driven by a Libertarian moral theory; at least so contracted to the physician, perhaps via the hospital or clinic, that eliminates any sense of obligation to the patient. Many speech language pathologists may hold that the patient has an intrinsic, inherent, inalienable right to such beneficence, the physician notwithstanding. This would suggest a Deontological or Kantian moral theory (see Introduction and

Chapter 2). However, in the absence of an absolute right, meaning there is an absolute and inalienable obligation to beneficence on the part of the patient by all those involved in the medical care, it becomes very problematic to define precisely what the speech language pathologist's obligation to beneficence is and to whom.

Obligation to beneficence

The question becomes what is the nature of the obligation to beneficence on the part of the speech language pathologist? Does the speech language pathologist also have a "contractual" obligation to provide beneficence to the patient independent of any obligation to the physician? To answer this question, one considers the nature of a contract. Any legally binding contract must have six elements: (1) offer, (2) acceptance, (3) consideration (exchange of goods), (4) mutuality of obligation, (5) competency and capacity, and (6) lawful purpose. Note that a contract need not be a written instrument. Finally, there is an implied requirement of good faith. For example, a misspelling in a contract cannot be used as a loophole to escape obligations, and when the execution of a contract is in dispute, all parties must be allowed an opportunity to redress any problems.

Does the speech language pathologist have a contract? There are several situations in which the speech language pathologist has a contract. In the course of caring for the patient, there is an exchange of goods. The speech language pathologist potentially provides the patient the goods of expert advice. In exchange, the patient, directly or through the proxy of insurance, gives the speech language pathologist money. Note that it is unlikely that the physician pays the speech language pathologist and thus there is no exchange of goods, in the monetary sense, between the physician and speech language pathologist. However, there may be a different form of contract between the physician and the speech language pathologist that will be considered later.

The contract between the patient and the speech language pathologist implies an obligation of good faith. By accepting payment directly or by proxy, the speech language pathologist agrees to abide by the contract except where it is unlawful. The patient may perceive that she has a right to expect that the speech language pathologist is going to act in the best interest of the patient. Whether it is the case depends on the implicit or explicit contract between the speech language pathologist and the patient. It is unlikely that the speech language pathologist and the patient came to mutual explicit understanding of the speech language pathologist's primary responsibility to the patient. In such cases, the standard practice of care or law would supervene. Indeed, as discussed later, some governmental laws or policies specifically prevent the speech language pathologist from making a diagnosis (perhaps only in its most restricted sense) and it would be reasonable that any obligation to beneficence that derives from making a diagnosis would not be owed to the patient. It is not clear that recommending the VFES would be prohibited in the sense that recommending the test necessarily is derivative of making the kind of diagnosis prohibited.

If the speech language pathologist is not prepared to act in the best interest of the patient, then the speech language pathologist may have an obligation to qualify the implied contract by applying limits or conditions prior to engaging in the conduct covered by the implicit or explicit contract. For example, the speech language pathologist can say that she or he will act in the patient's best interest except when it comes to arguing with a physician. It is unclear whether any patient or society would find this acceptable. In that case, advocating as robustly as necessary for the beneficence to the patient, such as providing the VFES, is an expectation that the society in general imposes on the implied contract between the patient and the speech language pathologist. A Libertarian moral theory holds that the greatest good is maximum freedom. In order to prevent anarchy, absolute freedom is limited by mutually agreed upon contracts. However, note that the expectation of the society, as described previously, imposes a further limit on the contract following from the Libertarian moral theory. The expectation of the society that the speech language pathologist acts first on the patient's behalf suggests a Deontological moral theory—the patient has an inalienable right to such beneficence. At least in the kind of society suggested earlier, neither the Libertarian nor the Deontological moral theories have hegemony. As will be seen repeatedly in the cases presented in Part 1, the ethics of everyday medicine are a complex amalgam of different moral theories, thereby creating a very complex ecology in which to effect justice.

In contrast to the speech language pathologist's obligation to beneficence to the patient and the responsibilities to just action such obligation implies, there is the fact that the physician was the agent between the speech language pathologist and the patient. In a sense, the physician has an implicit contract with the speech language pathologist. The physician allowed the speech language pathologist to see the patient initially. This is the good that the physician provides to the speech language pathologist as part of the implicit contract with the speech language pathologist. In exchange, there would appear to be some obligation to return some good to the physician.

This issue is what rights or privileges extend to the physician over the relationship between the speech language pathologist and the patient? As often the case where there is no explicit codified law or policy and there is no prior case precedent that exactly mirrors the case presented, one can look to metaphors elsewhere, such as the fiducial responsibilities of investment brokers to their clients (Chapter 4). The key is to determine how closely these metaphors relate to the case presented so as to avoid the Fallacy of Four Terms discussed in Chapters 3 and 5.

Another metaphor is the gag rules by which insurance companies penalize physicians for discussing certain aspects of care with patients (Liang, 1998). An insurer can "deselect" a physician who counsels patients in a manner unpleasant to the insurer without cause even if states pass laws to ban such practices. Perhaps the biggest deterrent to the use of gag rules is bad public press. Interestingly, Florida passed the "Firearm Owners' Privacy Act" prohibiting physicians asking patients about theirs and their family's use of firearms but was subsequently permanently enjoined by the US Court of Appeals for the Eleventh Circuit on February 16, 2017 in *Wollschlaeger v. Governor of Florida*. The physician could simply refuse to allow the speech language pathologist to ever be involved in the care of future patients.

Society, through the agency of government, also has a contract with the speech language pathologist. In this case, the exchange of goods is that the speech language pathologist is provided a "monopoly" on the practice of speech language pathology. Not just anybody can practice speech language pathology. In exchange for this "monopoly," the speech language pathologist agrees to abide by the contractual rights and obligations implicit and explicit in the speech language pathologist's license to practice.

Advocacy

The question becomes how would the obligation to beneficence on the part of the speech language pathologist to the patient, rather than the physician, be executed? The question could be reframed as advocacy, which implies a duty of one person on behalf of another. Borrowing from the extensive experience of the nursing profession's embrace of advocacy on the part of patients, a number of different definitions or conceptions of advocacy exist (Kalaitzidis and Jewell, 2015). Most acts of advocacy relate to what may be called policy advocacy where the profession argues for better healthcare, in its widest sense, in a community rather than for an individual. Individuals benefit as the community benefits. Extension to the case presented might be that the speech language pathologists work with the organization that supervises physicians, in general, to educate and advocate for VFES in future cases like the case presented. Individual future patients will benefit by a better educated physician staff. Further, as the individual physician is not singled out, there may be less defensiveness on his part, which consequently may change the attitude of the physician in the case presented. However, it is hard to see how such policy advocacy fulfills the obligation to the ethical principle of beneficence to the patient in the case presented.

Another form of advocacy is an information provision where the advocate attempts to provide the patient with necessary and sufficient information so that the patient or the patient's surrogate can make an informed decision, perhaps leading the patient to advocate on his or her own behalf. Patient advocacy is at the level of the individual patient.

There are a number of dimensions relating to patient advocacy. The speech language pathologist can advocate to advance the patient's interests (microlevel justice) or the patients' rights (macrolevel justice)—these two dimensions are not synonymous (Willard, 1996). In many ways, advocating for patients' rights is easier, as many organizations have adopted a systematic and explicit enumeration of patient rights, typically from nongovernmental organizations such as the American Hospital Association. For example, the US Congress considered a patient bill of rights, Bipartisan Patient Protection Act, in 2001, which was not passed. The American Speech and Hearing Association (ASHA) advocates a model for the Bill of Rights for People Receiving Audiology or Speech-Language Pathology Services [https://www.asha.org/policy/rp1993-00197.htm; The American Speech and Hearing Association (ASHA), 1994]. Among the rights enumerated were "The Right to receive a clear explanation of evaluation results; to be informed of potential or lack of potential for improvement; and to express their choices of goals and methods of service delivery" and "The Right to accept or reject services to the extent permitted by law." However, it is not clear to the degree to which these "rights" require the speech language pathologist, in the case presented, to be proactive in advocating these rights or reactive in response to a request initiated by the patient or the patient's surrogate. In the context of this case, the patient could ask the speech language pathologist for the results, interpretations, and implications of the pathologist's evaluation, and, in that case, the right of the patient would necessitate a clear and full account without prevarication, leading to the recommendation of a VFES according to the American Speech and Hearing position. But what if the patient or patient's surrogate does not ask? Can the speech language pathologist "coax" the patient to ask the speech language pathologist and, if so, how would this change the ethical implications?

An interesting question that will arise in Case 2 and discussed in Chapter 4 is whether the patient can demand a VFES be done. Note that the right stipulated by the ASHA specifies the patient's right to accept or reject services. This right is embedded within the notion of battery where the patient cannot be touched without the patient's permission. Note that this permission may be implicit, for example, a patient seeking the care of a physician or healthcare professional is expected to be touched in a professional manner. Merely seeking medical care, the patient provides implicit consent. In this case, the patient's right is clearly reactive, the patient chooses a course of action only after options are offered by the physician or healthcare professional. To be sure, the patient could be prompted by the speech language pathologist to ask for a VFES but this is likely to engender a negative response on the part of the physician and perhaps other healthcare professionals as described earlier.

Some in the nursing movement claimed a unique form of advocacy and demanded their role be codified [for many different and perhaps self-serving reasons (Willard, 1996)]. The apparent aggressiveness generated considerable backlash (Bird, 1994; Melia, 1994). In part, the most aggressive position argued was that nurses knew the patient in a unique way, which meant that the nurses were the most qualified, over and above physicians and other healthcare professionals (Salvage, 1987).

This author was in training during the ascendancy of the nursing advocacy movement and witnessed the resentment it generated. Physicians reacted as though they were accused of being less caring than nurses. This was reinforced by the nurses who claimed the nurses knew the patients better by virtue of the time actually spent with patients in the hospital. It was as though the quantity of time at the patient's side engendered a unique privilege to authority for the nurse—not discussed was the nature or quality of that time. The early discussion was not about caring because both physicians and nurses care for patients but in different and complementary ways. But there is only one patient and the system of justice must be in place to adjudicate differences between the nurse and all healthcare professionals and the physician. It is no less true between the physician and the speech language pathologist.

The centrality of diagnosis and who makes it

One could reasonably argue that even framing the case as it was presented presupposes a hierarchical relationship between the physician and the speech language pathologist and other healthcare professionals and, at the very least, an asymmetry of power. Why is it that the healthcare professional cannot just order a VFES? To be sure, the hospital or radiology department that would perform the VFES requires a requisition form to be signed by a physician. However, one could argue that this is only a bureaucratic rule to reinforce the physician's privileged position or, alternatively, a bottleneck in the delivery of healthcare in order to exert control. It also could be argued that a VFES must be taken in the total context of the patient—and the patient's ecology—and the physician could argue that this is beyond what could be expected reasonably on the basis of the typical training and experience of a speech language pathologist. Yet, this argument is problematic because it is very likely that the physician does not have the training or experience to fully appreciate the nature, need, benefits, and risks associated with a VFES and is, therefore, in a poor position to assess the potential impact of a VFES on the patient and the patient's ecology. However, the Wisdom of Solomon cannot be applied, the patient cannot be divided into two, one for the physician and one for the speech language pathologist. In the end, there has to be a decider, and the society has a vested interest in the decision about who is to be the decider.

Historically, there always have been multiple types of practitioners and the competition between them is unavoidable as they vie for the same patient. In many ways, the outcome of the competition was decided to a certain degree by defining scopes of practice. For example, the Hippocratic Oath in ancient Greece states that "But I [physician] will keep pure and holy both my life and my art. I will not use the knife, not even, verily, on sufferers from stone [thought to be urinary bladder stones], but I will give place to such as are craftsmen therein." Clearly, engaging in surgery would be beneath the physician. The unique power of physicians lies in physic, the term ancient physicians used to describe diagnosis based on their understanding of nature. The power of the physician rests primarily on the ability to diagnose and, from the diagnosis, to explain and prognosticate. Further, treatments follow from the diagnosis.

Throughout history, physicians have jealously guarded their prerogative to diagnose (Montgomery, 2018). In 1704, the British Parliament overruled the courts that had found an apothecarist liable for diagnosing, which heretofore was known as the exclusive privilege of physicians (Cook, 1990). Subsequently, apothecarists could diagnose but only in the strict context of the medicinals they were providing and even then could not charge the patient for the act of diagnosis. The act highlights the distinction between justice and law. If it were justice for apothecarists to diagnose, why shouldn't the apothecarists be paid for their efforts? The prohibition against payment may have been a means to control healthcare expenditures (even in 1704), not unlike the effect of insurers who, when denying coverage for care, insist they are not practicing medicine but only stating what they will cover in a contractual sense. Yet, the insurers often deny coverage, stating a lack of proven efficacy, but how is that different from what physicians and healthcare professional do or should do? The American Medical Association's Principle of Medical Ethics of 1912 stipulated:

> Sec. 4. It is the duty of physicians to recognize and legitimate patronage to promote the profession of pharmacy, on the skill and proficiency of which depends the reliability of remedies, but any pharmacist who, although educated in his own profession, is not a qualified physician, and who assumed to prescribe for the such, ought not to receive such countenance and support. Any druggist or pharmacist

who dispenses deteriorated or sophisticated drugs or who *substituted one remedy for another designated in a prescription ought thereby to forfeit the recognition and influence of physicians* [italics added].

(American Medical Association's Principle of Medical Ethics of 1912)

The hierarchical organization between physicians and healthcare professionals can be seen in practice laws. For example, only 23 states in the United States allow nurse practitioners to practice without direct supervision by a physician. Most of these states have large rural populations where access to healthcare, particularly from physicians, may be limited. The scarcity of resources likely affects who can practice medicine and ultimately establishes the system of justice for those patients (see the concept of moderate scarcity in Introduction). If the laws of Illinois are typical, the ability of speech language pathologists to see patients without referral is limited. The code states:

> (3) If the patient has a voice disorder or vocal cord dysfunction, he or she must be examined by a physician who has been granted hospital privileges to perform these procedures and the speech-language pathologist must have received from that physician a written referral and direct authorization to perform the procedure.
>
> (4) If the patient has a swallowing disorder or a velopharyngeal disorder, he or she must be examined by a physician licensed to practice medicine in all its branches and the speech-language pathologist must have received from that physician a written referral and direct authorization to perform the procedure.
>
> *(225 ILCS 110/9.3)*

Nearly all healthcare professionals have a limited power to diagnosis, in its widest connotation, in the strict context of their scope of practice. Indeed, the ethical physician recognizing her obligation to beneficence depends on the speech language pathologist's ability to diagnose, if only in the sense that the speech language pathologist "diagnoses" there is something wrong and the best way to evaluate the abnormality further is by using a VFES. There have been debates regarding more formal disease and disorder-specific diagnoses. The American Speech and Hearing Association maintains that "competent SLPs can diagnose communication and swallowing disorders but do not differentially diagnose medical conditions" and "use endoscopy, videofluoroscopy, and other instrumentation to assess aspects of voice, resonance, velopharyngeal function, and swallowing" (https://www.asha.org/uploadedfiles/sp2016-00343.pdf). The ability of speech language pathologists in Ontario, Canada to make a diagnosis is extremely curtailed, often necessitating some "gaming" of terminology in order to carry out their obligation to beneficence. However, Bill 200, Removing Barriers in Audiology and Speech-Language Pathology Act, 2018, has passed the first reading, although its ultimate fate is unsure. The bill states the following:

> The practice of speech-language pathology is the assessment, *diagnosis* [italics added], prevention and treatment of speech, language, communication, voice and swallowing dysfunctions or disorders to develop, maintain, rehabilitate or augment communication or swallowing functions.

Nevertheless, even just from a practical situation, the physician's diagnosis supersedes that of the healthcare professional in circumstances where a physician referral is necessary to authorize the care provided by the healthcare professional. Even for nurses in a hospital, there is a prior diagnosis for which the physician admits a patient. Even standing orders that can be executed by a nurse generally undergo prior review and authorization by the medical staff, consisting of physicians.

One rationale for physicians to have sole jurisdiction of diagnosis in the larger sense relates to the importance of a widely considered differential diagnosis, that is, a list of all reasonable possible diagnoses. In fact, the importance of a wide-ranging differential diagnosis is a matter of logic (Montgomery, 2018). An insufficient differential diagnosis risks the Fallacy of Limited Alternative (see Chapter 5), which is of the form *if (a inclusive OR b inclusive OR c) implies d is true and b and c are found false, then a must be true.* The fact is that, under conditions of an *inclusive OR* function, the truth or falsehood of *b* and *c* has nothing to do with the truth or falsehood of *a*. Further, one can remove *b* and *c* from the logical statement, leaving *if a implied d is true and d is true, then a is true.* However, this logical statement is invalid and is the Fallacy of Confirming the Consequence. The premise *d* could be true for any number of reasons other than premise *a*.

In the context of the case, it might be that the speech pathologist used the logic *if the patient has speech disorder a, then there should be an abnormality x in the VFES, there is an abnormality x in the VFES so the patient must have disorder a.* However, it is entirely possible that the patient may have a completely different speech disorder that would not be determined. The only way to be sure is to change the logic to *if and only if the patient has speech disorder a, then there should be an abnormality x in the VFES, there is an abnormality x in the VFES so the patient must have disorder a.* But the only way to be sure that the premise and proposition *if and only if the patient has speech disorder a* is true, and hence the conclusion is valid and true, is to rule out any other possible disorder, but this would depend on the range of disorders familiar to the speech language pathologist (Montgomery, 2018).

The problem is when the speech language pathologist considers only diagnoses common to his practice or encountered in his education and training (see Case 9). The speech language pathologist could miss other reasonable diagnoses that could account for the patient's problem. This is not to say that physicians do not inappropriately limit their differential diagnoses and thus miss the true cause as discussed in Case 9. However, one might reasonably conclude that a physician with 4 years of medical school after college and typically 4 years of residency after medical school would have broader experience and thus a greater range of reasonable diagnoses to include in the differential diagnosis. Consequently, any patient with any previously undiagnosed disorder should first be seen by a physician. Also, very importantly, any subsequent diagnosis by a healthcare professional should be in the context of the initial diagnosis provided by the physician.

It could be argued that healthcare professionals, for example, nurse practitioners, have sufficient education, training, and experience for common disorders and thus, within the context of these common disorders, could generate an initial diagnosis independent of a physician. However, this creates the critical question as to just how common does a disorder have to be that it is sufficient for a healthcare professional to make the initial diagnosis? Clearly, making a decision as to the cutoff is an ethical question. One consequence is that the patient who has an uncommon disorder may be misdiagnosed by the healthcare professional and thus be denied the care the patient would otherwise be entitled to. One could justify the misfortune of the patient with an uncommon disorder based on a Utilitarian moral theory that, with limited resources, greater access to a diagnosis for the common diagnosis justifies the failure of access to an uncommon diagnosis; a form of rationing (see the Oregon Healthcare Experiment in Chapter 4). However, for such a rationing program to be fair in a pluralistic modern liberal democracy, a sufficient number of citizens would have to agree either directly through referendums or by their political representatives. Such limitations to optimal care are frequent in everyday medicine, yet it is rarely the case that citizens have given their permission for such rationing.

It could be argued that the speech language pathologist is recommending a test, not making a diagnosis. Therefore, the issue of who has the right or privilege to diagnose the patient's swallowing problem is not the issue. However, such a suggestion would be disingenuous. The whole purpose of any test is to help establish a diagnosis, with the operative word "help." It has been a long tradition in medicine that laboratory tests do not diagnose, only physicians and healthcare professionals within the scope of their practice make diagnoses. The following is from the US Food and Drug Administration website on laboratory tests:

> Some laboratory tests are precise, reliable indicators of specific health problems, while others provide more general information that gives doctors clues to your possible health problems. *Information obtained from laboratory tests may help doctors decide* [italics added] whether other tests or procedures are needed to make a diagnosis or to develop or revise a previous treatment plan. *All laboratory test results must be interpreted within the context of your overall health and should be used along with other exams or tests* [italics added].
> *(https://www.fda.gov/medicaldevices/productsandmedicalprocedures/invitrodiagnostics/labtest/default.htm)*

One might ask what maleficence, that is, harm, is done by performing a VFES with the question presupposing that the risk of maleficence is minimal other than aspiration of the materials administered. However, the patient already is at risk and any incremental risk likely is minimal. However, the VFES does harm by its costs and relatively low risk of immediate medical complications. The costs can be harmful to those directly, such as the patient, or indirectly, such as society or insurer, who pays the cost. Further, there is the risk to other patients who would be displaced or have to wait further for their own VFES. However, there also is the risk of false positives, an abnormal test where there is no medical abnormality, leading to inappropriate treatments, and false negatives, leading to normal tests when there is a medical abnormality, to withholding appropriate tests.

The risk of a false positive or false negative is related to the specificity and sensitivity of the VFES and it can be reasonably said that the speech language pathologist is more of an expert in understanding the specificity and sensitivity of the VFES. However, the specificity and sensitivity of the VFES are only part of the considerations determining the value of the VFES. Both the specificity and the sensitivity measures have to be combined with the prior probability that the patient actually has the disorder for which the VFES is intended to diagnose or rule out. For example, a test that is 90% specific and 90% sensitive may have more false positives than true positives and more false negatives than true negatives if the probability of having the disease prior to the VFES is less than 10%. Thus, the value of the VFES depends on the prior differential diagnosis, which then engenders the aforementioned discussion regarding who should be making the differential diagnosis (Montgomery, 2018).

The discussion given previously regarding who is authorized to make what kind of diagnosis is the reason for the distinction maintained throughout this book between physicians and healthcare professionals. To be sure, both are involved in healthcare and both are professionals. The term physician as used in this text is not honorific—it does not denote any special privilege. Rather, it is to denote a unique responsibility as the decider. The acceptance of being a physician then is an acknowledgment of a responsibility commensurate with the authority and, therefore, a different obligation to beneficence. Further, the treatments offered in complying with the obligation to beneficence generally are at a higher risk of maleficence, or harm. This also adds to the different ethical challenges that physicians uniquely face.

The distinction between physician and healthcare professional can be understood by the metaphor to law. A distinction is maintained between a judge and lawyer, although most judges are lawyers. Both are involved in the dispensing of justice, but the judge is not the same as a lawyer and the term, judge, demarcates that difference. Thus, physician is to judge as healthcare professional is to lawyer, in the context of advising the physician. The analogy between physician and judge will be discussed more anon.

None of the aforementioned discussions about who should have the authority—to establish the differential diagnosis when *en route* to treatment—discounts the contributions that the speech language pathologist and other healthcare professionals make or qualifies or limits their importance. Indeed, the physician and, consequently, the patient would benefit greatly from the physician carefully considering the recommendations of the speech language pathologist. At the very least, the advice of the speech language pathologist helps expand the differential diagnosis in order to avoid the Fallacy of Limited Alternatives. This would seem so obvious that the physician considering the opinion of the speech language pathologist approaches the notion of a natural duty as part of a virtuous physician (see Introduction).

A physician can explicitly reject out of hand, the consultation of others, which would be a form of strong paternalism and hence perhaps unethical (see Case 5). Note that rejection or acceptance of consultation is not the same as accepting or rejecting the actual content of the consultation. The latter would not necessarily be unethical depending on the physician's analysis of the content. Rejecting the consultation itself would be unprofessional by most standards. The physician could also practice what is called soft paternalism by simply ignoring the consultation (Beauchamp and Childress, 2013).

Institutions, committees, and agencies to ensure ethics

Healthcare is conducted in a highly complex environment if for no other reason than the number of physicians, healthcare professionals, institutions, agencies, and insurers involved in nearly every aspect of everyday medicine. The advances in diagnostic and therapeutic interventions are increasingly complex, which necessitates a division of labor. The ability to positively affect the patient's difficulty in swallowing depends on understanding the pathophysiology, which is beyond the expertise of most physicians. It is for this reason that the legal notion of Captain of the Ship has been largely abandoned. Thus, it is not given that the physician is the only person with authority, the only shareholder. The integration across many disciplines requires complex administrative and logistical structures. Also, the time of an independent solo practitioner, physician, or healthcare professional, is past (see Chapter 4). Healthcare organizations are held accountable to the actions of all their employees, in both their obligation to beneficence and their obligation to nonmaleficence. The organization's stake in the outcomes of its patients necessitates a hierarchical supervisory structure. Many of these organizations have means for resolving ethical challenges, although the ones encountered nearly every day are rarely brought up (see Introduction).

In this author's experience, many ethics committees do not address issues arising from the professional and ethical competence of physicians and healthcare professionals. Indeed, some ethics committees demure from such cases. At times, the only opportunity is through risk management committees whose deliberations are often in the context of avoiding lawsuits and, if a lawsuit arises, to be able to mitigate its risks and losses. There are few things that increase the risk of a successful lawsuit than discord between physicians and healthcare professionals documented in the patient's medical record. Increasingly, institutions are recognizing that the risk management office is not the best instrument to deal with the challenges of everyday ethics and are developing alternative policies and procedures.

Other options include raising the concerns to the physician's supervisor. However, the perception by the speech language pathologist, as well as outside reviewers, is that it would be difficult for the physician's supervisor to avoid bias and self-interest in resolving the challenge. At least this is the presumption of many healthcare professionals.

There is the option of raising the concern with governmental regulatory authorities; however, the evidentiary and severity threshold for such reporting makes such actions a rather blunt instrument, particularly for ethical challenges experienced in everyday medicine. This author raised ethical concerns regarding the actions of another physician to the Pennsylvania Medical Society to urge private consultation with the physician only to be referred to the Pennsylvania State Board of Medicine, which would intervene only if a formal complaint was made. This seems to be an unacceptable "nuclear option," despite the lack of any alternatives. Clearly, one way to avoid dealing with a problem is making the dealings "radioactive."

Raising concerns to the physician's superiors or governmental agencies are not without risk to the whistleblower. This author is aware of a physician who followed the recommendations of the chair of her department concerned about poor care provided by another physician. The institution retaliated by attempting constructive dismissal by unilaterally changing the terms of employment to the point where continued employment was not acceptable to the whistleblower. An egregious example was the case of three nurses who anonymously reported a surgeon to the Texas Medical Board whereupon the surgeon complained to his friend, a sheriff, who subsequently filed criminal complaints against the nurses (Moser, 2009). Ultimately, courts found in favor of the nurses, although not until after considerable maleficence was administered to the nurses.

Resolving conflicting obligations to ethical principles

It could be argued that the nurses just described acted courageously, but given all the nurses had to go through, it would not be surprising or unreasonable that others, including the speech language pathologist in the case presented, would be unwilling to go so far. The operative concept is the obligation to beneficence. For example, the average person would not be expected to run into a burning house in an attempt to look for and save persons in the house. However, a firefighter would be expected to do so. The question is whether the speech language pathologist in this case is more like the average person or the firefighter? At the very least, the firefighter is contractually expected to enter the burning house, whether explicitly or implicitly in the professional standards. The question is whether the speech language pathologist is expected to take such actions that could be detrimental to her career.

As can be appreciated readily, the ethical challenge in this case is complex, with many competing interests, responsibilities, and authorities. These need to be sorted out (see Chapter 5). One approach is to identify all the stakeholders, defined as all those who reasonably might be affected by, however, the challenge is resolved, and shareholders, those having a reasonable potential to affect the resolution of the challenge. Some participants may be both stakeholders and shareholders (most are) but perhaps to varying degrees.

Once the shareholders and stakeholders are identified, the dynamics between the various stakeholders and shareholders can be defined in terms of the ethical principles of beneficence, autonomy, and nonmaleficence and, most importantly, the obligations to the ethical principles when using a Principlist approach; similarly, the obligations between parties also can be characterized in terms of the various Anti-Principlist approaches, such as the obligation to virtue (Chapter 5). Nearly always there is little doubt about the ethical principles that apply, for example, beneficence, nonmaleficence, and autonomy. Rather, any conflict arises from the contrasting obligations to those principles. Many times, the obligations to the ethical principles are compatible among the shareholders and stakeholders and the ethical issue is fairly easily resolved. However, as often occurs in ethical dilemmas or problems, the various obligations to the ethical principles do not mesh easily and indeed can be in conflict with one another.

To resolve conflicting obligations to the ethical principles in the ethical problem, one approach is to turn to the various moral theories. As already seen, the ethical differences between the physician and the speech language pathologist could be interpreted from a Libertarian moral theory. Thus, what the speech language pathologist can or cannot do is defined by the implicit or explicit contractual relationships, yet there are limits to the contract consequent to the necessary inclusion of other moral theories. A Deontological moral theory could hold that the patient has an inalienable right to be informed by the speech language pathologist irrespective of the physician's antagonism. An alternative form of the deontological argument is that the physician should respond positively to the speech language pathologist's recommendation because that is what virtuous physicians do. Yet, the limitations imposed in the setting of moderate scarcity of resources limit the application, hence rule, of the Deontological moral theory. A Utilitarian moral theory could argue that what should be done is what generates the greatest good and the physician's actions, no matter how arbitrary or capricious, is just by virtue of minimizing expenditure of resources for other purposes. These other purposes may provide a greater rate of return on society's investment of resources. The problem of all Utilitarian moral theories is the definition of good, which goes well beyond strictly financial concerns, such as a patient's intrinsic value held by the society. Just as deontological considerations limit the rule of Libertarian moral theories, deontological considerations also mitigate the rule of Utilitarian moral theories. An Egalitarian moral theory would hold that the physician's actions are not unjust as long as all patients received the same consideration, good or bad. Yet, again, most reasonable persons blinded to their self-interest would find the situation unacceptable. It may well be that these moral theories are insufficient or inappropriate in which case entirely different moral theories need to be created.

Much of the discussion given earlier was predicated on an implicit or explicit contractual relationship, which is transactional. Justice is considered in the agreements, policies, procedures, and laws among all the shareholders and is not driven by some inherent inalienable right on the part of any shareholder. Thus, the supervening moral theory is libertarian because the libertarian perspective holds that the greatest freedom creates the greatest good. One's freedom is not curtailed by some endowed inalienable right of some other (the Preamble of the Declaration of Independence notwithstanding, see Introduction). Taken to its logical extreme, libertarianism would be anarchy. What prevents anarchy is a mutual agreement to curtail one's own freedom in exchange for others doing similarly to the mutual benefit of all. One gives up his right to drive on any side of the road he wishes but does so only when everyone else gives up that right and all agree to drive on one particular side of the road. The physician and speech language pathologist give up some of their freedom to do anything they want in exchange for goods received when caring for patients and the privilege to practice in exchange for compliance to laws, regulations, and policies. Society ensures the protection of patients by regulating physicians and healthcare professionals, and the physicians and healthcare professionals receive benefit in the form of a type of monopoly where not everyone and anyone can practice medicine.

One could take a Deontological moral theory as supervening. The patient, merely by virtue of being human, has the right to a VFES and justice requires the physician to relent and authorize the VFES. Kant held that every person is an end (goal or purpose) unto themselves and should not be made a means to someone else's end. Thus, to deny the patient a VFES in order to save money, save face, or preserve or reserve resources for others is to make the patient a means to someone else's end and would be immoral and thus unjust. However, such a deontological approach is a slippery slope in the absence of infinite resources (see the notion of moderate scarcity in Introduction).

Utilitarianism is a moral theory that could be applied to the case, which holds that justice attempts to achieve the greatest good, however, good is defined. But it is just how good is defined that makes utilitarianism so problematic. As can be seen, the range of stakeholders can be vast. How then is one to appropriate the good for each shareholder into the good that drives justice? One could argue for egalitarianism in which the moral good is the equal provision of justice to everyone, whether the care be good, bad, or indifferent. Clearly, this is not helpful as the physician and the healthcare professional want to treat the patient differently. However, more importantly, another physician with another speech language pathologist may agree to conduct the VFES. Now the quality of care is determined by the luck of the draw as to which physician becomes the physician for the patient. A key concept in egalitarianism is that justice should not be a matter of chance or bias. Indeed, it is the responsibility of institutions that employ and governmental agencies that license physicians and healthcare professionals that all citizens have equal opportunities to appropriate healthcare.

Integration of care and distributed agency

The actions of the physician and the speech language pathologist in the case described appear superficially as a dispute between two parties and the dynamics dyadic between the physician and the speech language pathologist. However, this situation is far more complex. Despite what may be an intuitive response among all parties in grappling with the complexities, it may be to simply default to the physician. In other words, "the buck stops" with the physician. However, the complexity of modern healthcare delivery is increasing in complexity, making such deferral to the physician unworkable. In many ways, court rejection of the Captain of the Ship doctrine in medical liability is direct recognition of the complexity, but, perhaps more important, is the division and distribution of responsibility that follows the division of labor to meet the complexity of providing medical care. Court cases have recognized the unfairness of holding the physician responsible for the maleficence on the part of others over whom the physician has no influence or authority. Previously holding the physician liable as Captain of the Ship creates an incongruity between authority and responsibility. However, those court decisions were predicated on the ability to differentiate and compartmentalize responsibility and hence authority and liability.

The complexity of healthcare delivery is increasing, perhaps even to the point of demonstrating Chaos and Complexity, which, in principle, would make differentiation and compartmentalization of responsibility, authority, and liability virtually impossible (see Case 12 and Chapter 1). An interesting analogy is the development of artificial intelligence (AI), particularly machine learning, and its ethical implications (Taddeo and Floridi, 2018). Machine learning AI is remarkable on any number of accounts. Particularly, machine learning does not depend on humans to solve the computational problem and, strikingly, can solve problems in a manner that is unknowable and hence unchecked by any human. Further, machine learning AI has become ubiquitous and unrecognized in the daily lives of humans. The principle of autonomy would dictate that decisions based on unsupervisable AI are unethical, as it disrespects the human on which it operates. In a real sense, the person subjected to the AI is asked to merely take "AI's word" on it. Indeed, this is why the European Union passed laws instilling the right to explanation, whereby a person can demand that the AI system explains its decision (Goodman and Flaxman, 2017).

It may be possible for a false conclusion or an unacceptable ethical outcome to result from a situation that is so complex that any differentiation and compartmentalization of responsibility are impossible. Rather, the fault must be attributed to the entire healthcare system involved. The notion of a systems error as a cause of violation of nonmaleficence was highlighted by the finding of the US Institute of Medicine, where unintended human errors were the cause (Kohn et al., 1999). As these errors were unintended, they were not considered malpractice with the associated liabilities. Nevertheless, patients have received compensation for such adverse unintended consequences with the institution instead of individuals being held directly responsible and hence subject to civil law (Sohn, 2013).

Appreciation for the difference between unintended and unanticipated consequences is increasing (Bloomrosen et al., 2011). To be sure, unanticipated consequences likely are unintended but unintended actions may still be anticipated. In the event of a failure to take action on recognizable risks for unintended adverse effects, institutions may be held liable (Shavell, 1984). This leads to what constitutes a reasonable expectation to anticipate the risk of an unintended adverse effect. As a matter of practice, institutions are increasingly the force organizing and ordering the complex relations between all its parts, including the physician and the speech language pathologist. The days of sovereignty because the physician was the sole independent private practitioner are largely gone. Institutions are faced with the apportionment of authorities and responsibilities, for example, between the physician and the speech language pathologist (see Case 2).

As the notion of intended and unintended adverse effects underlies much thinking in terms of liability or culpability, the distinctions and implications of intended versus unintended harm may help in understanding the ordering of authorities and liabilities. Intended and unintended harms imply a sense of agency, as the entity responsible for the adverse effect must have had some agency by which to cause the harm or to prevent it. As seen from the preceding discussion of AI, is any harm caused by the physicians' and healthcare professionals' use of AI an intended or unintended harm? That a person was harmed demands some justice but from whom?

There must be some accountability necessitating the finding of agency so as to identify the part liable. However, in systems of sufficient complexity and robustness, there is distributed agency. The difficulty in grappling with distributed agency relates to the human's sense of human agency where agency is an anthropomorphic notion (Floridi, 2013). The anthropomorphism of the complex system makes understanding and dealing with such complex integrated systems difficult. Human rather than institutions or computer algorithms are the usual suspects, but such an approach risks failing to actually resolve the causes of injustice. A discussion of the ethical situation with AI is considered in Case 12. The interested reader is directed to Floridi (2013), Pagallo (2010), and Asaro (2007).

Physicians as judges

As there is but one patient, there must be one decider. The questions become what is the nature of the decider and who is authorized to be the decider in the era of specialization and division of labor? An analogy to the courts may be helpful. It has become increasingly clear that judges cannot be expected to be content experts in every case on which they are asked to judge. Even calling expert testimony often does not solve the problem for the judge as the experts can disagree. In nonjury trials, judges are then forced to choose between the differing expert opinions. In this case, a judge may not be a content expert but knows the rules of evidence, legal logic, and procedures and a not insignificant insight into human nature. The opinion of each expert can be judged on the consistency with the rules of evidence and the argument advanced by the expert evaluated on it logical or rational persuasiveness and the psychological assessment of the credibility of the experts.

Physicians could act similarly. The physician, not being an expert on VFES, can evaluate the rationale of the speech language pathologist based on the differential diagnosis offered by the speech language pathologist in the context of the physician's differential diagnosis. Further, the logic of the speech language pathologist's prediction of the types of findings that might result from the VFES might affect the physician's differential diagnosis and can be evaluated and acted upon in an appropriate matter. This adds a role to the speech language pathologist, which is the ability to convey his or her thinking in a manner accessible and acceptable to the physician, a role that may benefit from rhetoric in addition to logic. However, the speech language pathologist arguing some unique and independent privilege that supervenes on that of the physician is likely to be counterproductive, not only for the patient but also for future patients cared for by the speech language pathologist and the physician as discussed earlier. The opinions of the expert witnesses are not privileged such that they supervene over the considerations of the judge.

The analogy of the physician as a judge without expert content knowledge extends to accountability as well. Judges can be overruled and reversed. Judges can be reprimanded, impeached, and otherwise removed from the bench. Further, the actions of the judge, in most cases, undergo "sunlight [as] the best disinfectant" (Brandeis, 1914); in other words, are not hidden. However, in the case under discussion, there are reasons to keep deliberations in confidence, as discussed next. Yet, this risks making the disinfectant of sunlight unavailable when it could be helpful.

Settling out of court

No one is suggesting that the disagreement in the case presented between the speech language pathologist and the physician has to be settled in a court of law, ethics committee, or other forums where the physician and speech language pathologist are called forth to justify themselves. Even judges would discourage this and would prevail upon the parties to "settle out of court." Judges do so because often the complexities of the case exceed the precision of legal tools (see Chapter 5). Further, laws and ethics are contextualized and gain their force in this specific context of the case. By "settling out of court," the court does not set a precedent that might not be suited to future cases, as the court does not want to have to undo a legal precedent, stare decisis.

Clearly, it would be best if the dispute between the speech language pathologist and the physician could be "settled out of court," provided a just outcome can be achieved. In this case, "court" applies to any administrative proceedings such as within the hospital or governing medical organization. Further, it is important to realize that the relationship between the speech language pathologist and the physician may go well beyond this particular case and that it would be productive to develop methods to forestall future disagreements. The question then becomes how to "settle out of court?"

In many ways, the ethical principles or virtues still apply as well, but the "out of court" settlement also requires significant applications of psychology and sociology. The resolution of the dispute must recognize the autonomy of each party and an avoidance of maleficence (avoiding harm to reputations and standings). The psychology involves how to negotiate without causing a loss of face and defensiveness. The psychology stems from recognizing that a loss of face and standing is a realistic human concern. Further, both the speech language pathologist and the physician must be able to work affectively for other future patients, which will require respect from the entire medical staff for each.

What is clear is that most reasonable persons blinded to their self-interest would hold the speech language pathologist obligated to pursue the matter in a fair and reasonable way, but the paramount issue is the patient's benefit. Unless it is a crisis, a crisis mode using "nuclear options" should be avoided. Often it is best to take the discussion further but in a private and confidential matter. The discussion can be advanced by the mutual recognition that each party has the best interest of the patient at heart, thereby avoiding any adversarial appearance. For example, the speech language pathologist can start by stating that she or he sees their role in helping the physician in his or her care of the patient. The speech language pathologist's avoidance of defensiveness may help the physician avoid his or her own defensiveness. From there it is frequently possible to come up with a means to resolve the difference. For example, the speech language pathologist can ask the physician what would the physician need to resolve the difference?

If the speech language pathologist and the physician cannot come to a resolution in private and on their own, it may be possible to appeal to a third party respected by both the speech language pathologist and the physician. Because the third party has mutual respect for both the speech language pathologist and the physician, both may be willing to abide by the third party.

Should an attempt at a private conversation involving a mutually respected third party fail, many institutions have specific mechanisms to help resolve the problem. For example, it may be possible to talk to the physician's chairperson of their department or to the chief of staff. Many institutions also have bioethics committees to help resolve these issues in a confidential matter. Many institutions have morbidity and mortality rounds in which such issues can be discussed. In many circumstances, morbidity and mortality rounds are privileged and the proceedings may not be discoverable, meaning the record cannot be held against the medical professionals involved in a legal proceeding. If still unresolved, the speech language pathologist can obtain private legal counsel and other methods, such as open or anonymous referral to governmental agencies, can be pursued.

In the end, there is only one person that the speech language pathologist has to satisfy and that is himself or herself. The speech language pathologist needs to be able to sleep at night and a clear conscience is most conducive.

References

Akbar, U., Dham, B., He, Y., Hack, N., Wu, S., Troche, M., Tighe, P., et al., 2015. Incidence and mortality trends of aspiration pneumonia in Parkinson's disease in the United States, 1979–2010. Parkinsonism Relat. Disord. 21, 1082–1086. https://doi.org/10.1016/j. parkreldis.2015.06.020.

American Medical Association, 1912. Principles of medical ethics of the American Medical Association. Ann. Am. Acad. Pol. Soc. Sci. 101 (1), 260–265.

Asaro, P., 2007. Robots and responsibility from a legal perspective. In: Proceedings of the IEEE Conference on Robotics and Automation: Workshop on Roboethics, Rome, April 14th.

Beauchamp, T., Childress, J.F., 2013. Principles of Biomedical Ethics. Oxford University Press, Oxford.

Bird, A.W., 1994. Enhancing patient well-being: advocacy or negotiation? J. Med. Ethics 20 (3), 152–156.

Bloomrosen, M., Starren, J., Lorenzi, N.M., Ash, J.S., Patel, V.L., Shortliffe, E.H., 2011. Anticipating and addressing the unintended consequences of health IT and policy: a report from the AMIA 2009 Health Policy Meeting. J. Am. Med. Inform. Assoc. 18 (1), 82–90. https://doi.org/10.1136/jamia.2010.007567.

Brandeis, L., 1914. Other People's Money And How the Bankers Use It. Frederick A. Stokes, New York.

Cook, H.J., 1990. The Rose Case reconsidered: physicians, apothecaries, and the law in Augustan England. J. Hist. Med. Allied Sci. 45 (4), 527–555.

D'Amelio, M., Ragonese, P., Morgante, L., Reggio, A., Callari, G., Salemi, G., Savettieri, G., 2006. Long-term survival of Parkinson's disease: a population-based study. J. Neurol. 253, 33–37. https://doi.org/10.1007/s00415-005-0916-7.

Floridi, L., 2013. Distributed morality in an information society. Sci. Eng. Ethics 19 (3), 727–743. https://doi.org/10.1007/s11948-012-9413-4.

Giraldo-Cadavid, L.F., Leal-Leaño, L.R., Leon-Basantes, G.A., Bastidas, A.R., Garcia, R., Ovalle, S., Abondano-Garavito, J.E., 2017. Accuracy of endoscopic and videofluoroscopic evaluations of swallowing for oropharyngeal dysphagia. Laryngoscope 127, 2002–2010. https://doi.org/10.1002/lary.26419.

Goodman, B., Flaxman, S., 2017. European Union regulations on algorithmic decision-making and a "right to explanation". AI Mag. 38 (3), https://doi.org/10.1609/aimag.v38i3.2741.

Hassan, H.E., Aboloyoun, A.I., 2014. The value of bedside tests in dysphagia evaluation. Egypt. J. Ear Nose Throat Allied Sci. 15, 197–203. https://doi.org/10.1016/j.ejenta.2014.07.007.

Kalaitzidis, E., Jewell, P., 2015. The concept of advocacy in nursing: a critical analysis. Health Care Manag. (Frederick) 34 (4), 308–315. https://doi.org/10.1097/HCM.0000000000000079.

Karkos, P.D., Papouliakos, S., Karkos, C.D., Theochari, E.G., 2009. Current evaluation of the dysphagic patient. Hippokratia 13, 141–146.

Kohn, L.T., Corrigan, J.M., Donaldson, M.S., 1999. To Err Is Human: Building a Safer Health System. Institute of Medicine, Washington, DC.

Li, J., Wang, W., Yu, J., Zhu, L., 2016. Young children's development of fairness preference. Front. Psychol. 7, 1274. https://doi.org/10.3389/fpsyg.2016.01274.

Liang, B.A., 1998. The practical utility of gag clause legislation. J. Gen. Intern. Med. 13 (6), 419–421.

Martin-Harris, B., Jones, B., 2008. The videofluorographic swallowing study. Phys. Med. Rehabil. Clin. N. Am. 19 (4), 769–785.

Melia, K.M., 1994. The task of nursing ethics. J. Med. Ethics 20 (1), 7–11.

Miles, A., Allen, J.E., 2015. Management of oropharyngeal neurogenic dysphagia in adults. Curr. Opin. Otolaryngol. Head Neck Surg. 23, 433–439. https://doi.org/10.1097/MOO.0000000000000206.

Montgomery Jr., E.B., 2018. Medical Reasoning: The Nature and Use of Medical Reasoning. Oxford University Press, Oxford.

Moser, J., 2009. Texas nurses under fire for whistleblowing. Am. J. Nurs. 109 (10), 19. https://doi.org/10.1097/01.NAJ.0000361469.79160.a0.

Murphy, E.K., 2001. "Captain of the ship" doctrine continues to take on water. AORN J. 74 (4), 525–528.

O'Horo, J.C., Rogus-Pulia, N., Garcia-Arguello, L., Robbins, J., Safdar, N., 2015. Bedside diagnosis of dysphagia: a systematic review. J. Hosp. Med. 10, 256–265. https://doi.org/10.1002/jhm.2313.

Pagallo, U., 2010. Robotrust and legal responsibility. Knowl. Technol. Policy 23 (3/4), 367–379.

Rawls, J., 1999. A Theory of Justice, revised ed. The Belknap Press of Harvard University Press, Cambridge, MA.

Salvage, J., 1987. Whose side are you on. Sr. Nurse 6 (2), 20–21.

Shavell, S., 1984. Liability for harm versus regulation of safety. J. Leg. Stud. 2, 357–374.

Sohn, D.H., 2013. Negligence, genuine error, and litigation. Int. J. Gen. Med. 6, 49–56. https://doi.org/10.2147/IJGM.S24256.

Steele, C., Allen, C., Barker, J., Buen, P., French, R., Fedorak, A., Day, S., et al., 2007. Dysphagia service delivery by speech-language pathologists in Canada: results of a national survey. Can. J. Speech-Lang. Pathol. Audiol. 31, 161–177.

Surian, L., Franchin, L., 2017. Toddlers selectively help fair agents. Front. Psychol. 8, 944. https://doi.org/10.3389/fpsyg.2017.00944.

Suttrup, I., Warnecke, T., 2016. Dysphagia in Parkinson's disease. Dysphagia 31, 24–32. https://doi.org/10.1007/s00455-015-9671-9.

Taddeo, M., Floridi, L., 2018. How AI can be a force for good. Science 361 (6404), 751–752. https://doi.org/10.1126/science.aat5991.

The American Speech and Hearing Association, 1994. The protection of rights of people receiving audiology or speech-language pathology services. Task Force on Protection of Clients' Rights. ASHA 36 (1), 60–63.

Willard, C., 1996. The nurse's role as patient advocate: obligation or imposition? J. Adv. Nurs. 24 (1), 60–66.

Wirth, R., Dziewas, R., Beck, A.M., Clave, P., Heppner, H.J., Langmore, S., Hamdy, S., et al., 2016. Oropharyngeal dysphagia in older persons. From pathophysiology to adequate intervention: a review and summary of an international expert meeting. Clin. Interv. Aging 11, 189–208. https://doi.org/10.2147/CIA.S97481.

Case 2

Case 2—A physician requested a video fluoroscopic evaluation of swallowing (VFES) for his patient with Parkinson's disease who was experiencing difficulty with swallowing. A VFES involves videotaping an X-ray of the mouth and throat as the patient swallows radio-opaque solids and liquids of different consistencies. In this manner, the pathway of the food and liquids can be followed as the patient swallows. The physician was instructed to call a central outpatient coordination service to arrange for the outpatient study. The service informed the physician that they were unable to provide a VFES but would provide a "bedside" clinical examination. A clinical or "bedside" examination involves a speech-language pathologist observing and listening to the patient as the patient swallows liquids, looking for evidence of choking or gurgling or residue in the mouth following an attempt to swallow.[a]

Medical facts of the matter

The hospital-based clinic in which the physician practiced had no speech-language pathology department for a direct referral. Community speech-language pathologists do not have access to a radiology service for X-ray video recording. To obtain a VFES, the physician had to refer the patient to a central organization that controlled community-based evaluations. That program would only offer a clinical or bedside evaluation and then they would determine whether referral for a VFES was to be made. Such a referral for the VFES would involve a 13-month wait time.

The physician raised her concerns with the manager of the affiliated outpatient service (not a physician), indicating that the clinical examination did not have sufficient sensitivity, meaning the test could result in a high rate of false negatives (no abnormality found when the swallowing truly is abnormal). If the bedside or clinical evaluation was abnormal, the patient would have a VFES to determine the appropriate mode of treatment. Thus, there was nothing to gain by performing the clinical or "bedside" examination and would only delay definitive evaluation. Despite being presented with the scientific data that follows, the manager insisted that a clinical or bedside examination was sufficient and offered no further assistance. Appeals to the physician's department head were unhelpful. A direct appeal to the physician who directed a swallowing clinic located in a different hospital resulted in the patient being seen more quickly but only after a significant delay for the patient in the case described (also see companion Case 1 where the speech-language pathologist recommended the VFES while the physician refused).

---

[a] The case presented is hypothetical, although modeled on experience. No judgment is made relative to the correctness, appropriateness, or value of any actions described on the part of any of the participants. The purpose here is to describe and explore the variety of ethical considerations and consequences that depend on the perspective of different ethical systems.

The Ethics of Everyday Medicine. https://doi.org/10.1016/B978-0-12-822829-6.00002-3

Good test, bad test?

As a matter of fact, a "bedside" clinical examination is insufficient and indeed is counterproductive in the care of a patient with a high risk of aspiration. The clinical examination, like any other test, must be considered in terms of its Positive and Negative Predictive Values. The Positive Predictive value is the probability that a patient with an abnormality will be detected, while the Negative Predictive Value is the probability that a patient without an abnormality will be determined to be normal. The issue is made more pressing by the fact that pneumonia, presumably resulting from aspiration (food or liquids going into the lungs), is the leading cause of death in patients with Parkinson's disease. In one actuarial study, the incidence of death in hospitalized patients with Parkinson's disease due to aspiration pneumonia was 4.9% in 2010. The risk is nearly four times higher than hospitalized patients without Parkinson's disease (Akbar et al., 2015). However, it is highly likely that these results significantly underestimate the risk of aspiration and pneumonia for patients with Parkinson's disease in general.

The Positive Predictive Value is the probability that the test will recognize a true positive, in this case, someone at risk for aspiration and possible pneumonia. The Positive Predictive value is a function of the sensitivity of the test, which is the probability of the test being positive when applied to a sample of patients with known swallowing problems that risk aspiration. However, the specificity must be combined with prior probability, that is, the probability of aspiration risk in the population of patients of concern (note that the probability of these patients actually having a swallowing problem is unknown). The population likely will include those actually at risk and those who actually are not at risk, although, to the clinician, all are considered at risk. For example, if a test, such as a clinical examination, can detect 75% swallowing problems in all those known to have a swallowing problem, the sensitivity is 75%. This is based on studies suggesting that 25% of patients who aspirate are silent aspirators and would likely result in a false negative in a clinical examination. In a study of 16 patients with Parkinson's disease without symptoms of swallowing problems undergoing VFES, 4 patients (19%) had evidence of aspiration and an additional 14 patients (88%) had residual testing material in the vallecula (the space above the larynx in the throat) thought to be a risk for aspiration (Bird et al., 1994).

The sensitivity of a test is the probability that those with the problem will test positive. However, a physician in a clinic cannot know whether her patient has a swallowing problem that risks aspiration; consequently, the sensitivity of the test is not directly applicable to the patient of concern to the physician. Assume that among all patients for which a physician would be reasonably concerned, the actual probability that any of those patients actually has a problem swallowing risking aspiration is 80%. This means that the test, which is 75% sensitive, will identify 60% of those patients with a swallowing problem risking aspiration. However, the false negative rate will be 20%. If the risk of aspiration pneumonia and death is 20% for all patients with Parkinson's disease, then 4 out of the 100 patients who passed the "bedside" clinical evaluation will die. Presumably, these patients would have been discovered by the VFES and perhaps their deaths would have been prevented. Whether these deaths are ethically unacceptable will be discussed later.

One initial response might be to just say "that is the way things are and when does caring about things like this ever matter?" Martin Luther King Jr. said "A man dies when he refuses to stand up for that which is right. A man dies when he refuses to stand up for justice. A man dies when he refuses to take a stand for that which is true." It may well be that caring will not change things, but things will not change if one does not care. At the very least, it would be reasonable to assume that for reasonable persons or those ignorant of their own interests [John Rawls' Veil of Ignorance, from his work A Theory of Justice (Rawls, 1999)] such fatalism should not be the default position, as that can only perpetuate the bitterness of always failing. After careful analysis and deliberation, it may become clear that change from the condition described in this case is not possible or even desirable. Nonetheless, in this author's opinion, the bitterness of giving up can be replaced by the pride in trying. Indeed, as will be discussed, it may well be the duty of physicians and healthcare professionals to attempt change. Another response is to just condemn the situation described, arguing that what the physician orders should be done and what right does the clinic or hospital administration has to contradict the physician's orders? Alternatively, the physician could be condemned for giving the one patient everything that will "break the bank" and reduce care for others.

Gut reaction, stakeholders, and shareholders

In this author's experience, it would be highly unlikely that anyone would not have an immediate reaction to the situation described in the case presented, often called a "gut response." In actuality, the response is an affective (emotional or subjective) neuro-physiological response as discussed in Chapter 1. It is something that humans seem to be "wired" to feel and respond. Indeed, young children and animals respond to situations and circumstances that suggest a sense of justice (Chapter 1). These "gut reactions" can be counterproductive. Paraphrasing Claude Bernard, considered the father of modern physiology, "we are more often fooled by things we think we know than things we do not." Perhaps one of the first steps in the analysis of ethics is to "check your gut at the door" and check the prior positions, biases, and predispositions that one is likely to bring to the discussion. (These issues are critically examined in Chapter 3.) Better to think with your head (in the figurative sense) than your gut. As will be seen, a critical analysis will demonstrate that the immediate responses described here are likely to be counterproductive.

Another first impression is to see the problem described in the case as a problem between the physician and the administration about who gets to decide. In this case, the physician and administrators are shareholders in that they can impact whatever decision is reached. However, it is far more complex. Society, through its proxy in government, has a say by virtue of the laws, rules, and policies established to regulate medicine. Thus, the government is a shareholder as well. An initial impression may be that the argument centers on what beneficence to provide the patient, such as conducting the VFES. The patient is a stakeholder, meaning that the patient is affected by the decision regarding the VFES. However, there are many more stakeholders. The reputation and self-esteem of the physician and administrator are at risk as a consequence and thus they are stakeholders. Family, friends, and all those who care about the patient are stakeholders. Every current or future patient is a stakeholder as the precedent established in how the case presented is resolved may set a precedent for other patients, current and future.

To fully resolve the ethical problem, all shareholders and stakeholders must be identified and considered. Then the relationship among all pairs of shareholders and stakeholders can be analyzed on the dimensions depending on the ethical system to be utilized (see Chapter 2). Ethical principles, for example, beneficence, nonmaleficence, and autonomy are one dimension used primarily in Principlist ethical systems. The ethical principles relate to rules governing personal interactions irrespective of any notion of right or wrong; the latter is the purview of moral theories. Alternatively, the interactions can be based on laws or case precedents, with the latter being how similar situations and circumstances were judged just or unjust in the past. The interactions can be defined in terms of virtue, that is, what a good and just person would do.

From the Principlist ethical system, few would doubt the value or "truth" of the basic ethical principles, such as beneficence, nonmaleficence, and autonomy. Thus, the dimensions by which the ethical problem is to be resolved are related to the respective obligation to those ethical principles among all shareholders and stakeholders. Conflicts in terms of obligations can be identified and then resolved by selecting which of the moral theories will supervene—Egalitarian, Utilitarian, Libertarian, Deontological, or Kantian. Alternatively, providing the VFES is just something a virtuous healthcare organization naturally does without any appeal to ethical principles or moral theories, an example of virtue ethics and care ethics. Finally, VFES should (not) be offered because that was the decision in past similar cases. All seek justice as the final consequence of their actions although exactly what kind of justice is a matter for discussion.

Recognizing that there are many ethical systems, such as virtue, care, phenomenological, narrative, and feminist ethics (Chapter 2), the approach used here is based on Principlism, particularly the ethical principles of beneficence, nonmaleficence, and autonomy and the moral theories of Egalitarianism, Deontological or Kantian, Libertarianism, and Utilitarianism. Each will be defined as they are encountered in the discussion. Importantly, confining most of the discussion to these principles and moral theories should not be taken as suggesting these are exhaustive of all the types of ethical perspectives and approaches. These were selected because these principles and theories are the most "common coin" used in ethical discussions of everyday medicine. Further, perhaps based on past experiences, proceeding from these principles and theories seems to facilitate these discussions.

Cost (harm or maleficence) versus benefit (beneficence) analysis

Consider what happens to the patient with Parkinson's disease who has an abnormal "bedside" clinical examination. Presumably, this ultimately is done to help determine whether treatment is indicated. Depending on the exact nature of the abnormality, a variety of interventions can be brought to bear. Unfortunately, the clinical examination does not provide sufficient understanding of the pathophysiological mechanisms on which to base therapy if therapy is to be provided. This is not to say that some clinicians don't make treatment recommendations based on the clinical or bedside evaluation, but they would do so not based on knowledge of the pathological mechanisms involved. Further, for some patients, there is no safe manner by which to take oral foods and liquids and the safest treatment (although not necessarily the most desirable) is to feed the patient through a tube inserted into the stomach. There is no way a clinical or bedside can determine which treatment approach is optimal. Thus, there is the risk of underaggressive treatment, thereby exposing the patient to unnecessary risk or overaggressive treatment that unnecessarily compromises the patient's quality of life. Consequently, a positive result from a clinical examination would result in the patient requiring a VFES in any event, so the precise nature of the problem to be understood, the best treatment offered, and an understanding of what the future may hold (prognosis) can be offered. But appropriate therapy is delayed by intervening the bedside or clinical examination. If the bedside or clinical examination is "negative," a VFES will be ordered anyway given the insensitivity of the bedside or clinical examination. Nothing is gained, time is wasted, and the patient is at risk for longer than necessary.

A brief comment on the nature of maleficence is in order. Maleficence is the harm in the widest connotation. The term is not a value judgment such as good or bad or evil; it does not imply blame or incompetence. Almost invariably, every action in medicine risks some harm, for example, puncturing the skin to administer a vaccine. In the case presented, the patient could aspirate simply as a consequence of doing the test, even though the patient may be under constant supervision by a speech-language pathologist. It is only after determining the consequences of the exposure to maleficence that a value judgment can be made. For example, if the outcome is just, then the degree of maleficence may be appropriate. If the outcome is unjust, then the maleficence may be inappropriate and consequently blameful. However, if there is a prior value judgment, such that all maleficence is bad or evil, it will be more difficult to proceed even in those situations where the benefit would have outweighed any harm. Indeed, balancing beneficence and nonmaleficence is fundamental to every medical decision. Methods for striking a balance go back in history at least until Thomas Aquinas who addressed this issue in his Summa Theologica in 1485.

Financial cost is a form of maleficence. Someone has to pay for the tests, and money spent on such tests is money denied for other purposes, perhaps better purposes. In a cost analysis, the issue, in this case, is the incremental cost of performing a VFES in all patients with Parkinson's disease at risk. The incremental cost is only the additional cost of the VFES when the clinical examination was negative as all patients with a positive test would go on to have a VFES. Thus, the additional cost would be approximately 25% over all patients. One ethical question becomes whether the 25% increased cost is reasonable, realizing possible savings of 4 lives for every 100 tests. Assuming the VFES costs approximately US$1000, the additional expense for additional lives saved would be $4000 per 100 patients or $40 per patient.

This analysis is oversimplified but it is made to make a point. Additional costs would be the cost of treating those four additional patients. However, this could be offset by reducing the costs of treating pneumonia. Simply reducing admissions to the hospital by reducing the risk for pneumonia is likely to more than offset the cost of the increased number of VFES examinations and additional patients being treated. Thus, it would seem reasonable to obtain a VFES on patients with Parkinson's disease who are of concern to the physician, but this is not so simple. Perhaps an additional expenditure of $40 for every patient with Parkinson's considered to be at risk for aspiration pneumonia may be too much. In some sense, injustice would be created if the additional expenditure was made, creating an injustice of other stakeholders and shareholders. In that case, the physician may be unethical by insisting that the patient has a VFES. If it is judged that an additional expenditure of $40 is entirely justifiable and, therefore, would provide justice, then perhaps the clinic manager is acting unethically.

Who is paying which cost? For example, an outpatient VFES may be paid by one insurer while any hospitalization may be paid by a different insurer. Thus, an outpatient care insurer paying for the VFES does not realize the savings (return on the investment) by reducing hospitalization. The insurer for hospitalization realizes unearned savings by the other insurer paying for the VFES. A single insurer would be in a better position to benefit if the insurer covered the costs of VFES, as well as the cost of not conducting a VFES in terms of subsequent complications. If the physician was able to force, in some manner, the outpatient care insurer to pay for the VFES, then maleficence would have been done to the insurer who expends resources without a reasonable expectation of reward. It just may be that the outpatient care insurer has to suffer for the greater justice, a situation termed "telishment" (Rawls, 1955) for which numerous examples exist today, for example, taxing healthcare provider organizations rather than society at large to cover the cost of insurance.

As can be appreciated, there are many stakeholders (those affected by the decision whether to conduct the VFES) and shareholders (those in a position to influence whether a VFES is conducted). The number and types of stakeholders and shareholders go well beyond the ones identified thus far. For example, future patients are stakeholders because the decision made in the case presented becomes a precedent for deciding about future patients. The concept of *stare decisis*, do not deviate from past precedence, makes the decision in the present case very important (see *stare decisis* in Chapter 1). The concept of precedence in ethical decisions is analogous to *stare decisis* in law and reproducibility in biomedical research (Montgomery, 2019).

Each stakeholder and shareholder stands in some relation to all the others and that relationship can be analyzed in terms of the ethical principles—beneficence, nonmaleficence, and autonomy—and the obligation to those principles. As already seen, the obligations entailed by the relations often are in conflict as the obligation to beneficence to the patient (performing the VFES) creates maleficence for the insurer by the insurer having to pay the cost of the VFES. Note that the term maleficence means harm in any and every sense of the word harm and is not to be confused with malfeasance or malpractice. Maleficence does not take into account the intention. Thus, the harm of all stripes is considered maleficence, intended, unintended, unanticipated and that should have been anticipated, and anticipated. Intentional and unjustifiable maleficence is treated as malevolence. Sometimes maleficence is qualified by issues of intention, which creates problems for appreciating necessary and ethically acceptable causing of harm. For example, paying the cost and exposing the patient to X-rays are harmful even if they are not onerous.

A conceptual problem arises when maleficence is the primary focus rather than nonmaleficence. It may be more useful to consider maleficence as a failure of obligation to nonmaleficence. Thus, just actions may mitigate the obligation to nonmaleficence and result in justifiable maleficence in order to meet the greater obligation to the ethical principles of beneficence and autonomy. Just actions resulting from appropriate circumscription of the obligation to nonmaleficence do not defeat the ethical principle of nonmaleficence. Indeed, the Principle of Double Effect, such as discussed by Thomas Aquinas in his Summa Theologica (1485), can be considered management of the obligation to nonmaleficence. Thus, nonmaleficence is not qualified as a value judgment in itself, it is said to be contentless—containing no value but merely a statement of fact. The obligation to nonmaleficence can be estimated, at least in a qualitative sense. Thus, an intentional violation of the just obligation to nonmaleficence is the malevolent form of failure to the obligation of nonmaleficence. An unintentional violation is judged differently. But both intentional and unintentional violations only characterize the obligation and do not affect the standing of the principle.

The same approach is true of the other ethical principles of beneficence and autonomy. The principle of autonomy is not in conflict rather it is the obligation to autonomy that is in conflict, for example, between the respect for the position of the physician, the administrator of the clinic, and the entire system that created the situation in which the patient, physician, and administrator find themselves. It is important to note that none of the participants claim that the ethical principles of beneficence, nonmaleficence, and autonomy are invalid or do not apply to themselves or others. Rather, the relative obligations to those principles for one person are held to trump those for another.

Justice, ethical principles, moral theories, laws, case precedents, and virtues

Justice lies in the act. Justice may happen spontaneously and instantaneously. Justice can be the consequence of deliberative reconciling the conflicts among attempts to comply with the obligations to the ethical principles, moral theories, laws, case precedents, or virtue. For the sake of discussion here, the ethical principles include beneficence, nonmaleficence, and autonomy. Yet, when taken at face value, there is no inherent and obvious resolution given by attending to the ethical principles or the obligations to them. Indeed, comparisons between the specific obligations to the ethical principles for the individual stakeholders and shareholders almost become a matter of "comparing apples and oranges"; attempting to do so risks a category type error in epistemology (Ryle, 1949). It is not that the ethical principles do not apply. Rather, some means of prioritizing and balancing between the obligations to the ethical principles in conflict are required and usually require the selection of a moral theory. A moral theory is what determines right from wrong; it is normative meaning what should be done. To be sure, different moral theories may come to different conclusions as to what should be done; nonetheless, the purpose of a moral theory is to determine what should be done.

A moral theory is distinct from ethics, which relates to how humans interact. The origins of ethics are descriptive, what is being done that is just (or unjust), which forms the substrate for generalization, which is still descriptive of just (or unjust) actions. Note that how persons interact is the subject of ethics, but ethics, as used in this text (see Introduction), itself does not determine what is right or wrong. Ethical principles are critical to analyzing the conflict in the obligations to those principles that justice has to deal with. But moral theory is what determines whether the outcome is right (just) and not wrong (unjust). Thus, justice is determined by determining which moral theory or combination of theories supervenes.

From the perspective of a Deontological or Kantian moral theory, humans—the patient with Parkinson's disease—have the inherent right to beneficence and, more importantly, there is an obligation to the beneficence on behalf of the patient, in this case, performance of a VFES. Kant held that a human is an end to his- or herself; that is, the actualization or perfection of the person's own self is a right as most other Deontological moral theories (Kant, 1785). It would seem that the Kantian perspective would give priority to obligation to the principle of autonomy, which would then elevate the obligation to beneficence to the patient over the obligation to nonmaleficence to the insurer.

Kantian ethics goes beyond by requiring any ethical action to be tested by its universality, not in an empiric sense, but as a test of whether the ethical action would hold if applied universally. As will be seen frequently throughout this text, Deontological moral theories are difficult when the consequences are applied universally, particularly in the setting of the moderate scarcity of resources that necessitates conflict between those who control the resources and those that need the resources. Every patient with Parkinson's disease and considered at risk for aspiration just should have a VFES according to a Deontological moral theory, at least until the system goes bankrupt. To avoid bankruptcy in the situation or circumstance of the moderate scarcity of resources, the Deontological moral theory must be limited or bracketed by some other moral theory, such as Libertarian or Utilitarian moral theories.

Prioritizing the obligation to the principle of nonmaleficence toward the insurer, who has to pay, would make the patient the means to the insurer's ends, that is profit. By denying coverage or authorizing the VFES, the insurer preserves its profit, its beneficence, at the expense of the patient's beneficence. However, this perspective presumes that the human patient has some moral priority over the insurer. If one accords the insurer, as a corporation, the moral equivalence as the patient, then it is not clear that the Deontological or Kantian moral theories provide a clear and easy resolution. Indeed, corporations can own property, enter into contracts, and have, in certain circumstances, been afforded First Amendment rights in the United States (see Chapter 4).

One way to avoid the difficult conundrum of the relative moral standing of stakeholders and shareholders, in terms of autonomy, is to take the Libertarian perspective.

Here, the highest good is liberty, which, when taken to its logical extreme, obviates any obligations to beneficence and autonomy as first principles. Indeed, anarchy is the logical consequence. To avoid anarchy, humans negotiate among themselves to minimize the risk of experiencing maleficence and having some surety of beneficence. The obligation to the principle of autonomy is reduced to the terms of the contract. It may just be the case that the contract (explicit or implied) is where the administrator can simply dictate that the VFES will not be offered irrespective of the recommendations of the physician, speech-language pathologist, or patient. In the Libertarian perspective, such an incontestable edict is not impossible and not necessarily unjust. Indeed, it may be a form of telishment necessary for the society, as a whole, to function. The citizens submit to the telishment much in the manner of a Ulysses pact.

Generally, unilateral, nonnegotiable, and incontestable edicts are not well tolerated by humans (see Chapter 1). It is unlikely that anyone would find *laissez faire* capitalism, taken to its logical extreme, justice. Indeed, it is limited or bracketed by the concept that a contract in Anglo-American common law requires (1) an offer, (2) consideration, (3) acceptance, (4) a mutual intent to be bound by the contract, (5) lawful, and (6) the parties must be capable of entering into the contract. There is a further presumption of good faith, for example, innocent errors should not void the contract. The offer is the specifics or terms of the contract or agreement. Consideration is the exchange of things of value, goods, between each party. For example, the insurer exchanges financial coverage for healthcare expenses while the patient, or those responsible for the patient, offer to pay the premium. As another example, the physician, speech-language pathologist, and clinic offer to provide healthcare to the patient of an insurer in exchange for reimbursement, typically through an insurer. This is no less true when the insurer is a governmental agency. As can be appreciated, unilateral, nonnegotiable, incontestable edicts do not appear to satisfy the concept of a contract, at least in the ideal sense.

Whether by sympathy or good politics, most shareholders avoid such appearances of unfairness. Even Thomas Hobbes (1651), who supported an absolute sovereign, acknowledges revolution was justifiable if the sovereign fails to protect the life and limb of the citizens. Rather, the shareholder might say the test or treatment is not medically necessary or is experimental. When those rationales fail, the recourse is to merely state that a VFES is not a covered service. The result is an attempt to avoid the obligations to the ethical principle of beneficence to the patient, but this is difficult. What is a covered service is a complex question and attempts to define covered service have been difficult. For example, governments have attempted to explicitly state what was required to be a covered service. The 2010 Affordable Care Act in the United States listed (1) ambulatory patient services; (2) emergency services; (3) hospitalization; (4) maternity and newborn care; (5) mental health and substance use disorder services; (6) prescription drugs; (7) rehabilitative and habilitative services and devices; (8) laboratory services; (9) preventive, wellness services, and chronic disease management; and (10) pediatric services as covered services.

It is highly unlikely that any specified list will "drill down" to explicitly hold VFES as a covered service. The question then becomes what is the standard? Often, the standard is the "standard of care" typically described as what a similarly situated shareholder, such as a physician, would do. In other words, what is the practice of the shareholder's peers? However, increasingly, courts are suspect of such standards as these are inherently self-serving. Rather, the standard is becoming what a "reasonable" shareholder would do. But then, how does one define a "reasonable" shareholder (addressed in Chapter 3).

In this case, the administrator did not say that the VFES is not a covered service, which would be problematic in a single-payer universal system because at some point some patients are provided with a VFES. The administrator would have to explain why a particular patient was provided with a VFES in a timely fashion while the patient in the case under discussion was not offered a VFES in a timely fashion. Rather, the administrator held that the VFES was unnecessary, as a "bedside" clinical evaluation was equivalent and therefore refusal of the VFES was within the bounds of fairness. Perhaps the position of the administrator would be fair if it were true, but as discussed here, it was not. The question is whether the administrator is unethical or perhaps something more reprehensible? Perhaps not, it may be a prerogative afforded to administrators for the sake of greater justice that requires an efficient and financially stable healthcare system. Perhaps such administrators are necessary when the rest of society cannot come to some single system that would avoid the necessity of such arbitration. Perhaps society exchanges or surrenders its prerogative to the administrator in the manner of a Ulysses pact as discussed in Introduction.

The administrator need not say that VFES is not a covered service but only that it will take 13 months to actually have the VFES, thus the obligation to beneficence is qualified, the timing, and not the principle of beneficence. Estimating from survival analyses of patients with Parkinson's disease, 6% of patients die each year (Posada et al., 2011). It is likely that the survival rates are worse with patients known to have swallowing problems as described previously. Thus, it is likely that more than 6% of patients with Parkinson's disease will die while waiting for a VFES. The caveat is that the 6% who die while waiting may die for reasons that a timely VFES and subsequent treatment would not have circumvented. Nonetheless, it is prudent, if not also reasonable, that some patients with Parkinson's disease will die while waiting for a VFES who might not have died otherwise.

Failure to offer a timelier VFES is not necessarily an ethical lapse, but may be a moral failure depending on which moral theory supervenes on the ethical dynamics. For example, the patient, family, friends, caregivers, fellow citizens, physicians, healthcare professionals, insurers, and government agents are all stakeholders as they are affected by any decision to change the system to allow a timelier provision of a VFES. Clearly, there is potential beneficence to all the stakeholders. However, there also is maleficence to the shareholders, such as physicians, healthcare professionals, insurers, government agents, and other patients without Parkinson's disease, if only based on the added healthcare costs of increasing the infrastructure to allow more timely provision of a VFES. There also is the issue of autonomy as applied to all the shareholders and stakeholders. Thus, a conflict exists among the obligations to the ethical principles when parsed among all the shareholders and stakeholders that cannot be resolved solely on the ethical principles or the obligations to those principles. The resolution requires determining how the various moral theories supervene on the dynamics of the obligations to the ethical principles specified for all the shareholders and stakeholders.

The same cost analyses based on the moral theories described earlier for the cost of the VFES can be applied to the cost of the infrastructure necessary to provide a timelier access to VFES. Given the real risk of patient mortality during the waiting period, a Deontological or Kantian moral theory would argue that the patient has a right to a timely provision of a VFES. However, "moderate scarcity" (see Introduction) means that the obligation to provide a timely VFES is not absolute. Some prioritization is necessary. The problem becomes how to prioritize, but there is nothing within the Deontological or Kantian moral theory that readily leads to a prioritization. Thus, whatever deontological right and obligation exists, it must be limited or bracketed by the application of another moral theory. The implications of a Libertarian moral theory were discussed previously. The Utilitarian moral theory is onerous because it is difficult to determine the net good achieved by any specific possible resolution of the obligations to the ethical conflicts. For example, the good provided by obliging the beneficence provided to the patient in performing a VFES must be counterbalanced by the expenditure of healthcare resources, a situation analogous to "comparing apples to oranges."

The Egalitarian moral theory holds that what is right is that all persons receive the same consideration, whether good, bad, or indifferent. However, just as with the Deontological or Kantian moral theory, such equal consideration is not open-ended in the case of moderate scarcity. Further, complicating the ethical situations is what is meant by equal consideration. It cannot be that anyone and everyone wishing to undergo a VFES are entitled to have one performed because there has been another person for whom the VFES was performed. Similarly, in such an application of Egalitarian moral theory, it is likely intolerable that no one would have a VFES if not everyone was provided VFES. Clearly, some prioritization is necessary, presumably through some reasoned argument or algorithm. *Egalite* (equality) may be achieved by requiring anyone and everyone to be subjected to the same prioritization process regardless of the outcomes.

Egalitarianism is complicated when equity is considered in addition to equality. Equity suggests some compensation for those who cannot be equal to all those vying for equal treatment in the setting of moderate scarcity. For example, all citizens have an equal right to purchase health insurance irrespective of their financial ability to do so. This could be considered the equal treatment and hence just under an Egalitarian moral theory. Yet, it is not clear that reasonable persons or those operating under a veil of ignorance, thereby preventing self-interest, would find such equality, such as the right to purchase health insurance rather than receive healthcare, acceptable. In the case of equity, those unable to afford health insurance on their own would be supplemented in order to do so. Under a Libertarian moral theory, equality rather than equity may be just, but this may not be the case in a Deontological moral theory. Egalitarian and Utilitarian moral theories likely would not be decisive in adjudicating between equality and equity.

Further, it is not likely that the notion of equality is so narrowly defined. For example, how does providing or withholding a VFES compare with providing or withholding a magnetic resonance imaging (MRI) scan for stroke patients? It would seem unacceptable to provide stroke patients with an MRI but never allowing any patient with Parkinson's disease to have a VFES, but such situations may exist. Thus, the ethics of VFES must be viewed in a much larger context; indeed, when extended to its logical conclusion, judgments regarding a VFES must be considered in light of the provision of healthcare in its most general sense.

Consider the Oregon Health Plan where recognizing monetary scarcity, pairs of diagnoses and their respective treatments, were prioritized. The state government would decide a cutoff, where diagnoses/treatment pairs above the cutoff would be funded while those below would not. The cutoff was determined each year based on funds allocated by the state government (Oberlander et al., 2001). As of January 1, 2019, Parkinson's disease ranked 171 while stroke ranked 317. Thus, patients with stroke would be rationed out of care before those with Parkinson's. At least based on the assessment of utility by the Oregon Health Plan, there would be more utility in treating a patient with Parkinson's disease than a patient with a stroke.

Preferential treatment, implementing the obligation to beneficence, of a patient would seem unjust, at least if it is the Egalitarian moral theory that prevails. The concept of moderate scarcity (see Introduction) extends to the simultaneous consideration of scarcity across the entire spectrum of healthcare. It could be argued that providing full and excellent care to patients with stroke, while providing compromised care to patients with Parkinson's disease, violates the Egalitarian moral theory, primarily on the basis of a differential obligation to autonomy shown to patients with stroke compared to those with Parkinson's disease. Indeed, in circumstances of the case presented, patients in a hospital have timely access to VFES. This can be implemented by limiting speech-language pathologists to the hospital and even more limiting to a particular program in the hospital, such as the stroke service. Thus, a hospital-based physician cannot simply refer her patient who is not a stroke patient to the speech-language pathology service or department as none exists. As it is far more likely that a patient with an acute stroke will be in the hospital compared to a patient with Parkinson's disease, unless, for example, the patient with Parkinson's disease has pneumonia—the very eventuality the VFES was to help avoid—patients with strokes, not patients with Parkinson's disease, likely will have timely access to a VFES, assuming there was no effective alternative.

It is important to note that the use of an Egalitarian moral theory is not fundamental or foundational and thus of its own does not necessarily trump other moral theories. Libertarian, Utilitarian and Deontological, or Kantian moral theories may not trump any of the others. More typical, any one moral theory often has to be limited or bracketed by applying other moral theories. Nonetheless, justice demands action. Enforcement of justice requires some compelling agents. It is an empirical fact that by far the most frequent enforcement is predicated on the breaking of rules and laws, whether these are agreed to by all or forced on all. Thus, the breaking of rules and laws is considered a violation of contracts, the contractual duty of all citizens to follow the rules and laws, for example.

Ethics of healthcare systems

It just may be that the rules and laws allow the timely provision of VFES to patients with stroke and not to patients with Parkinson's disease. Under a Libertarian moral theory, such rules and laws may be acceptable if agreed to by all parties in the manner described previously for contracts. As one element of contracts is acceptance by all bound by the contract, patients with Parkinson's disease would have to agree that they would not receive a VFES, whereas a patient with a stroke would receive a VFES. However, the agreement by patients with Parkinson's disease may not be needed but this would require making the patient with Parkinson's disease somehow less than the patient with stroke. It may be that society's obligation to autonomy on behalf of the patient with Parkinson's disease is less than the obligation to patients with stroke. It is not clear how defensible such a position might be.

Another element is consideration; in the case presented, what considerations would be extended to the patient with Parkinson's disease for forgoing access to a VFES, for example, to achieve some balance of beneficence among all stakeholders? Perhaps it may be a value in knowing that somehow patients with Parkinson's disease are contributing to the public good by denying their own access to VFES examinations. This is a very problematic suggestion; it is not clear how such a choice could even be offered to the patient with Parkinson's disease. Even if patients with Parkinson's disease compromise their autonomy, that is, deserving respect in the consideration of a VFES, it may set an unacceptable precedence for patients without Parkinson's disease or stroke yet also in need of a VFES.

Clearly, confronting the patient with Parkinson's disease and all stakeholders associated with the patient, with such a choice is ethically and emotionally fraught, not to mention politically fraught. Thus, it could be that everyone without Parkinson's disease at the time simply agrees that no patient subsequently diagnosed with Parkinson's disease will have access to a VFES. Analogous situations have occurred. For example, a health insurer may merely state that the contract provisions hold that any current or future patients with Parkinson's disease will not have the cost of a VFES covered, essentially meaning that only the wealthy or those desperate enough to raise the funds will have access to a VFES. However, such a situation makes healthcare a right only in the sense that everyone has equal rights, if not abilities, to purchase healthcare and that the actual provision of healthcare is only a privilege extended through explicit or implicit contractual relations with insurers, private or public. The issue of equity would be immaterial.

The Oregon Experiment, described earlier, might be justice under the Utilitarian moral theory, that is, providing the greatest good, at least in the narrow context of the immediate concern. Such Utilitarian moral theory justification pales when one considers the amount of funds expended by the United States to free three whales trapped in the Arctic ice that mobilized the National Guard (https://www.nytimes.com/1988/10/18/us/unlikely-allies-rush-to-free-3-whales.html). However, the Oregon Experiment would fail under an Egalitarian moral theory in that, not every person is treated the same—those whose conditions did not meet the cutoff would not receive care. Perhaps more importantly, not everyone whose conditions did not meet the cutoff would be treated the same, as only those patients poor enough to qualify for Medicaid would be denied care, while those with other insurances and wealth would receive care.

From a Libertarian moral theory, the Oregon Experiment might be justified if all stakeholders voted to accept the terms of the Oregon Experiment, much as in the manner discussed for patients with Parkinson's disease regarding access to VFES. Such a vote did not take place among all stakeholders but rather only among the shareholders, those being the governor and legislature of the state of Oregon. It is an open question whether the actions of the shareholders truly reflected the interests of all citizens of the state, the stakeholders, and thus justifiable by a Libertarian moral theory. Perhaps the fairest approach is a citizen-wide lottery for healthcare where all take an equal risk regardless of circumstances or means.

It is not likely that the governmental representatives of the citizens have the knowledge, training, or experience to set the exact specifics of the allocation of healthcare to each patient, for example, whether any given patient should receive a VFES, nor should they be expected to. Yet, whether by providing health insurance or promulgating laws that shape health insurance by others, the government is the paramount shareholder. The only means to affect governmental policies is for the government to delegate authority to those who have the necessary expertise. Thus, it just may be that the government, explicitly or implicitly, delegates the authority, shareholder status, to the administrator to determine who has access to a VFES. As will be discussed further, in the case presented and assuming the only timely access to a VFES was through the hospital, the criteria for providing services effectively barred access to the VFES for the patient.

The critical question is whether the government transferred only authority to the administrator, even if only implicitly, thereby absolving the government of responsibility. It would seem that most reasonable persons if only by natural inclination (Chapter 1), would find this unacceptable. Most might expect that responsibility was also transferred to the administrator. Indeed, one perspective is that the administrator became an agent for the government for the administration of the social contract between the government and its citizens. Thus, all the aspects of a social contract with the government, as described previously, are now held among the administrator who is bound by the contract originally struck between the government and the stakeholders. Those aspects include an offer, consideration, acceptance, mutual intent to be bound by the contract, and the ability to enter the contract.

For the government, the offer may be in the form of protecting the interests of the citizens. The consideration is that the citizens provide legitimacy to the government. Acceptance is documented in the constitution or similar legal contracts. Mutual intent is that the government agrees to hold elections for the citizens to express their assent and that the citizens agree to be bound by the law and other legal instruments that result. However, mutual intent to be bound by the contract is reinforced primarily only at the time elections are held. For the hospital and clinic, the healthcare provider organizations, the offer by the administrator is the provision of care. Considerations include that the hospital and clinic will provide care, while the patient, directly or through a proxy, reimburses the healthcare provider organizations, allowing the healthcare provider organizations to function and its employees (and possible stockholders) to prosper financially. Both the healthcare provider organizations and the patients demonstrate acceptance by the fact that the healthcare provider organizations invite patients in and patients enter willingly, at the very least. In the United States, acceptance is essentially mandated, at least yearly, by fixed health insurance enrollment periods where insurance is increasingly a part of the healthcare provider organization—healthcare provider/insurer organizations. The consolidation of healthcare provider organizations provides patients with fewer options, thus essentially mandating acceptance. Similarly, mutual intent ultimately can only be enforced by the threat of legal action and the brief periods of time when patients can "vote with their feet." Even the ability to "vote with their feet" may be denied because of limited choice regarding preexisting conditions.

There are many ways governments, directly or indirectly, transfer their contract to healthcare provider/insurer organizations. These include establishing minimum requirements on insurers/healthcare provider organizations such as the intent in the US Affordable Care Act, described previously. The US Health Maintenance Organization Act of 1973 (Pub. L. 93-222 codified as 42 USC §300e) and the subsequent Employee Retirement Income Security Act of 1974 (ERISA) (Pub. L. 93-406, 88 Stat. 829, codified in part at 29 USC ch. 18) are also examples of a type of hegemony enjoyed by healthcare provider/insurer organizations with respect to mutual intent to be bound handed over by the government. Capping of healthcare expenditures whether regionally, such as block grants, or direct allocations to individual healthcare provider/insurer organizations or by healthcare provider/insurer organizations through capitated care directly challenges the mutual intent to be bound.

It is important to note that limitations on healthcare expenditures, such as through block grants or capitated care, are not new. For example, England and Wales began a process recognizable today to provide for the healthcare of the indigent beginning with the Act for the Relief of the Poor in 1597. The history of laws for the poor is fascinating if only for the many ways the government attempted to limit healthcare costs and the unintended consequences that are evident even today. For example, healthcare for the indigent typically was administered through individual parishes. Subsequently, some individual parishes were combined into a single jurisdiction without necessarily increasing the corresponding funding to account for the greater number of patients. Surgeons found themselves asked to pay for their own bandages, which had not been the case previously, a form of capitated care. The importance of these historical notes is to demonstrate that the structure of healthcare provision has fundamental principles that are transcendental—in other words, generalizations related to human nature and not the particular situations or circumstances of any particular time or place. Further, these fundamentals are often ethical principles and moral theories, both of which are derivative to the natural inclination to justice in humans (Chapter 1). As these fundamental principles and moral theories determined the provision of healthcare in the past, they will do so in the future. Hence, knowledge of the past can help deal with today and shape the future. As Sir Winston Churchill said, "The farther back you can look, the farther forward you are likely to see."

Ethics of clinicians in healthcare systems

Explicitly or implicitly and as a matter of practicality, the actual day-to-day conduct of healthcare delivery is largely in the hands of clinicians, even if largely constrained administrators, be they insurers, clinic managers, department heads, magistrates, or judges. However, the power of physicians to determine their work and destiny was never absolute. For example, Sir William Osler, the paragon of physician-hood, first worked in hospital wards until he gained a sufficient reputation to attract private patients. Even then Osler was consumed about the finances of his practice, which may explain some of his opposition to the reforms in medical education advanced by Abraham Flexner in the early 1900s. It may also explain his deference to fellow physicians, for example, failure to inform a surgeon that the hand amputated did not have cancer as was the surgeon's reason for the amputation (Bliss, 2007). The situation is true to this day, but a fuller discussion is beyond the purposes of this text. The book *Doctored: The Disillusionment of an American Physician* by Jauhar (2014) chronicles the many ways physicians are not in control.

Viewing the physician-patient relationship from a Libertarian moral theory avoids the problems described earlier when couching the physician-patient relationship in other moral theories (note that what will be discussed about physicians also applies to healthcare professionals, but physicians will be referred to for the sake of space). Thus, the Libertarian moral theory evolves quickly to a contractarian perspective. Thus, the contractual, implicit or explicit, relationship between the physician and the patient can be characterized based on the elements of a contract described earlier. Particularly important is the nature of the considerations or goods exchanged in the contract. Assuming that the patient believes the goods provided by the physician are "all diagnoses and treatments reasonable based on scientific medical knowledge," then it is unlikely the physician can knowingly have a mutual intent to fulfill the contract, as a good deal of the benefice a physician can provide is determined by the healthcare provider/insurer organization. As such, the contract would not be valid in the ideal sense of a contract, in which case the patient is allowed recompense.

Physician liability would not be the case if the patient's expectations for considerations from the physician are "all goods *allowable by the healthcare provider/insurer organization*." As the physician is a direct party to the contract, the physician has the responsibility to inform the patient as to the limits placed on considerations. This obligation is inherent in the element of contracts that require the parties of the contract to be capable of entering a contract. The physician cannot assure the provision of "all medical care reasonable based solely on scientific medical knowledge and capabilities." What the physician can assure is the provision of "all medical care reasonable based solely on scientific medical knowledge and capabilities allowable by the healthcare provider/ insurer organization of which the physician is a member" (hence, the physician has contractual obligations to the healthcare provider/insurer organization; see Chapter 4).

The limits of what the physician can assure immediately suggest a healthcare provider/insurer organization-patient relationship beyond that of the physician-patient relationship, of which many aspects have been discussed previously. Yet, this relationship would seem very different from the physician-patient relationship. To be sure, the administrator can claim intent and obligation to provide beneficence to the individual patient much the same as the physician, as well as adhering to patient autonomy. However, it is likely the obligation to beneficence that differs between the physician or clinician and the healthcare provider/insurer organization administrator. The physician must look at the patient in the eye and explain that the patient would not have timely access to a VFES. If the physician did not do so, then the physician violates the obligation to the principle of autonomy, making the needs of the patient subservient or the means to the ends of the healthcare provider/insurer organization. In addition, the physician would not be providing informed consent that requires disclosure of all reasonable alternatives, and presumably, reasonableness is based on scientific medical merit, not other considerations. By not having the patient's informed consent, either implicitly or explicitly, the physician or clinician does not have permission to treat the patient and to treat the patient risks being liable for technical battery. The obligation to beneficence to the individual patient on the part of the physician or clinician might suggest an obligation of the physician or clinician to advocate on behalf of the patient, perhaps counter to the interests of the healthcare provider/insurer organization (Chapter 4).

The concern here is the obligation to beneficence on the part of the administrator. It is highly unlikely that the exact candidacy for a VFES is covered explicitly in any governmental laws or policies or the policies and procedures of the healthcare provider/insurer organization. It comes down to the personal sensibilities of the administrator, advocacy by the physician or healthcare professionals notwithstanding. The question then is how does the administrator see the patient?

Some insight may be gained by considering the ethical dilemma of the runaway trolley car first described by Foot (1978) and extended by Thomson (1976). The original version involves a runaway trolley car hurtling toward five unsuspecting workers on a track. A person is standing next to a switch with which the person can divert the runaway trolley to another track and kill two workers on that track. Thus, the only choice is whether to throw the switch that will kill two workers or leave the switch alone in which case five workers would be killed. Most people would throw the switch in order to save five lives at the expense of two lives with a net gain of three lives.

The extension of the runaway trolley car dilemma involves the same scenario of the runaway trolley. However, there is no switch but the person is standing on a bridge over the track. Standing next to the person is a very large man wearing a heavy knapsack. The person knows that if she pushes the man off the bridge onto the track the man's body and knapsack would derail the trolley, thereby saving the five workers. The only option is to push the man off the bridge, as the observer is too light and her jumping onto the track would not deflect the trolley and an additional death would occur and none saved. The issue is expending one life, the large man on the bridge, but saving five on the track for a net gain of four lives or not pushing the man and five lives will be lost. Most people would not push the man off the bridge to save a net of four lives when just minutes before they were willing to throw a switch to kill two workers for a net gain of only three lives in the original problem.

One explanation for this paradox is the proximity of the agent, the person throwing the switch or pushing the man off the bridge, to the consequence, death of some persons. Somehow acting on an inanimate object, such as a switch at a far distance to the two workers who would be killed, is different than touching and pushing a real live man off a bridge and onto the tracks. In the former situation, the person is relatively removed from the consequence compared to the latter situation. Similarly, the administrator does not have to tell the individual patient directly that the patient will not undergo a VFES in a timely fashion, as this is left to the physician. The administrator is "distant" from the actual direct human consequences of the administrator's decision, unlike the physician. In a sense, the administrator deals with an "objectified" patient, thereby reducing the patient to something tractable for the operations of the administrator, thus reducing the dimensions by which the obligation of beneficence is understood. For example, the patient becomes a debit in the accounting and allocating of funds, which is not inconsequential if the "books must be balanced." Perhaps the administrator would not be effective if, while certainly sympathetic to the individual patient, the administrator engaged in sentimentality, which, according to Kant, would be ineffective sympathy. The sentimental administrator may well "break the bank" by providing a VFES.

The critical question becomes under what circumstance is the administrator's objectification of the individual patient just? A Kantian moral theory would argue that it is never just. As discussed earlier, arguments based on a Utilitarian moral theory are difficult. Further, moderate scarcity renders Deontological or Kantian and Egalitarian moral theories difficult to implement. Indeed, in the face of moderate scarcity, some form of prioritization is necessary and, most importantly, must be enforceable. Thus, the situation rapidly boils down to a contractual or Libertarian moral theory. In such cases, the consequence of objectification by the administrator must be analyzed and understood as they are critical to healthcare provision. Perhaps the least problematic is universal explicit acceptance by all shareholders and stakeholders, particularly the citizens, of the objectification of patients by administrators. Yet, this is highly unlikely to occur at the level of patients with Parkinson's disease and access to VFES. What then is the alternative?

Policing

Given the difficulty of anticipating each and every variation on the question of which patients should have timely access to a VFES, it is not likely that rules, laws, policies, and procedures will exist *a priori* (see Chapter 3). Further, any decisions by administrators by fiat immune to liability likely will not be tolerated. An option is an a posteriori system where departures from just behaviors on the part of all shareholders, in this case, administrators, are recognized, adjudicated, recompense extracted, and perhaps most importantly, an example to others to discourage unjust behaviors. However, such an a posteriori system would have to be 100% efficient to prevent future departures from ethical and just behaviors. Anything less than 100% efficiency will likely lead to compliance by "playing the odds." In other words, any a posteriori system takes on a policing function and deterrence is critical. Merely compensation for injury to the patient and other stakeholders consequent to departures from just behaviors would be insufficient. If the system allowed a large number of departures to escape detection and adjudication, the principle of autonomy would be violated, which would be immoral in Egalitarian and Deontological or Kantian moral theories.

It may well be ethical under Utilitarian moral theory as minimizing the loss of good that attends the payment of penalties, depending on what constitutes good and for whom. In a Libertarian moral theory, these apparent inequities may be acceptable just as the cost of doing business.

For deterrence to reduce the chance of future departures from just behaviors, the penalty must be greater than the cost of doing business. Indeed, the purpose of punitive damages and class action lawsuits was to moderate future actions on potential perpetrators of unjust behaviors by increasing penalties above the cost of doing business. To perceive punitive damages as a windfall for the plaintiff (and attorneys) is to miss the point, particularly because in many situations, it is only the threat of significant legal liability, perhaps by assessing punitive damages, that is operational in any meaningful way to curb future injustices.

For deterrence to work, the departures from just behaviors must be detected, but who is going to detect them? It is highly unlikely that most patients and their families, friends, and caregivers would recognize such departures. Thus, from an Egalitarian and Deontological or Kantian moral theory, leaving it to the patients and their families, friends, and caregivers to detect departures would be a violation of the obligation to autonomy and the principles of beneficence and nonmaleficence depending on the consequences. Unless the patients and their families, friends, and caregivers understand and agree that it is their responsibility to detect departures, it is hard to justify the situation under a Libertarian moral theory.

The question then becomes who has the obligation to detect departures from just behaviors? One could argue that it is up to the government, through its policing agencies, to detect departures. However, it is unlikely that there are sufficient police to ensure sufficient detection such that insufficient detection does not become capricious and undermine confidence in the system. Indeed, governmental agencies look to physicians and their professional societies to detect such departures. But this typically relates to individual physicians, not to healthcare provider/insurer systems. Indeed, the Employee Retirement Income Security Act of 1974 reduces health maintenance organizations' exposure to medical liability, such as malpractice, and effectively provides only redress for contractual disputes (Chapter 4). Thus, the default obligation would appear to fall to physicians and healthcare professionals. But is this realistic? Indeed, there are a great many ways physicians and healthcare professionals are directly or indirectly prevented, gagged, from alerting patients to departures from ethical and just behaviors on the part of healthcare provider/insurer systems (Brand et al., 1998). Moreover, there are risks to such advocacy, as discussed in Case 1.

A "get out of jail free" card?—Limiting obligation to beneficence

Interestingly, Canada has addressed this issue of physician liability related to a lack of access to care. The Canadian Medical Protective Association noted:

> The few legal cases touching on these issues signal that the courts are willing to consider the resources available to physicians when assessing whether the standard

of care was met. The courts have stated that an assessment of a physician's clinical care is not based on a standard of perfection, but rather on the standard of care that might reasonably be applied by a colleague in similar circumstances.

A court in Ontario, for example, has stated that "...a doctor cannot reasonably be expected to provide care which is unavailable or impracticable due to the scarcity of resources." [Mathura v Scarborough General Hospital, [1999] OJ No 47 at para 83 (Gen Div)]

Nevertheless, a physician is expected, *within those resource constraints* [italics added], to do the best he or she can for patients, and to act reasonably in such circumstances.
                      (https://www.cmpa-acpm.ca/en/advice-publications/browse-articles/2007/
                                  limited-health-care-resources-the-difficult-balancing-act)

The question is who benefits, as the physician certainly does but it is not clear that such a policy benefits the patient. Indeed, problems could arise if the policy appears to the physician as lessening any obligation on the part of the physician to advocate on behalf of the patient to receive the care in question. The result in the failure of the physician to do so may be unjust.

The situation just described would be unjust if there were an obligation on the part of the physician to advocate for the patient. The question is from whose perspective would the obligation arise? Certainly, the physician can hold the obligation her- or himself, which would be consistent with virtue ethics and can be seen in Thomas Percival's (1803) book on ethics (see Chapter 4). It is not clear that the insurer, governmental or private commercial, would consider this an obligation on the part of the physician. Yet, the patient may expect that the physician has an obligation to advocate as described previously.

Some Canadian provinces provide a mechanism for care not provided within the provinces or if requests for timely healthcare within the country would impose delays that might "result in death or medically significant irreversible tissue damage" (Flood and Epps, 2002). Yet, the difficult question is how much irreversible tissue damage and over what time period? Clearly, patients with Parkinson's disease are at an increased risk for death secondary to aspiration and pneumonia. But over what time frame? Is a 6% chance of dying within the next year sufficient to justify going to another country with the right to get it paid for?

"Not only must Justice be done; it must also be seen to be done" R v Sussex Justices, ex parte McCarthy

Another way this quote has been framed is that "not only must a trial be fair; it must look fair." Recall the discussion earlier regarding the Oregon Health Plan that rations not which people will or not receive healthcare provided they meet eligibility criteria, but just what healthcare they will receive. Thus, a patient can be said to be enrolled in the state's Medicare program, yet if the patient's only medical problems are those be-low the cutoff threshold, the patient essentially is not getting care and thus can be said to be rationed out of care. Yet, the results and experiences of the Oregon Health Plan have been mixed, not completely satisfying advocates or critics (Oberlander et al., 2001). Interestingly, Oberlander et al. (2001) wrote:

> Although the Oregon legislature has drawn a line across the list of prioritized services, that line has been rather fuzzy. Put simply, many Medicaid recipients continue to receive services that are supposedly excluded by the OHP. In large part, this can be attributed to the *reticence of providers when it comes to abiding by the rules* [italics added]. The OHP pays for all diagnostic visits and procedures even if treatment is not covered, and physicians have taken advantage of this loophole to provide uncovered medical services. Patients presenting with comorbidities have also been diagnosed with conditions covered by the list in efforts by physicians to secure patients' access to services below the line.

Thus, it would appear that even if the Oregon Health Plan may have been deemed just under the informing Utilitarian moral theory, the providers did not see it as just. It may be that those providers were held to an Egalitarian or Deontological (particu-larly Kantian) moral theory even though the state mandated a Utilitarian moral theory. Perhaps the reticence stems from the providers personally having to inform the patient regarding denial of care as discussed earlier for the runaway trolley car dilemma. Perhaps if the Oregon Health Plan was not only just, but also looked just, it would have not been undone, at least in part, by the providers who saw things differently.

Are new ethical principles and moral theories needed?

As can be appreciated from the previous discussions, there does not appear to be a clear answer to whether justice is served by the system acquiescing to the physician's demand that her patient with Parkinson's disease undergoes a VFES. No clear answer is evident even as the obligations to the ethical principles of beneficence, autonomy, and nonmaleficence were clarified. It also appears that no clear and universally ac-cepted solution was obtained when the obligations to the ethical principles were eval-uated by the Egalitarian, Utilitarian, Libertarian, and Deontological moral theories.

Perhaps new ethical principles and new moral theories are needed, but, if so, how will they be determined? In pluralistic modern liberal democracies, it is unlikely that the word of God or the dictates of an absolute sovereign will work. There seems to be little alternative than to appeal to the will of those to be subject to the new principles and moral theories, for example, the citizens. But how is the will of those to be subject determined? By asking, but how does one ask and exactly how should one ask? These issues are the concern of Part 2 of this book, particularly Chapter 3.

One has to admit that these issues in the case presented have been recognized, al-though to variable degrees and in one manner or another, since the time of Hippocrates in ancient Greece (c.460–375 BCE). The honest question is why have not more effective ethical principles and moral theories been developed? It is not simply that medicine is more complex or that there were not economical issues and governmental involvement. The first Poor Laws that also provided for the care of the poor were written in the early 1500s. Many efforts to curtail costs, such as capitated care, were used in England in the 1700s (Montgomery, 2018). There likely are a great many reasons but often it is valuable to look perhaps from a different perspective and starting point, one of probably many is offered in Chapter 1.

References

Akbar, U., Dham, B., He, Y., Hack, N., Wu, S., Troche, M., Tighe, P., Nelson, E., Friedman, J.H., Okun, M.S., 2015. Incidence and mortality trends of aspiration pneumonia in Parkinson's disease in the United States, 1979–2010. Parkinsonism Relat. Disord. 21 (9), 1082–1086. https://doi.org/10.1016/j.parkreldis.2015.06.020.

Bird, M.R., Woodward, M.C., Gibson, E.M., Phyland, D.J., Fonda, D., 1994. Asymptomatic swallowing disorders in elderly patients with Parkinson's disease: a description of findings on clinical examination and videofluoroscopy in sixteen patients. Age Ageing 23 (3), 251–254.

Bliss, M., Osler, W., 2007. A Life in Medicine. Oxford University Press, Oxford.

Brand, G.S., Munoz, G.M., Nichols, M.G., Okata, M.U., Pitt, J.B., Seager, S., 1998. The two faces of gag provisions: patients and physicians in a bind. Yale Law Pol. Rev. 17 (1). Article 6. Available from: http://digitalcommons.law.yale.edu/ylpr/vol17/iss1/6.

Flood, C.M., Epps, T., 2002. A patients' bill of rights: a cure for Canadians' concerns about Medicare? Inst. Res. Public Policy 3 (12).

Foot, P., 1978. The Problem of Abortion and the Doctrine of the Double Effect in Virtues and Vices. Basil Blackwell, Oxford. (Originally appeared in the Oxford Review, Number 5, 1967).

Hobbes, T., 1651. Leviathan or the Matter, Forme and Power of a Common-Wealth Ecclesiasticall and Civil. Yale University Press. (Reprint edition, 2012).

Jauhar, S., 2014. Doctored: The Disillusionment of an American Physician. Farrar, Straus and Giroux, New York.

Kant, I., 1785. In: Wood, A.W. (Ed.), Fundamental Principles of the Metaphysics of Morals. pp. 143–202. translated by Thomas K. Abbott in Basic Writings of Kant, Modern Library, New York, 2001.

Montgomery Jr., E.B., 2018. Medical Reasoning: The Nature and Use of Medical Knowledge. Oxford University Press, Oxford.

Montgomery Jr., E.B., 2019. Reproducibility in Biomedical Research: Epistemological and Statistical Problems. Academic Press, New York.

Oberlander, J., Marmor, T., Jacobs, L., 2001. Rationing medical care: rhetoric and reality in the Oregon Health Plan. CMAJ 164 (11), 1583–1587.

Percival, T., 1803. Medical Ethics, or a Code of Institutes and Percepts, Adapted to the Professional Conduct of Physicians and Surgeons. Cambridge University Press.

Posada, I.J., Benito-León, J., Louis, E.D., Trincado, R., Villarejo, A., Medrano, M.J., Bermejo-Pareja, F., 2011. Mortality from Parkinson's disease: a population-based prospective study (NEDICES). Mov. Disord. 26 (14), 2522–2529. https://doi.org/10.1002/mds.23921.

Rawls, J., 1955. Two concepts of rules. Philos. Rev. 64 (1), 3–32.

Rawls, J., 1999. A Theory of Justice, revised ed. Belknap Press of Harvard University Press.

Ryle, G., 1949. The Concept of Mind. University of Chicago Press, Chicago.

Thomson, J.J., 1976. Killing, letting die, and the trolley problem. Monist 59 (2), 204–217.

Case 3

Case 3—A 63-year-old male was seen in consultation requested by a family physician at which time the diagnosis of early and mild Parkinson's disease was confirmed. The consultant recommended the patient start selegiline in order to slow the progression of the disease. The consultant provided detailed information regarding the evidence to support slowing of disease progression, dosing information, a list of potential adverse effects, and medications that are contraindicated when combined with selegiline. The patient had no conditions or treatments that would contraindicate the use of selegiline and there was no reasonable expectation that the situation would change in the near future. There were no financial reasons to prevent the patient from receiving the medication. The consultant offered to help at any time and volunteered his cell phone number. The family physician refused to prescribe selegiline stating that was up to the consultant.[a]

Medical facts of the matter

The consultant expected the family physician to write the prescription. The letter from the consultant to the family physician included the following text:

Movement Disorders Unique Identification Code: XXXXXXXX

Dear Dr. XX:

We had the opportunity to see the abovenamed patient in the Clinic of the Division of Neurology at the XXXX. Enclosed, please find a copy of our evaluation and recommendations. PLEASE NOTE, a brief summary of the impressions and recommendations is given at the beginning of the note. The full impressions and recommendations are at the end of the note, often with supplemental information.

I will attempt to assist you by providing detailed recommendations that contain step-by-step recommendations specific to the patient for your use as well as more general information. Also, I am available by cell phone (XXX-XXX-XXXX) as well as e-mail (XX). With respect to e-mail, I cannot ensure confidentiality of the communication so I recommend that, for information communicated by e-mail, not to include any information that could identify the patient, such as name, date of birth, medical record number, or date of service. You may wish to use some unique and confidential code that will allow you to identify the patient in your e-mail to me and I can use in my response.

I look forward to working with you in the care of the patient. Please do not hesitate to contact our office if we can answer any questions or be of any further assistance.

Sincerely,

XXXX MD

Professor of Neurology

[a] The case presented is hypothetical, although modeled on experience. No judgment is made relative to the correctness, appropriateness, or value of any actions described on the part of any of the participants. The purpose here is to describe and explore the variety of ethical considerations and consequences that depend on the perspective of different relevant ethical systems.

The Ethics of Everyday Medicine. https://doi.org/10.1016/B978-0-12-822829-6.00003-5

My strategy for the management of patients with movement disorders, such as Parkinson's disease and tremor, may be different than what the patient, family physician, and the treating physician may have encountered with other movement disorders neurologists and general neurologists. My approach is borne out of a sense of responsibility to the future as well as present patients. The severe lack of movement disorders specialists risks prolonged delays in patients receiving specialized care, which has serious consequences. For example, a study of the US Medicare patient charts demonstrates a higher mortality rate in patient with Parkinson's disease not seen by neurologists, which translates to a 3% increased mortality rate in the first year the patient is not seen by a specialist and a 10% increased mortality if the patient is not seen within 2 years (Willis et al., 2011). These statistics are not a negative reflection on nonneurologist clinicians but rather reflect the failure of a system to enable nonneurologists to provide the same level of care. Canadian surveys show that 28% of patients wait at least 6 months before being seen by a neurologist (Kirby et al., 2016). As concerning as these statistics are, they do not begin to capture the preventable suffering of patients and all those who personally care for the patient. Unless there are significant changes in the manner in which care is provided, the de facto situation becomes a "first come, first serve" approach as movement disorders specialists in clinics become "saturated" and patients are turned away. This would not be good stewardship of the resources available.

The key is to maintain accessibility for patients to movement disorders neurologists. As it is unlikely that the supply of movement disorders neurologists will increase sufficiently to meet the needs, an effective partnership of shared and distributed responsibilities between the movement disorders neurologist and the treating physician, optimally the family physician needs to be established. The preferred relationship is one where the movement disorders physician's responsibility includes assisting the treating physician as the treating physician provides the day-to-day management (principal care) of the patient. In exchange, the movement disorders neurologist must provide the treating physician with sufficient tools to enable the treating physician and to be reasonably accessible at the moment the treating physician needs assistance.

A few family physicians have suggested that I conduct the initial management until the patient is "stable," meaning requiring less observation and intervention. Unfortunately, this is not effective because many patients, particularly those with Parkinson's disease, require frequent and ongoing evaluations and management changes. Parkinson's disease is a progressive disorder and the way patients respond to medications changes over time. Thus, the optimal treatment regimen is a moving target that requires relatively frequent and ongoing retargeting of the treatment regimen. Further, continuity of care is critical for effective management, which is best facilitated with the family physician or original treating physician.

The effort required of the family physician would be prescribing a 5-mg tablet of selegiline to take in the morning, checking one week later to see if the patient is tolerating the medication, and, if so, adding a second 5-mg dose of selegiline at noon. No further adjustments would have been necessary if the patient continued to tolerate the medication. To be sure, there is some controversy on whether monoamine oxidase type B (MAO-B) inhibitors are effective in slowing the progression of Parkinson's disease; however, that is not relevant here. The family physician did not disagree with the consultant's recommendations but simply refused to prescribe the medication.

The fact of the matter is that there is a severe shortage of neurologists if measured by the burden each neurologist carries and the delays in a referred patient being seen by the neurologist. In 2015, neurologists responding to a survey indicated they worked an average of 56 h/week, of which an average of 30 h was spent in direct patient care (Kirby et al., 2016). These figures are a little different compared to a survey conducted in 2002. What is different is that in 2002, academic and community neurologists saw an average of 15 and 30 new consultations per week, respectively. The number of new patients seen in 2012 was reduced by one-third compared to 2002. Thus, it is likely that neurologists have fewer new patient consultations despite long delays for new patients to be seen and are spending much more time seeing patients in follow-up. In 2002, 41% of academic and 55% of community neurologists could see a new patient within 8 weeks. By 2012, the percentage was reduced to 21% and 26% of community and academic neurologists, respectively (Kirby et al., 2016).

Stakeholders and shareholders

At the first level of ethical analysis, the stakeholders, those affected by the decision of the family physician to not provide the day-to-day management of the patient, include the patient, family physician, and consultant. The family physician and consultant are the shareholders, those in a position to affect decisions. However, as will be demonstrated, society, through its proxies, the government and insurers, also are stakeholders and shareholders. Society is a stakeholder because what happens to any individual citizen could happen to any other citizen. Therefore, each citizen's rights are only as protected as every other citizen's right, at least over the long run. The government, by virtue of the power vested in it by its citizens, is a stakeholder in that the government can be affected by what happens to its citizens who are patients and future patients. In a modern liberal democracy, governments can be turned out by the will of the citizens. In addition, the government is a shareholder by the enforcement of regulations related to licensure (power of the physician to practice). Further, the government enters into a "contractual" relationship with the physician where the goods of a monopoly given to the physician (not just anyone can practice as a physician) are exchanged for adherence to the policies and regulations of the government. The insurer is a shareholder, whether it is public or private, by establishing contractual agreements. The society also has a "contractual" agreement with physicians providing the goods of a subsidized medical education with the expectation of beneficence toward the members of society subsequently by physicians in return.

The point of the aforementioned discussion is that the typical distinction between shareholder and stakeholder as typically held in many discussions is ill-founded. Any ethics that presupposes the physician and healthcare professional as solely a shareholder and the patient solely as a stakeholder does not reflect the complex ecology in which the ethics of everyday medicine operates and, therefore, likely to allow injustices (see Introduction).

When there is a conflict among shareholders and stakeholders, the dimensions of the conflict need to be defined and some means to resolve the conflict found. A

Principlist approach would be to define the relationship between all the shareholders and stakeholders in terms of the obligation to the ethical principles of beneficence, nonmaleficence, and autonomy as few would hold that no party to the conflict expects these ethical principles. From the perspective of virtue ethics, most reasonable persons blinded to their self-interest would hold that virtues of kindness, generosity, and honesty, among others, are reasonably expected by all patients, although the obligations of shareholders to those virtues may be qualified. Certainly, all would agree that beneficence to the patient is important and, for the discussion here, beneficence would have been achieved by the prescription of selegiline to the patient. Further, both the consultant and the family physician believe that prescribing the medication was an obligation to which there were no constraints. Nonmaleficence is not an issue here; if the patient were to experience an adverse effect, it would have occurred regardless of who prescribed the medication. Thus, the issue seems to center on who has a greater obligation to beneficence in writing the prescription and providing the follow-up supervision. Further, the conflict also appears to involve considerations of autonomy of the family physician and consultant. The family physician appears to think that this is the consultant's responsibility, while the consultant believes the responsibility lies with the family physician.

Family physician's duty of care for the patient

The family physician already has an obligation to treat the patient's Parkinson's disease. By accepting the patient into the family physician's practice, there is a contractual, at least implicit, obligation to treat any and all illnesses that might befall the patient. An exception would be an arrangement whereby the family physician, prior to accepting the patient, specifically stated that the family physician will not provide any care for patients with Parkinson's disease. This is highly unlikely. Consequently, failure of the family physician to provide care for the patient's Parkinson's disease, which is reasonably expected of the family physician, could be considered an abandonment of care and unethical. However, in the case presented, it is not clear that the family physician abandoned the patient, at least with respect to the provision of the prescription to treat the patient's Parkinson's disease. The family physician may have believed that in referring the patient to the neurologist, there was an implicit transfer of care and thus responsibility fell to the neurologist with respect to the patient's Parkinson's disease. The issue of transfer of care is central. Further, there was no explicit transfer of care that established a type of contract agreed to by both the family physician and the neurologist. Consequently, the issue of an implicit transfer of care seems to hinge on the individual expectations of the family physician and the neurologist, which clearly are in disagreement. A central question becomes the nature and justification for the different expectations.

A potential defense might be that it is not reasonable or the standard of care for a family physician to provide management of patients with Parkinson's disease, at least the type of patient that is the subject of the case presented. The argument is that it is not the role that the family physician plays in healthcare. However, as noted by

Norman E. Bowie, the nature of a role is not determined by the individual but by the collective of those thought to play the same role (Bowie, 1982). Thus, other family physicians accepting the obligation that the family physician, in this case, declined to do argues that what the family physician was asked to do is not outside the role of the family physician (for a version of a role narrowed for the benefit of the physician, see Case 6).

To be sure, the family physician is not required to do anything that he or she does not think is in the patient's best interest nor is the family physician compelled to implement the consultant's recommendations; however, reasonable persons blinded to their own self-interest would expect that there is an appropriate reason in the patient's best interest for the family physician not to do so. An example may be a preexisting condition or circumstance that would contraindicate the consultant's recommendation. However, as described earlier, the risks of prescribing selegiline are relatively minimal, particularly as compared to any number of other mediations typically prescribed by family physicians. For example, in a clinical trial of another MAO-B, rasagiline, there were more complaints of adverse effects from patients taking the placebo (Olanow et al., 2009). Given the simplicity of the dosing, it is difficult to see where inexperience in prescribing selegiline constitutes a significant barrier.

There are a number of policies and regulations from governmental organizations and professional societies that bear on the issues in this case (note that the author is not an attorney and is not providing legal advice; references to laws and court actions are only exemplars of some ethical issues). However, there are considerable variations in the specific governing regulations and policies as they relate to family physicians, and primary care physicians more generally (Arya et al., 2017). The discussion here relates to the United States and Canada, where the author has practiced movement disorders as a subspecialty of neurology. One governing principle is whether a patient-physician relationship has been established. Establishment of a patient-physician relationship imparts an obligation on the physician connected to the patient by the established relationship. However, at best, such a definition is imprecise. Clearly, actions creating an orphaned patient, for whom there was a previously established patient-physician relationship, are tantamount to abandonment and are unethical. The Canadian Medical Protective Association, essentially the provider of medical liability protection, holds:

> In Québec, article 32 of the Code of ethics states that "A physician who has undertaken an examination, investigation or treatment of a patient must provide the medical follow up required by the patient's condition, following his intervention, unless he has ensured that a colleague or other competent professional can do so in his place." The College of Physicians and Surgeons of Newfoundland and Labrador has stated that in some circumstances, the failure to provide ongoing outpatient management for orphan patients might give rise to an allegation of misconduct.
>
> *(https://www.cmpa-acpm.ca/en/advice-publications/browse-articles/2008/*
> *follow-up-for-orphaned-patients)*

The College of Physicians and Surgeons of Nova Scotia holds "The consultant position initiating any diagnostic or therapeutic intervention bears the responsibility for any patient follow up, including notifying the patient of the results of investigations ordered" (The College of Physicians and Surgeons of Nova Scotia, Guidelines for physicians regarding referral and consultation, https://cpsns.ns.ca/wp-content/uploads/2017/10/Referral-and-Consultation.pdf). It is not clear whether merely examining the patient constitutes an obligation of continued care once the assessment and its implications have been discussed with the patient. Further, the policies appear to rely on the provision of treatment, which may be different from recommending treatment. Consider the implications of a curbside consult, for example, radiologists, pathologists, or on-call consultants, who respond to referring physicians but who may never encounter the patient directly. However, had the consultant prescribed the selegiline, he would have established a patient-physician relationship in the sense of now being obligated to continue to provide the principal care for the patient with respect to the selegiline.

It is important to note that the physician acting as a consultant likely is not immune to issues of medical liability for what the family physician may or may not do. The consultant may suffer liability vicariously as the referring physician will act, one way or another, on the consultant's advice. Had the patient experienced an adverse effect following the family physician's initiating the consultant's recommendations, the consultant could have been liable. Had the consult not made a recommendation or instructed the family physician that the consultant should have, and the patient has an adverse event because the family physician did not initiate what the consultant failed to recommend, the consultant may reasonably be held liable. Neither of these situations pertains to the case presented.

The issue here is who has the obligation to initiate and continue treatment. Interestingly, many court cases arising from a failure in the duty of care by consultants were adjudicated by contractual agreements (Abramson, 2008). In some of these cases, those dependent on the consultation, such as resident physicians at a teaching hospital or a crew aboard a ship at sea, could not be expected to act independently or be in a position to question the consultant.

An established patient-physician relationship constitutes one form of duty to care. Consultants could be required to act as though a patient-physician relationship was established by contractual arrangements between the consultant and the consultant's employer or the insurer by which the consultant is contracted. For example, a neurologist providing consultation to an emergency room physician was found to have an established patient-physician relationship as the bylaws where the neurologist practiced required the neurologist to accept telephone calls from the emergency room (Abramson, 2008).

As described previously, the College of Physicians and Surgeons of Nova Scotia holds that the consultant "initiating" a therapeutic intervention bears the responsibility for executing the therapeutic intervention. The question is whether the act of making recommendations to the referring physician constitutes "initiating." At least in the case of what is called "curbside" consultations, which are common, this does not appear to be the case as some courts have held that this would be too onerous and consequently stifle the important role that "curbside" consultations play. For example, curbside consultations generally are brief and often obviate the need and thus the expense of formal consultations (Grace et al., 2010), although at a high risk of inaccuracy (Burden et al., 2013). Interestingly, some studies of the use of curbside consultations argue that the savings incurred should be used to compensate the curbside consultant (Grace et al., 2010). One wonders whether this would constitute a contractual agreement, thereby implying a more intensive and extensive duty of care.

Some have suggested constraining the consultation procedure to specifically exclude discussion of the consultant's recommendations with the patient or the patient's representatives except by direct permission from the referring physician. Perhaps the notion is that not discussing the consultant's findings and recommendations with the patient avoid the establishment of a patient-physician relationship, thereby limiting the duty to care beyond the actual consultation (Cohn, 2003). However, the various cases described earlier would suggest such a strategy may likely fail.

To be sure, clear communication of expectations on the part of the family physician and the consultant is important. Often, a distinction is attempted between referral and consultation, with the former implying a transfer of care and the latter soliciting advice. Each could involve a different patient-(consultant) physician relationship with different duties to care. However, this distinction is often confused (American Academy of Family Physicians, Consultations, Referrals, and Transfers of Care: https://www.aafp.org/about/policies/all/consultations-transfers.html). It may well be that the family physician was making a "referral," expecting the consultant to take principal responsibility for the continuing treatment while the consultant understood "consultation." The consultant made this abundantly clear in his letter to the family physician in the case presented.

Family physician's duty of care of society

In a sense, the family physician did act on the consultant's recommendation by refusing to prescribe the selegiline. It is unlikely that this was a disregard of the ethical principle of beneficence.

Rather, the operative notion was the obligation to that beneficence. Perhaps the family physician felt the consultant had a greater obligation as the family physician had a referral in mind that implied a transfer of care with respect to management of the patient's Parkinson's disease. Ultimately, the consultant reluctantly prescribed the medication as otherwise the patient would have been deprived of the benefit to which the patient was entitled. Note that the family physician did not directly contact or discuss the situation with the consultant. The consultant learned of the family physician's refusal through the patient.

The question of what damage is done if the consultant prescribed the medication and thereby is obligated to see the patient more frequently compared to the situation if the family physician assumed the responsibility. Clearly, the consultant had in mind obligations to other and future patients given the scarcity of specialists in Parkinson's disease and the serious adverse effects of delays, such as increased mortality, as noted in the consultant's letter to the family physician. Some family physicians expect that the consultant would manage the treatment until such time that the patient is stable, which is not applicable to a progressive degenerative disorder such as Parkinson's disease. It would seem by his actions that the family physician may not have shared that same sense of obligation to other and future patients with Parkinson's disease. However, the question to be examined is whether society, through its proxies such as governments, charges the family physician with such obligations.

The question becomes is there any basis for an obligation on the part of the family physician toward other and future patients with Parkinson's disease? The College of Physicians and Surgeons of Ontario holds:

> Physicians are expected to make their patient's needs the first priority but accomplishing this requires a broader focus than the direct physician-patient relationship. In order to meet individual patient needs, physicians should consider their contributions to their individual patients, but also to their own practice, the community, and *the health care system* [italics added].
> *(https://www.cpso.on.ca/Policies-Publications/The-Practice-Guide-Medical-Professionalism-and-Col/Principles-of-Practice-and-Duties-of-Physicians)*

The position appears to suggest a required obligation to future patients, even if vague and not well regulated. However, there may be a contractual obligation as well in the manner by which the province pays family physicians as it impacts on the healthcare budget and thus support for other and future patients. Thus, Medicare in Ontario may expect the family physician to provide primary care for all of his patients to the extent reasonably possible, thereby minimizing expenses related to the consultant providing ongoing care.

The national Canadian healthcare system, administered by the provinces, was established in 1962. The "Saskatoon Agreement" established physicians as independent contractors paid on a fee-for-service basis. Dissatisfied with the lack of effectiveness and costs, provinces such as Ontario attempted to change the system by restructuring reimbursement and investing more heavily in primary care physicians such as family physicians (Marchildon and Hutchison, 2016). For example, mean payments to primary care physicians increased by 31% compared to all Ontario physicians of 25% between 2005 and 2009. The effort was highly successful as it reversed the decline in the number of graduating medical students entering primary care. Indeed, in 2014, 47.4% of physicians in Canada were primary care compared to 12.3% in the United States (Edwards et al., 2014).

Since 2002, the province has moved increasingly from fee-for-service reimbursement for primary care physicians to capitated and similar forms of reimbursement, such as salaried, which covers approximately 75% of family physicians in 2015. The government introduced new models for delivery of healthcare by family physicians, such as the Family Health Network, Family Health Organizations, and Family Health Teams that have three or more physicians. The latter include other healthcare professionals, such as nurse practitioners, dietitians, and others who are salaried while the physicians continue under capitation. Although studies are limited, the method of reimbursement may affect the professional behavior of primary care physicians.

Indeed, that is one of the intentions of insurers going to a capitated care model. For example, patients of fee-for-service primary care physicians, compared to capitated care physicians, are more likely to visit specialists (Gosden et al., 2000). It would seem that the capitated care model achieved the goal of reduced visits to specialists although not seeming to have an impact on hospitalizations c.2000. If the family physician, in this case, bought into the concept of capitated care to reduce healthcare expenditures and perhaps reduce pressures on specialists, the actions of the family physician requiring the consultant to provide primary care with regards to the selegiline seem inconsistent.

Evidence of the intent to have family physicians take responsibility for ongoing principal care is seen in the reimbursement regimens related to specialists. Reimbursements to specialists are graduated in the amount reimbursed. Specialists are limited to the number of higher reimbursing claims per patient per year regardless of the patient's need. This arrangement is particularly burdensome on specialists caring for patients with chronic progressing neurodegenerative disorders, such as Parkinson's disease, where patients need more rather than less care over time.

Certainly, there has been a concern with capitated care for what has been called "cream skimming" or "cherry-picking" whereby capitated care programs avoid the very sick and costly patients for which there is some suggestion of occurrence in an analysis of Ontario patients (Rudoler et al., 2015). This is not to say that the family physician's motivation, in the case presented, was to reduce his costs relative to his reimbursements in order to maximize his income. However, capitated care models have such a risk. Further, the intent of primary care reforms in Ontario was to encourage family physicians to take more responsibility for their patients, thereby lessening expenses by referrals to specialists. The question is whether the contract between the government through the Medicare program binds the family physician to the intent of the government's program.

Unfortunately, the difficult conditions that gave rise to the ethical problem, that is, moderate scarcity of specialist, care likely will get worse. There is an 11% shortage (percent of needed neurologists over those available) of neurologists in 2012 that is expected to increase to 19% by 2025 (Dall et al., 2013). The prevalence of parkinsonism for persons over the age of 45 for a metropolitan area of Ontario serving 2.3 million lives is 9937 patients as of 2016 [based on de Rijk et al. (1997) and 2011 Survey of Living with Neurological Conditions in Canada (https://www.statcan. gc.ca/pub/82-003-x/2014011/article/14112-eng.htm); Census Profile, 2016 Census, Statistics Canada (2016 http://www12.statcan.gc.ca/census-recensement/2016/dp-pd/ prof/details/page.cfm?Lang=E&Geo1=CSD&Code1=3525005&Geo2= CD&Code2=3525&Data=Count&SearchText=Hamilton&SearchType=Begins &SearchPR=01&B1=All&TABID=1; 2010/2011)]. There are only a few movement disorders specialists in this area, defined as neurologists with practices largely confined to movement disorders following fellowship training in movement disorders. Among movement disorders seen by such specialists, Parkinson's disease is the most prevalent. This means that the patient-to-Parkinson's disease specialist (assuming three Parkinson's disease specialists) ratio is 3312 to 1, which is nearly twice the average number of patients followed by a family practitioner in rostered capitated care programs.

To be sure, many patients are being seen by Parkinson's disease specialists in other cities nearby; however, it is likely not every patient with Parkinson's disease is being seen by a Parkinson's disease specialist.

Many patients have not been diagnosed and therefore are not receiving care. From community surveys, 24% of patients identified as having Parkinson's disease have not been diagnosed and hence not treated (de Rijk et al., 1997). Extending the 24% undiagnosed, there are an estimated 12,321 patients in the metropolitan area, at least theoretically. If those patients were to be diagnosed, the ratio of patient-to-Parkinson's disease specialists would be 4107 to 1.

It is likely that a large number of patients with Parkinson's disease are being managed by nonneurologists and that these patients have never seen a neurologist. In a study of practice in Ontario published in 2002, only 45% of Parkinsonian patients saw neurologists annually and the percentage of having only a single consultation was 59.5% over the 6 years (Guttman et al., 2002). This latter statistic is of concern where US Medicare data demonstrate accruing improvement with subsequent visits to a neurologist (Willis et al., 2012).

Significant health outcome disparities exist between care provided by neurologists and nonneurologists. In a retrospective cohort study based on a review of Medicare records in the United States, 13,489 out of 24,929 eligible incidents of Parkinson's disease cases were followed by a neurologist. There were 9112 Parkinson's disease-related hospitalizations, but the odds of admission, initial and repeated, were less when the patient was cared for by a neurologist—for psychosis [hazard ratio (HR) 0.71, 95% confidence interval (CI) 0.59–0.86), urinary tract infection (HR 0.74, CI 0.63–0.87), and traumatic injury (HR 0.56, CI 0.40–0.78)] (Willis et al., 2012). In addition to a reduction in mortality rates noted earlier, patients were less likely to be placed in a skilled nursing facility [odds ratio (OR) 0.79, 95% CI 0.77–0.82] and had a lower risk of hip fracture (OR 0.86, 95% CI 0.80–0.92) when treated by a neurologist (Willis et al., 2011).

Disparities also exist between movement disorders specialists and general neurologists. A retrospective examination of medical records at a US VA hospital demonstrated that adherence to indicators of quality of care in patients with Parkinson's disease was 78% for movement disorders specialists, 70% for general neurologist involvement, and 53% for nonneurologists. Indicators included treatment wearing off, assessments of falls, depression, and hallucinations (Cheng et al., 2007).

None of the aforementioned discussions suggests that family physicians or general neurologists cannot provide the kind of care provided by movement disorders specialists. Data suggest that the odds of optimal care are greater when advised by a movement disorders specialist, on average. However, the data do not inform about any particular family physician or general neurologist. Indeed, it is likely that there are family physicians and general neurologists every bit as capable as the best movement disorders neurologist. The difficult problem is how can the family physician and general neurologist assure themselves that they are practicing at a level comparable to a movement disorders specialist? It would seem to require a metacognitive skill and it is an open question whether medical training provides such metacognitive skills. Critically important, how would a patient, family member, caregiver, insurer, or governmental regulator be assured? It is likely that licensure requirements would vouchsafe the abilities of the family physician, as presumably, all family physicians have met such requirements. Thus, it is an open question whether failure to refer to a movement disorders specialist, whether by default or lack of access, is unjust and thus unethical.

The conflicts in the respective obligations to beneficence, that is, who is going to prescribe the selegiline and directly supervise its use, can be assessed based on a Deontological moral theory. It is the patient's just right to optimal treatment and therefore a right to be seen by a movement disorders specialist in a timely fashion. Failure to refer the patient would be unjust. Logistically, this would only be possible if the referring family care physician relieved the movement disorders neurologist of the burden of prescribing and directly managing the medications with the understanding that the movement disorders neurologist would be available in a meaningful way to assist by follow-up consultation. The implication is that the family physician has an obligation to facilitate the access of all patients with Parkinson's disease, present and future, to a movement disorders neurologist or, at the very least, to a neurologist.

How could the conflict be resolved based on an Egalitarian moral theory? The Egalitarian moral theory is violated if access to a movement disorders specialist is a matter of chance based on which family physician happens to care for the patient. The Egalitarian moral theory argues that justice should not be the luck of the draw and that all patients have equal access to a movement disorders specialist, even if not 100%. Perhaps the most egalitarian approach would be for patients with or suspected of having Parkinson's disease enter a lottery to determine whether they will be seen and followed by a movement disorders neurologist. It is very difficult to see how any society would accept a lottery for healthcare. One could argue that the Oregon Healthcare program is, in effect, a type of lottery, where access to medical coverage for the medically indigent is a matter of luck of getting a disease or disorder that is above the cutoff for funding (see Chapter 4). It is not clear that the odds of having a disease or disorder that makes the cutoff are distributed equally over all citizens. The Oregon Healthcare program may be just based on a Utilitarian moral theory where maximizing the good of the entire population is paramount and necessarily requires rationing.

The critical question is to define the good that any resolution would attempt to maximize. The presumption, actualized in the case presented, is that the patient was going to receive the good—prescription of the selegiline—regardless of who provided the prescription and direct follow-up. Generally, the family physician prescribing the selegiline and following up would not be any incremental good for the movement disorders neurologist as she would be maximally busy whether or not she prescribed the selegiline and provided the direct follow-up care. A potential good to the movement disorders specialist might be less bad feeling knowing that she would not have to turn away patients because access was improved by the family physician taking the burden of prescribing and directly supervising the medication.

A potential good of not prescribing and following up for the family physician whose income is capitated would be the same income but with less burden by having the movement disorders neurologist assume principal care. However, this would come at a cost to society where there is a net increase in expenditure on the patient's care due to having to reimburse the movement disorders neurologist for care that might otherwise be subsumed under the capitated reimbursement provided to the family physician. In a sense, the family physician may be being paid for not providing care, if it is accepted that the reimbursement policies were intended to reduce expenditures on specialist care. Note that in this situation or circumstance it is difficult for the family physician to justify transferring principal care to the specialist because the family physician believes that the government health program does not reimburse adequately.

The ethical dilemma posed is that, currently, the Egalitarian, Utilitarian, and Deontological moral theories do not seem to resolve the ethical conflict easily, leaving the Libertarian or contractual moral theory. Consequently, evolution in the way patients with Parkinson's disease are managed has to be based ethically on the contractual relationship among movement disorders specialists, neurologists, and family physicians. In Ontario, this would require government intervention. In a laissez-faire system of healthcare delivery the government likely still has to intervene as assumptions for the "invisible hand" of capitalism (Smith, 1776/1759) are not likely met. The assumption is an enlightened consumer free to make choices, which is problematic in healthcare (Le Grand, 2009). At the very minimum, a Libertarian moral theory likely must be tempered as a patient cannot be left with no healthcare due to contractual issues.

Whatever the response, it is hard to see that any solution in programs like those of Ontario, Canada does not require greater responsibility on the part of family physicians in the day-to-day management of patients with Parkinson's disease. In the United States, with the looming shortage of neurologists, particularly movement disorders specialists, a greater involvement of nonneurologists will be required. The question is the manner by which this will evolve, either with or without direct governmental intervention. For example, many of the measures requiring insurers to have a sufficient number of specialists in a sufficient range of specialties as implemented under the Patient Protection and Affordable Care Act of 2010 (the Affordable Care Act) are under threat.

Even if a Libertarian or contractual moral theory prevails and the government or insurer is the dominant shareholder, it is not clear how the government, for example, should exercise its shareholder status. An analogy is the relationship between the management of corporations (e.g., governmental agencies) and shareholders (citizens, particularly present and future patients). Ultimately, the shareholders are the deciders, for example, by voting out the corporate board or government and subsequently replacing management. However, there is concern that the shareholders, as a conglomerate, lack the understanding to manage the corporation effectively (Harris and Raviv, 2010). For example, most insurers, public or private, survey specific indicators of quality of care, such as diabetes, heart disease, and smoking cessation. It is unlikely that any insurer would "drill down" to the management of Parkinson's disease by the family physician. The questions become who and how would the quality of care to patients with Parkinson's disease be monitored so that any corrective action, if needed, could be instituted? Further, there is the question of the degree of obligation to beneficence (sympathy) a remote administrator can hold to the individual patient (see Case 2).

A patient in the United States who had essential tremor was seen as a second opinion and the patient asked that a report be sent to his treating physician. The consultant providing the second opinion agreed with the diagnosis but only cautioned about the use of topiramate, which could increase the risk of kidney stones as the patient had a history of mild kidney failure. The treating physician responded by letter to the consultant that the treating physician did not refer the patient, that he did not need the consultant's opinion, and that the treating physician was returning the consult. The consultant reached out to the treating physician to no avail. Concerned about the professional behavior of the treating physician, the consultant talked with the local medical society who referred the consultant to the state medical board. An attorney for the state medical board would only act if the consultant filed a formal complaint. The suggestion amounted to the "nuclear option," which likely would have inflamed the treating physician and lessen the probability of more professional conduct in the future.

Consultant's responsibility to the patient, society, and the family physician

A question now is what is the consultant's responsibility to the patient? As the consultant held the patient was being seen for consultation rather than a referral that did not imply a transfer of care and a different duty to care. The consultant believed there was a contractual obligation for the family physician to execute the treatment recommendations. The consultant's reasoning was logistical, that is, trying to maintain access to the consultant for other and future patients. Further, establishing such an obligation, consultants would increase the cost of care to society greatly and would be contrary to the intent of the government as discussed earlier. However, most reasonable persons blinded to their self-interest would hold that the obligation to beneficence to the patient was such that someone had to prescribe the selegiline and the consultant did that.

When viewed in the context of an obligation to society, the consultant has a contractual obligation to consider the effects of her actions on the health of society as described earlier as would be consistent with a Libertarian moral theory established by governmental policy, at least in spirit. Further, a Utilitarian moral theory would argue that the consultant should not prescribe the medication because by doing so the consultant accepts responsibility for principal care, which would limit the ability of the consultant to see other and future patients. A Deontological moral theory holds that there is an obligation to treat the patient regardless of who prescribes the medication. An Egalitarian moral theory would argue that the patient should receive selegiline as would any other patient. However, it is difficult to see how an Egalitarian or a Deontological moral theory would adjudicate between the family physician and the consultant as to who has the greater obligation.

It is not difficult to see that both the family physician and the consultant have a shared obligation to the patient. The family physician already has a responsibility to care for the patient irrespective of the patient's Parkinson's disease. In one sense, the family physician discharged that responsibility by referral to the movement disorders specialist. Further, by accepting the request for consultation, the consultant has an obligation to the family physician. The clear expectation is that the family physician is requesting assistance to ensure the best management of the patient. The consultant's responsibility is to assure that the family physician has the assistance the family physician needs. It is insufficient merely to confirm the diagnosis and to simply state "treat the patient with selegiline." However, in this case, the consultant provided the family physician with detailed information and step-by-step instructions and offered to receive the family physician's phone call at any time.

The consultant cannot take the Captain of the Ship defense holding that the family physician is the person ultimately responsible, thereby disavowing responsibility to the care of the patient (Case 1 and Chapter 4). The family physician requesting assistance likely means that the family physician lacks the knowledge and skill of the consultant with regard to Parkinson's disease or at least sufficient confidence. The family physician is dependent on the consultant to either convey to the family physician the knowledge and skill necessary to provide the care or the consultant providing the care directly. In this case, the consultant did the former and it is not clear whether it was necessary or desirable for the consultant to do the latter.

Some responses to the moderate scarcity of expert care

Central to the ethical problems posed by this case is the notion of moderate scarcity (see Introduction). If there were a surplus of movement disorders specialists, there unlikely would be as much of an ethical problem, assuming that the total cost of care is the same whether the cost consequent to treatment is by a movement disorders specialist or a family physician. In this author's experience, there have been a number of different responses to the scarcity. Many individual movement disorders specialists will simply close their practice to new patients, thus exacerbating the ethical challenge of access to optimal care. Closing a practice to new patients can be achieved explicitly or implicitly, the latter by not reserving any appointments for new patients with the available appointments filled by patients already cared for. For patients with Parkinson's disease, this is a likely scenario as patients with this progressive neurodegenerative disease tend to see the movement disorders specialist more frequently over time. Accessibility is driven by the mortality of those waiting or subsequent logistical impossibilities for patients yet alive.

Another option has been to engage "physician extenders" such as nurse practitioners and physician assistants. However, the utilization of nurse practitioners and physician assistants has been limited by practice laws and reimbursement regimens. Integration of nurse practitioners into Parkinson's disease clinics has been problematic, as management of these patients has not been a particular focus of education, training, or interest on the part of nurse practitioners generally and because physicians usually are not trained or experienced in working with nurse practitioners. For example, one study suggests little advantage in the clinical control of Parkinson's disease in practices that utilize nurse practitioners (Hurwitz et al., 2005). However, this study, although important, does not address the question here, that is, whether engagement of nurse practitioners can increase the access to expert level care. This has been the experience of the author. But to be effective, the nurse practitioner must be working with the movement disorders specialist rather than independently or within a Family Health Team as usually is the case in Ontario simply because the issue is freeing up the movement disorders expert.

Generally, most physicians feel an obligation to the patients to whom they already are committed. However, healthcare provider organizations are judged more often by access to new patients. In this author's experience, some organizations "ration" follow-up visits, for example, by requiring special permission for a patient to be seen sooner than 3 months. Other organizations limit the appointment slots available for follow-up patients. The next effect is to transfer the problem of access from new patients to follow-up patients. If the follow-up patients access to optimal care was diminished, the policy would be unethical from a Kantian (deontological) moral theory as a person is an end unto himself, not as means to someone else's end. In other words, the follow-up patient would be used as a means to increase accessibility for new patients. The situation would also be unethical based on an Egalitarian moral theory that holds all persons be treated the same. The limiting of access to follow-up patients to ensure access by new patients might be ethical from a Libertarian (contractualist) moral theory provided all parties freely agree, but it is unlikely this would ever be the case.

The situation of moderate scarcity would seem to pit the ethical obligations of the individual physician against the societal obligations of the healthcare provider organization. Note that both entities acknowledge the value of the ethical principle of beneficence to all patients but that the contention lies in the obligation of beneficence to the individual patient or to the collective of patients that constitutes society. This issue was taken up in greater detail in Case 2 related to the availability of video fluoroscopic evaluations of swallowing for patients with Parkinson's disease.

References

Abramson, M., 2008. Medical Malpractice Liability in the Information Age: The Evolution of the Physician-Patient Relationship on the New Healthcare Frontier. Available from: https://www.arbd.com/medical-malpractice-liability-in-the-information-age-the-evolution-of-the-physician-patient-relationship-on-the-new-healthcare-frontier/.

Arya, N., Gibson, C., Ponka, D., Haq, C., Hansel, S., Dahlman, B., Rouleau, K., 2017. Family medicine around the world: overview by region: The Besrour Papers: a series on the state of family medicine in the world. Can. Fam. Physician 63 (6), 436–441.

Bowie, N.E., 1982. 'Role' as a moral concept in health care. J. Med. Philos. 7 (1), 57–63.

Burden, M., Sarcone, E., Keniston, A., Statland, B., Taub, J.A., Allyn, R.L., Reid, M.B., et al., 2013. Prospective comparison of curbside versus formal consultations. J. Hosp. Med. 8 (1), 31–35. https://doi.org/10.1002/jhm.1983.

Cheng, E.M., Swarztrauber, K., Siderowf, A.D., Eisa, M.S., Lee, M., Vassar, S., Jacob, E., Vickrey, B.G., 2007. Association of specialist involvement and quality of care for Parkinson's disease. Mov. Disord. 22 (4), 515–522.

Cohn, S.L., 2003. The role of the medical consultant. Med. Clin. N. Am. 87, 1–6.

Dall, T.M., Storm, M.V., Chakrabarti, R., Drogan, O., Keran, C.M., Donofrio, P.D., Henderson, V.W., et al., 2013. Supply and demand analysis of the current and future US neurology workforce. Neurology 81 (5), 470–478. https://doi.org/10.1212/WNL.0b013e318294b1cf.

de Rijk, M.C., Tzourio, C., Breteler, M.M., Dartigues, J.F., Amaducci, L., Lopez-Pousa, S., Manubens-Bertran, J.M., et al., 1997. Prevalence of Parkinsonism and Parkinson's disease in Europe: The Europarkinson Collaborative Study. European community concerted action on the epidemiology of Parkinson's disease. J. Neurol. Neurosurg. Psychiatry 62 (1), 10–15.

Edwards, N., Smith, J., Rosen, R., 2014. The Primary Care Paradox: New Designs and Models. KPMG International Cooperative, Amsterdam, Netherlands. Available from: https://assets. kpmg.com/content/dam/kpmg/pdf/2014/12/primary-care-paradox-v3.pdf.

Gosden, T., Forland, F., Kristiansen, I.S., Sutton, M., Leese, B., Giuffrida, A., Sergison, M., Pedersen, L., 2000. Capitation, salary, fee-for-service and mixed systems of payment: effects on the behaviour of primary care physicians. Cochrane Database Syst. Rev. 2000 (3), CD002215.

Grace, C., Alston, W.K., Ramundo, M., Polish, L., Kirkpatrick, B., Huston, C., 2010. The complexity, relative value, and financial worth of curbside consultations in an academic infectious diseases unit. Clin. Infect. Dis. 51 (6), 651–655. https://doi.org/10.1086/655829.

Guttman, M., Slaughter, P.M., Theriault, M.E., DeBoer, D.P., Naylor, C.D., 2002. Parkinsonism in Ontario: physician utilization. Can. J. Neurol. Sci. 29 (3), 221–226.

Harris, M., Raviv, A., 2010. Control of corporate decisions: shareholders vs. management. Rev. Financ. Stud. 23 (11), 4115–4147.

Hurwitz, B., Jarman, B., Cook, A., Bajekal, M., 2005. Scientific evaluation of community-based Parkinson's disease nurse specialists on patient outcomes and health care costs. J. Eval. Clin. Pract. 11 (2), 97–110.

Kirby, S., Weston, L.E., Barton, J.J., Buske, L., Chauhan, T.S., 2016. Report of the Canadian Neurological Society Manpower Survey 2012. Can. J. Neurol. Sci. 43 (2), 227–237. https://doi.org/10.1017/cjn.2015.316.

Le Grand, J., 2009. Choice and competition in publicly funded health care. Health Econ. Policy Law 4 (Pt 4), 479–488. https://doi.org/10.1017/S1744133109990077.

Marchildon, G.P., Hutchison, B., 2016. Health reform monitor primary care in Ontario, Canada: new proposals after 15 years of reform. Health Policy 120 (7), 732–738.

Olanow, C.W., Rascol, O., Hauser, R., Feigin, P.D., Jankovic, J., Lang, A., Langston, W., et al., 2009. ADAGIO Study Investigators. A double-blind, delayed-start trial of rasagiline in Parkinson's disease. N. Engl. J. Med. 361 (13), 1268–1278. https://doi.org/10.1056/NEJMoa0809335. Erratum in: N. Engl. J. Med. 2011;364(19):1882.

Rudoler, D., Laporte, A., Barnsley, J., Glazier, R.H., Deber, R.B., 2015. Paying for primary care: a cross-sectional analysis of cost and morbidity distributions across primary care payment models in Ontario Canada. Soc. Sci. Med. 124, 18–28. https://doi.org/10.1016/j.socscimed.2014.11.001.

Smith, A., 1776/1759. An inquiry into the nature and causes of the wealth of nations. In: The Theory of Moral Sentiments. The Wealth of Nations.

Willis, A.W., Schootman, M., Evanoff, B.A., Perlmutter, J.S., Racette, B.A., 2011. Neurologist care in Parkinson disease: a utilization, outcomes, and survival study. Neurology 77 (9), 851–857.

Willis, A.W., Schootman, M., Tran, R., Kung, N., Evanoff, B.A., Perlmutter, J.S., Racette, B.A., 2012. Neurologist-associated reduction in PD-related hospitalizations and health care expenditures. Neurology 79 (17), 1774–1780. https://doi.org/10.1212/WNL.0b013e3182703f92.

Case 4

Case 4—A 72-year-old man followed by a movement disorders neurologist specializing in Parkinson's disease was admitted to a hospital because of fever and increased confusion. He was found to have a urinary tract infection and was treated with antibiotics. A general neurologist saw the patient in consultation while the patient was in the hospital. The consulting neurologist, without discussions with the treating movement disorders neurologist, markedly reduced the patient's dose of carbidopa/levodopa used to treat the patient's Parkinson's disease. The patient was discharged to a nursing facility. The movement disorders neurologist later was called by the patient's wife because the patient was unable to move, essentially confined to bed, and unable to care or feed himself. He was unable to converse with the facility staff. The movement disorders neurologist visited the patient in the nursing facility and restored the previous dose of carbidopa/levodopa. Within a week, the patient returned to his previous abilities to care for himself, get out of bed, and interact with the nursing staff, family, and friends.[a]

The medical facts

Parkinson's disease is associated with degeneration of a group of neurons deep in the brain in an area called the substantia nigra pars compacta. These neurons utilize a chemical, dopamine, to communicate with other neurons involved in the generation of movement. With degeneration of the neurons in the substantia nigra pars compacta, patients become slow in their movements and, in severe cases, slowed to the point of virtual immobility. Any and all movements can be affected, for example, of the throat muscles, making speech, eating, and swallowing very difficult. These patients are at serious risk for food or liquids entering the lungs to cause pneumonia.

One treatment is to replace the dopamine that has been lost due to the degeneration of neurons in the substantia nigra pars compacta. One cannot just give the patient dopamine as it cannot get into the brain. Consequently, patients are given levodopa, which is transported into the brain and then converted to dopamine. The benefits can be dramatic. Levodopa often is combined with a chemical, carbidopa, to reduce the adverse effects, such as nausea, that can be caused by the levodopa.

Review of the patient's hospital record did not find any explanation for the change in the patient's dose of carbidopa/levodopa. Perhaps the consulting neurologist thought that the carbidopa/levodopa was producing the increased confusion even though the patient had been on the original dose of carbidopa/levodopa for many months without difficulty. Further, it was more likely the confusion was caused by the urinary tract infection. Prior

---

[a] The case presented is hypothetical, although modeled on experience. No judgment is made relative to the correctness, appropriateness, or value of any actions described on the part of any of the participants. The purpose here is to describe and explore the variety of ethical considerations and consequences that depend on the perspective of different relevant ethical systems.

The Ethics of Everyday Medicine. https://doi.org/10.1016/B978-0-12-822829-6.00004-7

to hospitalization, the patient was treated with 1750 mg of levodopa per day. It may be that the consulting neurologist had never seen a patient on such a dose of levodopa and thought the dose excessive. However, in clinical trials of new treatments for advanced Parkinson's disease by experts, 1750 mg of levodopa per day was well within the range of the average dose (Weaver et al., 2012). If it were the case that the consulting neurologist was basing his or her opinion solely on the basis of his or her experience, then this is an example of physicians diagnosing and treating by pattern recognition based on their own idiosyncratic experiences rather than on scientific and medical facts and rational as discussed in Case 9. Arguably, the consulting neurologist was practicing suboptimal medicine but the issue here is whether it was unethical (see Introduction and Case 9). In this case, however, the issue goes beyond the question of suboptimal medical care provided by the consulting neurologist and to the ethical obligations of the consulting neurologist discussing the patient with the treating movement disorders neurologist prior to making changes in the patient's medications for the treatment of Parkinson's disease.

As an aside, there is a great reticence among physicians and healthcare professionals to be critical of another physician or healthcare professional (see Chapter 4). For example, following the publication of a report describing the disturbing prevalence of medical errors (Kohn et al., 2000) there has been an emphasis on errors as reflecting problems in medical "systems" (Wears and Sutcliffe, 2020). Even when there is a clear error on the part of an individual, there is a tendency to take issue with the act rather than the person. The implication is by assigning the error to the "system," the individual is somehow freed of responsibility and therefore accountability. It is not always the case that some form of anger toward a person making what should have been an avoidable error is not helpful, even virtuous, such as in righteous indignation (see Chapter 1). Often, it takes a "horror story" such as calling out a person's culpability to generate change (Iedema et al., 2009). Historically, the predominance of physician-centric ethics can be considered as sheltering physicians, perhaps inappropriately (see Chapter 4). In such situations and circumstances, it is not clear that the failure of the individual also is not a failure of the profession. This text will use the term "suboptimal" so as not to presuppose blame or censure, even though the latter terms may ultimately be the most productive of change.

One should not assume the case presented is an outlier or a rare anomaly and perhaps dismissed as an anomaly or merely "the cost of doing business." While there is a paucity of scientific studies on the problem of failure to consult among physicians involved in the care of patients, in this author's experience, it is not uncommon. One type of consultation is between the specialist and the nonspecialist, for example, between the family physician or primary care physician and a movement disorders neurologist (see Case 3). The problem is reflected in the disparities in the treatments used by each. For example, in the case of Parkinson's disease, family physicians and primary care specialists generally prescribe more levodopa and less dopamine agonist (agents that are not converted to dopamine but to which the brain responds as though the chemical was dopamine) compared to generalist neurologists, who similarly are more likely to use levodopa and less likely to use dopamine agonists compared to movement disorders neurologists. The concern is that early and continuous exposure to levodopa increases the risk of serious long-term complications (see Case 5).

Similarly, discrepancies regarding the care of similar patients between specialists and nonspecialists have been raised in the treatment of kidney failure (Smart et al., 2014) and heart disease (Ahmed et al., 2003). In a study of heart failure, evaluations of the function of the left side of the heart, the use of angiotensin-converting enzyme inhibitor medications, 90-day readmission, and 90-day mortality rates in hospitalized demonstrated that better care was provided by cardiologists compared to generalist physicians. Importantly, the best care was provided by a collaborative effort of cardiologists and generalist physicians.

If one were to discount logistical obstacles, the question becomes why the lack of collaboration when it is clear that it likely would be in the patient's best interest? Perhaps there is no perception of the need for consultations on the part of the consulting neurologist or other physicians or healthcare professionals principally involved in the patient's immediate care. If so, then who is responsible for the lack of perception, the person or the institutions and agencies responsible for the selection, education, training, and supervision of the physician or healthcare professional as discussed in Case 9? Perhaps it is hubris or insecurity that prevents consultations (see Case 3). Even if logistical problems prevent collaboration, the question is who is responsible for organizing healthcare delivery that would allow such logistical problems? However, even if there is an abject failure of consultations and collaborations, it is not a foregone conclusion that such a situation is unethical. It may be that a diversion of resources to prevent problems such as in the case presented would result in an injustice when all stakeholders and shareholders are considered, for example, conflicting with societies' healthcare priorities.

Shareholders, stakeholders, and obligations

The ethical analysis begins with the identification of shareholders, those in a position to affect the patient, and stakeholders, those affected by the decisions of the shareholders. At first, the obvious shareholder with respect to the management of the patient's Parkinson's disease while in the hospital fell to the consulting neurologist. The patient was the immediate stakeholder. Given a Principlist approach, all would agree that the ethical principles of beneficence, nonmaleficence, and autonomy are valued and are goals. Similarly, all the Anti-Principlist approaches would also agree that the best care of the patient, however, implemented and justified, is a top priority. Proceeding from a Principlist approach (not to discount the other approaches), the operative question is what is the obligation to these ethical principles? To be sure, the patient shares the same responsibilities of the obligations entailed by the ethical principles, such as paying the consulting neurologist, typically through the patient's proxy insurers. The larger concern here is the obligation of the consulting neurologist to the patient in the context of discussions with the patient's movement disorders neurologist.

To be sure, the consulting neurologist was acting from an obligation to beneficence, in this case, the reduction of confusion, agitation, and anxiety; although perhaps counterproductive in how to achieve the desired beneficence. The consulting neurologist also had an obligation to the ethical principle of nonmaleficence. However, the patient clearly was harmed by the consulting neurologist's actions because of the severe worsening of the patient's Parkinson's disease, thereby compromising the obligation to nonmaleficence. It is important to note that maleficence is the harm in the most general sense and does not necessarily imply intentional malfeasance. One could argue that the consulting neurologist's actions were a violation of the obligation to beneficence, that is, failure to provide adequate treatment to control the patient's symptoms. However, it is likely not a failure to the obligation of beneficence as the consulting neurologist was providing benefit to the extent of the consulting neurologist's knowledge. The issue here is the failure of the obligation to nonmaleficence as the harm was a direct consequence of an action taken by the consulting neurologist, an act of commission, rather than a failure to provide adequate treatment, an act of omission. The distinction often is subtle and often blurred, but this approach has heuristic value for a framework by which to address ethical problems.

It is unlikely that anyone would disagree that the ethical principle of autonomy does not operate. Rather, the issue is the obligation to the principle of autonomy, for example, as it relates to informed consent. Issues related to informed consent and paternalism as a frequent cause of violation of the obligation to autonomy embodied in informed consent are discussed in detail in Case 5. The question is whether the patient or the patient's proxy gave informed consent for the changes in the patient's medications for Parkinson's disease. It is a reasonable question whether the patient or the patient's proxy would have accepted such a precipitous reduction in the patient's medications had they been warned of the potential adverse effects. However, it is unknown whether informed consent was obtained, either explicitly or implicitly.

There is another sense of autonomy, which is the recognition that the patient has a chronic illness likely to subsequently involve the treating movement disorders neurologist. The risk is that the drastic worsening of the patient's condition following intervention by the consulting neurologist could cause the patient and caregivers to doubt the judgment of the movement disorders neurologist, thus potentially compromising future care. This risk has long been recognized by physicians as evident in Thomas Percival's (1803) "Medical Ethics; or a Code of Institutes and Precepts Adapted to the Professional Conduct of Physicians and Surgeons," as will be discussed in detail later. The dramatic change in medications could also cause the patient and caregivers to doubt the consulting neurologist. In any event, there is a net loss of confidence on the part of the patient and caregivers. Arguably, this could have been avoided by a conversation between the consulting neurologist and the movement disorders neurologist.

The question becomes why was there no communication prior to the act? Certainly, given the fact that most physicians are extremely busy, there may not have been time to communicate. However, there is a saying that "one makes time for what is important." If that be true, then the issue is why did not the consulting neurologist think it important to have a conversation with the treating movement disorders neurologist? There may be any number of reasons. It is possible that the consulting neurologist was completely confident in his or her decision and thus saw no need for a conversation with the treating movement disorders neurologist. That confidence may have been buoyed by the personal experience of the consulting neurologist, having never seen a patient on such high doses of levodopa, as discussed earlier. Even if the consulting neurologist did not think it important to discuss the patient with the treating movement disorders neurology, there is a matter of professionalism. Assuming that the code of ethics developed by Percival and the American Medical Association Code of Ethics of 1847 represent the necessary elements of professionalism, the following quotes from the 1847 code of ethics are interesting:

> § 4. A physician ought not to take charge of, or prescribe for a patient who has recently been under the care of another member of the faculty in the same illness, except in cases of sudden emergency, or in consultation with the physician previously in attendance, or when the latter has relinquished the case or been regularly notified that his services are no longer desired. Under such circumstances, no unjust and illiberal insinuations should be thrown out in relation to the conduct or practice previously pursued, which should be justified as far as candour, and regard for truth and probity will permit; for it often happens, that patients become dissatisfied when they do not experience immediate relief, and, as many diseases are naturally protracted, the want of success, in the first stage of treatment, affords no evidence of a lack of professional knowledge and skill.
>
> *American Medical Association (1848, p. 103)*

Other stakeholders are involved, such as the treating movement disorders neurologist who had to resume care for the patient. Failure by the consulting neurologist in terms of the problems produced by the unagreed changes in medications imposes a significant burden on the treating movement disorders neurologist. The burden is not only the necessity to expend additional resources to restore beneficence to the patient but also to repair the trust in physicians, including trust in the treating movement disorders neurologist as well. The issue of the importance of trust in medical institutions and, necessarily, the agents of those institutions—physicians and healthcare professionals—raises serious ethical concerns in the practice of everyday medicine, which will be addressed in more detail later. Damaging that trust represents a failure in the obligation to the ethical principle autonomy afforded to the movement disorders neurologist. The movement disorders neurologist, being human, may see the actions of the consulting neurologist as a repudiation of the care provided by the movement disorders neurologist; certainly, the patient and the patient's family, friends, and caregivers may see it as such. It is not hard to see that such a slight would generate resentment that might further limit collaboration, even just a subsequent discussion between the movement disorders neurologist and consulting neurologist as to what should have happened. As can be seen, the ripple effect of the action of the single consulting neurologist spreads, affecting a great many persons.

Principle of double effect

One tension in the case presented is the risks associated with the large reduction in the patient's dose of carbidopa/levodopa. On the one hand, the consulting physician thought that the reduction would lessen the patient's confusion and agitations, which is a clear benefit. On the other hand, such a reduction could result in marked worsening of the patient's symptoms of Parkinson's disease. Indeed, experiences in the past of purposeful withholding of anti-Parkinson's disease medications in hopes of restoring better responsiveness were called "drug holidays." These were abandoned because of the severe associated risks (Toru et al., 1981; Friedman et al., 1985). The question becomes how to balance the benefits and risks of maleficence? Clearly, the benefits and risks of maleficence do not range over the same dimension. It is not as though two units of benefit outweigh one unit of harm. As occurs frequently in the practice of everyday medicine, physicians and healthcare professionals have to compare "apples to oranges."

One approach is based on the Principle of Double Effect expounded by Aquinas (1485) where four principles were offered: (1) the nature of the act that must be morally good or at least indifferent, (2) the harm possible from the act cannot be the means by which one produced a morally good effect, (3) the motives of the actor must be to do good, and (4) the potential bad effect must not be disproportionately bad compared to the intended good effect. In this case, the act was to relieve the patient of the confusion and agitation, clearly a good effect. The possible harm, worsening of the patient's parkinsonism, was not a necessary condition to produce the good. In other words, the act might not necessarily produce the harm—the patient's symptoms may not have worsened. It is likely that the consulting physician was not aware of the magnitude of the risks with the marked reduction in the carbidopa/levodopa dose.

The critical question, however, is whether the consulting physician *should have known* about the probability and the seriousness of adverse effects. Clearly, the consulting physician acted on the patient's levodopa dose; thus, it would seem that the physician had an obligation to have sufficient knowledge of the potential benefits as well as risks, as one would not allow a person to drive a car unsupervised if the person had not demonstrated sufficient working knowledge of how to drive a car. Even if the consulting physician need not have known, the consulting physician, at the very least, might have realized the limits of his or her knowledge and experience. The contrast between the consulting neurologist's opinion of the appropriate doses of levodopa and that of the treating movement disorders neurologist perhaps was a "red flag." Those acting to provide beneficence are expected to act reasonably. Indeed, Good Samaritan laws have been enacted to encourage persons to provide help in emergencies. However, in *Cleary v. Hansen* in 1981, the Canadian court ruled:

> Even during an attempt to assist someone in an emergency, the law expects reasonable care to be exercised even though the standard is relaxed to a certain extent. The court does not expect perfection, but rescuers must be sensible. They, like anyone else, must weigh the advantages and the risks of their conduct. Their conduct, too, however laudable, must measure up to the standard of the reasonable person in similar circumstances.
>
> *(Cited in https://www.kcyatlaw.ca/can-sue-good-samaritan/)*

Physicians have been held to the standard of what a similarly situated physician would do under the circumstances (https://www.kcyatlaw.ca/can-sue-good-samaritan/). If the expectation is that the consulting physician should have known, then the fourth principle of the Principle of Double Effect, the potential bad effect must not be disproportionately bad compared to the intended good effect, is violated and the action may not be justified—but is it unjust? There may be certain circumstances under certain prevailing moral theories where the consulting physician's lack of knowledge may not be unjust. For example, there may be logistical barriers to the consulting physician acquiring the knowledge, such as no movement disorders neurologist was available. The utilitarian costs of removing such barriers or to insist the necessary infrastructure may exceed the costs of other obligations, assuming a moderate scarcity of resources and competing societal interests.

Balancing obligations to the individual against to the institution

Returning to the issue of obligations to the ethical principle of autonomy, the consulting physician has an obligation to the treating movement disorders neurologist, who now must deal with the consequences of the consulting physician's action. These include not only the medical issues, such as restoring beneficence in terms of relieving the Parkinsonian symptoms, but also perhaps having to deal with a now skeptical patient, family members, friends, and caregivers. To be sure, the skepticism may be applied to both the consulting and the movement disorders neurologists. However, the skepticism could extend to every physician and healthcare professional. It is interesting that the American Medical Association (AMA) Code of Ethics of 1847 stipulated "All discussions in consultation should be held as secret and confidential. Neither by words nor manner should any of the parties to a consultation assert or insinuate, that any part of the treatment pursued did not receive his assent. The responsibility must be equally divided between the medical attendants—they must equally share the credit of success as well as the blame of failure" (American Medical Association, 1848, p. 21).

The question becomes does such secrecy risk undermining the healthcare of other patients, now and in the future? While this may constitute a violation of the ethical principle of autonomy in the interests of physicians and healthcare professionals, the consequence could be an undermining of the obligation to beneficence for patients that requires some trust in physicians and healthcare professionals. Note that the treating movement disorders neurologist likely will have to address the actions of the consulting neurologist with the patient and patient's family, friends, and caregivers. The question becomes what is the obligation to the autonomy of the movement disorders neurologist to the consulting neurologist in that situation?

The issue of obligations relative to the ethical principle of autonomy between physicians figures heavily in the first code of ethics of the AMA adopted in 1847 (American Medical Association, 1848). As will be seen, those ethics could be viewed as protective of the medical establishment, perhaps to the detriment of others, including patients. However, it is important to realize that Rationalist/Allopathic physicians in 1847 were under great pressure from competition with the Empiric healthcare practitioners, such as homeopaths, of which there were more practitioners than Rationalist/Allopathic physicians (Montgomery, 2018). Indeed, the American Institute of Homeopathy was established in 1844 and the creation of the American Medical Association in 1847 has been seen as a reaction. The AMA code of ethics prohibited fraternalization with homeopaths and other Empirics by members of the AMA. In the code of ethics, it is written: "A physician should practice a method of healing founded on a scientific basis; and he *should not voluntarily professionally associate with anyone who violates this principle* [italics added]." Those that violated the principle of science were "the crooked devices and low arts, for evidently selfish ends, the unsupported promises and reckless trials of interloping empirics, whose very announcements of the means by which they profess to perform their wonders are, for the most part, misleading and false, and, so far, fraudulent." The US Supreme Court ruled the AMA code of ethics in violation of antitrust laws [*Wilk v. American Medical Association*, 895 F.2d 352 (7th Cir. 1990)], and the prohibition against fraternization with Irregulars and Empirics, in this case chiropractors, was removed from the code of ethics.

In an important way, the consulting neurologist was also a consultant to the movement disorders neurologist who had to continue the care of the patient after the patient was discharged from the hospital. The obligations of the consulting neurologist were specified by the AMA Code of Ethics of 1847, where it was written:

> In consultations, the attending physician [movement disorders neurologist in the present case] should be the first to propose the necessary questions to the sick; after which the consulting physician should have the opportunity to make such farther inquiries of the patient as may be necessary to satisfy him of the true character of the case. Both physicians should then retire to a private place for deliberation.
>
> *AMA Code of Ethics (1847, p. 18)*

> But no statement or discussion of it should take place before the patient or his friends, except in the presence of all the faculty attending, and by *their common consent* [italics added]; and no opinions or prognostications should be delivered, which are not the result of previous deliberation and concurrence.
>
> *AMA Code of Ethics (1847, p. 19)*

> In consultations, the physician in attendance should deliver his opinion first; and when there are several consulting, they should deliver their opinions in the order in which they have been called in. *No decision, however, should restrain the attending physician from making such variations in the mode of treatment, as any subsequent unexpected change in the character of the case may demand* [italics added]. But such variation, and the reasons for it, ought to be carefully detailed at the next meeting in consultation. The same privilege belongs also to the consulting physician if he is sent for in an emergency, when the regular attendant is out of the way, and similar explanations must be made by him at the next consultation.
>
> *AMA Code of Ethics (1847, pp. 19–20)*

> A physician who is called upon to consult, should observe the most honourable and scrupulous regard for the character and standing of the practitioner in attendance: the practice of the latter, if necessary, should be justified as far as it can be, consistently with a conscientious regard for truth, and no hint or insinuation should be thrown out which could impair the confidence reposed in him, or affect his reputation.
>
> *AMA Code of Ethics (1847, p. 22)*

It is relatively straightforward to see that the consulting neurologist may have violated the code of ethics as stipulated by the AMA in 1847. The repercussions of such a violation likely are none. In such cases, ethical rules become inspirational rather than aspirational, thus depending on the virtue of the consulting neurologist. The consulting neurologist could not have discharged his or her responsibilities to the attending physician as there was no communication with the attending physician. Indeed, the treating movement disorders neurologist was unaware of what transpired until contacted by the patient's family after the patient was discharged to the nursing facility.

Whether this makes the consulting neurologist's actions unjust requires further analysis. For a Libertarian/Contractualist moral theory to prevail there must be some sort of contract, either explicit or implicit. As seen repeatedly, the expectation is that the consulting neurologist followed the standard of care. If the standard of care is defined as what similarly situated physicians and healthcare professionals would do, then it is unlikely that the consultant in the case presented deviated from the standards of care today. Similarly, this standard of care also addresses the Egalitarian moral theory as all patients would be treated similarly if their physicians followed the standard of care. In other words, the standard of care would level the care provided to all patients.

If one modifies the standard of care to be what a reasonable physician or healthcare professional similarly situated would do, then one invites a Utilitarian moral theory. A reasonable physician or healthcare professional would reason as to what is the greatest good, despite the problematic nature of defining the greatest good. Perhaps the time and effort expended to discuss the case with the treating movement disorders neurologist would have resulted in a loss of good to other patients for whom the consulting neurologist was responsible. It may be that the consulting neurologist has a Deontological moral duty to the movement disorders neurologist, but again the situation and circumstance of moderate scarcity make enforcing a Deontological duty difficult.

From the aforementioned discussion, it would seem that a unique relationship exists among the physicians that would seem to be different from the general public. It may be that physicians hold a special and exalted place that commands respect for physicians, not only from the public but also from its members. In the AMA Code of Ethics of 1847, it is written:

> Art. II. —Obligations of the public to physicians.
> § 1. *The benefits accruing to the public, directly and indirectly, from the active and unwearied beneficence of the profession, are so numerous and important, that physicians are justly entitled to the utmost consideration and respect from the community* [italics added]. The public ought likewise to entertain a just appreciation of medical qualifications; —to make a proper discrimination between true science

and the assumptions of ignorance and empiricism,—to afford every encouragement
and facility for the acquisition of medical education, —and no longer to allow
the statute books to exhibit the anomaly of exacting knowledge from physicians,
under liability to heavy penalties, and of making them obnoxious to punishment for
resorting to the only means of obtaining it.

AMA Code of Ethics (1847, pp. 29–30)

There clearly is a sense of entitlement in the code of ethics on behalf of physicians. The entitlement appears based on two factors, the beneficence that physicians bestow onto members of the public and their monopoly on the science necessary for the best practice of medicine, although in truth there was relatively little difference in the practice of medicine at the time between Rationalist/Allopathic physicians and the Empirics, as both practice mostly Galenic medicine (Montgomery, 2018). Indeed, patients were jeopardized more often by excesses of the Rationalist/Allopathic physicians (the origin of the pejorative adjective of "heroic" medicine) than from the practice of the homeopaths. Other elements may be contributing to the notion of entitlement, for example, survival of the demanding competitive gauntlet that physicians must succeed to become physicians; the presumption of self-sacrifice on the part of physicians toward their patients; and a shared sense of threat from the public, such as posed by unfair claims of medical malpractice.

The notion of an obligation consequent to receiving the entitlement of physician-hood is well established through history. In many ways, the AMA Code of Ethics of 1847 is largely due to Thomas Percival's book on medical ethics (King, 1982), with the difference likely related more to historical context than to fundamental ethics or morality. In his book, Percival (1803) wrote:

XXII I. The *Esprit de Corps* [italics in the original] is a principle of action found in
human nature, and when duly regulated, is both rational and laudable. Every man
who enters into a fraternity engages, by a tacit compact, not only to submit to the
laws, but to promote the honour and interest of the association, so far as they are
consistent with morality and the general good of mankind. A physician, therefore,
should cautiously guard against whatever may injure the general respectability of his
profession; and should avoid all contumelious representations of the faculty at large;
all general charges against their selfishness or improbity; and the indulgence of an
affected or jocular skepticism, concerning the efficacy and utility of the healing art.

(pp. 43–44)

The Hippocratic oath also speaks to a special bond within physician-hood, particularly through the vehicle of teaching as written:

To hold my teacher in this art equal to my own parents; to make him partner in
my livelihood; when he is in need of money to share mine with him; to consider
his family as my own brothers, and to teach them this art, if they want to learn it,
without fee or indenture; to impart precept, oral instruction, and all other instruction
to my own sons, the sons of my teacher, and *to indentured pupils who have taken the
physician's oath* [italics added], but to nobody else.

(Hippocratic oath translation in https://en.wikipedia.org/wiki/Hippocratic_Oath)

There clearly is a sense of brotherhood (and today sisterhood) among physicians. To get a sense of this brotherhood, at least as it stood in 1896, an extended quote from Dr. George McNaughton to the Brooklyn Medical Society is interesting. Dr. McNaughton was quoted as saying:

> In preface, I shall attempt to tell a story which in a small way illustrates the brotherhood which exists between physicians. A few years ago, on a passenger car on the Pennsylvania Railroad, each seat was occupied by two persons, save one, and in the seat was a man who had placed his luggage in the other half of his seat with the evident intention of making people believe that the occupant had stepped out for a moment. Presently a gentleman entered the car, discovered the one seat not occupied by a passenger, and politely asked if this seat were taken. The reply was that it seemed to be. One word led to another until finally the argument became heated, much to the amusement of the passengers. At this juncture, the conductor appeared and addressed the man who was standing as doctor. The seated man looked somewhat surprised and abashed, and at once asked his opponent if he were a physician, and was answered in the affirmative. The selfish individual apologized, said that he also was a physician, and beg the gentleman to be seated. They began to converse, and it is fair to believe that they very pleasantly swapped—truth until they reached their destinations. This fairly represents the selfishness of men, or some men, and at the same time the fellowship found among physicians.
>
> *(Anonymous Report, 1896)*

While perhaps representative of the ethos at the time, it is not clear that such brother- and sisterhood persists today in any similar way. In 1950, approximately 75% of physicians belonged to the AMA. By 2011 the percentage dropped to an estimated 15% (Collier, 2011). While many factors have been suggested, such as a general decrease in interest in large organizations, some believe that medicine has become just a job rather than a calling, with the transition to more corporate-styled medicine where physicians are employees, so the change in perception may be justified. Nevertheless, as described later, viewing the practice of medicine as a job rather than a calling does not seem hospitable to developing the type of virtue held by past physicians. It is important how a calling is defined that distinguishes it from strictly vocational or career concerns. Dik and Duffy (2009) define calling as "a transcendent summons, experienced as originating beyond the self, to approach a particular life role in a manner oriented toward demonstrating or deriving a sense of purpose or meaningfulness and that holds other-oriented values and goals as primary sources of motivation."

Virtue and the physician and healthcare professional

It would seem that medicine perceived as a calling as just defined would be most conducive to the development of virtuous physicians and healthcare professionals (Case 7 and Chapter 4) with the expectation that virtue extends to relationships between physicians and healthcare professionals. Indeed, the argument could be made that how physicians treat each other is part and parcel of how they treat patients. The question is does viewing medicine as a calling rather than a job lead to better patient care and physician and healthcare professionals' sense of well-being (Goodin et al., 2014; Westerman, 2014)? Evidence shows that perceiving work as a calling has benefits (Dik et al., 2012).

Such a sense of entitlement likely risks paternalism and, as seen, the roots of paternalism run extremely deep in history. Indeed, historical analysis demonstrates that the notion of informed consent, such as involved in the practice of everyday medicine, is only a recent development (see Introduction). In the absence of an obligation to informed consent, paternalism flourishes (see Case 5). But the question is open whether a sense of entitlement necessarily results in injustice? Indeed, it may just be that those entitled are in the best position to be beneficent (Douthat, 2018). In 1803, Thomas Percival wrote:

> To a young physician, it is of great importance to have clear and definite ideas of the ends of his profession; of the means for their attainment; and the comparative value and dignity of each. Wealth, rank, and independence, with all the benefits resulting from them, are the primary ends which he holds in view; and they are interesting, wise, and laudable. But knowledge, benevolence, and active virtue, the means to be adopted in their acquisition, are all still higher estimation. And he has the privilege and felicity of practicing an art, even more intrinsically excellent in its mediate then and its ultimate objects. The former, therefore, have claim to uniform pre-eminence.
>
> *(pp. 40–41)*

The admonition is to be a virtuous physician, who, by being virtuous, provides beneficence and avoids maleficence even if autonomy becomes secondary (see Virtuous Ethics, Chapter 2). It is not necessarily the case that actions of providing beneficence and avoiding maleficence, even at the cost of autonomy, are unjust (Case 5). Perhaps if virtue becomes a habit that is embedded in practice then it represents a form of ethical procedural knowledge. According to Immanuel Kant, instilling just behaviors by the generation of habits may be a better guarantee of just actions than any declarative rational analysis (Chapter 2). Perhaps if the consulting neurologist had the virtuous habit of communicating with the movement disorders neurologist, the maleficence might have been avoided.

The question arises as whether the duty of the movement disorders neurologist to the medical profession limits discussions with the patient and the patient's family, friends, and caregivers? The concern becomes whether the patient or the patient's representatives have the right to know, as it is highly unlikely that the patient's family was not suspicious of some error in care. Throughout history, physicians and healthcare professionals have held knowledge in confidence or secrecy. The Hippocratic Oath stipulates secrecy, as it states "And whatsoever I shall see or hear in the course of my profession, as well as outside my profession in my intercourse with men, if it be what should not be published abroad, I will never divulge, holding such things to be holy secrets." In 1803, Thomas Percival wrote "No physician or surgeon, therefore, should *reveal* occurrences in the hospital, which may injure the reputation of any one of his colleagues; except under restrictions contained in this exceeding article." The requirement was that the complaint or concern had to be vetted by a physician in the institution. Percival continues, writing "But neither the subject matter of such references [contentions between physicians], nor the adjudication, should be communicated to the public; as they may be personally injurious to the individuals concerned and can hardly fail to hurt the general credit of the faculty" (Percival, 1803, p. 46).

There appears to be a paucity of studies examining the incidence of medical errors that are reported, perhaps the most tangible example of discussions among physicians and healthcare professionals. Most studies take a descriptive and qualitative approach, such as surveying the attitudes of patients, physicians, and healthcare professionals (O'Connor et al., 2010; Perez et al., 2014). In a survey of physicians in a tertiary hospital in Saudi Arabia, at least 41% did not report an error as they did not believe it was their responsibility. An error with more serious consequences was more likely to be reported (Alsafi et al., 2011). Yet, the incidence of medical errors would seem very high. The US Institute of Medicine estimated that between 44,000 and 98,000 Americans die each year from preventable errors in hospitals (Kohn et al., 2000). The incidence of nonfatal medical errors is likely much higher. While there likely are a great many factors, the track record of protecting those reporting medical errors provides little confidence (see Case 7).

An important factor in approaching a colleague who may have made a medical or ethical error is the fear of repercussions. There have been a great many efforts and continuing efforts to develop methods that would encourage discussion of medical errors. In 1803, Thomas Percival emphasized the importance of discussions among physicians and surgeons, writing "To advance professional improvement, a friendly and reserved *intercourse* should subsist between the gentleman of the faculty, with a free communication of whatever is extraordinary or interesting in the course of their hospital practice. An *account of every case or operation* which is rare, curious, or instructive should be drawn up by the physician or a surgeon, to who is charge it devolves, and entered into a register kept for the purpose, but open only to physicians and surgeons of the charity [italics in the original]" (Percival, 1803, p. 15).

One such approach is morbidity and mortality conferences, in which possible errors are discussed. In many jurisdictions, shield laws protect discussions from legal discovery such as might be used in a legal complaint. Interestingly, such conferences are almost exclusively in the domain of surgery and anesthesiology, rarely in internal medicine. Indeed, this author cannot recall there ever being such a conference in any neurology program with which this author was affiliated. Even when held, the conferences seldom addressed medical errors but more often presented interesting or educational case reviews (Orlander et al., 2002).

In the absence of any personal relationship between pairs or groups of physicians and healthcare professionals that generates trust and respect to discuss problems, there are very limited resources. For example, a physician returned another physician's report with a curt letter saying he did not refer the patient nor did he need the consultation. The author of the report had sent the report to the physician at the request of the patient. The author of the report was concerned that this action and repeated similar actions might not be in the patient's best interest and sought the assistance of the local medical society. The author of the report referred the physician to the state medical board, who suggested a discussion with the board's attorney, who said no action would be taken without a formal complaint. A similar situation arose when a physician raised concerns about an error on the part of a colleague to the chairman of the department. The chairman stated he would only intervene if a formal complaint was raised and then would convene a review committee. Another situation occurred when a physician, concerned about the actions of another physician, was directed to report to the risk management department. These instances demonstrate that discourse among some physicians and institutions resorts to one of two options: the "nuclear" option, potentially injurious and acrimonious to all, or to let the concerns slide.

"We are in this together to make it work"

With the corporatization of medicine, it is not surprising that the "manufacturing mentality" would become the operative force in the organization of healthcare. Two key concepts in the rise of modern manufacturing are (1) the concept of interchangeable parts and (2) the assembly line. Interchangeable parts required components of the system to be identical in the relevant factors, for example, the bore of the cannon had to be exactly the same among all cannons so that a single projectile could be fired from any cannon and any projectile could be fired from the same cannon. Prior to that, each cannon would have to have its own unique projectile, which greatly limited the economies of scale.

The interchangeability of parts allowed for the development of the assembly line, where any part from a set of the identical parts could be assembled in any device, such as a car. This means that efforts used to build the car could be organized in a piecemeal fashion where the specific stage in the assembly occurred at a specific and different point on the conveyer belt on which the partial car was moved between assembly points. The assembly line offered enormous efficiencies beyond the economies of scale. For example, division of the task of building a car could be divided among the different assembly points where the worker did not need to understand the building of the entire car but rather only those aspects directly related to the component that was to be assembled at that point. The implications for such a division of labor are seen in Cases 3 and 6.

Interchangeability is the key concept. As applied to the ethics of everyday medicine, the concept of interchangeability applies to both patients and physicians and healthcare professionals. With respect to physicians and healthcare professionals, patients and thus society may want perfect interchangeability. Otherwise, the quality of the healthcare provided would vary depending on the particular physicians and healthcare professionals the patient happens to engage. Thus, the quality of healthcare would depend on which physicians and healthcare professionals can be accessed for healthcare in particular healthcare delivery systems. Licensing to practice medicine and credentialing for the privilege to practice medicine are means to ensure homogeneity among clinicians to ensure interchangeability.

Interchangeability also applies to patients. Central to the Rationalist/Allopathic system of medicine is reductionism where the extreme variability in the manner in which patients present themselves for care can be reduced to an economical set of principles. Each individual patient is merely some combination among the elemental principles, which allows for a "homogenization" of patients. The homogenization of patients is increased by the use of randomized control trials in evidence-based medicine where the findings relate to the average patient, the *l'homme moyen* (see Chapter 4). Every patient subject to the recommendations of randomized controlled trials in evidence-based medicine is not considered as an individual but as a single construct, the *l'homme moyen*. As discussed in Chapter 4 and in Montgomery (2018, 2019), such an approach risks the Fallacy of Four Terms and associated medical errors.

Even if treating the patient as merely one instantiation of the average patient may not be unjust, even if it produces medical errors. It may merely be that changing the healthcare system to be fully responsible for the idiosyncrasies of the individual patient diverts limited resources from other priorities. The critical question is whose priorities would suffer? If the priorities that suffer were those that the citizenry all agreed took precedent, then leaving the healthcare system as is may be just. If the priorities that suffer are profits of commercial insurers, then other questions are raised. It could be that pecuniary priorities of the commercial insurer are necessary inducements for the insurer to provide a just distribution of healthcare. Note that just distribution does not mean that every patient has a right to healthcare but perhaps only that every patient has a right to purchase healthcare insurance. The justice of the commercial insurer's actions must be parceled out in the obligations to the ethical mechanisms, such as principles or virtues and weighed by the prevailing moral theories (see Introduction and Chapter 5).

The interchangeability of physicians and healthcare professionals is the central issue in the case presented. The critical question is how interchangeable are the physicians and healthcare professionals engaged in the case presented? What are the dimensions of interchangeability? Note that most governmental regulatory agencies provide a single license to all physicians regardless of how the physicians organize their practices, such as becoming a movement disorders neurologist, at least in the United States. In Canada, there is a single qualifying examination and passage is a necessary but not a sufficient condition. To obtain a medical license, the physician must pass a certifying examination from either the College of Family Physicians of Canada or the Royal College of Physicians and Surgeons of Canada (http://www. healthforceontario.ca/UserFiles/file/PRC/recruitment-essentials-licensing-en.pdf). In the United States, physicians can obtain board certification in various medical specialties. These latter specializing requirements would suggest that physicians are not interchangeable, at least in some way.

Other forces have the effect of enforcing interchangeability. For example, having a family or primary care physician as the gatekeeper for referral to other physician and healthcare professionals may enforce a type of interchangeability (see Case 3). When a referral is not made, the effect is that the family or primary care physician is holding himself or herself as equivalent to the specialty for which a referral could have been considered. In this sense, the family or primary care physician is held interchangeable with the specialist. As discussed, strong economic forces, such as penalties for what might be considered excessive referral to specialists, are driving the notion that the family and primary care physicians are interchangeable with the specialist and thus no reason for and many reasons against referral to the specialist or other healthcare professionals. Similar considerations hold in the case presented where the consulting neurologist held himself to be interchangeable with the patient's treating movement disorders neurologist.

A significant problem in the United States is the balkanization of healthcare delivery as evidenced by the rise in financial burden by added costs when care is provided by "out of network" physicians and healthcare professionals. However, if a specific healthcare insurer does not have a sufficient range of physicians and healthcare professionals, then one either accepts the additional costs of "out of network" care or one forgoes care, which is problematic in urgent situations or circumstances. The Patient Protection and Affordable Care Act of 2010 (Obamacare) required sufficient access to a range of different healthcare services. However, the range of services can be very limited (see Case 2). In effect, such situations and circumstances suggest a de facto interchangeability of physicians and healthcare professionals within the network as those outside the network. Consequently, the insurer could argue that there is no justification not to apply out of network penalties. The net effect is to control the patient's access to any physician or healthcare professional, thereby exerting leverage on physicians and healthcare professionals to join the network on terms more favorable to the insurer (see Chapter 4). Note that the last situation may not be unjust if it is deemed as an acceptable manner to retain resources for other societal priorities in the setting of moderate scarcity.

One could reasonably argue that in situations and circumstances where the general physician is interchangeable with the specialist, then there is no loss of beneficence, no violation of nonmaleficence, and autonomy is unaffected. Thus, deferral of referral is not unjust. However, it is likely that most reasonable persons blinded to their self-interests would hold unjust where the refusal to refer occurred in situations or circumstances in which the general physician and the specialist or healthcare professional are not interchangeable. It would seem that recognition of the situation or circumstance in which the general physician and the specialist or healthcare professional are not interchangeable and thus justice would demand referral is up to the discretion of the general physician. As discussed in Case 9, it is clear that such discretion does not always lead to justice.

Similar considerations are obtained when the government is the insurer. For example, in Canada, healthcare insurance is regulated by each province and some services in one province are not covered by the insurance of other provinces (https://www. ontario.ca/page/ohip-coverage-across-canada). The presumption is that all reasonable and urgent healthcare needs provided in one province obviate any need of the patient to seek healthcare in another province. The exception would be reasonable urgent needs. What is reasonable and urgent may be up for debate, such as ambulance services, which generally presume an urgent situation or circumstance.

There are other situations or circumstances where forces seek to deny interchangeability in a peculiar way. For example, in capitated care, the economic incentive to refer difficult or time-demanding patients to others reduces the physician's costs, in terms of time and distraction from less demanding patients, and as the reimbursements are fixed, a net profit is achieved solely by reducing costs (see Case 3). Whether this is unjust requires consideration of the entire complex ecology surrounding the actions. For example, the capitated physician cannot be effective and help others if the physician is bankrupt. Thus, at a microlevel or at the level of the dyadic relationship between the physician and patient, off-loading the provision of care to another physician may be understandable. However, in the macrolevel or global sense, the obligations to the patient must be met by someone; if not the family or primary care physician, then it is by the specialist—unless forgoing of care is an acceptable option. There will not be any net savings to society as a whole if physicians "play hot potato" with patients (see Case 10).

Given the incredible complexity of everyday medicine, it may not be possible to "administrate" the consistent provision of justice based on a necessarily reduced schema of policies, regulations, laws, or reliance on the virtue of administrators (see section Ethics as an Exercise in Chaos and Complexity in Chapter 1). Some "wiggle room" is necessary to accommodate the complexities and idiosyncrasies of human beings. The importance of "wiggle room" has been noted in studies of patient safety. Wears and Sutcliffe (2020, p. 69) wrote:

> Perrow (a sociologist) is even more pessimistic, asserting that since designs, equipment, procedures, supplies, operators, and work environments almost contain hidden flaws, small failures are inevitable and complex interactions will make these failures both unexpected and incomprehensible to the operators. In systems with a little slack [author – wiggle room], the small incidents will occasionally turn into large accidents "because the nature of the system itself; they are system accidents, and are inevitable, or 'normal' for these systems."

Such "wiggle room" is seen in prosecutorial discretion in bringing criminal charges, judicial discretion in sentencing, and the ability of legal juries to disregard the law, as discussed in Chapter 1, should following the letter of the law prove too inflexible and risk injustice. In the context of the discussion of the case presented, wiggle room can be created by lessening the restrictions or obstacles for consultation and providing the physician with the time and resources to discuss the particular case with other physicians. This author remembers as a medical student, physicians knew each other and their particular interests and expertise. Physicians were very willing to refer certain patients to those physicians with particular interests and expertise. Even if not constrained by policies and time limitations, it is not clear that physicians today really know each other, particularly with the breakdown of the "community" of physicians. Wiggle room can be accomplished by providing physicians and healthcare professionals to attend to the needs of the patient that cannot be confined to the spatial-temporal confines of the clinic or bedside visit and not limited by reimbursement schemes. In the case presented, perhaps the consulting neurologist would have talked with the patient's treating physician had the consulting neurologist had time. The question in the ethics of everyday medicine is whether if such communication was logistically possible, would physicians today avail themselves of it in the patient's best interest?

References

Ahmed, A., Allman, R.M., Kiefe, C.I., Person, S.D., Shaneyfelt, T.M., Sims, R.V., Howard, G., DeLong, J.F., 2003. Association of consultation between generalists and cardiologists with quality and outcomes of heart failure care. Am. Heart J. 145 (6), 1086–1093.

Alsafi, E., Bahroon, S.A., Tamim, H., Al-Jahdali, H.H., Alzahrani, S., Al Sayyari, A., 2011. Physicians' attitudes toward reporting medical errors: an observational study at a general hospital in Saudi Arabia. J. Patient Saf. 7 (3), 144–147. https://doi.org/10.1097/PTS.0b013e31822c5a82.

American Medical Association, 1848. Code of Ethics of the American Medical Association. T.K. and P.G. Collins, Philadelphia, PA.

Anonymous Report, 1896. Twenty-fifth anniversary of the organization of the Brooklyn Pathological Society. Brooklyn Med. J. 10, 56.

Aquinas, T., 1485. Principles of double effect. In: Summa Theological.

Collier, R., 2011. American Medical Association membership woes continue. CMAJ 183 (11), E713–E714. https://doi.org/10.1503/cmaj.109-3943.

Dik, B.J., Duffy, R.D., 2009. Calling and vocation at work: definitions and prospects for research and practice. Couns. Psychol. 37 (3), 424–450.

Dik, B.J., Eldridge, B.M., Steger, M.F., Duffy, R.D., 2012. Development and validation of the Calling and Vocation Questionnaire (CVQ) and Brief Calling Scale (BCS). J. Career Assess. 20 (3), 242–263.

Douthat, R., 2018. Why we miss the wasps: their more meritocratic, diverse and secular successors rule us neither as wisely nor as well. The New York Times (December 5).

Friedman, J.H., Feinberg, S.S., Feldman, R.G., 1985. A neuroleptic Malignantlike syndrome due to levodopa therapy withdrawal. JAMA 254 (19), 2792–2795.

Goodin, J.B., Duffy, R.D., Borges, N.J., Ulman, C.A., D'Brot, V.M., Manuel, R.S., 2014. Medical students with low self-efficacy bolstered by calling to medical specialty. Perspect. Med. Educ. 3 (2), 89–100. https://doi.org/10.1007/s40037-014-0110-7.

Iedema, R., Jorm, C., Lum, M., 2009. Affect is central to patient safety: the horror stories of young anaesthetists. Soc. Sci. Med. 69 (12), 1750–1756. https://doi.org/10.1016/j.socscimed.2009.09.043.

King, L.S., 1982. Medicine in the USA: historical vignettes. V. The 'old code' of medical ethics and some problems it had to face. JAMA 248 (18), 2329–2333.

Kohn, L.T., Corrigan, J.M., Donaldson, M.S., Institute of Medicine (US) Committee on Quality of Health Care in America (Eds.), 2000. To Err is Human: Building a Safer Health System. National Academies Press, Washington, DC.

Montgomery Jr., E.B., 2018. Medical Reasoning: The Nature and Use of Medical Knowledge. Oxford University Press, Oxford.

Montgomery Jr., E.B., 2019. Reproducibility in Biomedical Research: Epistemological and Statistical Problems. Academic Press, New York.

O'Connor, E., Coates, H.M., Yardley, I.E., Wu, A.W., 2010. Disclosure of patient safety incidents: a comprehensive review. Int. J. Qual. Health Care 22 (5), 371–379. https://doi.org/10.1093/intqhc/mzq042.

Orlander, J.D., Barber, T.W., Fincke, B.G., 2002. The morbidity and mortality conference: the delicate nature of learning from error. Acad. Med. 77 (10), 1001–1006.

Percival, T., 1803. Medical Ethics; or a Code of Institutes and Precepts Adapted to the Professional Conduct of Physicians and Surgeons. Printed by S. Russell, for J. Johnson, Saint Paul's Church Yard, and R. Bickerstaff, Strand, London.

Perez, B., Knych, S.A., Weaver, S.J., Liberman, A., Abel, E.M., Oetjen, D., Wan, T.T., 2014. Understanding the barriers to physician error reporting and disclosure: a systemic approach to a systemic problem. J. Patient Saf. 10 (1), 45–51. https://doi.org/10.1097/PTS.0b013e31829e4b68.

Smart, N.A., Dieberg, G., Ladhani, M., Titus, T., 2014. Early referral to specialist nephrology services for preventing the progression to end-stage kidney disease. Cochrane Database Syst. Rev. 6 (6), CD007333https://doi.org/10.1002/14651858.CD007333.pub2.

Toru, M., Matsuda, O., Makiguchi, K., Sugano, K., 1981. Neuroleptic malignant syndrome-like state following a withdrawal of antiparkinsonian drugs. J. Nerv. Ment. Dis. 169 (5), 324–327.

Wears, R.L., Sutcliffe, K.M., 2020. Still Not Safe: Patient Safety and the Middle-Managing of American Medicine. Oxford University Press, Oxford.

Weaver, F.M., Follett, K.A., Stern, M., Luo, P., Harris, C.L., Hur, K., Marks Jr., W.J., et al., 2012. Randomized trial of deep brain stimulation for Parkinson disease: thirty-six-month outcomes. Neurology 79 (1), 55–65. https://doi.org/10.1212/WNL.0b013e31825dcdc1.

Westerman, M., 2014. Reflections on having a 'calling' as a medical doctor. Perspect. Med. Educ. 3 (2), 73–75. https://doi.org/10.1007/s40037-014-0112-5.

Case 5

Case 5—A neurologist does not prescribe dopamine agonists to patients with Parkinson's disease but rather prescribes levodopa, although there is concern about greater long-term risks with early and sole use of levodopa. The neurologist does not think that the risk of long-term complications associated with levodopa outweighs the additional cost for the dopamine agonists. The neurologist does not mention the possibility of treatment with dopamine agonists to the patients.[a]

The medical facts of the matter

Parkinson's disease is a neurodegenerative disorder. The symptoms include tremor, slowness of movement, stooped posture, and difficulty with balance and walking. Muhammad Ali (http://www.youtube.com/watch?v=jvVbgnCiyo0) and Michael J. Fox have Parkinson's disease. The disease is caused by the degeneration of brain cells (neurons) deep in the brain in an area called the substantia nigra pars compacta, in addition to other areas. Neurons in this region make and use a chemical called dopamine to communicate with other neurons. When the neurons degenerate, various regions of the brain experience a loss of information associated with the loss of dopamine.

A medication called levodopa can be used to replace the dopamine that has been lost from the brain, much like giving insulin helps patients with diabetes when their pancreas no longer makes insulin. The levodopa is absorbed into the brain and then is converted to dopamine. Dopamine agonists are another type of medication that can be used to treat Parkinson's disease. The dopamine agonists are absorbed into the brain but are not converted to dopamine. Instead, the dopamine agonist molecules "fool" the brain into thinking they are dopamine and the brain responds accordingly.

Both levodopa and dopamine agonists are very helpful in patients with Parkinson's disease. Levodopa is more powerful, meaning that more patients may find sufficient benefit from levodopa than with dopamine agonists. However, many patients find sufficient relief with dopamine agonists alone. Levodopa is less expensive than the dopamine agonists. However, there are long-term complications. After a few years, patients who take these medications can develop severe involuntary movements, called dyskinesia. Indeed, many videos of Michael J. Fox show him having these uncontrollable movements (http://www.youtube.com/watch?v=ECkPVTZlfP8). Research demonstrates that the risk of developing involuntary movements and other complications, such as wearing off, where the benefit is not persistent, occurred in 28% of patients taking pramipexole, a dopamine agonist, compared with levodopa (51%)

---

[a] The case presented is hypothetical, although modeled on experience. No judgment is made relative to the correctness, appropriateness, or value of any actions described on the part of any of the participants. The purpose here is to describe and explore the variety of ethical considerations and consequences that depend on the perspective of different relevant ethical systems.

The Ethics of Everyday Medicine. https://doi.org/10.1016/B978-0-12-822829-6.00005-9

(hazard ratio, 0.45; 95% confidence interval, 0.30–0.66; $P < 0.001$) (Parkinson Study Group, 2000). To be sure, the mean improvement in symptoms, as measured by the Unified Parkinson's disease Rating Scale, was greater in the levodopa group than in the pramipexole group at 23.5 months (9.2 points vs 4.5 points; $P < 0.001$). However, two quality-of-life measures showed essentially no difference up to 20 months between patients beginning on pramipexole compared to those beginning on levodopa. A reasonable conclusion is that up to 20 months there is little difference in the quality of life but with a significant reduction in the risks of involuntary movement with the preferential use of dopamine agonists over levodopa. Very similar results were obtained with another dopamine agonist, ropinirole (Rascol et al., 2000).

The purpose here is not to debate the relative superiority, however, that is defined, of dopamine agonists versus levodopa. Rather, it is about whether and how to involve patients in the actual treatment decisions. At the very least, it can be said that dopamine agonists are a reasonable alternative to levodopa in the initial treatment of patients with Parkinson's disease; indeed, dopamine agonists are the treatment of first choice recommended by a great many experts. Experts advocating initial treatment with levodopa argued, "Finally, dopamine agonist monotherapy is not nearly as efficacious as levodopa. This is why only a very limited number of patients could be maintained on agonist monotherapy for more than a few years in any of the published clinical trials to date" (J. Eric Ahlskog, reply to Montgomery, 2004). The counterpoint is that there are some patients, perhaps many, for whom the dopamine agonist is sufficiently efficacious and, if for those patients there is a reduced risk of long-term adverse effects, perhaps the wiser option would be to try the dopamine agonists. Importantly, one cannot know *a priori* which patient will find satisfactory improvement with less involuntary movements at the time the decision is made to start symptomatic therapy. Initial use of dopamine agonists at the very least deserves consideration. It is the nature of that consideration that is at the heart of the case presented.

Some have argued that the lower risk of complications, such as involuntary movements, was a consequence of less aggressive treatment for those on dopamine agonists. However, these critics would have to explain how it is in a double-blinded study where the physician and the patient were not told which medication they were given that the treating physicians knew which patients were taking dopamine agonists and then be less aggressive in the treatment of those on dopamine agonists. Yet, data clearly demonstrate a reduced risk of complications during the time frame in which dopamine agonists are equally effective as demonstrated in the quality-of-life measures. Ahlskog is quite right; initially, comparable quality-of-life measures fall behind levodopa but after 20 months. But note the value judgment. Why is comparable efficacy as measured by quality-of-life indices and with lower risk of complications for 20 months not worthwhile? Perhaps it may be that the greater costs of dopamine agonists over levodopa do not offset the advantage. But doesn't that depend on who is paying the bill rather than who is prescribing it (see Chapter 4)? Is the neurologist preempting the patient's discretion?

Countering the argument, the author wrote "Finally, if it is the patient's or the patient's proxy's right to decide how the patient is treated, then our responsibilities are to present all sides reasonably. If so, the debate as to whether a physician should prescribe dopamine agonists or levodopa is misdirected" (Montgomery, 2004). In response, Roger Albin wrote "Finally, debate about initial treatment of PD is appropriate and necessary. Patients do not have *unlimited rights* [italics added] to decide how they are treated. The generally accepted formulation of medical ethics identifies four crucial principles: respect for autonomy, nonmaleficence, beneficence, and concern for justice" (Beauchamp and Childress, 2001). Application of the last three principles involves paternalistic behavior. Beauchamp and Childress (2001), however, specify that none of the basic principles have priority. According to Albin, "Assigning priority to autonomy is questionable philosophy and poor clinical practice" (Roger Albin, reply to Montgomery, 2004).

Autonomy

The central issue involved in this case, like a great deal of many others in part 1, revolves around autonomy. The position of Albin and likely Ahlskog that the obligation to beneficence, in terms of greater efficacy of levodopa, and to nonmaleficence, lower short-term adverse effects of levodopa, allows paternalistic trumping of the obligation to autonomy. Note that it is highly unlikely that Albin and Ahlskog do not believe in autonomy. Indeed, the author knowing them can attest to their respect for patients, the fundamental basis of autonomy. Rather, as suggested by Albin, the difference is the degree of obligation to autonomy, but in this case, autonomy is the patient's right to be informed of the option, not necessarily to demand a particular treatment.

Albin cited the work of Tom L. Beauchamp and James F. Childress (2001). In their seventh edition, Beauchamp and Childress (2013, pp. 220–223, 228) provided a number of situations where trumping autonomy by paternalism is justified by best interests, by consent, and by prospective benefit. In the case of justification by best interests, the situation is when patients cannot make their preferences known, which is not the case here. The second situation is by consent, where patients doubt their ability to always act in their best interests and consent to others overriding their autonomy (e.g., a Ulysses pact discussed later). This is not likely the case here. With respect to the third situation, Beauchamp and Childress (2013, p. 221) wrote:

> *Paternalism justified by prospective benefit* [italics in the original]. Accordingly, the justification of paternalistic actions that we recommend places benefit on a scale with autonomy interests and balances both: As a person's interests in autonomy increase and the benefits for that person decrease, the justification of paternalistic action becomes less plausible; conversely, as the benefits for a person increase and that person's autonomy interests decrease, the justification of paternalistic action becomes more plausible. Preventing minor harms or providing minor benefits while deeply disrespecting autonomy lacks plausible justification, but actions that prevent major harms or provide major benefits while only trivially disrespecting autonomy have a plausible paternalistic rationale.

The argument just presented by Beauchamp and Childress (2013) is a "slippery slope." It would appear that a particular person's disinterest in his or her autonomy would justify disregard of the obligation to autonomy. Extending that logic, a person not making the effort to vote in an election would seem to offer justification for disregarding voting rights. Further, the interest in one's autonomy would vary according to the perceived power to exercise that autonomy (see Case 14). One could see any number of measures to defeat persons' interest in autonomy, such as "poll taxes" and having to travel unreasonable distances, often at the expense of losing pay to exercise their autonomy inherent in their vote. Similar concerns attend the sliding scale of beneficence and nonmaleficence. A distinction must be made between justice at a micro- and macrolevel (see Chapter 1). What may be nonmaleficence to an individual may be a greater beneficence to society. Yet, in a pluralistic modern liberal democracy, macrolevel justice is predicated on sufficient microlevel justice among the populace.

An interesting argument is whether the second situation, paternalism by consent, can be invoked. One example may be the Ulysses pact. Ulysses (the Roman version of the Greek Odysseus) wanted to hear the beautiful song the sirens sing but which lured men to their death. Ulysses put wax in the ears of his crew while he had himself lashed to a pole and instructed his crew not to free him. Thus, Ulysses was able to hear the sirens' song but not be lured to his death. In some cases, the patient may not want to make an informed consent, instead asking someone else to make the decision, and, not infrequently, it is the physician or healthcare professional. Often, this is in the form of "you (the physician or healthcare professional) tell me what to do" or "I'll (the patient or the patient's surrogate) do what you tell me to do." At times it is more subtle, such as "what would you do if you were me?"

The Ulysses scenario is relatively less problematic. The crew knew when and why there was an inability of Ulysses to exercise autonomy (when listening to the sirens' song), and presumably once the sirens were past, Ulysses would be able to exercise autonomy. However, what if the crew could not be sure when the sirens were past? If the crew could never know, then they could never be sure that Ulysses was capable of exercising autonomy when asked to be released. In that case, the last autonomous act would take precedent, and that last autonomous act was the command to be lashed to the pole and, consequently, Ulysses would never be released.

The argument may be that the patient just cannot understand the medical issues sufficiently as to render an informed opinion. In this case, it would seem that the patient explicitly or implicitly consents to the physician with greater knowledge to make the decision. Often, the patients are presumed to be unable to understand sufficiently to fully participate in informed consent, perhaps because the patient likely did not complete medical or healthcare professional schools. However, young children as young as 9 years of age can understand major surgeries sufficient to provide informed consent (see Case 13). Yet, the presumption that the patient cannot understand or be led to understand implicitly and preemptively leads to failure to the obligation to autonomy, resulting in paternalism. In a survey of hospital-based outpatient physicians in their use of ethics committees, factors related to the use or nonuse of ethics committees were explored (Orlowski et al., 2006). Many surgeons felt that members of the ethics committee could not grasp the issues involved in the surgical cases and thus the surgeon was entitled to exercise paternalism. Perhaps, in the case presented, the physician did not think the patient capable of understanding the risks and benefits, and the patient's not objecting or asking for alternatives constituted an implicit consent to act paternalistically. Yet, if the patient is unaware of alternatives, how could he or she ask?

Informed consent

In many ways the issues surrounding informed consent illustrate the issue of autonomy and thus are worth investigating. However, autonomy is not just about informed consent, it is a matter of respect of which respecting the right to informed consent is a component (see Case 13). The discussion of informed consent here is not that of human research, emergencies, or when the patient is not competent as these are not relevant to the current case. A public health emergency, medical emergency, patient waiver (essentially a consent to not give consent for the treatment), "therapeutic privilege" when obtaining the consent could harm the patient, and patient incompetence are recognized exceptions to the necessity of informed consent (Cocanour, 2017).

Beauchamp and Childress (2013, pp. 110–125) identified a number of different forms of consent, including implicit or implied, presumed, and explicit, with the latter written or verbal. An example of implicit or implied consent is that when a patient sees a physician or healthcare professional, the patient can reasonably expect that he or she will be asked a number of questions and be examined. The mere presence of the patient in the clinician's examining room implies consent for the history and physical examination. However, in certain circumstance, such as examining genitalia, breast, or rectum, explicit consent may be required or best practice, particularly in circumstance where a reasonable patient may not expect such an examination. In surgery, the patient consents to an operation but does not provide explicit consent to the standard procedures involved. For example, the patient would not have to give explicit consent as to the nature of the suture material. Further, often there are a large and complex constellation of physicians and healthcare professionals involved in the care of a particular patient. There is the reasonable presumption that personal health information would be exchanged freely among those involved in delivering the care (with certain notable exceptions). Indeed, these presumptions are codified in the US Health Insurance Portability and Accountability Act of 1996 (Pub.L. 104-191, 110 Stat. 1936, enacted August 21, 1996 section 45 CFR 164.506).

There are four fundamental criteria for informed consent; the patient must be (1) competent; (2) of legal age for consent, although minors may have a right to assent or even consent if deemed sufficiently mature, see Case 13; (3) adequately informed; and (4) able to make uncoerced decision (Cocanour, 2017). It is important to appreciate that such medical informed consent is no different from the elements of any valid contract between people (and corporations) in most pluralistic modern democracies. These elements include: (1) offer and acceptance; (2) consideration, typically an exchange of goods; (3) capacity to enter into a contract; (4) good faith (e.g., disallowing innocent typographical errors to invalidate a contract); and (5) lawful purpose. The elements of a valid contract have existed for hundreds of years and so it is striking that medical informed consent should have gained appreciation essentially since the 1970s. The interesting question is why did the prior relationship between patient and physician or healthcare professional appear to be unlike a valid contract? A valid contract implies a balancing of power between those making and those accepting the contract. Historically (and to this day), the power between the patient and the healthcare delivery system has not been balanced (see Chapter 4).

The elements of a valid contract, and thus medical and research informed consent, are derivative of the fundamental right to bodily integrity and freedom of thought (conscience). These may not be deontological rights but they are de facto rights if there is a universal commitment to enforcement. Universality can be established if a sufficient number of reasonable persons blinded to their self-interest make it so.

In the case presented, the critical issue is the patient being adequately informed. Information provided by the patient usually includes: (1) diagnosis, if known; (2) nature and purpose of the recommended treatment; (3) burdens, risks, and expected benefits; and (4) all reasonable options, including forgoing any treatment. The last element, information regarding all reasonable options, is central to the case presented. Recently added is any potential conflict of interest on the part of those involved in executing the care, for example, whether the physician has a financial interest in the use of levodopa.

Technical battery versus negligence

Accepting the caveat that the author is neither an attorney nor giving legal advice, what follows is the author's understanding of the legal issues. Battery is a criminal offense in which a person is "touched" without his or her consent. Historically, early cases of failed medical informed consent were based on battery but subsequently rely primarily on negligence (Cocanour, 2017). However, there continues to be a risk for cases of battery based on failed informed consent (Alpers, 2004). The discussion here is not one of law or legal procedures. Rather, the discussion of law or legal decisions is to explore more general issues in the ethics of everyday medicine in which court actions are exemplars. According to the Canadian Medical Protective Association, "A physician may be liable in assault and battery when no consent was given at all, when the treatment went beyond or deviated significantly from that for which the consent was given, or if consent to treatment was obtained through serious or fraudulent misrepresentation in what was explained to the patient" (https://www.cmpa-acpm.ca/en/advice-publications/handbooks/consent-a-guide-for-canadian-physicians#assault).

The issue of battery in the context of informed consent sometimes is referred to as technical battery. It appears relatively unambiguous in cases involving surgery, such as the surgeon conducting a mastectomy when only a biopsy was consented to. Commenters have suggested that technical battery only applies to surgical procedures, perhaps derivative of the original conception of battery as "violation of the bodily integrity." Indeed, this conception was supported in *Pennsylvania Superior Court Boyer v. Smith* (345 Pa. Superior Ct. 66 (1985) 497 A.2d 646). Wendy O'Connor of *Physicians News Digest* wrote "Thus, at present, informed consent is not required for procedures such as clamping a wound, inserting a drain following surgery, the use of forceps during natural childbirth, the nonsurgical administration of medication, chiropractic manipulation, angiogram, intravenous administration of antibiotics or other nonexperimental prescription medication, or the oral administration of prescription drugs" (https://physiciansnews.com/2015/10/19/informed-consent-what-every-pennsylvania-physician-needs-to-know/). However, the Arizona Supreme Court was reported to uphold a claim of battery based on injecting fentanyl intravenously for a magnetic resonance imaging procedure when the patient specifically requested a different agent to which the providers had previously agreed on (Alpers, 2004).

The ethical implications persist, regardless of how violation of bodily integrity occurs, and the distinction between surgical and nonsurgical, while perhaps a legal distinction, may not be a distinction ethically. In some ways, the distinction may be analogous to the paradox of the runaway trolley dilemma discussed in Introduction and in Case 2. The personal proximity to the action that determines the outcome greatly determines the degree of perceived maleficence, throwing a switch that results in two lives lost and five saved is somehow less culpable than pushing a single person onto railway tracks to save five persons, although losing one. Surgery is highly personal, prescribing medications, even if having a higher risk of maleficence than surgery, seems less personal and hence less culpable. Thus, nonsurgeon physicians may be held to the same notion of battery as surgeons as described in the Arizona Supreme Court case referred to previously.

A crucial distinction between battery and negligence is the standards used. Battery often is found on prima facie evidence, evident on the face of the act, and is considered true unless proven otherwise. Negligence generally occurs when there is a departure from a standard of care. Those claiming injury due to negligence must prove, usually by expert opinion, that there was a departure from the standard of care. However, defining the standard of care as what similarly situated physicians would do is ethically problematic. As Erich Fromm (1955) wrote, "The fact that millions of people share the same vices does not make these vices virtues, the fact that they share so many errors does not make the errors to be truths, and the fact that millions of people share the same mental pathology does not make these people sane."

Informing in informed consent

A particularly troublesome notion for informed consent is the nature and extent of the information regarding all reasonable options. There have been three standards: (1) the reasonable professional, (2) the reasonable patient, and (3) the subjective standards. The professional standard was established in 1957 in the case of *Salgo v. Stanford* (Cocanour, 2017). Subsequently, one standard was the information a reasonable patient might require; this was taken as that information which, had the patient been made aware, would have affected his or her consent. The subjective standard recognizes the uniqueness of the individual patient. Quoting from Cocanour (2017),

> In *Cobbs v. Grant and Wilkinson v. Vesey*, the decisions were more in line with a subjective standard. Whether a patient should proceed with a therapy requires reference to the values of that patient and thus are not exclusively medical determinations. From Cobbs, the scope of the physician's communications to the patient, must be measured by the patient's need and that need is whatever information is material to the decision.

This standard still is patient-centric but refers to the particular patient rather than the generic or average patient (see Chapter 2). The key is what is reasonable information the physician is required to disclose is what is reasonable and relevant to the patient, not the physician. It is not likely that the physician in the case presented met that requirement. From the Canadian Medical Protective Association, essentially providers of medical liability coverage:

> Although obtaining a valid consent from patients has always involved explanations about the general nature of the proposed treatment and its anticipated effect, the Supreme Court of Canada, over two decades ago, imposed a more stringent standard of disclosure upon physicians. The adequacy of consent explanations is to be judged by the "reasonable patient" standard, or what a reasonable patient in the particular patient's position would have expected to hear before consenting...

> Throughout these and other legal judgments which have been rendered in more recent years, there is repeated reference to the need to disclose "material" risks to patients. However, there can be some understandable uncertainty as to what in fact does constitute a "material" risk. One court has defined it as follows:

> *"A risk is thus material when a reasonable person in what the physician knows or should know to be the patient's position would be likely to attach significance to the risk or cluster of risks in determining whether or not to undergo the proposed therapy"* [italics in the original].
> *(https://www.cmpa-acpm.ca/en/advice-publications/handbooks/ consent-a-guide-for-canadian-physicians#assault)*

Thus, the particular circumstances of the patient are an important determinant of materiality (see discussion of phenomenological ethics in Chapter 2).

The question arises as to how frequent do discussions of alternative treatments occur in patient-physician/healthcare professional interactions? In a study of patient-physician/healthcare professional conversations regarding the use of antiretroviral therapy for HIV stratified for complexity, discussion came up 0 out of 2 (0%) occasions of the time when the decision-making was basic, 2 out of 12 (17%) occasions of intermediate complexity, and 5 out of 7 (74%) occasions when the decision was complex (Callon et al., 2017). In another study of 1057 recorded and analyzed discussions over a much more varied clinical context in 1999, alternatives were discussed in 5.5% of basic decision-making, 15.8% of intermediate complexity, and 29.5% of complex cases (Braddock et al., 1999). Ideally, the discussion of alternatives should have occurred in 100% of the cases of all complexities. The question is why did only the most complex cases engender a discussion of alternatives? Perhaps this may reflect the relative lack of confidence on the part of the physician/healthcare provider in the complex cases. One wonders whether the discussion on alternatives may reflect a way of sharing the burden of doubt and perhaps liability.

What is reasonable?

As evident from the preceding discussion, much depends on what constitutes reasonable. How would one recognize what is reasonable or perhaps what is unreasonable? The latter may be epistemically more tractable. Another interesting question is how can physicians and healthcare professionals be taught what is reasonable if that what is reasonable cannot be defined (Meno's paradox described by Plato in the Socratic dialogues)? Perhaps there is no economical set of principles by which every situation of reasonableness can be deconstructed, analyzed, and reduced. In other words, the virtually infinite varieties of situations and circumstances that involve concerns regarding reasonableness are not variations on a set of economic principles as a Principlist approach would hold. Rather, variety is diversity where each case is considered as de novo and treated on its own merits, where perhaps virtue ethics may be useful (Chapter 2).

One approach to defining what is reasonable comes from Immanuel Kant, who held that for any proposition or action to be moral, the advocate must accept that it be universal. In 1785, Kant wrote:

> The shortest way, however, and an unerring one, to discover the answer to this question whether a lying promise is consistent with duty, is to ask myself, Should I be content that my maxim (to extricate myself from the difficulty by a false promise) should hold good as a universal law, for myself as well as for others? and should I be able to say to myself, "Every one may make a deceitful promise when he finds himself in a difficulty for which he cannot otherwise extricate himself"?
>
> *Kant (2001, p. 161)*

Such a position may be too philosophically strong, meaning the position would disallow propositions that otherwise would be accepted as moral. For example, exceptions to the obligation for informed consent noted earlier would be a Kantian argument that informed consent is not a universal law. Lacking in a moral status derivative of being universal, insistence on informed consent is potentially unreasonable.

The problematic nature of the necessity of universality is precisely an argument against Principlism (Chapter 2). However, the question is what alternative does not invite a slippery slope to absolutism or solipsism? One could take a consensus approach, such as the inherent in Reflective Equilibrium (Chapter 3), such that there is a level of consensus that the proposition approaches universality; once the threshold is crossed, the proposition is effectively universal and the proposition held reasonable. However, short of universal agreement in the consensus, the proposition is still at risk for the Fallacy of Four Terms and therefore at risk for injustice (Chapter 3).

Perhaps another method to determine reasonableness of any proposition is to determine to what degree the proposition represents a formal or informal logical fallacy. Examples of formal logical fallacies ubiquitous in medical reasoning include: (1) Fallacy of Four Terms, (2) Fallacy of Confirming the Consequence, (3) Fallacy of Confirming a Disjunctive, and (4) Fallacy of Limited Alternatives (Montgomery, 2019a). An example of the Fallacy of Confirming the Consequence would be *if moral premise **a** implies behavior **b** is true and behavior **b** is demonstrated, then moral premise **a** is true and, for purposes here, moral premise **a** is considered reasonable.* This is invalid, meaning that the truth or falsehood of *moral premise **a*** cannot be established by the argument. The demonstration of *behavior **b*** could have arisen from any number of other premises besides *premise **a***. Interestingly, arguing from Principlism, a putative a deductive process, is an example of the Fallacy of Confirming the Consequence (Chapter 1).

Hard paternalism, soft paternalism, and paternalism by disinformation

Paternalism essentially occurs when the shareholder superimposes the shareholder's value onto the stakeholder, thereby trumping the stakeholder's value. Hard paternalism is when the trumping is overt. The stakeholder has expressed his or her values but is overruled by the shareholder's values. Soft paternalism occurs when the stakeholder is not provided an opportunity to state his or her values in a fully informed way. Had the neurologist simply said "I am not going to prescribe dopamine agonists to you" it would be an example of hard paternalism. As the neurologist did not mention the possibility of dopamine agonists, the patient was unaware that this was an option and therefore had no sense of the value of dopamine agonists; hence, soft paternalism. At least with hard paternalism, the patient could have voiced his or her concern to the contrary. There was no option in response to soft paternalism.

Another form of paternalism is by disinformation. In this case, had the neurologist simply said "dopamine agonists have a higher rate of side effects and therefore I will not prescribed the medications," this would be paternalism by disinformation. The statement by the neurologist is not a fair statement of the issues related to a risk of maleficence, that being adverse effects, as discussed previously.

Implicit in paternalism by disinformation is the creation of a type of straw man argument. Alternatives other than the physician's preferred position are characterized in such terms that it is highly unlikely any reasonable patient would find the alternatives, the straw man, acceptable. The patient then defaults to the physician's position. The straw man can be subtle. In the early 2000s, tissue plasminogen activator (TPA) was found to reduce the disabilities associated with stroke, provided the activator was administered within hours after the onset of the stroke. This time requirement introduced serious logistical issues and ultimately resulted in the establishment of stroke code teams on call, typically at large academic medical centers staffed with residents and fellows. The author recalls neurologists at local community hospitals stating publicly that they would not institute such a rapid response. Their rationale was not that they couldn't administer TPA, but that it was too dangerous. One wonders whether the logistical issues were the real reason. To be sure, TPA had significant risks, which, viewed in isolation, the straw man, would seem too dangerous. However, when counterbalanced by the potential benefit, TPA is considered the treatment of choice in the appropriate circumstance that can only be when there is a timely response.

Suasion, for (persuasion) and against (dissuasion)

Paternalism by disinformation can be accomplished by rhetoric, the art of suasion—persuasion when an action is desired, dissuasion when the action is to be prevented. Note that rhetoric consists of logical reasoning (*logos*), appealing to emotion (*pathos*), and personal appeals of the one attempting the suasion (*ethos*), in the Aristotelian tradition (for an excellent review, see Dubov, 2015). Currently, rhetoric most often relates to appeals to emotions and the personal characteristics and positions of the one attempting the suasion. While physicians and healthcare professionals would like to think they operate solely on logic and rationality, *logos*, the fact is appealing to emotion and personal attributes to the person seeking to sway is far more common. Indeed, studies of implicit bias, held here as unreasonable, suggest that physicians are no less biased than the general public, despite many years of training in rational medicine (Fitzgerald and Hurst, 2017).

Many have argued that rhetoric may be useful and this possibility will be discussed later. The focus here is the risks associated with rhetoric; risks, although prevalent, are rarely recognized. One approach to recognizing rhetorical risks on the part of physicians and healthcare professionals is to see what is present in the patient that could be played upon by the rhetoric of the physician or healthcare professional (Shaw and Elger, 2013). One frailty of human nature is the difficulty of balancing short-term risks with long-term benefits. In economics, the phenomenon is referred to as the time-discounted utility theory—rewards promised after some time delay have less value than immediate rewards (Frederick et al., 2002). It is likely that similar phenomena also operate in consideration of immediate and long-term adverse effects. In the case described, it is possible that patients might overvalue the greater short-term risks of the dopamine agonists over the long-term risks for levodopa. The neurologist in the case presented may be playing to the time-discounted utility phenomenon.

There are a number of biases, sometimes referred to as heuristic devices, which include loss/gain framing bias, where a possible loss is given more valence than a potential gain; availability bias, where the most recent experience has greater valence than the past; sunk cost effect, where previous failed efforts diminish enthusiasm for potential efforts; order effect, where the first option presented is taken as greater preference or credibility; bandwagon effect or going with whatever the majority do; and omission bias, where errors of commission, that is, direct results of an intervention, are considered worse than the adverse consequences of inaction, although in reality, inaction is an active "no" in most cases. This problem is particularly poignant in surgical therapies (see Montgomery, 2015). The sunk cost effect is seen, for example, in the infrequent consultations with movement disorder experts by patients with essential tremor. After having exhausted all reasonable pharmacological therapies, patients seem less than enthusiastic at the potential for deep brain stimulation (author's experience). Another bias is impact bias, where future potential emotional states are estimated inaccurately (Blumenthal-Barby and Krieger, 2015).

Interestingly, physicians and healthcare professionals are more likely to have these biases than patients—in a systematic review, 80% of medical personnel compared to 61% of patients had inherent biases. It is not difficult to see where the inherent biases of patients can be leveraged by the inherent biases of the physician or healthcare professional in an implicit manner, yet conflict with rational (presumed to be ethically most sound) decisions. In fact, patients may be influenced because their own biases resonate with those of the physician or healthcare professional.

Examples of the impact of inherent biases include when a physician or healthcare professional says to the patient "you have a 10% chance of dying as a result of the treatment" and leaves it at that. The patient may look at the treatment much differently if the physician or healthcare professional said "you have a 90% chance of surviving the treatment" (Keller and Siegrist, 2009). This is an example of "framing" effects, which play on the inherent biases of loss/gain framing and order effect (Levin et al., 1998).

Modes of thinking, both in the patient and in the physician or healthcare professional, feed paternalism by disinformation. All humans are subject to the power of an intuitively appealing narrative that often trumps fact or logic (Johnson-Laird, 2006). A physician said to a patient that the patient's very high first morning blood glucose level was "the body's way of getting ready for the coming day." The patient took this as in indication not to be concerned and, consequently, felt no motivation to correct the high blood glucose level. Whether the physician's statement is a contrivance of dissuasion or an expression of the physician's mode of thinking, the fact is it was not an act of beneficence and indeed may be maleficent to the patient. The impact of analogies and metaphors on paternalism by disinformation is discussed more fully later. Unfortunately, logical or reasoning errors in physicians and healthcare professionals often provide an opportunity for rhetoric to have an ethically counterproductive effect.

The necessity, indeed inevitability, of pathos and changing the patient's mind

The recent trend in biomedical ethics has been a greater influence of autonomy for the patient, as it should be. The challenge of obligation to autonomy is seen when autonomy is thought synonymous with absolute acceptance of the patient's position when that position is counterproductive from a medical perspective. In some cases, it seems that any attempt to change the patient's position is seen as imposing or manipulating the patient, thereby violating patient autonomy (see Case 14). Consequently, some ethicists have argued that the purpose of suasion was to "clarify" the patient's position (Dubov, 2015). However, this seems disingenuous. For whom is the patient's position being clarified? Is it for the benefit of the physician or healthcare professional, perhaps to clarify the thinking of the physician or healthcare professional, particularly as it relates to the patient's needs for phenomenological authenticity (see Chapter 2)? If so, it would seem that the effort would default to the patient's unmodified position nonetheless and the conundrum persists. Is it for the benefit of the patient? If so, is not the intention then to change the patient's mind? One is back to where one started.

Some ethicists have sought to differentiate the changing of a patient's mind by suasion from coercion and manipulation (Dubov, 2015). However, the distinction attempted may be a distinction only in appearance. Rather, in terms of the patient, suasion may be a continuum and in continuums, only the extremes are recognizable. Hard paternalism, clearly coercive, may be at one extreme, soft paternalism at the other, with paternalism by disinformation in the middle. The latter two would qualify as manipulation but is not the middle position manipulation? Some have argued that suasion is not deception, suggesting that suasion is the "art of making the truth apparent" (Dubov, 2015) but such an approach begs the question, naively, that there is a truth out there that can be rendered apparent for the patient and by implication independent of the patient. The presumption makes suasion an epistemic tool to "discover" some ontological truth. It would be a relief if such was possible but so far it does not seem to be the case.

If the intent is to make the truth apparent in order to change the patient's mind, with truth equated with certainty, it would seem that Aristotle's *logos* (logic) would be the most optimal. Yet, the notion of a "deductive" ethics is very problematic (Chapter 1). Even if a deductive logic, as an epistemic tool, were possible, it is not clear how effective it would be. Immanuel Kant would have been skeptical of the role of (deductive) reason in ethical behavior, having wrote:

> In the physical constitution of an organized being, that is, a being adapted suitably to the purposes of life, we assume it as a fundamental principle that no organ for any purpose will be found but what is also the fittest and best adapted for that purpose. Now in a being which has reason and a will, if the proper object of nature were its *conservation*, its *welfare*, in a word, its *happiness*, then nature would have hit upon a very bad arrangement in selecting the reason of the creature to carry out this purpose. For all the actions which the creature has to perform with a view to this purpose, and the whole rule of its conduct, would be far more surely prescribed to it by instinct [author—taken here as that amendable to *pathos* and *ethos* rather than *logos*], and that end would have been attained thereby much more certainly than it ever can be by reason [author—taken here as that amendable by *logos*]. Should reason have been communicated to this favored creature over and above, it must only have served it to contemplate the happy constitution of its nature, to admire it, to congratulate itself thereon [author—taken here as examples of post-hoc declarative commentary on the consequence of procedural knowledge, see Chapter 1], and to feel thankful for it to the beneficent cause, but not that it should subject its desires to that weak and delusive guidance, and meddle bunglingly with the purpose of nature. In a word, nature would have taken care that reason should not break forth into *practical* exercise [author—taken as procedural knowledge],

nor have the presumption, with its weak insight, to think out for itself the plan of happiness, and of the means of attaining it. Nature would not only have taken on herself the choice of the ends, but also of the means, and with wise foresight would have entrusted both to instinct [italics in the original].

Kant (2001, p. 153)

Indeed, there may only be "instinct" and what is called reason is a post hoc commentary in an attempt to explain (Chapter 1). In any event, reason, as in *logos* or logic, may be impotent in changing the patient's mind. Aristotle noted that persons cannot change their minds unless they are predisposed to do so. Note that the predisposition cannot be just some prior logic as then the concept would be circular as that logic would require a prior logic ending only in instinct. This leaves only *pathos*, emotion, and *ethos*, authority, as the motivators of change.

Aristotle went further to suggest the impotence of *logos*, logic, to motivate change, for example, a change in the patient's position. He wrote "The contemplative intellect contemplates nothing practicable, and says nothing about what is to be pursued or avoided. But even when it contemplates something of the kind, it does not straight away command pursuit or avoidance, e.g., it often thinks of something fearful or pleasant, but it does not command fear, although the heart is moved, or, if the object is pleasant, some other part" (De Animia, 432b27–433a1). Hume shared a similar (although not exactly the same) position, writing in his "Treatise on Human Nature,"

It would be tedious to repeat all the arguments, by which I have proved [Book II. Part III. Sect 3.], that reason is perfectly inert, and can never either prevent or produce any action or affection, it will be easy to recollect what has been said upon that subject. I shall only recall on this occasion one of these arguments, which I shall endeavour to render still more conclusive, and more applicable to the present subject.

Reason is the discovery of truth or falsehood. Truth or falsehood consists in an agreement or disagreement either to the real relations of ideas, or to real existence and matter of fact. Whatever, therefore, is not susceptible of this agreement or disagreement, is incapable of being true or false, and can never be an object of our reason. Now it is evident our passions, volitions, and actions, are not susceptible of any such agreement or disagreement; being original facts and realities, complete in themselves, and implying no reference to other passions, volitions, and actions. It is impossible, therefore, they can be pronounced either true or false, and be either contrary or conformable to reason.

(https://ebooks.adelaide.edu.au/h/hume/david/treatise-of-human-nature/B3.1.1.html)

What then of dispassionate logical discussions for informed consent with patients? Does the question even make sense if the patients are not swayed by logic, ultimately? Aristotle was not so pessimistic and did believe that reason could at least influence actions. Reason may not be the engine that propels the patient but it may determine or, at least, influence the direction the patient takes. Yet, reason alone is not sufficient.

When value is set in the context of *pathos* (emotion) and *ethos* (the person attempting suasion), and the subsequent power of *pathos* and *ethos* to motivate, the notion of informed consent takes on a new perspective. As noted previously, the Supreme Court of Canada stipulated "*A risk* [as discussed in informed consent] *is thus material when a reasonable person in what the physician knows or should know to be the patient's position would be likely to attach significance* [author—taken here as value] *to the risk or cluster of risks in determining whether or not to undergo the proposed therapy*" [italics in the original] (https://www.cmpa-acpm.ca/en/advice-publications/hand-books/consent-a-guide-for-canadian-physicians#assault). It would appear to argue for a patient subjective standard. However, if the patient subjective standard centers on value as defined earlier, then is not the controlling variable *pathos* and perhaps *ethos*? If the common coin, between physicians and healthcare professionals in informed consent, is a matter of subjectivity and emotions and given the relatively privacy (perhaps necessarily private, according to Willard Van Orman Quine, 1960) of one's own thoughts and the difficulty of "translating" them to something understandable by others (indeterminacy of translation) and difficulties of reference (inscrutability of reference), the notion of informed consent becomes rather porous. The use of narrative, feminist, care, and phenomenological ethics is a valuable attempt at understanding this problem and could contribute to the ethics of everyday medicine (Chapter 2). Note that this is not to say that such narrative, feminist, care, or phenomenological ethics are not rigorously justifiable. Rather, the justice would derive from empirical observations of the consensus of reasonable persons blinded to their self-interest (see Chapter 3).

Metaphor

Some ethicists have argued for the use of analogies and metaphors to assist patients' understanding of disease and treatment as necessary in giving informed consent as discussed in Case 14 (Mabeck and Olesen, 1996; Sweeney et al., 2001; Álvarez et al., 2017). Indeed, metaphor is central to biomedical research and the practice of medicine (Montgomery, 2019a, 2019b). A particularly powerful and widely used metaphor by both patients and physicians and healthcare professionals is the metaphor to the mechanical. The body is a machine, disease is a breakdown of the machine, and treatments fix what is broken (Mabeck and Olesen, 1997). Thus, the physician or healthcare professional conveying a "mechanical" explanation of the patient's disease or treatment finds a ready partner in the patient's intuitive understanding as the body as a machine and disease as a mechanical disfunction.

The problem is what if the body is not a machine, at least the kind of machine conveyed in the metaphors by the physicians and healthcare professionals? For example, physicians and healthcare professionals, due to complex and historical factors, generally conceive of Parkinson's disease as a relative deficiency of dopamine. Consequently, the natural inclination shared by both patients and physicians and healthcare professionals is to replace the dopamine, for example, by using levodopa, which is converted to dopamine upon entering the brain. The question is whether this shared metaphor becomes an obstacle to perhaps better treatments, which do not intuitively convey a notion of dopamine replacement, for example, deep brain stimulation? This is a problem of the ethics of everyday medicine because controlled trials of deep brain stimulation compared to the best medical therapy demonstrate the clinical superiority of deep brain stimulation. Yet, most experts believe that deep brain stimulation is underappreciated and therefore underutilized to the detriment of many patients. Perhaps such underutilization would be acceptable if the patients were fully informed and decided not to pursue deep brain stimulation. However, in the author's experience, one of the most common complaints after successful deep brain stimulation surgery is "why did my doctor wait so long to recommend deep brain stimulation?" To be sure, this is the question of patients who noted successful results, a very different question may be most asked by those who fail. However, the large majority of patients do well.

The metaphor "War on Cancer" is an interesting example of the positive and negative effects of metaphor. To be sure, the metaphor of a "War on Cancer" spurred the creation of the National Cancer Institute, many foundations, and research efforts. However, its negative effects on patients have become an increasing concern (Hines, 2014; Malm, 2016). Very often the patients who fought the war against disease are described as survivors (winners, champions, and other adjectives) and those who do not as victims. The roles have an implicit bias that could result in overtreating the patient, which is clearly an ethical concern. There are a number of cancers for which watchful waiting or active surveillance is argued to be the preferred management, thus obviating or delaying aggressive treatments (Brito et al., 2014; Stavrinides et al., 2017). For these cancers, the "War on Cancer" appears to be more of a truce or a mutual nonaggression pact. Yet, how willing is a physician, healthcare professional, or patient to call a truce? Consider the issue of mammograms in women under age 40 without risk factors (Montgomery, 2019a) and prostate-specific antigen testing in older men. The war slogan of "shock and awe" or the use of overwhelming force to achieve decisive victory early may not be the best metaphor for some of these cancers. Clearly, the use of such metaphors by physicians and healthcare professionals carries ethical risks in the practice of everyday medicine.

Metaphors can be perceived as a means of incorporating value of a deductive-like approach. The structure of the metaphor appears to provide the logic, while the content of the metaphor contains the value, particularly as an emotive driving force. A metaphor in everyday medicine may be of the form *disease or condition A is to treatment B* (proposed but unfamiliar and not yet accepted by the patient) *as disease C is to treatment D* (familiar to and accepted by the patient). The metaphor contains the target domain, that is, *disease A* and its *treatment B*, and the source domain *disease C* and its *treatment D*. An understanding of *disease A* and *treatment B* is borrowed from the prior understanding of *disease C* and *treatment D*. If successful, the patient accessed with *disease A* and accepts *treatment B* based on the patient's understanding of *disease C* and its *treatment D*. The metaphor can be restated as an argument in propositional logic, which would be *if a (disease A) implies b (treatment B) is true and c (disease C) implies d (treatment D) is true and a (disease A) implies c (disease C) is true, then d (treatment D) implies b (treatment B) is true*. If one accepts that *a (disease A) implies c (disease C)*, which would be the case if *a (disease A)* and *c (disease C)* were the same, for all intents and purposes, then *a = c*, then the argument can be restated as *if a implies b is true and a implies d is true, then d implies b is true*. However, this is the Fallacy

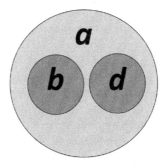

Fig. C5.1 Set theory representation of the Fallacy of Pseudotransitivity, which is of the form *if a implies b is true and a implies d is true, then d implies b is true.* Set *b* is wholly contained within set *a*; consequently, anything true of set *a* will be true of set *b*. Similarly, set *d* is wholly contained within set *a* and therefore everything that is true of set *a* will be true of set *d*. However, set *b* is not within set *d*; consequently, what may be true of set *d* is not necessarily true of set *b* and the argument is invalid and the metaphor fails.

of Pseudotransitivity and can be appreciated by Set theory as shown in Fig. C5.1. As can be seen, set *b* is wholly contained within set *a*; consequently, anything true of set *a* will be true of set *b*. Similarly, set *d* is wholly contained within set *a* and therefore everything that is true of set *a* will be true of set *d*. However, set *b* is not within set *d*; consequently, what may be true of set *d* is not necessarily true of set *b* and the argument is invalid and the metaphor fails.

The metaphor may be further understood as a syllogism; an example of a valid syllogism follows:

Argument 5.1

Major premise: *All humans* (bridging term) *are mortal* (major term)
Minor premise: *Socrates* (minor term) *is a human* (bridging term)
Conclusion: *Socrates* (minor term) *is mortal* (bridging term)

The set of humans is wholly contained in the set of mortal things. Everything that is true of mortal things is true of humans (in the domain of longevity). The set Socrates is wholly contained in the set humans, which, necessarily, is contained in the set of mortal things. Therefore, it is necessarily true that Socrates is mortal. The metaphor of the aforementioned discussion can be restructured to a syllogism as shown here in Argument 5.2:

Argument 5.2

Major premise: *Cancer, as the enemy* (bridging term), *must be battled* (major term)
Minor premise: *Patient A* (minor term) *has cancer, as the enemy* (bridging term)
Conclusion: *Patient A cancer* (minor term) *must be battled* (bridging term)

Argument 5.2 is valid and its conclusions certain if the bridging term, cancer, in the major premise is the same as the bridging term in the minor premise. If the bridging terms in the two premises are not exactly the same, then the argument is invalid

(Fallacy of Four Terms, a major, a minor, and two versions of the bridging term) and the truth or falsehood of the conclusion cannot be determined from the argument alone. As can be seen, low-grade prostate cancer is not the same as anaplastic small cell carcinoma. Yet, such an argument, metaphor, may be made by the physician or healthcare professional to the patient. As the argument is invalid, its use does not ensure informed consent, and it must be taken that informed consent was not obtained.

The certainty of Argument 5.2, and hence, its utility in obtaining informed consent, depends on how close are the two bridging terms in the major and minor premises? This question has far-ranging implications. For example, the Fallacy of Four Terms figures importantly in the crisis of irreproducibility in biomedical research (Montgomery, 2019b). Thus, the question becomes how does one assess the ethical "reproducibility" of the arguments made to patients (Chapter 3)? This is particularly difficult when the two terms are not quantifiable or differ in their dimensions (having to compare apples and oranges).

One approach is to consider the Epistemic Risk, which is made up of the Epistemic Distance between the two forms of the bridging term and the Epistemic Degrees of Freedom, for example, how many conceptual twists and turns are necessary to get from one version to the other version of the bridging term? For example, rodent models of human diseases are often used to predict whether a particular treatment will be effective in humans. The argument comes down to the presumption that what is true of the rodent having received the treatment will be the same as for humans who would receive the treatment. This depends on how similar rodents are to humans relative to the disorder being treated. The metabolism of glucose in rodents may be very similar to humans; thus, any differences (Epistemic Distance) between rodents and humans relative to glucose metabolism may be relatively small and the Epistemic Risk of the Fallacy of Four Terms in using the animal model likewise small. However, cognition in rodents may be of a very different kind than humans (Epistemic Degrees of Freedom); thus, predicting the effects of treatment on cognition in humans may be very difficult and the Epistemic Risk of the Fallacy of Four Terms may be very high (Montgomery, 2019b). Similarly, if an ethical principle, moral theory, law, or precedent derived from Caucasian males in the source domain for the metaphor by which to make inferences that involve African-American females, then the Epistemic Risk will be quite high.

Physicians and healthcare professionals should exercise wisdom in their use of metaphor if, for no other reason, it makes a difference to patients. In the case presented, the neurologist's argument would seem to be that shown here:

Argument 5.3

Major premise: Patients should not be exposed to drugs (major term) that have the most adverse effects (bridging term).
Minor premise: Dopamine agonists (minor term) have the most adverse effects (bridging term).
Conclusion: Patients should not be exposed (major term) to dopamine agonists (major term).

This argument is at risk for the Fallacy of Four Terms. The bridging term in the major premise, *the most adverse effects*, may not be the same as *the most adverse effects* in the minor premise. *Adverse effects* in the major premise make no distinction with regards to the specific treatment, it is a general statement. However, the notion of *adverse effects* in the minor premise is specific to dopamine agonists and may not be the same as the general notion of *adverse effects* in the major premise. In this particular case, if one considers adverse effects in the near term, then dopamine agonists have a higher rate than levodopa. However, long-term adverse effects are fewer with dopamine agonists. Thus, there are two types of adverse effects resulting in two versions of the bridging term and four terms resulting in a fallacious argument.

Physicians and healthcare professionals can also engage in paternalism by disinformation by allowing the patient or patient's surrogate to operate under false or counterproductive metaphorical reasoning, such as what follows:

Argument 5.4

Major premise: *Craniotomy surgery* (bridging term) *is neurosurgery* (major term)
Minor premise 1: *Craniotomy surgery* (bridging term) *is very dangerous* (minor term)
Conclusion 1: *Neurosurgery* (major term) *is very dangerous* (major term)
Minor premise 2: *Deep brain stimulation* (minor term 2) *is neurosurgery* (now bridging term)
Conclusion 2: *Deep brain stimulation* (minor term) *is very dangerous* (major term)

The extended argument is the Fallacy for Four Terms; specifically, there are two versions of *neurosurgery*, which in the major premise is different than the term in minor premise 2. Deep brain stimulation requires placing a burr hole in the skull approximately 14 mm in diameter and inserting a series of tubes on the order of 1.5 mm in diameter through the brain. This is very different than a craniotomy, which requires the removal of relatively much larger pieces of the skull typically as a preliminary to more invasive manipulations of the brain. This is not to say that deep brain stimulation does not have risks; however, the risks are rare, as surgeries go, although the consequence can be severe. However, to equate deep brain stimulation with a craniotomy is invalid and any suasion based on the invalid argument is not informed consent (Montgomery, 2015).

Can a patient demand a treatment?

It appears to be a relatively settled ethical principle or judgment that patients or their surrogates have the right to refuse any treatment. This is a negative right. There also is the notion of a positive right, the right of a patient to a treatment desired by the patient even if it does not comport with the position of the physician or healthcare professional. Also, there is a growing consensus that physicians can refuse patients' requests for medical treatments deemed futile, even given the problematic issue of defining futile (Drane and Coulehan, 1993; Müller and Kaiser, 2018; Pope, 2018). However, there may be the temptation to hold anything that a patient requests, which the physician or healthcare professional disagrees with, as medically futile; otherwise why would the physician or healthcare professional refuse the patient's request? To be sure, there are degrees of futility. It is possible that the neurologist in this case felt that the notion of a superior benefit from dopamine agonists over what could be achieved with levodopa was the equivalent of futile. However, such a presumption risks abuse given the asymmetry in power between the patient and the physician or healthcare professional. Where are the checks and balances?

On this subject, Brett and McCullough (1986) wrote:

> Stated another way, the foundation of the clinical encounter is a *specified body of knowledge and expertise about what is beneficial for patients* [italics added]. When a patient seeks to exercise a positive right to an intervention, a necessary condition is that there is either an established or a theoretical medical basis for the patient's request. If that necessary condition has been satisfied, the patient's unique circumstances and stated reasons for wanting the intervention should guide the final decision-making process.

The critical issue is who establishes the "specified body of knowledge and expertise about what is beneficial for patients?" If this is solely at the discretion of the physician or healthcare professional, there is a significant risk of paternalism, which, in addition to violation of the obligation to autonomy, may possibly violate the obligation to beneficence (see Chapter 4). Perhaps more importantly, the insurer or government may exercise a preemptive shareholder position in what is beneficial for patients. The admonition that the "*physician knows or should know to be the patient's position would be likely to attach significance to the risk or cluster of risks in determining whether or not to undergo the proposed therapy*" [italics in the original] (https://www. cmpa-acpm.ca/en/advice-publications/handbooks/consent-a-guide-for-canadian-physicians#assault) discussed previously was in the context of asserting a negative right. It is not clear whether the same admonition would apply to the patient asserting a positive right.

To be sure, there are a great many persuasive arguments that physicians and healthcare professionals should not simply defer to the patient. Indeed, such deferral may constitute a violation of obligations to the ethical principles afforded to other stakeholders and shareholders. Moderate scarcity requires some form of prioritization. However, this is a very difficult topic, as prioritization runs counter to the Egalitarian and Deontological moral theories and could be counter to a Utilitarian moral theory as well. However, attempts of prioritization based on a Libertarian or contractual moral theory only work to the degree that the other moral theories are superseded. In any event, it would seem that labeling a patient's request as without medical value and therefore futile is an easy out but not necessarily just.

Patients may not be able to demand unapproved treatments, for example, the situation with laetrile for cancer treatments. Advocates argued to force the US Food and Drug Administration (FDA) to not interfere with the interstate distribution of laetrile but were denied by the US Supreme Court, *United States v. Rutherford, 442 U.S. 544* (1979). Further, a physician or healthcare professional may refuse to provide a legally available treatment for which the physician or healthcare professional sees as a moral objection; however, in most cases, that physician or healthcare professional is expected to transfer the patient's care to someone else without those objections.

None of these circumstances pertain in this case. The dopamine agonists were approved by the US FDA and the neurologist did not hold that dopamine agonists were futile or immoral. Rather, the neurologist believed that the relative advantages and disadvantages favored the use of levodopa.

The question is what recourse was available to the patient? The patient could see another physician, but this presupposes the patient knew about the options, which the patient in the case presented did not. Even if the patient knew about dopamine agonists, it is not clear that the patient's outcome would be different with another neurologist. Perhaps the patient could enlist the help of an ombudsman or healthcare advocate through the healthcare provider organization, insurer, or governmental organization. Paid healthcare advocates or case managers are increasingly being retained by patients. The necessity of patients having to expend their own funds for such advocacy raises serious ethical questions of whether the various systems are failing the patients. These systems may have contractual obligations that should make outside paid advocacy unnecessary. Although unlikely to succeed, the question arises whether a legal complaint alleging the neurologist failing to provide the standard care could be considered?

Paternalism by lack of alternatives

Traditionally and typically, unresolvable disputes between what the patient wants and what the physician or healthcare professional is willing to provide involved transfer of care to those physicians or healthcare professionals who would not object to the patient's request. This presumes that the patient's requests are not illegal or violations of supervening policies, rules, and contracts. But what if every physician and healthcare professional reasonably available to the patient, for example a geographic area, share in the same refusal? Such a situation cannot be taken as vouchsafing the position of the physicians and healthcare professionals. For example, consolidation of healthcare provider systems leaving only religious affiliated or governed in certain areas could nullify legal, contractually valid, and otherwise obtainable reproductive health rights.

Another example is healthcare provider systems where patients are required to use the resources available only through the insurer. These systems may lack the options that the patient seeks even if the physician or healthcare professional would be willing to prescribe them. These systems are sometimes referred to as narrow networks. In a survey of medical plans in 2015 through the marketplace under the Patient Protection and Affordable Care Act (ACA), 15% lacked physicians in at least one specialty. The survey also demonstrated that 26% did not cover out-of-network services, while 78% of the remaining required at least 50% cost sharing with the patient for out-of-network care (Dorner et al., 2015). Patients seeking to have their requests are forced to go out of the system, potentially with significant financial consequences.

There is the temptation to see the use of an out-of-network provider or service as a single event to meet a single need, for example, a knee replacement. However, it would be exceptionally rare for any medical care not to have some risk of a complication that would require further care. The question then becomes whether the financial responsibility falls back onto the network, the out-of-network system, or the patient? The author was unable to find relevant studies in the literature. However, one can look to other precedents. In the United States, there are provisions in Medicare and under the ACA that require insurers to cover general medical costs associated with participation in some clinical trials (Martin et al., 2014). For this precedent to be informative, the clinical trials would have to be viewed sufficiently similar to out-of-network care so that the patient's network would be responsible for all follow-up care. The polices of Medicare and ACA were intended to spur participation in clinical trials. It is unlikely that this sentiment is prevalent among network providers as it would be seen as abetting out-of-network services. Another analogy is medical tourism and whether home insurers have a responsibility to cover medical complications for treatments conducted in other countries (Snyder et al., 2011; Crooks et al., 2013). Given the magnitude of medical tourism and the paucity of studies regarding the impact on aftercare provided at home, it is unclear whether that lack of discussion is good or bad news. It would appear that out-of-network care and medical tourism may not be a panacea for paternalism by limited alternatives.

References

Alpers, R.L., 2004. *Duncan v*. Scottsdale Medical Imaging, Ltd.: Restoring a patient's right to sue for battery. Ariz. L. Rev. 46, 843.

Álvarez, I., Selva, L., Medina, J.L., Sáez, S., 2017. Using root metaphors to analyze communication between nurses and patients: a qualitative study. BMC Med. Educ. 17 (1), 216. https://doi.org/10.1186/s12909-017-1059-0.

Beauchamp, T.L., Childress, J.F., 2001. Principles of Biomedical Ethics, fifth ed. Oxford University Press, Oxford.

Beauchamp, T.L., Childress, J.F., 2013. Principles of Biomedical Ethics, seventh ed. 228 Oxford University Press, Oxford, pp. 220–223.

Blumenthal-Barby, J.S., Krieger, H., 2015. Cognitive biases and heuristics in medical decision making: a critical review using a systematic search strategy. Med. Decis. Mak. 35 (4), 539–557. https://doi.org/10.1177/0272989X14547740.

Braddock 3rd, C.H., Edwards, K.A., Hasenberg, N.M., Laidley, T.L., Levinson, W., 1999. Informed decision making in outpatient practice: time to get back to basics. JAMA 282 (24), 2313–2320.

Brett, A.S., McCullough, L.B., 1986. When patients request specific interventions: defining the limits of the physician's obligation. N. Engl. J. Med. 315 (21), 1347–1351.

Brito, J.P., Hay, I.D., Morris, J.C., 2014. Low risk papillary thyroid cancer. BMJ 348, 3045. https://doi.org/10.1136/bmj.g3045.

Callon, W., Saha, S., Wilson, I.B., Laws, M.B., Massa, M., Korthuis, P.T., Sharp, V., et al., 2017. How does decision complexity affect shared decision making? An analysis of patient-provider antiretroviral initiation dialogue. Patient Educ. Couns. 100 (5), 919–926. https://doi.org/10.1016/j.pec.2016.12.013.

Cocanour, C.S., 2017. Informed consent: it's more than a signature on a piece of paper. Am. J. Surg. 214 (6), 993–997. https://doi.org/10.1016/j.amjsurg.2017.09.015.

Crooks, V.A., Turner, L., Cohen, I.G., Bristeir, J., Snyder, J., Casey, V., Whitmore, R., 2013. Ethical and legal implications of the risks of medical tourism for patients: a qualitative study of Canadian health and safety representatives' perspectives. BMJ Open 3 (2), e002302. https://doi.org/10.1136/bmjopen-2012-002302.

Dorner, S.C., Jacobs, D.B., Sommers, B.D., 2015. Adequacy of outpatient specialty care access in marketplace plans under the affordable care act. JAMA 314 (16), 1749–1750. https://doi.org/10.1001/jama.2015.9375.

Drane, J.F., Coulehan, J.L., 1993. The concept of futility. Patients do not have a right to demand medically useless treatment. Counterpoint. Health Prog. 74 (10), 28–32.

Dubov, A., 2015. Ethical persuasion: the rhetoric of communication in critical care. J. Eval. Clin. Pract. 21 (3), 496–502. https://doi.org/10.1111/jep.12356.

Fitzgerald, C., Hurst, S., 2017. Implicit bias in healthcare professionals: a systematic review. BMC Med. Ethics 18 (1), 19. https://doi.org/10.1186/s12910-017-0179-8.

Frederick, S., Loewenstein, G., O'Donoghue, T., 2002. Time discounting and time preference: a critical review. J. Econ. Lit. 40, 351–401.

Fromm, E., 1955. The Sane Society. Rinehart & Company, Inc, New York, NY.

Hines, I., 2014. The war on cancer: time for a new terminology. Lancet 383 (9932), 1883. https://doi.org/10.1016/S0140-6736(14)60907-7.

Johnson-Laird, P., 2006. How We Reason. Oxford University Press, Oxford.

Kant, I., 2001. Fundamental principles of the metaphysics of morals. In: Wood, A.W. (Ed.), Basic Writings of Kant. The Modern Library, New York.

Keller, C., Siegrist, M., 2009. Effect of risk communication formats on risk perception depending on numeracy. Med. Decis. Mak. 29 (4), 483–490. https://doi.org/10.1177/0272989X09333122.

Levin, I.P., Schneider, S.L., Gaeth, G.J., 1998. All frames are not created equal: a typology and critical analysis of framing effects. Organ. Behav. Hum. Decis. Process. 76 (2), 149–188.

Mabeck, C.E., Olesen, F., 1996. Methaphors and understanding of diseases. Ugeskr. Laeger 158 (51), 7384–7387.

Mabeck, C.E., Olesen, F., 1997. Metaphorically transmitted diseases. How do patients embody medical explanations? Fam. Pract. 14 (4), 271–278.

Malm, H., 2016. Military metaphors and their contribution to the problems of overdiagnosis and overtreatment in the "war" against cancer. Am. J. Bioeth. 16 (10), 19–21. https://doi.org/10.1080/15265161.2016.1214331.

Martin, P.J., Davenport-Ennis, N., Petrelli, N.J., Stewart, F.M., Appelbaum, F.R., Benson 3rd, A., 2014. Responsibility for costs associated with clinical trials. J. Clin. Oncol. 32 (30), 3357–3359. https://doi.org/10.1200/JCO.2014.57.1422.

Montgomery, E.B., 2004. Slowing Parkinson's disease progression: recent dopamine agonist trials. Neurology 62 (2), 343 (Author reply 343–345).

Montgomery Jr., E.B., 2015. Twenty Things to Know about Deep Brain Stimulation. Oxford University Press, Oxford.

Montgomery Jr., E.B., 2019a. Medical Reasoning: The Nature and Use of Medical Knowledge. Oxford University Press, Oxford.

Montgomery Jr., E.B., 2019b. Reproducibility in Biomedical Research: Epistemological and Statistical Problems. Academic Press, New York.

Müller, R., Kaiser, S., 2018. Perceptions of medical futility in clinical practice. A qualitative systematic review. J. Crit. Care 48, 78–84. https://doi.org/10.1016/j.jcrc.2018.08.008.

Orlowski, J.P., Hein, S., Christensen, J.A., Meinke, R., Sincich, T., 2006. Why doctors use or do not use ethics consultation. J. Med. Ethics 32 (9), 499–502.

Parkinson Study Group, 2000. Pramipexole vs levodopa as initial treatment for Parkinson disease: a randomized controlled trial. JAMA 284 (15), 1931–1938.

Pope, T.M., 2018. Medical futility and potentially inappropriate treatment: better ethics with more precise definitions and language. Perspect. Biol. Med. 60 (3), 423–427.

Quine, W.V.O., 1960. Word and Object. MIT Press.

Rascol, O., Brooks, D.J., Korczyn, A.D., De Deyn, P.P., Clarke, C.E., Lang, A.E., 2000. A five-year study of the incidence of dyskinesia in patients with early Parkinson's disease who were treated with ropinirole or levodopa. N. Engl. J. Med. 342 (20), 1484–1491.

Shaw, D., Elger, B., 2013. Evidence-based persuasion: an ethical imperative. JAMA 309 (16), 1689–1690. https://doi.org/10.1001/jama.2013.2179.

Snyder, J., Crooks, V.A., Johnston, R., Kingsbury, P., 2011. What do we know about Canadian involvement in medical tourism? A scoping review. Open Med. 5 (3), e139–e148.

Stavrinides, V., Parker, C.C., Moore, C.M., 2017. When no treatment is the best treatment: active surveillance strategies for low risk prostate cancers. Cancer Treat. Rev. 58, 14–21. https://doi.org/10.1016/j.ctrv.2017.05.004.

Sweeney, K.G., Edwards, K., Stead, J., Halpin, D., 2001. A comparison of professionals' and patients' understanding of asthma: evidence of emerging dualities? Med. Humanit. 27 (1), 20–25. https://doi.org/10.1136/mh.27.1.20.

Case 6

Case 6—A 78-year-old female with a long history of multiple medical problems complained of excessive shortness of breath (dyspnea) after modest exertion and fatigue. The patient e-mailed her physician regarding the recommendation of an angiogram (an X-ray procedure where a "dye" is injected into the heart and shows on the X-ray). The patient was concerned that the angiogram might be abnormal but may not be the cause of the shortness of breath and fatigue. The cardiologist stated that he was very busy and suggested e-mailing specific questions. The cardiologist went on to write that his responsibility was to review the cardiovascular issues, which he did, implying that he was not responsible for other noncardiac medical issues, as this was the standard of the discipline. He went on to say that it was the patient's decision regarding proceeding with the angiogram and just to let him know.[a]

The medical facts of the matter

The patient first sought help for her shortness of breath (dyspnea) and fatigue from her family physician who subsequently referred the patient to a cardiologist. The cardiologist limited his investigation to what might be the causes of coronary artery disease. However, there are many causes beyond coronary artery disease that could be inducing the dyspnea and fatigue; some are shown in Table C6.1 that follows. Indeed, coronary artery disease accounts for only less than 20% of the causes of shortness of breath and fatigue. Thus, the coronary arteriogram likely would not have demonstrated the most likely causes, and even if the angiogram was abnormal, there would be no way that the cardiologist could know that the abnormal blood flow through the heart was the cause. Confounding matters is the fact that just by virtue of the patient's age, the angiogram likely would be abnormal even if the blood flow through the heart had nothing to do with the patient's shortness of breath. Further, there were many cases where the patient had multiple reasons for dyspnea.

Most importantly, more than half of the patients had more than one condition for the causes of shortness of breath and the patient in the case presented had multiple poossible causes. Thus, the demonstration of an abnormality of coronary artery disease on the cardiac angiogram could not indicate which of the disorders was causing the patient's symptoms. Even if the patient underwent invasive coronary artery angiography, the probability of detecting a clinically significant abnormality in blood flow to the heart, as detected by the invasive fractional flow reserve, is 0.71% or 71%—the sensitivity of the invasive angiogram. However, the probability of a false-negative

---

[a] The case presented is hypothetical though modeled on experience. No judgment is made relative to the correctness, appropriateness or value of any actions described on the part of any of the participants. The purpose is to describe and explore the variety of ethical considerations and consequences that depend on the perspective of different relevant ethical systems.

The Ethics of Everyday Medicine. https://doi.org/10.1016/B978-0-12-822829-6.00006-0

Table C6.1   Patient characteristics from 247 dyspnea patient records reviewed by 4 expert internists.

Characteristic		
Total male patients (%)	115	(46.6)
Average age (SD)	69.9	(14.9)
Final diagnosis	**No.**	**(%)**
Heart failure	51	(20.6)
Chronic obstructive pulmonary disease/ bronchitis/asthma	59	(23.8)
Pneumonia	47	(19.0)
Malignancy	16	(6.5)
Pulmonary embolism	14	(5.7)
No diagnosis	3	(1.2)
Other (e.g., meningitis, cholecystitis, anemia, septic shock, pneumothorax, viral infection, hyperventilation, hepatorenal syndrome, bronchiolitis, pancreatitis)	57	(23.1)
Relevant comorbidity[a]	116	(51)
Heart failure	18	(15.5)
Atrial fibrillation	15	(12.9)
Pneumonia	13	(11.2)
Chronic obstructive pulmonary disease/ bronchitis/asthma	20	(17.3)
Renal insufficiency	11	(9.5)
Ischemic heart disease without heart failure	11	(9.5)
Malignancy	8	(6.9)
Hypoglycemia	9	(7.8)
Anemia	5	(4.3)
Other (e.g., aorta valve stenosis, meningitis, septic shock, viral infection, lung fibrosis, urinary tract infection, liver failure, HIV, hypertension, pulmonary, embolism, thrombosis)	74	(63.8)

[a] There could be more than one diagnosis per patient.
From Zwaan, L., Thijs, A., Wagner, C., van der Wal, G., Timmermans, D.R., 2012. Relating faults in diagnostic reasoning with diagnostic errors and patient harm. Acad. Med. 87(2), 149–156. https://doi.org/10.1097/ACM.0b013e31823f71e6.

angiogram is 0.34% or 34% (Danad et al., 2017). Assuming that 20.6 out of 100 patients with shortness of breath have heart disease (from Table C6.1), 14.6 persons would be identified by the coronary angiogram (assuming that the heart disease is due to coronary artery disease) whereas 6 would be failed to be identified. However, in 79.4 patients with shortness of breath not due to coronary artery disease, 26.9 will have a false positive test on the coronary artery angiogram. Thus, it is more likely that the patient would be diagnosed falsely based on the angiogram, that is not diagnosed accurately. A false-positive diagnosis of clinically significant coronary artery disease could lead to inappropriate treatment, placing the patient at the risk of harm—violating the obligation to the ethical principle of nonmaleficence. Note that this is not to say that coronary angiography is a poor test, but if shortness of breath is the only problem, an angiogram may not be a good test.

A coronary artery angiogram in the patient presented may not be a good test because the prior probability of heart disease causing the patient's symptoms is so low. However, the prior probability could be increased, thereby increasing the diagnostic value of the coronary angiogram, if alternative causes were evaluated and ruled out. However, this would require the physician to consider other noncardiac causes thoroughly, which, apparently, this cardiologist was not inclined to do. Whether the actions in the case presented are unethical, and therefore unjust, they are addressed subsequently and depend greatly on the supervening moral theory. It just may be that the cardiologist acted ethically and justly. If, however, the conduct was unethical and unjust, then where does the responsibility lie? It may not be with the cardiologist, as will be discussed later, and thus, the cardiologist personally may be acting within the ethical bounds of his discipline.

What is being asked of the physician and what is being given in response?

The patient clearly is concerned about the diagnostic utility of the cardiac angiogram as she recognizes that she has many medical problems that could be the cause of her shortness of breath and fatigue. The patient seems uncomfortable with the presumption that the cause relates to her coronary arteries. The patient is concerned about a "false" positive that could lead to inappropriate treatments, some of which are invasive and of relatively high risk. Indeed, the risk of a false positive is substantial as described previously. Essentially, the patient consulted the cardiologist for concerns about the shortness of breath and fatigue and it is unlikely that the patient consulted the cardiologist solely about the status of her coronary arteries. What the physician offered was an assessment of the patient's coronary arteries.

To be sure, the patient expects, and the cardiologist seeks to fulfill the obligation to beneficence, which is an improvement in the patient's shortness of breath and fatigue. The questions become what is the nature of the obligation to providing beneficence, what does the patient expect, and what does the physician feel obligated to provide? The physician clearly limits his obligation to beneficence as it relates solely to the coronary arteries, with improvement in the shortness of breath and fatigue as a corollary benefit. Indeed, the cardiologist holds that he has no obligation to look at anything other than the coronary arteries. Clearly, there is a conflict in what the patient expects, what the cardiologist feels obliged to do, and what the cardiologist's sense of his obligations to beneficence is—it is not that the ethical principle of beneficence does not apply. Further, differences in the expectations of beneficence do not defeat the principle of beneficence.

To be sure, other ethical systems are relevant here (see Chapter 2). The patient's expectations may be based on the patient's sense of herself, as would be informed by phenomenological ethics. Further, the patient's sense of herself is embedded in the narrative of her life and her expectations of herself and the physician may be best understood in terms of narrative ethics. Finally, the asymmetry of power between the wants of the patient and the offers by the physician can be fleshed out by feminist ethics. Finally, it just may be that a virtuous physician would want to address the wants of a patient who entrusts herself to the physician that can be illuminated by virtue ethics.

The ethical principles of beneficence and nonmaleficence

Virtually every medical decision involves the risk of violations of the ethical principle of nonmaleficence. In the case of a cardiac angiogram, the risk of a major complication is on the order of 2% of patients undergoing the procedure and the risk of death is on the order of 0.08% (Tavakol et al., 2012). However, the question becomes whether the risks of the diagnostic can be dissociated from the consequences of a completed safe angiogram. For example, if the successfully completed angiogram demonstrates abnormal blood vessels that are falsely thought to be the cause of the patient's symptoms (false positive), then the risks of subsequent treatment must be considered. If the patient goes on to have coronary artery bypass graft surgery, the risk of major complications, for example, the risk of myocardial infarction (heart attack), is 3.4% (Kunt et al., 2005) and the risk of a stroke is 1% (Bakaeen et al., 2014). Note that it is reasonable to include the risks of subsequent treatment, such as coronary artery bypass graft surgery, as the patient likely would have already agreed implicitly or explicitly to undergo treatment if the angiogram suggested a treatable condition; otherwise, why would a reasonable patient undergo the angiogram?

There is considerable precedent for considering the risks of what follows from the diagnostic test. For example, women under the age of 40 without risk factors that increase the chance for breast cancer are discouraged from having mammograms. It is not because of the risk of having the mammogram as this is relatively minor. Rather, the risk of having the mammogram is the potential consequence of having a false-positive mammogram, such as invasive biopsies. Clearly, the recommendation regarding undergoing the diagnostic test, in this case, a mammogram, considers the consequences from actions taken after the test, not just the risks of the mammogram. Similar positions have been taken regarding testing for prostate-specific antigen in older males.

For the sake of discussion, it is likely that the cardiologist fully appreciated the risks associated with the coronary angiogram and conveyed that information to the patient. However, it cannot be assumed that the cardiologist also included the risks of possible subsequent treatments, which could be considered necessary for the patient to give fully informed consent for the angiogram, depending on the moral theory that prevails in this case. In obtaining informed consent, either implicitly or explicitly, the countervailing risks should not be restricted only to those associated with the test offered, at least according to the precedents described earlier. There are further concerns as the risks of a false positive extend beyond subsequent cardiac treatments. There may be other causes of the patient's complaints of shortness of breath and fatigue that may not be explored with the convenient finding of coronary artery disease that is irrelevant to the patient's symptoms. The situation was described beautifully by Leo Tolstoy, who wrote:

> The combination of causes of phenomena is beyond the grasp of the human intellect. But the impulse to seek causes is innate in the soul of man. And the human intellect, with no inkling of the immense variety and complexity of circumstances conditioning a phenomenon, any one of which may be separately conceived of as the cause of it, snatches at the first and most easily understood approximation, and says here is the cause.
>
> *Tolstoy (1998)*

The aforementioned problem is an example of the Fallacy of Affirming a Disjunctive, which is of the form *if (a inclusive or b inclusive or c) implies d is true and d is true, then (a inclusive or b inclusive or c) is true and if a is true, then b or c is false*. In this case, *a* is coronary artery disease as demonstrated by the angiogram, *b* is anemia, *c* is a pulmonary disease, and *d* is the symptoms of shortness of breath and fatigue. There is nothing about the truth or falsehood of *a* that affects the truth or falsehood of *b* or *c* (at least with the usual sense of the use of "or" as in the "inclusive" rather than "exclusive or"). Even if one did not even consider *b* or *c*, the argument is invalid as it reduces to *if a implies d is true and d is true, then a is true* or, in the context of this case, *if coronary artery disease implies shortness of breath and fatigue, the patient has shortness of breath and fatigue, then the patient has coronary artery disease*, which is the Fallacy of Confirming the Consequence (Montgomery, 2019a). This reasoning is invalid as there may be any number of reasons *d* is true even if *a* is false; in the context of this case, there may be any number of reasons why the patient has shortness of breath and fatigue other than coronary artery disease.

Obligation to the ethical principle of autonomy

The cardiologist believes that he has fulfilled his obligation to autonomy to the patient, stating that it was up to the patient to decide whether to undergo a cardiac angiography. It is a reasonable question to ask whether the response by the cardiologist is a good faith effort to meet his obligation to the patient's autonomy. While perhaps not explicitly intended, the effect could have been to short circuit further discussion. Whether intended or not, the question is whether a reasonable patient would have reacted to the cardiologist's response such that the patient would feel discouraged from pursuing her concerns other than to acquiesce to the cardiologist's recommendation. If the latter was the case, then should not the cardiologist have been aware of the potential effect and at least be concerned that it could be a violation of the obligation to autonomy toward the patient, not to mention a failure to obtain informed consent?

An Egalitarian moral theory would argue that there is not a failure in the obligation if the patient is treated the same as every other patient; in this case, every patient is limited to the physician's sense of obligation. But it is not clear that this is the optimal solution to the question of obligation. It is unlikely that society would tolerate the risk of such failures for every patient. A Utilitarian moral theory would ask what is the cost of fully complying with the obligation for every patient relative to the benefits of just complying with the cardiologist's recommendations? It could be that every physician fully complying with the obligation to autonomy, for example, addressing each patient's concern exhaustively, would exhaust the resources and bankrupt the system (see Case 2). A Deontological moral theory would say that there is no debate, the cardiologist "just has to" fully fulfill the obligation to autonomy no matter the consequences. Similarly, virtue ethics would suggest that a virtuous physician would fully address the patient's concerns. As usual, the problem with Deontological moral theories, including this version of virtue ethics, arises in how to ensure compliance, particularly in the situation or circumstance of the moderate scarcity of resources.

The Libertarian moral theory would argue that there is no obligation; however, to avoid anarchy, the parties come to some agreement that is acceptable to all parties that likely contains enforcement provisions. Note that ultimately all moral theories depend on how they will be enforced, which, in modern liberal democracies, typically involves a contract, implicit, or explicit, and whether a form of contract is established by a statue or common law precedence. The resolution to the issue of obligation to autonomy can be considered in the light of a contract (Contractualism). The elements of a contract include (1) an exchange of agreed-upon goods, (2) each party is free to enter into the contract, (3) each party is capable of entering into a contract, (4) there is a good faith effort to fulfill the terms of the contract, and (5) lawful purpose.

The case presented can be analyzed in terms of a contract. It cannot be assumed that each party entered the agreement freely. Scarcity of goods and those who provide them, such as cardiologists, may constrain options. For example, within reasonable logistics, the cardiologist in the case may be the only cardiologist available. Further, the asymmetry of knowledge, which could be wielded as a tool of power or coercion, and the inherent structure of the encounter place the patient at a significant disadvantage (see feminist ethics in Chapter 2). There may not be agreement on the goods to be exchanged. In this case, the patient and the family physician desire the good of assessment of the cause of the patient's shortness of breath and fatigue. The cardiologist instead offers an assessment of the patient's coronary arteries. The good received by the cardiologist is payment for his services.

Regardless of the differences in the goods expected, by the time the cardiologist started seeing the patient, an implied contract was established. To be sure, many other parties are implicitly involved in the contract, including the insurers, healthcare delivery systems, regulatory agencies, government, and society (see Case 2). Once in a contract, the expectation is a good faith effort to fulfill the contract. In the e-mail exchange, the patient sought to clarify the obligation, in the sense of a good to be received from the physician. The patient specifically addressed the concern of a false positive. The cardiologist did not address the concern and although volunteering to meet with the patient and the patient's family member, it may be seen reasonably as a half-hearted effort and possibly not in good faith.

The cardiologist's statement that he was busy and it was up to the patient to decide seems to provide relatively little room for discussion and could be taken by the patient as an ultimatum or, at the very least, a *fait accompli*. At the very least, it could be considered insensitive, violating the obligation to beneficence or for the physician to be virtuous in the sense of good of reassuring the patient. However, it cannot be assumed that such an obligation to this beneficence exists and therefore the cardiologist may not be obliged (see Case 7 and Chapter 4). Nevertheless, if the cardiologist did not address reasonable concerns on the part of the patient, then there is doubt that the patient could provide fully informed consent. Proceeding with the angiogram could be considered a technical battery (see Case 5).

Ethical responsibilities of the system

This case can be contrasted with Case 9 where the physician falsely diagnosed a patient with dystonia as being psychological. In that case, the physician did not purposely confine the range of possible diagnoses as did the cardiologist in this case. From the discussion given earlier, it would be reasonable to ask whether the actions of the cardiologist would be considered good medical practice, although not necessarily unacceptable medical practice in the sense of being unethical or unjust. Further, if situations described in this case are a rare exception, it may just not merit any significant expenditure of effort to prevent, at least from a Utilitarian moral theory. It just may be the "cost of doing business."

The key here is how rare are such circumstances? In a study of possible diagnostic errors, Zwaan et al. (2012) studied the diagnostic efforts of 72 physicians examining dyspnea (shortness of breath), as in the patient described in this case, using a chart review of 247 patients over 9 months. Suboptimal cognitive acts, perhaps a more palatable euphemism for errors of reasoning, were found in 66% of the patients. In 34% of the patients, diagnostic errors were found, most often leading to patient harm. The most significant factors in diagnostic error were occurred during the history and physical exam, where these types of errors were most likely to lead to patient harm (Zwaan et al., 2012). A failure to consider alternative diagnoses was identified as a cause of the diagnostic error.

A great concern is whether circumstances of this case are institutionalized in medical education and the delivery of healthcare. A study was conducted where a series of cases were given to a number of specialists, with each specialist receiving the same cases. When given the same clinical case, the specialists were more likely to diagnosis a condition within rather than outside their specialty (Hashem et al., 2003). In other words, many physicians given the same case made different diagnoses depending on the specialty of the physician. Perhaps the situation described in the case presented is not rare.

It is highly unlikely that the failure to consider a wider range of diagnoses within the differential diagnosis was intentional, making it more likely that the cardiologist took this as the standard of care as indicated in his response, which may be correct. Further, given the ubiquity of the problem, many other physicians and healthcare professionals also do so. Yet, professional organizations specifically caution about such a narrow prospective and the subsequently limited sense of obligation. The European Society of Cardiology Education Committee created a curriculum for the training of cardiologists in 2013. The history-taking objectives included "symptoms of any co-morbidities [disorders other than cardiovascular]... family history (cardiovascular and other diseases)... several symptoms of cardiovascular disease and features that differentiate them from non-cardiovascular conditions... and clinical manifestations and treatment of the comorbidities often associated with cardiovascular disease" (Gillebert et al., 2013). The question arises that, despite good intentions, the implementation may be insufficient.

More generally, diagnostic errors are a large concern given the prevalence of errors significantly increasing the risk of death (National Academies of Sciences, Engineering, and Medicine, 2015). Interestingly, a report by the National Academies of Sciences, Engineering, and Medicine (2015) generated goals for improving diagnosis and reducing diagnostic error, including:

Facilitate more effective teamwork in the diagnostic process among healthcare professionals, patients, and their families

Enhance healthcare professional education and training in the diagnostic process

Ensure that health information technologies support patients and healthcare professionals in the diagnostic process

Develop and deploy approaches to identify, learn from, and reduce diagnostic errors and near misses in clinical practice

Establish a work system and culture that supports the diagnostic process and improvements in diagnostic performance

Develop a reporting environment and medical liability system that facilitates improved diagnosis by learning from diagnostic errors and near misses

Design a payment and care delivery environment that supports the diagnostic process

Provide dedicated funding for research on the diagnostic process and diagnostic error

Interestingly, the goals do not appear to address the individual responsibility of the physician or healthcare professional. Rather, these seem to be "systems" errors or limitations, where "systems" appear to apply to a larger aggregate acting in some joint manner. The closest to individual responsibility is the goal to "Facilitate more effective teamwork in the diagnostic process among healthcare professionals, patients, and their families." Nevertheless, the National Academies squarely places responsibility on the shoulders of healthcare delivery systems, whether private or public, and educational systems. The responsibility necessarily invokes questions of ethics and thus the ethical responsibilities of these systems (see Cases 2 and 9).

Lapses in medical reasoning

The analysis of the current case suggests that no system or teamwork would have prevented the perhaps lapse in medical reasoning in the case presented. Interestingly, in the discussion of the article by Zwaan et al. (2012) described previously, they considered possible sources of diagnostic error, writing, "For example, different information-processing strategies (e.g., pattern recognition and hypothetico-deductive reasoning) could be manipulated to see which one leads to the most SCAs [suboptimal cognitive acts]." A detailed epistemological and historical analysis demonstrating the various factors contributing to failures in medical reasoning have been provided (Montgomery, 2019a). This author demonstrated that medical reasoning necessarily involved the use of logical fallacies and that their injudicious uses result in medical errors. Similar issues were obtained in biomedical research where an injudicious use of necessary logical fallacies results in irreproducibility in research (Montgomery, 2019b).

There are many examples where various disciplines limit the scope of their responsibilities such as in the case presented. This author volunteered to conduct an educational session on the diagnosis and initial treatment of Parkinson's disease for residents in internal medicine. The instructors responded that any discussion of Parkinson's disease was excluded as the available seminars were to be devoted to epilepsy, stroke, and infections, despite that Parkinson's disease in older people is more common, even after demonstrating that patients with Parkinson's disease in that area were not receiving state-of-the-art-and-science care. One suspects that coronary artery disease was a frequent topic of discussion.

There may be a more insidious action that affects such limitations, that is the process of medical education itself, was the introduction of Scientism into the medical school curriculum following the Flexner report on the status of medical education in the early 1900s (Montgomery, 2019a). One factor in the rise of "scientific" medicine was remarkable advances in correlations of pathology with disease symptomology and physical findings, particularly consequent to the work of Sir William Osler in the late 1800s and early 1900s. The subsequent mode of reasoning in the predominant Rationalist/Allopathic tradition was to provide reason from hypotheses of the pathology to the expected history and physical findings, thus providing a route to diagnosis. The result was advocacy for the hypothetico-deductive approach that suffers from the Inverse Problem, where multiple pathologies can result in the same history and physical findings, as described later. Never fully embraced, it appears that the predominant mode of reasoning is pattern recognition, which may be even more subject to the Fallacy of Limited Alternatives (described later and in Case 9). Even adoption of the problem- or case-based medical education does not address the Fallacy of Limited Alternatives when the problem or case is limited to a single exemplar in isolation.

Perhaps due to the remarkable advances in clinical-pathological correlations on the part of physicians such as Sir William Osler, pathology typically is the means to introduce clinical medicine to students. Indeed, much of the teaching of normal anatomy, physiology, and biochemistry is to serve as background to the pathological correlates. Typically, a pathology is described and explicated and the associated symptoms and signs of the pathology are listed. Rarely, other pathologies or diagnoses that can lead to the same symptoms and signs are discussed contemporaneously. Often, the other pathology or diagnosis is discussed another time and again in isolation. Very often it is left to the student to link the discussions based on the symptoms and signs (see Fig. C6.1 that follows). As the clinician is presented only with the set of signs and symptoms, the clinician must reason backward from what has been learned to the possible

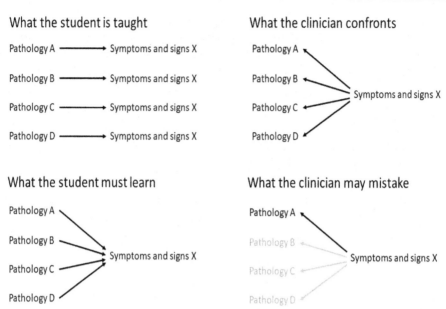

Fig. C6.1 Since the ascendency of Rationalist/Allopathic medicine with the success of the cell theory and histopathology, pathology has been the mode of introduction for students in medicine (Montgomery, 2019a). Very often instruction begins with a description of specific pathologies that are linked to symptoms and signs (What the student is taught). However, different pathologies may give rise to the same symptoms and signs. As the pathologies and the associated symptoms and signs are taught in isolation, it is often up to the student to create the linkage between pathologies (differential diagnosis) associated with the same symptoms and signs that constitute the clinical presentation (syndrome) (What students must learn). However, patients present with symptoms and signs, not with pathological diagnoses. The clinician must infer in reverse from how the clinician was taught, where the same set of symptoms and signs may indicate more than one pathology or diagnosis—the Inverse Problem (What the clinician confronts). If the clinician is unaware of alternative diagnoses, the clinician will default to the diagnosis the clinician knows. Note that the symptoms and signs will still correlate with the putative pathology, thus giving false confidence that the pathology the clinician knows is correct when it is truly false (What the clinician may mistake).

pathologies, but there is nothing in the set of symptoms and signs that point to a single pathology—the Inverse Problem. The Inverse Problem is compounded by a failure of the clinician to consider all possible pathologies, whether due to ignorance or by constraints on the clinician's obligation to do so.

Interestingly, the report on diagnostic errors by the National Academies tacitly, if unknowingly, recognized the Inverse Problem, writing "the committee sought to develop a definition of diagnostic error that reflects the iterative and complex nature of the diagnostic process, as well as the need for a diagnosis to convey more than simply a label of a disease" (National Academies of Sciences, Engineering, and Medicine, 2015). Yet, the National Academies continued, writing "The term 'health problem' is used in the definition because it is a patient-centered and inclusive term to describe a patient's overall health condition," suggesting that their reasoning was based on patient-centric ethics rather than on an epistemic logical perspective. Further, the National Academies failed to address the Inverse Problem when they wrote:

> In addition, the lack of focus on developing clinical reasoning and understanding the cognitive contributions to decision making represents a major gap in education within all healthcare professions. Proposed strategies to improve clinical reasoning include instruction and practice on generating and refining a differential diagnosis, *generating illness scripts* [italics added], developing an appreciation of how diagnostic errors occur and strategies to mitigate them, and engaging in metacognition and debiasing strategies.
>
> *National Academies of Sciences, Engineering, and Medicine (2015)*

Essentially, illness scripts are a form of pattern recognition, which only exacerbates the Inverse Problem and the Fallacy of Limited Alternatives (Montgomery, 2019a). Now the number of alternatives that should be considered is limited by the number of illness scripts. Using a hypothetico-deductive approach may at least allow reasonable alternatives to be derived from an economical set of foundational principles, such as physiology and pathophysiology. There are a potentially infinite number of patient manifestations, and as the illness scripts are tied to the manifestations, as argued by the National Academies, then an infinite number of illness scripts potentially exist, which would produce impossible logistical and reasoning demands. Simply reducing the illness scripts by prioritization or neglect will still result in some patients not receiving the full obligation of the ethical principles of beneficence, nonmaleficence, and autonomy that follow from a correct diagnosis (see Case 9).

One might hope that a single physician with sufficient general knowledge could supervene on the specialists to ensure that the patient receives the appropriate range of considerations (see Case 3). In the mid-1970s there was a push to establish and support Health Maintenance Organizations [Health Maintenance Organization Act of 1973 (Pub. L. 93-222 codified as 42 USC §300e); Employee Retirement Income Security Act of 1974 (Pub.L. 93-406, 88 Stat. 829, enacted September 2, 1974, codified in part at 29 USC ch. 18)] in the United States, which then would exercise control over referrals to specialists (see Chapter 4). In the late 1990s, in the United States and Canada, there was a move to shift training toward the production of primary care or family physicians (see Case 3). Funding for postgraduate education typically used to train specialists was reduced. Medical school curriculums were changed to emphasize primary care and family medicine. Indeed, specialists or academic physicians at times were barred from mentoring incoming medical students out of fear that the specialist would "entice" the student into specialty training, ignoring the large concerns of students in recouping educational expenses and quality of life related to the location of practice affecting the choice of general or specialized medical practice. Despite any number of efforts, there continues to be a shortage of primary care or family physicians (Association of American Medical Colleges, The Complexities of Physician Supply and Demand: Projections from 2016 to 2030, https://aamc-black.global.ssl.fastly.net/production/media/filer_public/85/ d7/85d7b689-f417-4ef0-97fb-ecc129836829/aamc_2018_workforce_projections_up-date_april_11_2018.pdf). Even if there were more primary care and family physicians and associated nurse practitioners and physician assistants, it is not clear that these clinicians would assume the supervening role described previously (see Case 3).

Justice

It can be reasonably argued that allowing the situation, such as that described in the case presented, represents a violation of the obligation to beneficence and nonmaleficence on the part of those responsible for the education, training, and supervision of physicians and healthcare professionals. To a very large degree, the responsibility of ensuring fulfillment of the obligation to beneficence and avoiding maleficence falls on governmental agencies in their proxy role for the citizens of modern liberal democracy. It is reasonable to further argue that the maximum beneficence and nonmaleficence were not assured and have not been afforded to the patient in the case presented here. However, it is critical to realize that the situations described may not be unethical and, therefore, not unjust. The cardiologist may not have any liability for the state of affairs just described.

Even if the responsibility for diagnostic errors lies in the individual physician or healthcare professional, the ubiquity and seriousness of such errors would seem to require a response from the system ultimately enforced by society through its government and health insurer proxies. Combined with the system's errors and limitations, the onus is even greater on the governmental and insurer proxies to ensure the fulfillment of the ethical obligation of beneficence and nonmaleficence. Yet, the obligation is not open-ended, if only due to the presence of moderate scarcity of resources that necessitates prioritization. Required prioritization removes, or at least mitigates, Deontological moral theories as to the supervening theory. Under a Deontological moral theory, a patient just has the right to full beneficence, nonmaleficence, and autonomy that should result in full beneficence, nonmaleficence, and autonomy being received by the patient. However, the nature of the rights is open to interpretation. One could argue that every person has the right to good health insurance or, alternatively, that every person has the right to purchase good health insurance but not necessarily good health care, as often is the situation in the United States. However, in the case presented, the patient had insurance. The only issue then is whether the insurance allows the physician to limit his obligation to the ethical principles.

The two interpretations of "rights," the right to healthcare or the right to purchase insurance for healthcare, can be explicated on Aristotle's notion of actuality and potentiality. The state where every person has the right to purchase good health insurance would mean to say that every person has the potential for good healthcare. If the potential to purchase, directly or indirectly, good health insurance results in having good health insurance, then the potentiality is actualized. The relationship between potentiality and actuality is complex and interesting. Aristotle would argue that potentiality is a derivative of actuality in that potentiality has within it a force to drive toward actuality. Potentiality will achieve actuality unless prevented from doing so by something external to the potentiality. In this case, there is the potentiality of beneficence and nonmaleficence in every person to extend to every other person, and the nature and the force of that potentiality are something innate in every person; people are wired to behave in just manners (see Chapter 1). In this sense, actuality is the obligation that derives from the potentiality.

In an ideal situation, every person would have the right to purchase good health insurance and would have the means to do so. Thus, potentiality immediately becomes actuality. However, in the real world, there is a disconnect between potentiality and actuality by virtue of external competing forces and moderate scarcity of resources. It would be an error to argue that the lack of actuality, all persons not receiving good healthcare, means that there is no potentiality, all persons could be provided good health insurance that actualizes into good healthcare. This is similar to the frequent argument in ethics that instances where conduct does not comport with the ethical principles of beneficence, nonmaleficence, and autonomy means that these principles do not exist or are not operative.

It would seem from the experiences and discussions described here that a patient's likelihood of obtaining a correct diagnosis often depends on chance or luck—the specialty the patient is referred to and the specialist ultimately consulted. In other words, a correct diagnosis becomes a lottery, which would not be unethical in an Egalitarian moral theory as long as everyone was offered the same chance at winning the diagnosis lottery and all agreed previously to be held by such a lottery—a necessity in modern liberal democracies. The current situation may be ethical under a Utilitarian moral theory where the necessary education, training, and procedural checks to assure adequate and appropriate care are too resource expensive and, in moderate scarcity, would reduce other more desired goals—assuming sufficient "votes" in favor of such an ethic such that the limitations would be imposed on all.

A purely Libertarian moral theory would be mute on the ethical situation described here were it not necessary to avoid anarchy. Consequently, contracts are established among the shareholders and stakeholders, and the ethical status of the case described would be determined by the contract established. However, it is unlikely that any contract, implicit or explicit, direct or indirect, would specify to the level of detail involved in the case presented (see Case 2). Consequently, it would be open to debate as to the standards to be upheld and how they are enforced. To be sure, if the standard of care is what similarly situated physicians would do, then the physician, in this case, did not do anything different from what most specialist physicians do. In the case presented here, the cardiologist was consistent with the standard of care. If the standard of care is what a similarly situated *reasonable* physician would do, then the standard may be quite different. Often, one looks to the standards established by professional societies, but these can be biased in favor of their member physicians and healthcare professionals (see Chapter 4). It is not clear that government agencies are likely to establish the standards of care at the level of detail in the case presented.

Ultimately, it may come down to the standards a physician or healthcare professional seeks to impose on himself or herself—what it means for the individual to be virtuous. Establishing such virtues is the aim of virtue ethics (see Chapter 2). But it may lack the force to ensure that every particular patient receives the standard of care even from a physician or healthcare professional not inclined to be virtuous; hence the limitations of virtue ethics.

References

Bakaeen, F.G., Chu, D., Kelly, R.F., Holman, W.L., Jessen, M.E., Ward, H.B., 2014. Perioperative outcomes after on- and off-pump coronary artery bypass grafting. Tex. Heart Inst. J. 41 (2), 144–151. https://doi.org/10.14503/THIJ-13-3372.

Danad, I., Szymonifka, J., Twisk, J.W.R., Norgaard, B.L., Zarins, C.K., Knaapen, P., Min, J.K., 2017. Diagnostic performance of cardiac imaging methods to diagnose ischaemia-causing coronary artery disease when directly compared with fractional flow reserve as a reference standard: a meta-analysis. Eur. Heart J. 38 (13), 991–998. https://doi.org/10.1093/eurheartj/ehw095.

Gillebert, T.C., Brooks, N., Fontes-Carvalho, R., Fras, Z., Gueret, P., Lopez-Sendon, J., Salvador, M.J., et al., 2013. ESC core curriculum for the general cardiologist (2013). Eur. Heart J. 34 (30), 2381–2411. https://doi.org/10.1093/eurheartj/eht234.

Hashem, A., Chi, M.T., Friedman, C.P., 2003. Medical errors as a result of specialization. J. Biomed. Inform. 36 (1–2), 61–69.

Kunt, A.S., Darcin, O.T., Andac, M.H., 2005. Coronary artery bypass surgery in high-risk patients. Curr. Control. Trials Cardiovasc. Med. 6, 13.

Montgomery Jr., E.B., 2019a. Medical Reasoning: The Nature and Use of Medical Knowledge. Oxford University Press, Oxford.

Montgomery Jr., E.B., 2019b. Reproducibility in Biomedical Research: Epistemological and Statistical Problems. Academic Press, New York.

National Academies of Sciences, Engineering, and Medicine, 2015. Improving Diagnosis in Health Care. The National Academies Press, Washington, DC.

Tavakol, M., Ashraf, S., Brener, S.J., 2012. Risks and complications of coronary angiography: a comprehensive review. Global J. Health Sci. 4 (1), 65–93. https://doi.org/10.5539/gjhs.v4n1p65.

Tolstoy, L., 1998. War and peace, vol. 4, Part 2, Section 1 (A. Maude, L. Maude, Trans.). In: Gifford, H. (Ed.), Oxford World's Classics. Published June 25th by Oxford University Press.

Zwaan, L., Thijs, A., Wagner, C., van der Wal, G., Timmermans, D.R., 2012. Relating faults in diagnostic reasoning with diagnostic errors and patient harm. Acad. Med. 87 (2), 149–156. https://doi.org/10.1097/ACM.0b013e31823f71e6.

Case 7

Case 7—A nurse asked a physician to stop sending copies of the physician's evaluation and recommendations to patients because the patients would call and ask questions.[a]

Medical facts of the matter

The physician, in the case presented, typically would send his patients or their legal representatives copies of the same evaluations and recommendations the physician would send to the patient's family or primary-care and other physicians. The physician did so for several reasons: (1) to be sure there were no errors in the information obtained from the patient or the patient's representative; (2) upon receipt of the report, the patient could be reasonably confident that the treating physician, for example, the family physician, likewise received a report; the patient or the patient's representative could call the treating physician in order to initiate any agreed-upon treatments without having to wait until the patient was seen again by the treating physician; and (3) so that the patient would be fully informed. A distinction has to be made between the record created by the physician or healthcare professional, here called the medical record, and what are called personal health records, which are created and managed by individuals such as patients (Tang et al., 2006). Typically, the personal health records do not contain all the information in the medical record. In this case, the physician shared the medical record generated from the visit with the patient or the patient's representative.

Although definitive studies are lacking, there is considerable evidence for concern about errors in the medical record. One indication is patient-generated requests for amendments to the medical record. In one study of patient requests, only 0.2% requested amendments. As the threshold for actually requesting an amendment likely is quite high, 0.2% or 2 out of 1000 likely is a significant underestimation. Among requests by 181 patients, inaccuracies generated 78% of requests; of those, 50% were granted, suggesting an important risk of inaccuracies (Hanauer et al., 2014). Importantly, the relatively low incidence of patients receiving copies of their medical records suggests that any burden imposed on physicians and healthcare professionals would be modest. In an interview of 998 patients recently discharged from the hospital, 23% had at least one adverse event; however, only 11% had at least one adverse event identifiable from the medical record, of which 1.1% were considered serious and preventable, suggesting that adverse effects were not be recorded in the medical record (Weissman et al., 2008). Compounding the problem is the use of dictation

---

[a] The case presented is hypothetical, although modeled on experience. No judgment is made relative to the correctness, appropriateness, or value of any actions described on the part of any of the participants. The purpose here is to describe and explore the variety of ethical considerations and consequences that depend on the perspective of different relevant ethical systems.

The Ethics of Everyday Medicine. https://doi.org/10.1016/B978-0-12-822829-6.00007-2

and templates using preexisting scripts and "drop-down" menus and automated completions in the electronic medical record. These technologies risk encouraging a "one-and-done" mentality in physicians and healthcare professionals who then do not double check the data entry (David et al., 2014). Indeed, it is not uncommon to see disclaimers on medical records that state "dictated but not read." While the intent may be to expedite transmission of the medical record by bypassing delays for the author to review the note, often the note that is "dictated but not read" remains the only documentation. Further, even if there is a follow-up report subsequent to the author's review, it is not clear how such an amended report would be brought to the attention of other physicians and healthcare professionals involved in the patient's care. The expectation that the receiving physician or healthcare professional will look repeatedly for the final report may be unrealistic. In any event, the "dictated but not read" has the full force of the official or final report as the expectation that the receiving physician or healthcare professional will act on the basis of the "dictated but not read" report.

The second reason for the physician sending the patient a copy of the medical record was to encourage the patient or the patient's representative to contact the treating physician to quickly initiate agreed to recommendations. In this author's experience, patients typically are seen on a prearranged basis, typically at 3- to 6-month intervals. It is unusual for patients to be seen sooner than their previously arranged follow-up. Thus, a patient seen by a consultant just after being seen by the treating physician may have to wait 3–6 months before the consultant's recommendations are acted upon. Studies of this phenomenon do not appear to have been conducted. However, other evidence may bear on the issue. For example, of patients discharged from the emergency department following mild traumatic brain injury, only 44% were seen within 3 months, while many patients continued to have symptoms. Among those with more moderate to severe symptoms, only 52% were seen in follow-up following the injury (Seabury et al., 2018). In a review of Medicare fees for service billings in patients readmitted to the hospital within 30 days of the initial discharge, 50.2% showed no evidence of a visit to the physician's office between initial discharge and readmission. Among those readmitted within 30 days of a surgical procedure, the reason for readmission in 70.5% of patients was medical (Jencks et al., 2009).

The third reason was to allow the patient and the patient's family, friends, and caregivers to be fully informed after the visit with the physician or healthcare professional. Studies demonstrate that patients immediately forget between 40% and 80% of medical information provided immediately after the information was conveyed (Kessels, 2003). Almost half of the information thought remembered is incorrect (Anderson et al., 1979). Further, in complex cases, a greater amount of information lowers the amount recalled correctly (McGuire, 1996).

It would appear that providing patients or their representatives with written copies of the medical record would be a relatively inexpensive and nonpainful way of helping ensure best execution of quality of the medicine recommended. The objection that the materials would be too difficult for the patient or the patient's representative to understand likely does not hold in the era of ubiquitous Internet-based computer search engines. Virtually any medical term can be looked up and then understood by most competent adults. Indeed, studies have shown that children as young as 9 years can understand complex surgical issues (see Case 13). To suggest that the patient in the case presented cannot understand the copy of the medical report provided suggests paternalism, a possible violation of the obligation to patient autonomy. Further, the objection fails if physicians and healthcare professionals actually are required to explain to the patient or to the patient's representative fully and in terms the patient or the representative could understand (see Case 5). The subsequent copy of the medical record then serves as a reminder, not a source of novel and de novo information. Indeed, the written reminder is often more effective than continued verbal information (Blinder et al., 2001).

Who owns the medical record?

The issue of who owns the medical record has been fraught with arguments perhaps since the time of Hippocrates, where confidentiality or silence, depending on one's perspective, was paramount. The debate appears to be dichotomized into actual ownership, such as owning a piece of property with the attendant property rights versus unfettered access by patients even if the medical record is owned by others, such as the physicians and healthcare professionals of the healthcare delivery systems. The issues have become more pronounced in the era of big data (medical) and genetics, where a simple genetic test for one problem can be exploited scientifically and commercially (Kish and Topol, 2015; Rumbold and Pierscionek, 2016).

The Canadian Medical Protective Association, essentially providers of medical liability coverage, pointed out that the Supreme Court of Canada ruled in *McInerney v. MacDonald* (1992) that

> …the information in the medical record belongs to the patient. However, the physical record belongs to the person or organization responsible for its creation, that is, the hospital or a physician in private practice…. Patients have a right to see the content of their record at any time and for any reason, subject to certain exceptions (e.g., if there is likelihood of harm to the patient). Any inappropriate notations can be embarrassing to, or even grounds for litigation against, the writer.

In the United States, only New Hampshire stipulates that the patient owns the information in the medical record; 18 states stipulate that the hospital owns the information; and the remaining states are mute on the issue (http://www.healthinfolaw.org/comparative-analysis/who-owns-medical-records-50-state-comparison). The privacy rule under the Health Insurance Portability and Accountability Act of 1996 (HIPPA) stipulates that healthcare providers must provide medical records requested by patients within 30 days. Yet, some patients are charged fees that may be prohibitive (US Government Accountability Office, Medical Records: Fees and Challenges Associated with Patients' Access, Report to Congressional Committees, May 2018, https://www.gao.gov/assets/700/691737.pdf). A number of challenges in patient access to medical records have been identified, particularly in the era of the electronic medical record (Beard et al., 2012). Further, a study demonstrated significant noncompliance with federal regulations in the provision of medical records (Lye et al., 2018).

An act of kindness, responding to an unvoiced need

As the primary issue here is patients' access to their medical records, it appears that it is a forgone conclusion that patients have the right to access, if not ownership. The ethical question, how humans treat each other, of whether to do so appears settled. So, what is at issue in the case presented here? Certainly, the physician's patients could have petitioned for a copy of the medical record, although the time delays would not have met the physician's concern regarding the timeliness of follow-up on any recommendations. Rather, the issue seems to be the proactive stance of the physician in ensuring that nearly every patient or the patient's representative receives a copy of the medical report in the timeliest manner possible. Most likely, the nurse in the case presented would not object to answering the questions put by the patient or the patient's representative, but only when asked. The nurse was not anxious to encourage such questions. The nurse viewed the physician's actions as inviting questions that the nurse would have to answer.

The immediate particular situation has both the nurse and the patient as stakeholders, as they are affected by the actions of the physician, the shareholder. It is unlikely that both would not see answering the patient's questions as an act of beneficence, that is, responding to the obligation to beneficence or what a virtuous physician or nurse would do. The question becomes what is the obligation to such beneficence or virtue and whether that obligation requires the proactive provision of a copy of the medical record to the patient?

Countering the nurse's obligation to the ethical principle of beneficence is an obligation to nonmaleficence. In this case, the harm is done to the nurse having to divert resources to answering the patient's questions, particularly when such expenditures of resources do not appear to be reimbursed. Indeed, the salary of the nurse likely comes from revenues generated by the physician and what goes to the nurse does not go to the physician. However, it may be impractical to develop alternative methods to support the nurse. To be sure, patients and their representatives may appreciate access to nurses answering their questions, but it is not clear there is any "marketing" value. The situation of moderate scarcity of physicians makes "voting with one's feet" likely ineffective. Historically, beneficence to the patient has not been a driving force (see Chapter 4).

The questions posed to the nurse likely take place outside the standard clinic visit where services provided during the visit render the visit reimbursable. To be sure, if the nurse's efforts were reimbursable and the nurse or the proxy healthcare provider sought reimbursement, then a contractual relationship can be asserted to exist and the nurse has an obligation to answer the questions, under a Libertarian/Contractualist moral theory.

Whether such a contractual relationship actually exists, the nurse may not perceive it to exist. The insurer could claim that the reimbursements paid covered all aspects of service related to the incident of care regardless of whether the related care was provided inside or outside of the clinic visit. This concept is clear in reimbursement for surgical care when any additional related care conducted within 30 days of the surgery is held to already have been paid by the prior reimbursement for the surgery. These are considered covered under the reimbursement for the actual surgery, referred to as bundling. However, the obligation is not open-ended, as after 30 days, it would appear that the clock starts over. Similar bundling for nonsurgical care is under consideration (Hackbarth et al., 2008). Capitated care is a clearer example of bundling. Perhaps if the patient calls more than 30 days after the visit, or whatever time interval is considered standard of care, then the nurse may not be obligated to take the patient's call even if virtue would suggest otherwise. In any event, it appears that any obligation is only initiated when the patient calls—no calls, no obligation.

An Egalitarian moral theory would appeal to whether the actions requested of the nurse and the nurse's response were considered standard of care, as then the nurse has an obligation to respond. A Utilitarian moral theory would be difficult to apply because it cannot be determined easily what greater good would be achieved by the nurse discouraging patient questions. To be sure, numerous studies have demonstrated that well-informed patients have better outcomes, but it is unclear whether these are cost effective, as suggested by analyses of the timeliness of follow-up evaluations following hospital discharge, assuming that such follow-ups provide for addressing questions asked by patients (Javorsky et al., 2014; Jackson et al., 2015).

The interesting question is whether there is an obligation to beneficence on the nurse toward the patient based on the patient's expectations? The idea that patients can demand or otherwise directly place obligations on physicians and healthcare professionals is unclear and doubtful (see Chapter 4 and Case 11). It would appear the only obligation to beneficence where the physician or healthcare professional proactively encourages patients' questions would be driven by a Deontological moral theory—there just is an obligation based on an inherent right of the patient. However, as often stated, enforcing a Deontological moral theory quickly devolves into Contractualism, inviting a Libertarian moral theory, in the context of moderate scarcity. One alternative, and perhaps the only alternative, is an appeal to virtue on the part of the nurse (see Chapter 2).

The appeal to virtue has been a long tradition but it is from an interesting perspective, that is, the virtue that benefits the physician rather than the virtue that is of benefit to patients. Benefits to patients seem to be a by-product (see Chapter 4). An extended quotation from Sir William Osler, long held the paragon of physicianhood, from his lectures to medical students, nurses, and practitioners of medicine, is instructive:

> My message is chiefly to you, Students of Medicine, since with the ideals entertained now your future is indissolubly bound. The choice lies open, the paths are plain before you. *Always seek your own interests, make of a high and sacred calling a sordid business, regard your fellow creatures as so many tools of trade, and, if your heart's desire is for riches, they may be yours; but you will have bartered away the birthright of a noble heritage, traduced the physician's well-deserved title of the Friend of Man, and falsified the best traditions of an ancient and honourable Guild* [italics added]. On the other hand, I have tried to indicate some of the ideals which you may reasonably cherish. No matter though they are paradoxical in comparison with the ordinary conditions in which you work, they will have, if encouraged, an ennobling influence, even if it be for you only to say with Rabbi Ben Ezra, "what I aspired to be and was not, comforts me." *And though this*

*course does not necessarily bring position or renown, consistently followed it will at
any rate give to your youth an exhilarating zeal and a cheerfulness which will enable
you to surmount all obstacles* [italics added]—to your maturity a serene judgment
of men and things, and that broad charity without which all else is nought—to your
old age that greatest of blessings, peace of mind, a realization, maybe, of the prayer
of Socrates for the beauty in the inward soul and for unity of the outer and the inner
man; perhaps, of the promise of Bernard, "pax sine crimine, pax sine turbine, pax
sine rixa" (peace without crime, peace without turmoil, peace without quarrel).

Osler (1939, pp. 42–43)

Osler's statement "regard your fellow creatures as so many tools of trade" is striking. Yet, the advancement of physicians often followed a pattern where patients in
different classes were used. Osler wrote:

He should stick closely to the dispensaries. A first-class reputation may be built
up in them. Byrom Bramwell's *Atlas of Medicine* [italics in the original] largely
represents his work while an assistant physician to the Royal Infirmary, Edinburgh.
Many of the best known men in London serve ten, fifteen, or even twenty years in the
out-patient departments before getting wards. Lauder Brunton only obtained his full
physicianship at St. Bartholomew's after a service of more than twenty years in the
out-patient department.

Osler (1939, p. 145)

The traditional career path was from working in dispensaries, often for the poor,
then advancing to the wards. Once a sufficient reputation was obtained, the physician
could go into private practice where the fee was proportional to the reputation. Indeed,
Osler's biographer, Michael Bliss (2007), described Osler's preoccupation with money
and perhaps, in the thinking of this author, such a concern was the basis for Osler's
resistance to physicians being salaried, as recommended in the Flexner Report, which
revolutionized medical education and, with it, medical practice (Flexner, 1910).

At the very least, those patients who could afford it could "vote with their feet" and
thus could reward the virtuous physician or healthcare professional according to the
benefit bestowed to the patient. The relative mobility of physicians, healthcare professionals, and patients at the very least established a modicum of a system of checks and
balances. Osler wrote:

As the practice of medicine is not a business and can never be one, the education of
the heart—the moral side of the man—must keep pace with the education of the head.
Our fellow creatures cannot be dealt with as man deals in corn and coal; "the human
heart by which we live" must control our professional relations. After all, the personal
equation has most to do with success or failure in medicine, and in the trials of life the
fire which strengthens and tempers the metal of one may soften and ruin another.

Osler (1939, p. 350)

Yet, how does one reconcile the statement just given with Osler's statement "regard
your fellow creatures as so many tools of trade?" Perhaps this is a tacit acknowledgment
that patients, at least then, could vote by transferring to another physician. Thus, if for
no other reason than retaining patients, physicians did have to treat their patients with

some modicum of deference and rationalize such deference as a display of virtue. However, any cynicism might be tempered by the degree by which patients received better care. The contrast to the current delivery of healthcare is striking. Patients rarely get to choose their physicians and healthcare professionals in a meaningful way, whether due to managed care, where patients are bartered by healthcare delivery systems, or due to scarcity. For example, in many areas of Canada, patients shudder at the thought of changing their family physician for fear of being unable to find another physician, at least in this author's limited experience.

Whether or not beneficence to patients affected by a self-serving sense of virtue is just or unjust is a different question. Even posing such a question presupposes or begs the question that the relevant perspective is that of the physician or healthcare professional (Chapter 4). It would seem to matter less to the patient whatever motivates the appropriate care. There is no question, or should not be, that betterment of the patient, the ethical principle of beneficence or virtue, is a goal but perhaps not exclusively. The concern becomes defining betterment and the resources required to provide betterment with the means to achieve betterment only secondary. If betterment is the diagnosis of the physical basis of disease and disorder, even if in general or abstracted terms, and then to affect the physical basis of disease and disorder, modern medicine has been remarkably successful, assuming the correct diagnosis. If so, at least theoretically if not practically, modern Rationalist/Allopathic medicine has achieved the goals that are possible today and provides confidence that the same will happen with the new goals in the future. The practice of medicine, by this standard, is just, regardless of which moral theory is chosen.

The problem is whether patients justifiably expect more. Surveys demonstrate that many patients do not fully voice their medical concerns (Tjia et al., 2008; Low et al., 2011). In a survey of Canadian patients, the most frequent complaint is access and sufficient time with physicians.

Patients may need more in the sense that being made whole (see phenomenological ethics in Chapter 2) and just having the physical basis of disease and disorder recognized and the physical basis affected by treatment is insufficient. Indeed, physicians, since the ancient Greeks, recognized that they provided more than just treatment, which most often was left to apothecaries and barber/surgeons. Physicians jealously guarded their exclusive right to make diagnoses and intervened legally and politically to prevent others from making diagnoses (Montgomery, 2019) (see Case 1). Wherein lies the power in diagnosis in the absence of treatment? It lies in understanding the past and present, in understanding why disease happened, and what is likely to happen in the future. It feeds into the narrative every human creates for his or herself. It is not a sentimentality to appreciate that, at the very least, understanding provides peace of mind for the patient. Perhaps the patient is deserving of aequanimitas (calmness, patience, equanimity) just as William Osler wished for medical students, nurses, and medical practitioners as quoted earlier. The lack of understanding results in fear in patients, and the pain associated with fear is described beautifully by John Donne in 1952 in his aphorism that follows:

> I observe the Phisician, with the same diligence, as hee the disease; I see hee fears, and I feare with him, I overrun him in his feare, and I go the faster, because he makes his path slow; I feare the more, because he disguises his fear, and I see it with more sharpnesse, because hee would not have me see it...

He [Phisician] knowes that his feare shall not disorder the practise, and exercise of his Art, but he knows that my fear may disorder the effect, and working of his practise.

Donne (1952)

There should be little question that calmness, patience, and equanimity are beneficent. Indeed, such beneficence is the basis for palliative care medicine. The effectiveness of palliative care in a setting of incurable illness argues for the effectiveness of similar approaches in everyday medicine (Kavalieratos et al., 2016). Perhaps it can be as simple as proactively providing patients with a copy of their medical records. One could ask whether the provision of such care aimed at providing calmness, patience, and equanimity to the patient is just and failure to provide such care is unjust. Again, the debate is not of the ethical principle of beneficence or virtue, rather it is the obligation to beneficence or virtue.

Determining whether a failure to provide calmness, patience, and equanimity to the patient is unjust depends on the prevailing moral theory. A Deontological moral theory would hold that the obligation is fundamental and inherent, just as the right to life, liberty, and the pursuit of happiness is an inalienable right, a right of Enlightenment natural law as enshrined in the Declaration of Independence, but interestingly, such sentiment does not have the force of law (see Introduction). Such deontological notions seem only quaint in the current postmodernist world (Hodgkin, 1996; Harrison, 1997; Raithatha, 1997), particularly under the dominance of *laissez faire* capitalism (McKee and Stuckler, 2012). An Egalitarian moral theory is moot if no patient is provided calmness, patience, and equanimity, as all are treated the same. A Utilitarian moral theory would question the net effect on the universal good of fulfilling an obligation to provide the beneficence of calmness, patience, and equanimity, particularly in the setting of moderate scarcity of resources. However, in this case, the cost of merely copying the patient the medical record would seem to pale; however, the perceived cost is the subsequent resources expended by the nurse in answering the questions that would not have been incurred had the patient or the patient's representatives not received a copy of the medical record. A Libertarian moral theory would argue that there is no obligation other than that which is contracted. All moral theories lead to Contractualism as the only means of enforcement of justice in the behaviors of recalcitrant physicians and healthcare professionals.

Once again, it is demonstrated that what appears to be inconsequential in the ethics of everyday medicine belies the depth of the ethical and moral questions. It is the lack of perceptiveness on the part of physicians and healthcare professionals that gives the appearance of inconsequentiality.

Be careful what you ask for, you just might get it— Ascendency of reductionist medicine

Rationalist/Allopathic medicine has become dominant since the early 1900s. While Rationalist/Allopathic physicians have claimed the mantle of Scientism since the early 1800s to distinguish themselves from other practitioners, it was not until the mid-1900s that modern medical science proved the Rationalist/Allopathic physicians correct at least with respect to the pathophysiology of the disease and the subsequent development of effective treatments (Montgomery, 2019). The success of medical technology—whether by drugs, biologics, or devices—would appear to have justified, at least in retrospect, the parsimonious attention to the patient's multifold complaints typically given by Rationalist/Allopathic physicians.

At the heart of Rationalist/Allopathic medicine is the concept that the variety of patient manifestations, such as their complaints, are just variations on an economical set of general principles (Montgomery, 2019). The idiosyncrasies of every patient could be understood as just some unique combination of general principles, and thus the Rationalist/Allopathic physician or healthcare professional only needs to consider the patient so as to determine the general principles that were operative. Indeed, focusing on the patient's idiosyncratic complaints may distract the physician or healthcare professional.

The logistics necessary in the face of moderate scarcity of medicine throughout history placed premiums on efficiencies on the part of physicians and healthcare professionals. In situations in the past where there was the expectation that little treatment could be offered, the physician or healthcare professional could provide some solace in the form of understanding—the diagnosis—and expectations for the future—the prognosis. However, with advances in medical science and medical technology, the ability to intervene on the physical basis of disease and disorder began to dominate the patient-physician/healthcare professional relationship. Use of the latest medical technologies was expected by patients and their representatives, and physicians and healthcare professionals responded, such as prescribing antibiotics for clearly viral upper respiratory infections. Success in this regard in the face of logistical pressures, whether demographic or financial impositions of scarcity, put priority on the diagnosis and treatment of the physical basis (see Case 10 and the discussion on chronic pain that follows). The admonition "careful what you wish for, you might just get it" seems appropriate. It just may be the case, from Utilitarian and Libertarian moral theories, that patients' emotional, psychological, or spiritual comfort is secondary—unless there is a change in reimbursement. This is not to say that emotional, psychological, and spiritual comfort do not contribute to physical well-being, as this certainly is not the case. The issue is how to pay for it that appears to be just for all shareholders and stakeholders.

An interesting situation helps illustrate the problematic nature of treating patients' complaints that are not clearly derivative of diagnosable physical injury is chronic pain. As pointed out by Meldrum (2016), treating chronic pain, particularly with opioids, has always been contentious and any consensus often vacillating. There clearly have been times where patients were made to suffer pain excessively due to excessive reticence in prescribing by physicians. Note that this is likely not because of any failure on the part of physicians and healthcare professionals to recognize the beneficence of relieving pain, but rather concerns over addiction and cognitive complications due to opioids. Thus, an obligation to the ethical principle of nonmaleficence tempered what might be an obligation to the ethical principles of beneficence and autonomy on behalf of the patient. The obligation to the ethical principle of beneficence in reducing pain was mitigated by these concerns. However, such concerns would not seem appropriate in the case of a terminal patient suffering excruciating cancer pain.

As Meldrum pointed out, the promotion of opioids, such as OxyContin, as having a low addiction potential that reduced the barrier to a widespread prescription of opioids, was thought to lead to the opioid crisis (Meldrum, 2016). Further, the requirement by the US Joint Commission requiring every patient's pain be assessed as any other vital sign, such as temperature, height, weight, blood pressure, and pulse, was thought to contribute to the rise in opioid use with recent calls to stop the requirement (Meldrum, 2016; Baker, 2017, https://www.jointcommission.org/assets/1/6/Pain_Std_History_Web_Version_05122017.pdf). However, it is important to note that there were pain management programs as alternatives to opioids; an extended quote from Meldrum (2016) is illuminating:

> The dimensions of the problem were and are immense. An estimated *25 million* [italics added] adult Americans, according to the most recent data, suffer daily from pain, and *23 million others* [italics added] suffer from severe recurrent pain, resulting in disability, loss of work productivity, loss of quality of life, and reduced overall health status. The best-known alternative to opioids is a multidisciplinary team approach involving reliance on physical and psychological therapies, including cognitive-behavioral therapy, relaxation and pain coping skills training, and self-hypnosis. *While such methods can be highly successful, many third-party payers regard them as too costly* [italics added]; insurance coverage is usually inadequate, and only major medical centers can support such programs. *Fewer than 200,000 patients currently participate in multidisciplinary treatment* [italics added].

Meldrum went on to write "But as the CDC guidelines demonstrate, patients with severe chronic pain will still need opioids, and physicians will be called on to prescribe it, albeit with more caution and, as Foley and Portenoy wrote 30 years ago, the physician's 'intensive involvement.'" Yet, it is likely the same physician's intensive involvement was thought too expensive by insurers in the past.

References

Anderson, J.L., Dodman, S., Kopelman, M., Fleming, A., 1979. Patient information recall in a rheumatology clinic. Rheumatol. Rehabil. 18 (1), 18–22.

Baker, D.W., 2017. The Joint Commission's Pain Standards: Origins and Evolution. The Joint Commission, Oakbrook Terrace, IL.

Beard, L., Schein, R., Morra, D., Wilson, K., Keelan, J., 2012. The challenges in making electronic health records accessible to patients. J. Am. Med. Inform. Assoc. 19 (1), 116–120. https://doi.org/10.1136/amiajnl-2011-000261.

Blinder, D., Rotenberg, L., Peleg, M., Taicher, S., 2001. Patient compliance to instructions after oral surgical procedures. Int. J. Oral Maxillofac. Surg. 30 (3), 216–219.

Bliss, M., 2007. William Osler: A Life in Medicine. Oxford University Press, New York.

David, G.C., Chand, D., Sankaranarayanan, B., 2014. Error rates in physician dictation: quality assurance and medical record production. Int. J. Health Care Qual. Assur. 27 (2), 99–110.

Donne, J., 1952. Meditation VI. In: Coffin, C.M. (Ed.), The Complete Poetry and Selected Prose of John Donne. Modern Library, New York, pp. 421–422.

Flexner, A., 1910. Medical Education in the United States and Canada: A Report to the Carnegie Foundation for the Advancement of Teaching (PDF) (Bulletin No. 4). The Carnegie Foundation for the Advancement of Teaching, New York City, 346.

Hackbarth, G., Reischauer, R., Mutti, A., 2008. Collective accountability for medical care—toward bundled Medicare payments. N. Engl. J. Med. 359 (1), 3–5. https://doi.org/10.1056/NEJMp0803749.

Hanauer, D.A., Preib, R., Zheng, K., Choi, S.W., 2014. Patient-initiated electronic health record amendment requests. J. Am. Med. Inform. Assoc. 21 (6), 992–1000. https://doi.org/10.1136/amiajnl-2013-002574.

Harrison, J., 1997. Medicine, postmodernism, and the end of certainty. Doctors have a duty to remain true patient advocates. BMJ 314 (7086), 1044–1045.

Hodgkin, P., 1996. Medicine, postmodernism, and the end of certainty. BMJ 313 (7072), 1568–1569.

Jackson, C., Shahsahebi, M., Wedlake, T., DuBard, C.A., 2015. Timeliness of outpatient follow-up: an evidence-based approach for planning after hospital discharge. Ann. Fam. Med. 13 (2), 115–122. https://doi.org/10.1370/afm.1753.

Javorsky, E., Robinson, A., Boer Kimball, A., 2014. Evidence-based guidelines to determine follow-up intervals: a call for action. Am. J. Manag. Care 20 (1), 17–19.

Jencks, S.F., Williams, M.V., Coleman, E.A., 2009. Rehospitalizations among patients in the Medicare fee-for-service program. N. Engl. J. Med. 360 (14), 1418–1428. https://doi.org/10.1056/NEJMsa0803563. (Erratum in N. Engl. J. Med. 2011;364(16):1582.

Kavalieratos, D., Corbelli, J., Zhang, D., Dionne-Odom, J.N., Ernecoff, N.C., Hanmer, J., Hoydich, Z.P., et al., 2016. Association between palliative care and patient and caregiver outcomes: a systematic review and meta-analysis. JAMA 316 (20), 2104–2114. https://doi.org/10.1001/jama.2016.16840.

Kessels, R.P., 2003. Patients' memory for medical information. J. R. Soc. Med. 96 (5), 219–222.

Kish, L.J., Topol, E.J., 2015. Unpatients—why patients should own their medical data. Nat. Biotechnol. 33 (9), 921–924. https://doi.org/10.1038/nbt.3340.

Low, L.L., Sondi, S., Azman, A.B., Goh, P.P., Maimunah, A.H., Ibrahim, M.Y., Hassan, M.R., et al., 2011. Extent and determinants of patients' unvoiced needs. Asia Pac. J. Public Health 23 (5), 690–702. https://doi.org/10.1177/1010539511418354.

Lye, C.T., Forman, H.P., Gao, R., Daniel, J.G., Hsiao, A.L., Mann, M.K., de Bronkart, D., et al., 2018. Assessment of US hospital compliance with regulations for patients' requests for medical records. JAMA Netw. Open 1 (6), e183014. https://doi.org/10.1001/jamanetworkopen.2018.3014.

McGuire, L.C., 1996. Remembering what the doctor said: organization and adults' memory for medical information. Exp. Aging Res. 22 (4), 403–428.

McKee, M., Stuckler, D., 2012. The crisis of capitalism and the marketisation of health care: the implications for public health professionals. J. Public Health Res. 1 (3), 236–239. https://doi.org/10.4081/jphr.2012.e37.

Meldrum, M.L., 2016. The ongoing opioid prescription epidemic: historical context. Am. J. Public Health 106 (8), 1365–1366. https://doi.org/10.2105/AJPH.2016.303297.

Montgomery Jr., E.B., 2019. Medical Reasoning: The Nature and Use of Medical Knowledge. Oxford University Press, Oxford.

Osler, W., 1939. Aequanimitas: With Other Addresses to Medical Students, Nurses and Practitioners of Medicine. Blakiston's.

Raithatha, N., 1997. Medicine, postmodernism, and the end of certainty. Postmodern philoso-
  phy offers a more appropriate system for medicine. BMJ 314 (7086), 1044.
Rumbold, J., Pierscionek, B., 2016. Why patients shouldn't "own" their medical records.
  Nat. Biotechnol. 34 (6), 586. https://doi.org/10.1038/nbt.3552. Reply to Why patients
  shouldn't "own" their medical records. Nat. Biotechnol. 2016;34(6):586–587. http://dx.
  doi.org/10.1038/nbt.3615.
Seabury, S.A., Gaudette, É., Goldman, D.P., Markowitz, A.J., Brooks, J., McCrea, M.A.,
  Okonkwo, D.O., et al., 2018. Assessment of follow-up care after emergency depart-
  ment presentation for mild traumatic brain injury and concussion: results from the
  TRACK-TBI study. JAMA Netw. Open 1 (1), e180210. https://doi.org/10.1001/
  jamanetworkopen.2018.0210.
Tang, P.C., Ash, J.S., Bates, D.W., Overhage, J.M., Sands, D.Z., 2006. Personal health records:
  definitions, benefits, and strategies for overcoming barriers to adoption. J. Am. Med.
  Inform. Assoc. 13 (2), 121–126.
Tjia, J., Givens, J.L., Karlawish, J.H., Okoli-Umeweni, A., Barg, F.K., 2008. Beneath the sur-
  face: discovering the unvoiced concerns of older adults with type 2 diabetes mellitus.
  Health Educ. Res. 23 (1), 40–52.
Weissman, J.S., Schneider, E.C., Weingart, S.N., Epstein, A.M., David-Kasdan, J., Feibelmann, S.,
  Annas, C.L., et al., 2008. Comparing patient-reported hospital adverse events with medical
  record review: do patients know something that hospitals do not? Ann. Intern. Med. 149
  (2), 100–108.

Case 8

Case 8—The patient is a 67-year-old female with a 15-year history of idiopathic Parkinson's disease. Her disease is currently manifested by motor fluctuations where she fluctuates from having severe slowness of movement and tremor to periods where she has severe involuntary uncontrollable movements. She has tried a wide range of treatment regimens, including all reasonable commercially available medications. The patient heard about the possibility of deep brain stimulation and inquired of her neurologist whether she would be a good candidate for it. The patient's physician dismissed the patient, saying "you are not bad enough."[a]

Medical facts of the matter

Deep brain stimulation (DBS) involves permanently placing an array of electrodes deep in the brain to provide ongoing electrical stimulation of the brain for a variety of neurological and psychiatric disorders (Montgomery, 2015). For Parkinson's disease, many studies have demonstrated that DBS provides remarkable benefits when patients have failed all other reasonable treatments. Head-to-head comparisons of DBS to best medical therapy demonstrated that DBS is more effective with fewer adverse effects after 6 months (Weaver et al., 2009). A summary of the results of a randomized control trial from the abstract follows.

> Patients who received deep brain stimulation gained a mean of 4.6 h/d [hours per day] of on time without troubling dyskinesia compared with 0 h/d for patients who received best medical therapy (between group mean difference, 4.5 h/d [95% CI, 3.7–5.4 h/d]; $p < .001$). Motor function improved significantly ($p < .001$) with deep brain stimulation vs best medical therapy, such that 71% of deep brain stimulation patients and 32% of best medical therapy patients experienced clinically meaningful motor function improvements (> or = 5 points). Compared with the best medical therapy group, the deep brain stimulation group experienced significant improvements in the summary measure of quality of life and on 7 of 8 PD quality-of-life scores ($p < .001$). Neurocognitive testing revealed small decrements in some areas of information processing for patients receiving deep brain stimulation vs best medical therapy. At least 1 serious adverse event occurred in 49 deep brain stimulation patients and 15 best medical therapy patients ($p < .001$), including 39 adverse events related to the surgical procedure and 1 death secondary to cerebral hemorrhage.
>
> *Weaver et al. (2009)*

---

[a] The case presented is hypothetical, although modeled on experience. No judgment is made relative to the correctness, appropriateness, or value of any actions described on the part of any of the participants. The purpose here is to describe and explore the variety of ethical considerations and consequences that depend on the perspective of different relevant ethical systems.

The Ethics of Everyday Medicine. https://doi.org/10.1016/B978-0-12-822829-6.00008-4

For all adverse effects, 83% experienced adverse effects, primarily as a consequence of stimulation. Nearly, all were resolved within 6 months, suggesting resolution with appropriate DBS programming (Weaver et al., 2009). Similar benefits and safety were found in a controlled trial in Europe (Deuschl et al., 2006). The study reported "Serious adverse events were more common with neurostimulation than with medication alone (13 percent vs 4 percent, $p < 0.04$) and included a fatal intracerebral hemorrhage. The overall frequency of adverse events was higher in the medication group (64 percent vs 50 percent, $p = 0.08$)."

Evidence demonstrates that the patient in the case presented had a high probability of achieving a significant improvement in her quality of life with DBS. To be sure, there are risks involved with DBS. With respect to perioperative risks, there is a 2%–3% chance of severe and permanent disability during and following the surgery. There is a 0.7% chance of death, pooling mortality in the two studies cited earlier. However, it is important to put the risks in context by comparison to other surgical procedures considered standard and accepted treatments. In carotid artery endarterectomy to reduce the risk of stroke, the combined risk of stroke and death ranged from 1% to 6% (Paraskevas et al., 2016).

What does "you are not bad enough" mean?

The patient likely would benefit from DBS in terms of improvement of her motor functions. It is unlikely that the reasoning on the part of the physician was not that the patient would not benefit. Thus, the hesitancy implied by the physician's statement must mean that the degree of improvement did not exceed the cost associated with DBS, with the term "cost" considered in its widest context.

One notion of cost is in terms of the risks associated with DBS, where the potential benefits must outweigh the risks. Perhaps this is the basis of the physician's reticence. One can consider a ratio of benefit (b) to risk (r) such that a ratio (b/r) of < 1 would mean that the risks are greater than the benefits. Thus, if DBS produces 10 units of benefit but is only associated with 5 units of risk, the b/r would be 2, or the benefit is twice the risk. Thus, DBS would be offered—assuming that a ratio of 2 was above the accepted cutoff for the b/r ratio for a clinically meaningful b/r ratio. In attempting to measure the benefit, one needs to appreciate what is to be gained. In the case presented, the benefit is proportional to the degree of disability. Perhaps when the patient's disease worsens sufficiently, then the benefit-to-risk ratio will be sufficient to warrant the physician to recommend DBS.

How is a unit of benefit and a unit of risk calculated? How is 1 unit of benefit equal to 1 unit of risk? If the abolishment of Parkinsonian tremors is 1 unit of benefit and a 1% risk of paralysis on one side of the body is worth 1 unit of risk, then is 1 unit benefit = 1 unit of risk? If a 50% chance of paralysis on one side of the body is worth 1 unit of risk, the b/r still would be equal to 1 as when the risk was 1%. Note that one cannot value paralysis itself because it has not happened and may not happen—it is the risk that must be considered.

At this level, units of benefit and units of risk are analytically incommensurable and comparing them is like comparing apples to oranges. Yet, if the basis of the physician's statement that the benefit does not outweigh the risks, then the physician is making such a calculation.

Even if benefit and risk are not analytically and quantitatively measurable, and thus incommensurable, one still can attempt a qualitative comparison by empiric induction, much in the manner of Reflective Equilibrium (see Chapter 3) (Moro et al., 2009). A comparison could be achieved by a consensus of experts or judges. It merely may be that experts reviewing a sufficient number of cases, where there was consensus that DBS surgery was thought indicated and where DBS was thought not indicated, would allow for such a metric. If the patient in the case presented fell into the group that the consensus of experts held DBS was not indicated, then the physician's decision would be vindicated, although not necessarily on the basis that the patient was "not bad enough."

Even under ideal circumstances, a number of problems are associated with the Reflective Equilibrium method for translating a descriptive (empiric review of a number of cases) to a normative claim (that which should be done). The problem with qualitative consensus is how sharp is the distinction between those for whom DBS was thought indicated and those not? What if there is some overlap between the groups and the patient in the case presented falls into the overlap? To which group does the patient belong? How easy is it to assign the patient in the case presented to either of these groups?

There are circumstances where the consensus of a group of experts is sought. Many centers providing DBS require vetting of individual patients for their candidacy for DBS surgery by a committee of experts. In this author's experience, committees rarely base their recommendations on the ratio of benefit to risk, which likely is to be highly idiosyncratic to the individual patient such that extrapolation from generalities is problematic. Rather, the question addressed is whether the patient has been treated with all reasonable alternatives, as these have an impact on informed consent (see Case 5). Are there some mitigating circumstances, either in favor of or adverse to, in recommending DBS? Are there unusual risk factors that might require prior and proactive preparations or contingency planning? Is the diagnosis consistent with being potentially treatable by DBS correct?

Even if such consensus conferences were available, it likely would not have been of use here. Typically, the physician would have to first refer the patient to the committee for said committee to provide guidance. In the case presented, it is unlikely that the physician would have made the referral. To be sure, the physician could have simply referred the patient for a second opinion but that would require some sense of duty or the need to refer for a second opinion. Note that given the moderate scarcity of resources (see Introduction), it merely may not be an unlimited or unqualified right of every patient for a second opinion. Alternatively, there may be some self-realization that the physician, in the case presented, may be wrong in his or her assessment. It merely may be that the physician feels that he or she is sufficiently knowledgeable and wise and that there are no alternatives save the beliefs the physician already has. Indeed, in a study of general neurologists, 87.8% thought they knew the selection criteria for DBS, while on further study, only 59.2% actually could describe the major criteria (Lange et al., 2017). While such observations suggest perhaps overconfidence in the neurologist's ability to decide if a patient is an appropriate candidate for DBS, it also suggests overconfidence in the ability to decide who is not a candidate. The latter is evidenced by the underutilization of DBS for patients with Parkinson's disease (Lange et al., 2017).

Who is paying the cost?

One version of "cost" could be the actual financial cost of providing DBS. Studies have demonstrated that DBS for patients with advanced Parkinson's disease is very cost effective compared to best medical (nonsurgical) treatment (Pietzsch et al., 2016). However, to pose this as an issue for the physician in the case presented presumes that it is the physician's prerogative to consider financial costs in the threshold to providing standard and accepted healthcare, such as DBS (see Case 5). Note that this is not to say that the physician does not bear some responsibility when considering financial costs but only in the setting of ceteris paribus (all other things being equal) and even then, it is a discussion that should occur with the patient or the patient's representative for there to be informed consent. It is another issue altogether (ethically speaking) to expect the physician or healthcare professional to discount effective therapies, by hard paternalism of refusing to offer DBS, based on the relative costs.

Interestingly, there was a move popular in the late 1970s and early 1980s to send physicians copies of itemized expenses their hospitalized patients were incurring. The idea was to encourage "frugality." Historically, it does not appear that financial cost considerations affect the medical decision-making of physicians and healthcare professionals (Glickman et al., 1994; Ballard et al., 2008; Schweiger et al., 2014; Greco et al., 2015; Van der Wees et al., 2017; Hammond et al., 2018). However, some physicians have argued that the physician does have to consider the patient's costs of treatment, for example, prescribing levodopa instead of dopamine agonists for patients with Parkinson's disease (see Case 5). Recommending less expensive treatments while failing to mention more expensive treatments would be an example of soft paternalism, which would be of questionable ethics based on a possible violation of the duty to the ethical principle of autonomy to the patient. Unless the treatment options are absolutely identical in every way, the discretion should be offered to the patient consistent with the obligation to the ethical principle of autonomy to the patient as that is what a virtuous physician would do. The physician or healthcare professional cannot know, without asking, how much expense a patient would be willing to bear and thus the soft paternalism would preempt the patient's choice.

The argument could be that the insurer is paying the cost. But if the physician or healthcare professional is considering what is in the best interests of the insurer, whether commercial or governmental, then the physician or healthcare professional could be at a conflict of interest. Such a conflict of interest need not be unethical or unjust depending on the moral theory that supervenes (see Chapter 4). In a Libertarian moral theory, there is nothing immoral in the physician having such a conflict of interest as the maxim is maximal freedom. However, as most applications of Libertarian moral theories require a contract to avoid anarchy, then conflicts of interest would be adjudicated based on the contract. For example, there is nothing to prevent a patient to agreeing that the physician could have a conflict of interest. In a way, a patient joining a particular insurer may be agreeing to only those treatments covered by the insurer and agreeing that the choice of treatments offered by the physician must comport with those allowable by the insurer. In this way, the physician is serving the interests of the patient as well as those of the insurer. There is no conflict. The elements of the contract can still be fulfilled, an exchange of agreed to goods where the patient receives care, although limited within the scope of the insurance, and the physician or healthcare professional gets paid. Each are able to enter a contract and do so freely. Both engage in a good faith effort to execute the contract. There is nothing in the contract that prevents the physician from a conflict as any conflict is obviated by the patient's willingness to limit his or her choice in treatments.

As a practical manner, it is highly unlikely that any relationship among patient, physician, and insurer would be so confined or constrained. It is likely that reasonable persons blinded to their self-interests would find the situation unacceptable. In a tacit agreement, most insurers allow patients to go out of the formulary (list of medications paid for by the insurer) or out of system. Patients receive treatments outside of the contract with the insurance organization reimbursing the out-of-system physician and healthcare professional, and healthcare delivery system. Medical tourism is one way of obtaining care outside the constraints of the typical insurer, although medical tourism is fraught with ethical concerns when the patient returns to the insurer. Thus, it is hard to see how a physician or healthcare professional could escape the necessity of at least discussing these alternatives irrespective of who is actually paying (see Case 5).

It is possible that the physician or healthcare professional is concerned about societal costs in the setting of moderate scarcity of resources in which providing treatment to one person may result in denying treatment to another. In this case, the patient who is denied treatment because another patient underwent DBS means that the DBS was "paid for" by the patient that was denied treatment. Specialists in DBS have expressed those concerns; however, such concerns may need to be considered in context, such as DBS provided in the United States compared to Canada (Bell et al., 2011). If someone is to be denied treatment so that another may receive treatment, then it is not at all clear that physicians and healthcare professionals could or should be the ones deciding.

It is not clear that physicians and healthcare professionals independently make the kinds of choices in allocating treatment to some and denial to others. For example, in the United States, some studies suggest that, ceteris paribus (all other things being equal), patients with different insurers are treated differently. However, such an analysis is difficult because the choice of insurers is affected by other factors that influence healthcare, such as socioeconomic, ethnic, and gender status (Chan et al., 2014). So, it is not at all clear if the physician in the case presented is basing his decision on whether the patient is "not bad enough" because of financial costs, resource allocations, or risk assessments. But what then is it based on?

Physician-centric estimation of risk

One possible explanation is that the risks and costs, broadly defined, factored in the physician's decision actually relate to the physician rather than the patient. This is not unheard of; as argued in Chapter 4, the history of medical ethics has largely centered on persons and institutions other than the patient. For example, until the 1960s and 1970s, the practice of medicine was largely determined by the individual independent physician. Thus, it would not be surprising that the ethical systems governing the ethics of everyday medicine would be physician-centric. Since the 1970s, commercial and governmental agents have largely shaped and constrained the boundaries of everyday medicine, resulting in what may be called business/government-centric ethics.

The rejoinder could be that the assessment of risks and costs is not physician-centric but is driven by factors outside of the control of physicians. The risks are "objective," where the costs are decided by others such as insurers, governmental agencies, and society. The physicians and healthcare professionals are merely responding to outside factors. However, if this was completely true and the sole determinant, then there should be synonymy between the estimations of cost and risk by physicians and healthcare professionals and those measured scientifically and objectively. But is this the case?

Estimation of adverse effects

A survey of neurologists polled their estimates of risks associated with DBS. The estimate for dysarthria was 7.0%, which is in the range of expert opinion. However, 18.2% estimated the risk of permanent harm between 6% and 15%, which is well above those reported by expert centers. Regarding the risk of intracranial bleeding, 10.5% of neurologists estimated the risk to be higher than 15%, where the rate is $< 4\%$ for expert centers (Lange et al., 2017).

Clearly, there is a tendency on the part of many neurologists to overestimate the risks of DBS. But this phenomenon is not confined to DBS. In a study of emergency room physicians, Gandhi et al. (2018) demonstrated that emergency room physicians generally correctly estimated costs 14% of the time. Physicians overestimated the cost of more cost-effective items and underestimated the cost of low-yield tests.

In an analysis of a physician-centric ethical framework, the relationships in which physicians and healthcare professionals are embedded can be analyzed on the ethical principles of beneficence, nonmaleficence, and autonomy on behalf of the physician or healthcare professional. The obligations to those ethical principles were defined by the supervening moral theory. However, physicians and healthcare professionals are foci of the analysis of the ethics of everyday medicine just as are patients, insurers, governmental officials, and citizens. A significant defect of many prior ethical analyses is a failure to attend to the ethical situation confronted by the physician or healthcare professional beyond the relationship to the patient and, in doing so, the analyses missed perhaps the most critical element (see Introduction). An alternative relevant ethical system may be virtue ethics. Good ethics is what virtuous physicians and healthcare professionals do.

The abandonment of fee-for-service reimbursement for managed care and capitated reimbursement has greatly altered the "contractual" landscape on which physicians and healthcare professionals operate. It is reasonable to suggest that the great changes in the United States, Canada, and many European countries, particularly since the 1960s and 1970s, have been dramatic shifts of liability of financial risk—from a shared burden, in the manner of a cooperative such as Dallas, Texas teacher organized health insurance (Buchmueller and Monheit, 2009) to one that is commoditized and negotiated. However, the commoditization and negotiated health insurance for some patients has been operating ever since the 1600s, such as the Poor Laws in England (Montgomery, 2019).

Managed care and capitation placed the financial risks squarely on healthcare delivery systems, which in turn, directly or indirectly, placed the risks on the shoulders of physicians and healthcare professionals. In many ways, such transfers were overt. The response of many healthcare delivery systems was to consolidate so as to increase their bargaining power with insurers and to negotiate contracts with physicians and healthcare professionals that held the physicians and healthcare professionals, directly or indirectly, responsible for the financial success of the healthcare delivery system. Again, one should not automatically presume that this arrangement is unethical or unjust. Such decisions can only be made in the context of the supervening moral theory. Commercial or private healthcare delivery systems cannot operate at a deficit, as that would risk failure and potentially stranding patients. Similarly, there may well be a limit to how much a deficit government-based healthcare delivery system can accumulate and thus a limit to how much these entities will allow physicians and healthcare professionals to contribute to the debit side of the ledger.

Unintentional lack of transparency and the effects on justice

The ecology of shareholders and stakeholders judged ethical and just must at least include transparency as necessary in a modern liberal democracy, particularly those relying on the "Invisible Hand" of *laissez faire* capitalism (see Case 3 and Chapter 4). Lack of transparency can be intentional, such as "gag" rules that were applied, even if only implicitly, by managed care organizations, or can be unintentional. One way of unintentional lack of transparency is ignorance, where important factors and principles are not recognized by the parties involved. Frequently, these opacities are unintended consequences of well-intentioned efforts.

It is important to emphasize the relationship of intention to ethics and justice, for example, the notion of a crime versus an unethical or unjust act. One component is the intention on the part of the perpetrator of the unethical or unjust act, *mens rea* in legal terminology. An intentional unethical act often can be readily recognized as unjust and perhaps even a crime. It becomes more problematic when the unethical act is unintentional. But if unintentional unethical acts are "forgiven" or "given a pass," then it is the person who suffered the unethical act who "pays the bill." Indeed, reducing personal responsibility on the part of the physician or healthcare professional may be one motivation to discuss medical errors as "systems" errors. However, the justification often given is to relieve some fear on the part of physicians and healthcare professionals, allowing them to more fully participate in remediation and prevention efforts. It is unclear how successful these efforts have been (Wears and Sutcliffe, 2020).

The diagnosis-related group (DRG) method or reimbursement of hospitals may be an example of unintentional lack of transparency that could be argued to limit patient care, which again may not be unethical or unjust. However, to come to some conclusion of the ethics and justice of DRGs, it is important to assess how DRGs affect patient care, particularly unintentionally limiting care in a manner that reasonable persons blinded to their self-interests would find unacceptable. Further, an ethical analysis of DRGs is relevant to any method of prospective payment, such as capitated care. The intent of the DRG program was "[create a] DRG prospective payment system, Medicare pays hospitals a flat rate per case for inpatient hospital care so that efficient hospitals are rewarded for their efficiency and inefficient hospitals have an incentive to become more efficient" (https://oig.hhs.gov/oei/reports/oei-09-00-00200.pdf). Typically, the reimbursement is based on the average length of stay (for inpatients) associated with specific diagnostic groups. As a government report stated,

> The average standardized charge for each DRG is calculated by summing the charges for all cases in the DRG and dividing that amount by the number of cases classified in the DRG. Statistical outliers, those cases outside three standard deviations of the average charge for each DRG, are eliminated. The average charge for each DRG is re-computed and then divided by the national average standardized charge per case to determine the weighting factor.
>
> *(https://oig.hhs.gov/oei/reports/oei-09-00-00200.pdf)*

The operating presumption is that basing reimbursement on average length-of-stay reimbursements should be fair, assuming that enough patients within a DRG are treated by a specific institution. A patient staying in the hospital longer than the average length of stay would cause a financial loss that could be made up by another patient whose stay is shorter. However, in this author's experience, the DRG length of stay was a target that was to be met with for every patient. Whether intended or not, soon some physicians and healthcare professionals were modifying their practice more to comport with the DRG's average length of stay than what might be in the patient's best interest. Note that if the distribution of lengths of stay is normal (following a bell curve), 50% of patients will stay longer than the average and 50% will stay shorter. However, if every patient stays the average or less, then 50% of patient who should have stayed longer will be discharged prematurely. Again, such actions are not necessarily unethical or unjust, depending on which moral theory is to supervene.

Perhaps one measure of premature discharges, perhaps related to the undue influence of DRGs, is the readmission rate; however, this is problematic as it may represent only extreme cases. The assumption is that an excessive readmission rate is taken as evidence of premature discharge of a patient. However, it is unknown how many patients are discharged prior to optimal treatment but whose conditions are not sufficient to warrant readmission. How pervasive and significant the effects of physicians emphasizing length of stay, as possibly evidenced by the readmission rate, are difficult to assess and the results of various studies have been mixed. Not many studies have directly examined whether the length of stay in a DRG-based system of reimbursement increases the risk of premature discharge. In part, the complexity and confounds, such as comorbidities, demographics, and socioeconomic status, make such analyses very difficult. For example, just increasing the variance in any measure, as would occur in more complex situations, would make it more difficult to determine the effects of the length of stay on readmission rates. In a study of treatment of head and neck cancers, shorter lengths of stay increased the risk of readmission (odds ratio and [95% confidence interval] 1.34 [1.22–1.48]) (Puram and Bhattacharyya, 2018). However, a number of other studies failed to demonstrate any effect of shorter lengths of stay on readmission rates (Rich and Freedland, 1988; Morse et al., 2013; Weissenberger et al., 2013; Kim et al., 2016; Dexter et al., 2017).

It is interesting and perhaps valuable to better analyze the effects of DRG average length of stay, at least for the potential of misunderstanding, as the implications are relevant to any prospective payment system based on the benchmarks such as the average length of stay. Assume that the physician or healthcare professional takes the average length of stay as the target for the management of each and every patient and that the target has a significant effect on the physician's or healthcare professional's discharge from care. What are the implications from a statistical perspective? For example, if every patient used to calculate the DRG associated average length of stay was exactly the same, for example, 7 days, then the physician or healthcare professional discharging the patient after 7 days likely would be appropriate for each and every patient. What if the range of length of stay was from 6 to 8 days with a mean of 7 days (assuming a normal distribution within the data set)? Then, on average, some patients may be discharged 1 day earlier than the average, making the likely impact relatively small. If the range is from 1 to 13 days, a significant number of patients may be discharged many days earlier than would be optimal and likely would experience a greater rate of harm, a violation of the obligation to the ethical principle of nonmaleficence. Thus, more variance in the data set used to determine the DRG average length of stay creates a greater impact on the consequences of discharging every patient at the average length of stay, assuming earlier the discharge means greater harm. Perhaps the government excluding "outliers" in the DRG length of stay calculations, as described earlier, may be a problem.

This author could not find any studies that examined and reported the variance in the calculation of the DRG average length of stay with the exception that patients more than three standard deviations from the average were excluded as described previously. Note that this is not to say that there are no such studies but they would be relatively hard to find using traditional literature search engines. The exception was a study of the lengths of stay in patients with 17 dermatologic conditions in an area in Germany that reported mean and standard deviations in the lengths of stay (Wenke et al., 2009). Median and interquartile scores were also reported, but for the sake of discussion, means will be considered as they were not much different than the median for many conditions. In order to combine experiences across different diagnostic categories, the standard deviation was used to normalize the standard error by dividing the standard deviation by the square root of the sample size. The standard error was divided by the mean and then multiplied by 100 to generate the relative standard error of the mean.

As can be seen in the figure shown here, there is a broad range of relative standard errors of the mean. To appreciate the significance, in the case of lymphomas, consider that the relative standard error of the mean was 7.46 and the associated mean was 9.2 days with a standard deviation of 11.6 (the caveat is that the median was 5 days, suggesting a skewed deviation, and therefore the relative standard of the error must be taken with caution). In this case, the standard deviation is greater than the mean. Targeting solely the length of stay would result in many patients discharged prematurely. In the case of psoriasis, the mean length of stay was 15.6 days with a standard deviation of 7.2 days or half the mean. The relative standard error of the mean was 1.13. In the case of psoriasis, discharging all patients at 15.6 days would result in relatively fewer patients being discharged prematurely. To be sure, the treatment of psoriasis may be very different than the treatment of lymphoma where treatment of the former is more regimented compared to the latter. The point here is not to comment on dermatological practices but only to illustrate the problematic nature of using the average length of stay to guide treatment (see Montgomery, 2019) (Fig. C8.1).

The discussion just described makes no claim to the ethical status or justice of using the average length of stay or other such metric as a significant factor in healthcare

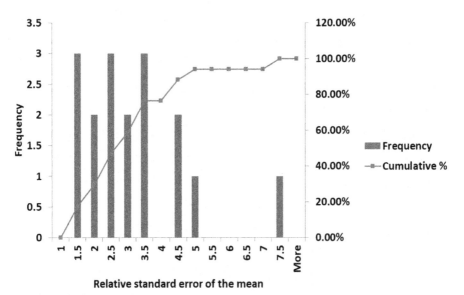

Fig. C8.1 Reanalysis of data presented for lengths of stay for 17 dermatological disorders studied by Wenke et al. (2009).

decisions by any shareholder. There is, at least, the potential for such an approach to affect the consequences (clinical outcomes) and thus the ethical status can be judged in terms of the consequences in the context of moral theories. For example, from a Utilitarian moral theory, if the resources saved by early discharge are less than the resources expended for any subsequent remedial treatment, such as readmission (disregarding any pain or suffering from insufficient treatment), then targeting the average length of stay, while perhaps not fully fulfilling the obligations to the ethical principles of beneficence, nonmaleficence, and autonomy to the individual patient, would not be unethical and, therefore, not unjust. A Deontological moral theory would hold that the average length of stay should not have an effect if it is the right of every patient being hospitalized as long as necessary to achieve optimal treatment. From an Egalitarian moral theory, as long as every patient is treated the same, that is, all patients are discharged at or prior to the average length of stay and sufficient numbers of persons (e.g., citizens who presumably are shareholders) agree (directly or via their proxy), then the use of average length of stays would not be unethical or unjust. The Libertarian moral theory would settle on a contract with unethical and justice behaviors based solely on violation of the contract.

Ultimately, all moral theories resolve down to Contractualism in modern liberal democracies. This is not to say that such contracts cannot be regulated consistent with other moral theories, such as a Deontological moral theory. Indeed, in most pluralistic modern liberal democracies, there are constraints on contracts. The elements of most socially acceptable contracts include (1) an exchange of agreed upon goods, (2) each party is free to enter into the contract, (3) each party is capable of entering into a contract, (4) there is a good faith effort to fulfill the terms of the contract, and (5) lawful purpose. Justification of the elements of a contract cannot come from a Libertarian moral theory, as Contractualism is not part of a Libertarian moral theory but a necessity to avoid anarchy. Further, the elements do not appear to abide any qualification, thereby reducing the normative effects of Egalitarian or Utilitarian moral theories.

Contracts rely on a mutual exchange of goods with the parties entering into such contracts freely. However, there is a subtle difficulty in DRG-based systems. The "ICD-10-CM/PCS MS-DRG v34.0 Definitions Manual" wrote:

> When clinicians use the notion of case mix complexity, they mean that the patients treated have a greater severity of illness, present greater treatment difficulty, have poorer prognoses and have a greater need for intervention. *Thus, from a clinical perspective case mix complexity refers to the condition of the patients treated and the treatment difficulty associated with providing care. On the other hand, administrators and regulators usually use the concept of case mix complexity to indicate that the patients treated require more resources which results in a higher cost of providing care. Thus, from an administrative or regulatory perspective case mix complexity refers to the resource intensity demands that patients place on an institution. While the two interpretations of case mix complexity are often closely related, they can be very different for certain kinds of patients* [italics added].
> For example, while terminal cancer patients are very severely ill and have a poor prognosis, they require few hospital resources beyond basic nursing care.

In the past, there has sometimes been confusion regarding the use and interpretation of the DRGs because the aspect of case mix complexity measured by the DRGs has not been clearly understood. *The purpose of the DRGs is to relate a hospital's case mix to the resource demands and associated costs experienced by the hospital* [italics added]. Therefore, a hospital having a more complex case mix from a DRG perspective means that the hospital treats patients who require more hospital resources but not necessarily that the hospital treats patients having a greater severity of illness, a greater treatment difficulty, a poorer prognosis or a greater need for intervention.

(https://www.cms.gov/ICD10Manual/version34-fullcode-cms/fullcode_cms/Design_ and_development_of_the_Diagnosis_Related_Group_(DRGs)_PBL-038.pdf)

This quote suggests a difference in perspective and expectations between physicians and healthcare providers and the healthcare delivery systems. Such differences, if they are not resolved, raise the question of whether the terms of a good contract can be met, thus placing stakeholders at risk. Should the differences ultimately place the patient at risk, then where does the responsibility lie? Do physicians and healthcare professionals understand the statistical nuances of average length of stay or other similar benchmarks and, if not, should they? If they should, then how much responsibility falls on the institutions responsible for education, training, and supervision?

Primum non nocere (first, do no harm)

Primum non nocere has long been held as a maxim in medicine. The implication is that considerations of obligation to nonmaleficence take precedence over the obligation to beneficence, such that all other things being equal, any benefit of the doubt would be in favor of not doing harm (stronger obligation to the ethical principle of nonmaleficence). If, hypothetically, a medical action might produce 5 units of beneficence and 5 units of maleficence, the action is not taken. As discussed previously, there is no such metric where all beneficence and maleficence can be placed on the same rational number dimension. Yet, all shareholders and stakeholders come to some sense of relative value, even if only qualitatively. This is not to suggest that such qualitative assessments cannot be robust (see Chapter 3).

Balancing between beneficence, in the case presented improvement in the patient's quality of life with DBS, and maleficence, the surgery and its attendant risks, was addressed by Thomas Aquinas in his principle of double effect ("Summa Theologica," written in 1265–1274). Four principles were offered: (1) the nature of the act that must be morally good or at least indifferent, (2) the harm possible from the act cannot be the means by which one produced a morally good effect, (3) the motives of the actor must be to do good, and (4) the potential bad effect must not be disproportionately bad compared to the intended good effect. Applying these principles to the case presented, the maleficence necessitated by the surgery cannot be the means by which the beneficence is obtained. This clearly is not the case as the maleficence has no causal relation to the beneficence. In other words, the production of beneficence does not absolutely necessitate producing injury other than those associated with routine and uncomplicated surgery. The crux appears to be that the maleficence, invading the body, producing pain, and placing the patient at risk, cannot be disproportionately greater than the benefit to be gained. The entire discussion up to this point has been an attempt to understand the proportionality.

The question is what might happen if there is a bias against maleficence that is not derivative of the actual treatment? Of concern is the notion of omission bias, that is, when everything is equal, errors of commission, acting, are somehow intrinsically worse than errors of omission, not acting. Omission bias is a large factor in childhood vaccinations (Asch et al., 1994) but can be observed in a variety of scenarios (Aberegg et al., 2005). Interestingly, children as young as 7 or 8 years old demonstrate omission bias (Hayashi, 2015), suggesting that perhaps humans have an innate "wired" predisposition to a sense of justice that is perceived as containing omission bias (see Chapter 1). Indeed, functional magnetic resonance imaging studies demonstrated different patterns of brain activation between moral problems that involve intentional harm compared to those that may be associated with passive harm (Schaich Borg et al., 2006; Cushman et al., 2012).

The power of omission bias can be seen in the ethical dilemma of the runaway trolley car first described by Philippa Foot in 1967 (Foot, 1978) and extended by Judith Thomson in 1976. The original version involves a runaway trolley car hurtling toward five unsuspecting workers on a track. A person is standing next to a switch with which the person can divert the runaway trolley to another track and kill two workers on that track. Thus, the only choice is whether to throw the switch that will kill two workers or leave the switch alone in which case five workers would be killed. Most people would throw the switch in order to save five lives at the expense of two lives with a net gain of three lives.

The extension of the runaway trolley car dilemma involves the same scenario of the runaway trolley. However, there is no switch but the person is standing on a bridge over a track. Standing next to the person is a very large man wearing a heavy knapsack. The person knows that if she pushes the man off the bridge onto the track the man's body and knapsack would derail the trolley, thereby saving the five workers. The only option is to push the man off the bridge, as the observer is too light. The question comes down to expending one life but saving five for a net gain of four lives or not pushing the man and losing five lives. Most people would not push the man off the bridge to save a net of four lives when just minutes before they were willing to throw a switch to kill two workers for a net gain of only three lives in the original problem.

One explanation for this paradox is the proximity of the agent, the person throwing the switch or pushing the man off the bridge, to the consequence, death of some persons. Somehow acting on an inanimate object, such as a switch at a far distance to the two workers who would be killed, is different than touching and pushing a real live man. In the former situation, the person is relatively removed from the consequence compared to the latter situation. From the perspective of the potential victim, someone is going to die whether the switch is thrown or the person is pushed off the bridge. Thus, from the victim's perspective, there is very little distinction between death as a consequence of an act of commission or an act of omission.

The question is who benefits from acts of omission versus acts of commission? In the dilemmas presented, the ethical question is one for the person throwing the switch or pushing the man off the bridge. Applying these dilemmas to the physician in the case presented, it may be that the physician was more uncomfortable with an act of commission, that is, instigating the patient undergoing DBS surgery and the possibility of harm to the patient, than an act of omission by denying a referral for DBS. To the degree that the analogy applies, the resolution of the ethical challenge centered on the physician, not the patient. In other words, the physician's actions were physician-centric rather than patient-centric.

Omission bias appears to be present ever since childhood. Thus, it is likely that physicians and healthcare professionals bring those biases with them to their education, training, and practice. If the risks of omission bias affecting the obligations to the ethical principles of beneficence and autonomy are to be mitigated, then a proactive approach is necessary with the presumption that unless specifically "treated," omission bias will persist. Yet, even a cursory review of the curriculum for professional training suggests very little opportunity to address inherent omission bias.

To be sure, most professional schools make efforts to teach ethics but the question is whether what they teach is relevant to the ethics of everyday medicine (see Introduction). Further, there is concern that ethical and moral judgment actually may deteriorate during medical school and postgraduate training as students are affected by the "hidden curriculum," often an implicit transfer of ethical and moral predispositions from their experiences in everyday medicine (for a review, see Yavari, 2016). An examination of suggested core curriculum in ethics by professional societies would appear to cover most, if not all, ethical concerns, at least as principles and topic domains (Doyal and Gillon, 1998; ww.acgme.org/acWebsite/RRC_120/120_prIndex.asp; www.acgme.org/acWebsite/RRC_120/120_pifindex.asp). Yet, drilling down to specifics is difficult. In a survey study of teaching of ethics in 17 institutions, 17 taught end-of-life issues; 9, beginning of life; 8, patient autonomy; 4, research ethics; 3, distributed justice; 1, Hippocratic oath; 1, breaking bad news; and 1, ethical principles as discussed by Beauchamp and Childress (2013). The following table shows topics from a survey of United States and Canadian medical schools (Soleymani Lehmann et al., 2004). The topics certainly suggest that issues specific to the ethics of everyday medicine (see Introduction) may be covered but there is no assurance of this. In terms of texts, most deal with what might be considered "headline" medical ethics rather than cases specific to the ethics of everyday medicine. An exception is the text "Ward Ethics: Dilemmas for Medical Students and Doctors in Training," edited by Kushner (2001). Tsai (2017) suggested that the following are examples for constructing educational materials for ethics:

> *medical graduates/generalists: truth telling, confidentiality, end-of-life ethics, and futile treatment* [italics added]. At the same time, ethics dilemmas in genetics, organ donation and transplantation, and cloning are relevant for specialists/scientists. Regarding hospital administrators and clinical leaders, the relevant issues are balancing care quality and efficiency; improving access to care; and allocating limited medical resources. `

Interestingly, the illustrative case presented in the chapter, a middle-aged male with a history of epilepsy driving a taxi, is not uncommon in the everyday practice of medicine.

Topics and percentages from a survey of United States and Canadian medical schools when each topic was addressed.

Topic	Response number (percentage)
Consent (informed consent, proxy consent, etc.)	87 (96)
End-of-life issues	84 (92)
Confidentiality	84 (92)
Truth telling	82 (90)
Ethical issues relating to student status	68 (75)
Allocation of scarce resources	68 (75)
Assessing patient competence	65 (71)
Access to care	64 (70)
Managed care	62 (68)
Financial incentives/conflicts of interest	55 (60)
Genetic testing and screening	50 (55)
Reproductive technologies	49 (54)
Research ethics	49 (54)
Pediatrics/neonatal care	48 (53)
Medical errors	45 (49)
Abortion	42 (46)
Role of ethics committees	38 (42)
Other	32 (35)

From Soleymani Lehmann, L., Kasoff, W.D., Koch, P., Federman, D.D., 2004. A survey of medical ethics education at U.S. and Canadian medical schools. Acad. Med. 79(7), 682–689.

The priority of specific topics can be considered from studies evaluating the ethics of physicians, healthcare professionals, and students of the respective disciplines. Some studies are based on case vignettes. In one study, the topics included (1) infertility counseling in a couple with intellectual disabilities, (2) a patient with a massive brain hemorrhage, (3) an 82-year-old woman with delirium, (4) HIV positivity, and (5) a depressed man with a malignant lump (Hébert et al., 1990). Another used the vignette of a 10-year-old boy in respiratory distress facing possible lifetime mechanical ventilation (Langer et al., 2016). Anorexia, suicidal ideation, and confidentiality of a college-aged woman were concerns (Kaufman et al., 1991). Another study of medical students' understanding of ethics included the following vignettes (Goldie et al., 2004):

> Request for withdrawal of treatment by a competent, paralyzed patient who has required a ventilator to keep her alive for 3 years and who has no hope of recovery.

> Whether to inform a patient with poorly controlled epilepsy, who is opposed to abortion and birth control, about a new medication that carries a 10% risk of severe birth defects.

> How to respond to a seriously injured patient who requires immediate surgery and blood transfusion, but due to her religion (she is a devout Jehovah's Witness) will not consent to transfusion, thereby greatly reducing her chance of survival.

Whether to report an HIV-positive prostitute who refuses to refrain from acting in ways that could transmit the virus to her clients.

Whether to refer a 15-year-old Catholic patient for a termination without her parents' consent.

How to respond to the request for information about her prognosis by an intelligent, terminally ill 12-year-old patient with leukemia whose parents are adamant she should not be informed of her terminal status.

Whether, as the only doctor on a remote island, to accept an inappropriate invitation to dinner with a patient of the opposite sex.

Whether to "blow the whistle" on a colleague who disregards the wishes of a patient to be resuscitated because she had signed a previous advance directive asking not to be resuscitated.

Whether to report a taxi driver, recently diagnosed with epilepsy and who continues to drive, to the authorities.

None of the aforementioned discussion suggests that the topics and the vignettes are not relevant to or would not be encountered in the ethics of everyday medicine, only that they would be relatively infrequent, not every day. Further, even if principles could be extracted from the experiences associated with the topics and the vignettes actually addressed, it is not clear how generalizable the experiences would be to more common issues in the ethics of everyday medicine, such as omission bias. Still, the relative lack of instruction in the ethics of everyday medicine not covered by current efforts may not be unethical or unjust. Those determinations depend on applications of the moral theories and deciding which moral theory supervenes, as discussed previously.

Fear of obligation

The issue of omission bias, as related to surgical treatments, would seem overrated given the continued and remarkable advances in invasive therapies. However, the situation may be different for treatments such as DBS. From the preceding discussion, because the ethical dynamics seem to be physician-centric, the question becomes what is it about the omission bias that benefits the physician even if not the patient? DBS is somewhat unique in that it is not a neurosurgical treatment. The benefits only accrue once the stimulator is turned on and adjusted. Generally, this should be done by a physician or healthcare professional who is an expert in brain stimulation. However, at least for patients with Parkinson's disease, the stimulation is not adjusted in isolation but rather in concert with other treatments, particularly medications. Thus, optimal postoperative DBS management requires expertise in all forms and aspects of treatment, medical as well as surgical. Thus, it would seem that the primary and ongoing responsibility would fall on the shoulders of the neurologist. Thus, unlike ablative or resecting surgeries, DBS requires ongoing commitment from the neurologist.

From a principled analysis, it would seem by education and training that neurologists would be the ones to look to for provisions of postoperative DBS programming. Yet, this author has been unable to find any statement from an organization that neurologists, collectively and individually, should be charged with the responsibility of providing postoperative DBS care (except for those issues clearly surgical, urgent, and just happen to befall the neurologist). Further, in over 25 years of involvement with DBS as a neurologist, this author has not encountered any neurology resident training program that requires the resident to become knowledgeable, experienced, and capable of providing postoperative care for patients who have undergone DBS. A survey of neurologists' understanding of DBS was sent to 7722 neurologists but only 4.2% responded (Shih and Tarsy, 2011). While there may be any number of reasons for the low response rate and no definitive explanation can be offered, it is at least possible, if not probable, that the response rate reflects the lack of interest or commitment of neurologists, as a whole, to DBS.

Assigning motivations to any person's actions, whether acts of commission or omission, is difficult at best. One could focus on the actions as they relate to ethical principles and moral theories and judge accordingly. It may be unnecessary and perhaps counterproductive to force physicians and healthcare professionals to be "pure of heart as well as deed." However, as always, in matters of ethics, ultimately the question resolves to enforcement. For those not "pure of heart," they may need to be compelled to be "pure of deed" at the very least. From a Contractualism perspective, there are grounds for such compulsion based on the notion that society has granted physicians and healthcare professionals unique status in return for the physician and healthcare professional attending to the needs of society. Yet, physicians and healthcare professionals are members of society and to compel them when similarly situated persons are not compelled is difficult and risks renting the fabric of society. Alternatively, inducements can be offered; yet, the moderate scarcity of physicians and many healthcare professionals may allow them to require more than society would be willing to pay unless prevented. Often the effectiveness of such inducements relates to what motivates physicians and healthcare professionals, who are as human as anyone else, unless virtue ethics can be successful (see Introduction and Chapter 2).

The questions become just what is the obligation of the physician in the case presented and if the actions of the physician are found unethical and, therefore, unjust, what could the reason be? In Case 3, the question was addressed as to who has the responsibility for ongoing day-to-day management of a patient with Parkinson's disease—the family physician or the movement disorders neurologist? At least one view is to assign the responsibility to who initiated care. For example, if the movement disorders neurologist wrote the prescriptions for medications used to treat the patient, then the movement disorders neurologist likely would be expected to continue that obligation.

The question becomes whether the physician or the healthcare professional, in the case presented, obligates her- or himself to provide postoperative DBS management should the physician or healthcare professional make the referral? If so, then one way to avoid the obligation is to not offer the possibility of DBS, a form of soft paternalism (see Case 5). Interestingly, this author has never heard a physician declining to offer DBS and saying to the patient that it was because the physician did not want to provide the follow-up care. Perhaps it is easier to just say to the patient "you are not bad enough."

This author experienced a similar phenomenon in the late 1990s with the introduction of thrombolytic therapy for stroke using tissue plasminogen activator (tPA). Even at that time there was little doubt about the efficacy of reversing stroke disability to a worthwhile degree, provided that tPA was given under the appropriate circumstances. The appropriate circumstances required urgent evaluation by the appropriate physician and neuroimaging to exclude prior associated bleeding in the brain. The logistics at that time required the ready availability of expert personnel, something academic medical centers staffed by residents and fellows generally could provide more readily, unlike community-based neurologists. Interestingly, when asked why community-based neurologists did not provide tPA treatment, some stated because it was too dangerous, which could only be the case if the conditions necessary for safe use were not afforded, for example, by the physician and the healthcare provider system the physician was affiliated with.

The physician in the case presented could have objected to the obligation of providing postoperative DBS management on the basis of lack of education and training. However, this rings hollow given the plethora of continuing educational programs available (and most physicians are required to participate). However, such an answer only partially shifts the responsibility to those charged with educating, training, and supervising the physician in the case presented. Again, this is not to say that a failure to refer a patient for DBS is unethical and thus unjust, although the soft paternalism in the physician's response would seem undesirable. Rather, the obligations to the ethical principles involved in the case need to be considered in light of the prevailing moral theory.

As can be readily appreciated in this case and in many of the cases presented, the ethical dynamics are far from dyadic—just between the patient and the physician or healthcare professional. The dynamics include all the permutations and combinations between all the shareholders and stakeholders. However, if the situation is to be resolved, it is important to note that the moderate scarcity that drives justice has to be fully appreciated in all its nuances. For example, the moderate scarcity of neurologists (see Case 3) is compounded when neurologists opt out of providing care that one might expect the patient to have a right to.

References

Aberegg, S.K., Haponik, E.F., Terry, P.B., 2005. Omission bias and decision making in pulmonary and critical care medicine. Chest 128 (3), 1497–1505.

Asch, D.A., Baron, J., Hershey, J.C., Kunreuther, H., Meszaros, J., Ritov, I., Spranca, M., 1994. Omission bias and pertussis vaccination. Med. Decis. Making 14 (2), 118–123.

Ballard, D.W., Reed, M.E., Wang, H., Arroyo, L., Benedetti, N., Hsu, J., 2008. Influence of patient costs and requests on emergency physician decision making. Ann. Emerg. Med. 52 (6), 643–650. https://doi.org/10.1016/j.annemergmed.2008.03.001.

Beauchamp, T., Childress, J.F., 2013. Principles of Bioethics, seventh ed. Oxford University Press, Oxford.

Bell, E., Maxwell, B., McAndrews, M.P., Sadikot, A., Racine, E., 2011. Deep brain stimulation and ethics: perspectives from a multisite qualitative study of Canadian neurosurgical centers. World Neurosurg. 76 (6), 537–547. https://doi.org/10.1016/j.wneu.2011.05.033.

Buchmueller, T.C., Monheit, A.C., 2009. Employer-sponsored health insurance and the promise of health insurance reform. Inquiry 46 (2), 187–202.

Chan, A.K., McGovern, R.A., Brown, L.T., Sheehy, J.P., Zacharia, B.E., Mikell, C.B., Bruce, S.S., et al., 2014. Disparities in access to deep brain stimulation surgery for Parkinson disease: interaction between African American race and Medicaid use. JAMA Neurol. 71 (3), 291–299. https://doi.org/10.1001/jamaneurol.2013.5798.

Cushman, F., Murray, D., Gordon-McKeon, S., Wharton, S., Greene, J.D., 2012. Judgment before principle: engagement of the frontoparietal control network in condemning harms of omission. Soc. Cogn. Affect. Neurosci. 7 (8), 888–895. https://doi.org/10.1093/scan/nsr072.

Deuschl, G., Schade-Brittinger, C., Krack, P., Volkmann, J., Schäfer, H., Bötzel, K., Daniels, C., et al.German Parkinson Study Group, Neurostimulation Section, 2006. A randomized trial of deep-brain stimulation for Parkinson's disease. N. Engl. J. Med. 355 (9), 896–908. Erratum in: N. Engl. J. Med. 2006;355(12):1289.

Dexter, F., Epstein, R.H., Dexter, E.U., Lubarsky, D.A., Sun, E.C., 2017. Hospitals with briefer than average lengths of stays for common surgical procedures do not have greater odds of either re-admission or use of short-term care facilities. Anaesth. Intensive Care 45 (2), 210–219.

Doyal, R., Gillon, R., 1998. Medical ethics and law as a core subject in medical education. BMJ 316, 1623–1624.

Foot, P., 1978. The Problem of Abortion and the Doctrine of the Double Effect in Virtues and Vices. Basil Blackwell, Oxford.

Gandhi, R., Stiell, I., Forster, A., Worthington, J., Ziss, M., Kitts, J.B., Malik, R., 2018. Evaluating physician awareness of common health care costs in the emergency department. CJEM 20 (4), 539–549. https://doi.org/10.1017/cem.2017.43.

Glickman, L., Bruce, E.A., Caro, F.G., Avorn, J., 1994. Physicians' knowledge of drug costs for the elderly. J. Am. Geriatr. Soc. 42 (9), 992–996.

Goldie, J., Schwartz, L., Mcconnachie, A., Morrison, J., 2004. The impact of a modern medical curriculum on students' proposed behaviour on meeting ethical dilemmas. Med. Educ. 38 (9), 942–949.

Greco, M., Zangrillo, A., Mucchetti, M., Nobile, L., Landoni, P., Bellomo, R., Landoni, G., 2015. Democracy-based consensus in medicine. J. Cardiothorac. Vasc. Anesth. 29 (2), 506–509. https://doi.org/10.1053/j.jvca.2014.11.005.

Hammond, D.A., Chiu, T., Painter, J.T., Meena, N., 2018. Nonpharmacist health care providers' knowledge of and opinions regarding medication costs in critically ill patients. Hosp. Pharm. 53 (3), 188–193. https://doi.org/10.1177/0018578717739005.

Hayashi, H., 2015. Omission bias and perceived intention in children and adults. Br. J. Dev. Psychol. 33 (2), 237–251. https://doi.org/10.1111/bjdp.12082.

Hébert, P., Meslin, E.M., Dunn, E.V., Byrne, N., Reid, S.R., 1990. Evaluating ethical sensitivity in medical students: using vignettes as an instrument. J. Med. Ethics 16 (3), 141–145.

Kaufman, J., Glantz, L.H., Grodin, M.A., Bersoff, D.N., 1991. Case vignette: Niki goes to school—autonomy, control, and psychiatric hospitalization. Ethics Behav. 1 (4), 273–281.

Kim, T.H., Park, E.C., Jang, S.I., Jang, S.Y., Lee, S.A., Choi, J.W., 2016. Effects of diagnosis-related group payment system on appendectomy outcomes. J. Surg. Res. 206 (2), 347–354. https://doi.org/10.1016/j.jss.2016.08.024.

Kushner, T.K. (Ed.), 2001. Ward Ethics: Dilemmas for Medical Students and Doctors in Training. Cambridge University Press, Cambridge. https://doi.org/10.1017/CBO9781316036501.

Lange, M., Mauerer, J., Schlaier, J., Janzen, A., Zeman, F., Bogdahn, U., Brawanski, A., et al., 2017. Underutilization of deep brain stimulation for Parkinson's disease? A survey on possible clinical reasons. Acta Neurochir. (Wien) 159 (5), 771–778. https://doi.org/10.1007/s00701-017-3122-3.

Langer, T., Jazmati, D., Jung, O., Schulz, C., Schnell, M.W., 2016. Medical students' development of ethical judgment: exploring the learners' perspectives using a mixed methods approach. GMS J. Med. Educ. 33 (5), Doc74.

Montgomery Jr., E.B., 2015. Twenty Things to Know about Deep Brain Stimulation. Oxford University Press, Oxford.

Montgomery Jr., E.B., 2019. Medical Reasoning: The Nature and Use of Medical Knowledge. Oxford University Press, Oxford.

Moro, E., Allert, N., Eleopra, R., Houeto, J.L., Phan, T.M., Stoevelaar, H., International Study Group on Referral Criteria for DBS, 2009. A decision tool to support appropriate referral for deep brain stimulation in Parkinson's disease. J. Neurol. 256 (1), 83–88. https://doi.org/10.1007/s00415-009-0069-1.

Morse, R.B., Hall, M., Fieldston, E.S., Goodman, D.M., Berry, J.G., Gay, J.C., Sills, M.R., et al., 2013. Children's hospitals with shorter lengths of stay do not have higher readmission rates. J. Pediatr. 163 (4), 1034–1038.e1. https://doi.org/10.1016/j.jpeds.2013.03.083.

Paraskevas, K.I., Kalmykov, E.L., Naylor, A.R., 2016. Stroke/death rates following carotid artery stenting and carotid endarterectomy in contemporary administrative dataset registries: a systematic review. Eur. J. Vasc. Endovasc. Surg. 51 (1), 3–12. https://doi.org/10.1016/j.ejvs.2015.07.032.

Pietzsch, J.B., Garner, A.M., Marks Jr., W.J., 2016. Cost-effectiveness of deep brain stimulation for advanced Parkinson's disease in the United States. Neuromodulation 19 (7), 689–697. https://doi.org/10.1111/ner.12474.

Puram, S.V., Bhattacharyya, N., 2018. Identifying metrics before and after readmission following head and neck surgery and factors affecting readmission rate. Otolaryngol. Head Neck Surg. 158 (5), 860–866. https://doi.org/10.1177/0194599817750373.

Rich, M.W., Freedland, K.E., 1988. Effect of DRGs on three-month readmission rate of geriatric patients with congestive heart failure. Am. J. Public Health 78 (6), 680–682.

Schaich Borg, J., Hynes, C., Van Horn, J., Grafton, S., Sinnott-Armstrong, W., 2006. Consequences, action, and intention as factors in moral judgments: an FMRI investigation. J. Cogn. Neurosci. 18 (5), 803–817.

Schweiger, M.J., Giugliano, G.R., Cook, J.R., Mayforth, J., Rothberg, M.B., 2014. Does knowledge of the cost of cardiovascular tests influence physician ordering patterns? Hosp. Pract. (1995) 42 (4), 46–52. https://doi.org/10.3810/hp.2014.10.1141.

Shih, L.C., Tarsy, D., 2011. Survey of U.S. neurologists' attitudes towards deep brain stimulation for Parkinson's disease. Neuromodulation 14 (3), 208–213 (Discussion 213). https://doi.org/10.1111/j.1525-1403.2011.00350.x.

Soleymani Lehmann, L., Kasoff, W.D., Koch, P., Federman, D.D., 2004. A survey of medical ethics education at U.S. and Canadian medical schools. Acad. Med. 79 (7), 682–689.

Thomson, J.J., 1976. Killing, letting die, and the trolley problem. Monist 59, 204–217.

Tsai, T.C., 2017. Twelve tips for the construction of ethical dilemma case-based assessment. Med. Teach. 39 (4), 341–346. https://doi.org/10.1080/0142159X.2017.1288862.

Van der Wees, P.J., Wammes, J.J.G., Jeurissen, P.P.T., Westert, G.P., 2017. Stewardship of primary care physicians to contain cost in health care: an international cross-sectional survey. Fam. Pract. 34 (6), 717–722. https://doi.org/10.1093/fampra/cmx077.

Wears, R., Sutcliffe, K., 2020. Still Not Safe: Patient Safety and the Middle-Managing of American Medicine. Oxford University Press, New York.

Weaver, F.M., Follett, K., Stern, M., Hur, K., Harris, C., Marks Jr., W.J., Rothlind, J., et al.CSP 468 Study Group, 2009. Bilateral deep brain stimulation vs best medical therapy for patients with advanced Parkinson disease: a randomized controlled trial. JAMA 301 (1), 63–73. https://doi.org/10.1001/jama.2008.929.

Weissenberger, N., Thommen, D., Schuetz, P., Mueller, B., Reemts, C., Holler, T., Schifferli, J.A., et al., 2013. Head-to-head comparison of fee-for-service and diagnosis related groups in two tertiary referral hospitals in Switzerland: an observational study. Swiss Med. Wkly. 143, w13790. https://doi.org/10.4414/smw.2013.13790.

Wenke, A., Müller, M.L., Babapirali, J., Rompel, R., Hensen, P., 2009. Development of lengths of stay and DRG cost weights in dermatology from 2003 to 2006. J. Dtsch. Dermatol. Ges. 7 (8), 680–687. https://doi.org/10.1111/j.1610-0387.2009.07029.x.

Yavari, N., 2016. Does medical education erode medical trainees' ethical attitude and behavior? J. Med. Ethics Hist. Med. 9, 16.

Case 9

Case 9—A 47-year-old female was seen with complaints of neck pain and a twisting of her head on her neck. The symptoms began 4 years previously. Each time she sought medical care, she was told that it was psychological. Other than the pain and twisting in her neck, she did not have other complaints and her past medical history was unremarkable. To be sure, the pain and twisting were stressful and an embarrassment and occasionally she would admit to being depressed. However, any episodes of feeling depressed lasted only a day or two. Her health otherwise had been excellent. Her only medication was a selective serotonin reuptake inhibitor antidepressant, which she took only after persistent entreaties by her past physician. After seeing her latest neurologist, she was diagnosed as having idiopathic cervical dystonia and her symptoms were resolved with intramuscular injections of botulinum toxin.[a]

Medical facts of the matter

The medical facts of the matter in the case presented are straightforward. The patient was misdiagnosed, evidenced as having been diagnosed as a psychogenic disorder. The patient's excellent and continuing response to the intramuscular injections of botulinum toxin and the lack of other causal factors typically associated with psychogenic disorders clearly weigh in favor of the diagnosis of idiopathic cervical dystonia. The likelihood of cervical dystonia is increased by the fact that the prevalence of cervical dystonia, particularly in middle-aged women, is not rare. It also is a fact of the matter that the patient suffered, hence subjected to maleficence, and was deprived of benefit, failure in the obligation to the ethical principles of beneficence, nonmaleficence, and autonomy for a number of years. The physical and emotional tolls cannot be overestimated. Yet, the question remains whether the past treatment was unjust and hence unethical.

Failure of the obligation to the ethical principle of beneficence

The patient had to endure years of symptoms without relief before the correct diagnosis was made. While regrettable, it did not appear to cause any substantial ethical concerns. Indeed, difficult diagnoses are encountered not infrequently and physicians

---

[a] The case presented is hypothetical, although modeled on experience. No judgment is made relative to the correctness, appropriateness, or value of any actions described on the part of any of the participants. The purpose here is to describe and explore the variety of ethical considerations and consequences that depend on the perspective of different ethical systems.

The Ethics of Everyday Medicine. https://doi.org/10.1016/B978-0-12-822829-6.00009-6

and healthcare professionals are not omniscient. Ultimately, the patient got help and, as this appears to be a relatively infrequent occurrence, it would not appear necessary to respond in any radical way with the admonition "if it ain't broke, don't fix it." But is the reality assumed so reassuring?

In a mail survey, the prevalence rate of cervical dystonia was estimated to be 390 cases per 100,000 adults with less than half receiving treatment (Jankovic et al., 2007). In a moderately large city such as Hamilton, Ontario, Canada, where the population is 536,915 (in 2016), there are 240,120 persons age 45 and older. Thus, an estimated 934 persons have cervical dystonia. If 50% were not diagnosed, then 468 patients are at risk of sharing the misfortune of the patient described in this case. Is this acceptable as the cost of doing business in healthcare? Clearly, this is a question of ethics, and as these patients are in the system of everyday medicine, it is a question for the ethics of everyday medicine.

From a survey of 50 patients in a movement disorders clinic, 49 completed a questionnaire. Only 10% of patients were diagnosed with cervical dystonia after their initial contact with a physician. The patients were seen by an average of three physicians before receiving an accurate diagnosis, and the average time between symptom onset and correct diagnosis was 6.8 years, although a substantial number of patients did not seek a diagnosis within 6 months of symptom onset (Bertram and Williams, 2016). Nevertheless, a very large number of patients had long delays in receiving the correct diagnosis and, consequently, appropriate treatment.

Not receiving the beneficence, that is, effective treatment, would be a violation of the obligation to beneficence to the patient usually expected of physicians and healthcare professionals, and increasingly of the insurers that have a major influence on the everyday practice of medicine. The question becomes how to assess the obligation? Viewed in the traditional dyadic relationship, that between the patient and the physician, the issue would seem relatively straightforward to many patients, physicians, and healthcare professionals. Perhaps most would view that it was unfortunate (perhaps a euphemism for unjust without have to say "unjust") but "it just happens." Such a response would not appear to justify any wholesale effort to change the everyday practice of medicine. Indeed, lacking is any moral outrage, which is often the prerequisite for social change (Iedema et al., 2009).

It merely may be the case that not having an obligation to provide beneficence in the case of diseases such as cervical dystonia is justifiable. Note that a failure to provide beneficence is not the same as committing maleficence. Thus, it is important to disambiguate failure in the obligation to provide beneficence, which is not the same as committing maleficence. From a Libertarian/Contractual moral theory perspective, one could argue that, unless stipulated in a contract, there is no obligation. However, the elements of a contract require good faith. In other words, the disallowance of cervical dystonia cannot be "in the fine print" or just implicit. Thus, some insurers have stipulated such exclusion, typically through an exclusion of preexisting disease, mental health problems, and infertility. It is highly unlikely that cervical dystonia would be on the radar for exclusion.

Exclusion of the obligation to beneficence in providing relief to patients with cervical dystonia could be just based on Utilitarian moral theory. It is possible that a greater good can be achieved by diverting scarce resources from treating cervical dystonia, and hence the prerequisite diagnosis, to other conditions. However, what would be the rationale if other similar disorders requiring equivalent resources were provided? Assuming there are similar disorders, then something other than utility would be necessary to exclude patients with cervical dystonia. Further, in a modern liberal democracy, all citizens, perhaps through their representatives, would have to agree prior to any exclusion. As discussed in Case 2, there is the question of who is responsible for the resources, for example, payment? If the patient agrees to pay for those resources, then the Utilitarian moral theory would be less relevant. Note that this option is preempted by the physician's failure to diagnose.

The notion of resources can be expanded to include changes in the healthcare system, as well as physician education and training, as will be discussed later. It just might be that training physicians sufficiently to accurately diagnose cervical dystonia in a timely fashion would divert resources that would not be justifiable from the Utilitarian moral theory perspective. However, the same objection applies in singling out cervical dystonia if equivalent education and training resources are provided to other comparable diagnoses. A consensus of reasonable persons blinded to their self-interests would hold; it should not be a matter of inertia or "that's the way we have always done it." If this is the accepted position, then no new treatments and no new diagnoses would ever become part of everyday medicine. In this case, the responsibility of the physician failing to make the diagnosis, and thus providing the beneficence of proper care, may also fall to the entire system of physician education and training and healthcare delivery, as discussed later. Yet, failure to make the correct diagnosis as a prerequisite for appropriate care, while a failure of the obligation to beneficence to the patient, may not be unjust and therefore unethical.

Failure of the obligation to the ethical principle of nonmaleficence

Then there is the ethical principle of nonmaleficence and the obligation to that principle. It is not a matter of the time it takes to be diagnosed with the interim suffering; it also is the consequence of being misdiagnosed, particularly as a psychogenic disorder. The fact is that the patient was given a diagnosis, just the wrong one. Indeed, only 31% of patients with cervical dystonia were initially referred to a neurologist, meaning a diagnosis other than cervical dystonia was being entertained (Bertram and Williams, 2016). More critical is the fact that 17 of the 49 patients (35%) were diagnosed as a psychiatric or psychological disorder (Bertram and Williams, 2016). The question becomes what damage is done by an incorrect diagnosis of a psychiatric or psychological disorder? Studies of patients given a diagnosis of psychogenic nonepileptiform seizures suggest a number of stressful responses: self-doubt, hostility to physicians and healthcare professionals, and stigma (Wyatt et al., 2014b; Robson et al., 2018). One wonders whether the patient could ever again have confidence in physicians and healthcare professionals once informed of the correct diagnosis. Similarly, receiving the diagnosis of psychogenicity can damage self-worth and self-trust, which, in addition to being emotional jarring and painful, can also affect the ability of the patient to act in a fully autonomous manner (McLeod and Sherwin, 2000). Despite the plethora of studies regarding the nature of true psychogenic disorders, there is very little information regarding the impact of a false diagnosis of a psychogenic disorder.

Failure of the obligation to the ethical principle of autonomy

The concept of autonomy is complex (see Case 13). Autonomy includes a notion of self-determination, but not exclusively. For example, one respects the dead, who have no self-determination, and, indeed in many places, desecrating the dead is a punishable offense. Autonomy is also linked to a sense of self, particularly a wholeness of self—an integrity of the unitary notion of self (see Phenomenological Ethics in Chapter 2). Consequently, the notion of self-agency is fundamental to one's autonomy and it is that sense of self-agency that should command respect, autonomy, from others.

The question then becomes what is the effect on the patient when she is told that her problems are psychological? What does such a diagnosis mean to the patient, particularly in the context that the patient's symptoms are real and the patient herself would have no reason to think otherwise? Persons enjoy what is called epistemic privilege related to first-person knowledge of themselves. Despite popular sayings, no one can know what the person is thinking and feeling more than the person herself. The first-person perspective is unique to the person holding it. To be sure, the person can attempt to describe or explain her first-person knowledge but it would be a translation. As discussed in Case 5, Willard Van Orman Quine demonstrated that there is no way to assure that two or more persons have the exact same meaning for the vocabulary in common. This results in an inscrutability of reference and thus an inherent indeterminacy of translation (Quine, 1960). Further, most persons believe that what they think and feel, they actually think and feel. Thus, the patient has no one to believe that she is feeling the pain in her neck or the twisting of her head.

Next, consider the dissonance when the patient's epistemic privilege encounters the physician's statement that indeed the patient's epistemic privilege is false or unreliable. Statements issued by physicians are often taken as fact by patients even when they are not (Smith, 1996). In this case, the fact is that examination shows the head to be turned to the left and back and the sternocleidomastoid on the right is hypertrophied. The statement that the condition is psychogenic is an inference; it cannot be a fact as it is incorrect. Nonetheless, the structure and dynamics of the patient-physician/healthcare professional interaction predispose the patient to consider the inference as fact. A disequilibrium results when the physician's fact is counter to the patient's own facts known to her by her epistemic privilege.

The first-person perspective likely is important to the patient's sense of self, which allows a valuation of experience past, present, and, importantly, future (Hermans, 1999). The sense of self over past, present, and future time can be considered as narrative (see Narrative Ethics in Chapter 2) and, in many ways, is similar to William James' notion of "stream of consciousness." The question is what effect does an incorrect diagnosis of a psychogenic disorder have on the patient's ongoing self-narrative, which in one instant changes the past, present, and future as it alters the self-narrative?

The conflation of the physician's diagnostic inference and fact is exacerbated if the physician does not invite the patient's response or, indeed, discourages the patient's response implicitly. Most physician-patient/healthcare professional interactions are in the form of questions and answers where the physician or healthcare professional asks the questions, not the patient (West, 1983). Even when questions are asked, questions from men are more likely viewed as rational, which often is not the case with questions posed by women patients (Smith, 1996, p. 212).

Patients can be prevented from fuller discussions by implicit actions of the physician/ healthcare professional, for example, after delivering the diagnosis, the physician can prevent the patient's response by writing a referral to a psychologist or psychiatrist and then walking toward the door. Indeed, the structure of the patient-physician/healthcare professional interactions often follows a script where transitions between the sequential operations are controlled by the physician or healthcare professional (see the proceduralization of the ethics of everyday medicine in Introduction). There is history taking, followed by a physical examination, then a review of laboratories, and then pronouncement of the diagnosis as though it was a verdict. Note that this scripted performance or procedure is expected by the patient as well as by the physician or healthcare professional (Smith, 1996). Even then it is not clear on what evidentiary grounds the patient might challenge the physician's diagnosis.

Perhaps more than coincidental in the case presented was that the patient diagnosed as having a psychogenic or conversion syndrome was female. The literature would support that a large majority of patients diagnosed with psychogenic movement disorders are women (Jankovic et al., 2006; Bhatia and Schneider, 2007), although one study suggests that it may not be a large majority (Kamble et al., 2016). There is the temptation to suggest that women are biologically predisposed to psychogenic or conversion symptoms. But it also is possible that women are more vulnerable to having the diagnosis applied to them. The problem is more invidious because the patient actually had cervical dystonia and would not have met the diagnostic criteria for a psychogenic or conversion syndrome. Thus, the wobbliness of any doubt in the diagnosis fell to the presumption of psychogenicity. How is this possible if there is not a different autonomy for any particular gender? If this is a bias that violates the obligation to the ethical principle of autonomy, then whence its origin, the physician or elsewhere? Is the physician a sexist? To be sure, the physician likely would be aghast at the notion that he could have acted in such a gender-based bias.

Considerable literature shows that male and female patients are treated differently with respect to pain, where women's pain is underappreciated and often has to be reported as more severe than males in order to be recognized (Bertakis et al., 2004; Blum et al., 2004; Stålnacke et al., 2015; Falk et al., 2016; Razmjou et al., 2016). In one study, two sets of two types of pain, one headache and the other abdominal, were presented to 120 primary care physicians (Colameco et al., 1983). The two cases in each set were identical except for personal pronouns that indicated male or female gender. The primary care physicians thought the female patients more emotional and less authentic or ill than the male patients. However, studies of the assessment of care of acute coronary syndromes do not indicate significant gender bias (Blum et al., 2004; Wyatt et al., 2014a; Nestler et al., 2017; Ruane et al., 2017).

The fact is women are not weak in any meaningful way, although they may be weakened by the social context. The increasing standing of women as combatants in the military is one piece of evidence. However, the paradox is that the weakness or emotionality of women need not even be an issue as long as physicians perceive them as such, particularly where the social organization of everyday medicine facilitates such perceptions.

Failure of medical reasoning

The error of commission in applying a false diagnosis of psychogenicity is a result of misreasoning abetted by the social dynamics of the female patient-physician/health-care professional relationship. The reasoning follows from the often necessary use of logical fallacies (Montgomery, 2019). The reasoning is structured as *if patient has (disease A or disease B or disease C), then the patient would have manifestations D is true; the patient has manifestation D is true; therefore, the patient has disease A or disease B or disease C*. The argument is extended to *disease A and disease B do not have manifestation D, therefore the patient must have disease C*. Thus, the argument resolves to *if patient has disease C, then the patient would have manifestation D is true; the patient has manifestation D is true; therefore, the patient has disease C*. The first and last arguments are examples of the Fallacy of Confirming the Consequence, which is of the general form *if a implies b is true and b is true, then a is true*. The argument is invalid, meaning that its conclusions are indeterminant. It is possible that *b* is true for reasons other than *a*. The disjunction *(disease A or disease B or disease C)* is the differential diagnosis.

The logical fallacy also includes the Fallacy of Affirming a Disjunctive, for example, *if (a or b or c)* [the disjunctive] *is true*, one cannot say whether it is *a*, *b*, or *c* that is making the disjunctive *(a or b or c)* true. But note that from the Fallacy of Confirming the Consequence, the disjunctive *(disease A or disease B or disease C)* is held as true (falsely) but within the disjunctive, the truth or falsehood of *disease A* or *disease B* has no bearing on the truth or falsehood of *disease C*. To hold that the falsehood of *disease A* or *disease B* somehow makes *disease C* true is the Fallacy of Limited Alternatives. When applied to probability, the notion is somehow the probability of *disease A* or *disease B* affecting the probability of *disease C* is the Gambler's Fallacy (Montgomery, 2019). Note that the preceding discussion assumes that *diseases A, B,* and *C* are independent probabilities.

The argument with absolute certainty is of the form *if and only if disease C causes manifestation D, the patient has manifestation D; therefore, the patient has disease C*. Consider the situation where *disease C* is psychogenicity and *manifestation D* is the patient's symptoms and signs in this case. If *manifestation D* is inconsistent with anatomical and physiological facts, then the cause cannot be physical (organic), leaving by default psychogenicity; the argument is valid and the conclusions are certain. This would be the strong criterion for establishing psychogenicity, *disease C*. But what if *manifestation D* is just weird? Thus, weirdness becomes the criterion for establishing psychogenicity. But what is the nature of weirdness as used in the present case, as certainly the patient's manifestations were not inconsistent with anatomical and physiological facts? Nor were the patient's symptoms and signs weird to the last neurologist.

Very often the weirdness of the manifestations that leads to the diagnosis of psychogenicity is really unfamiliarity. The physician may have never seen cervical dystonia before or, if the physician had seen patients with cervical dystonia previously, he did not recognize it. It is likely that the physician has seen things he might describe as weird, perhaps associating weirdness with psychogenicity. Thus, the basis for the physician diagnosing psychogenicity is the fact that he has never seen such manifestations in a patient with a known physical or organic disease, that is, a patient with cervical dystonia. The physician implicitly or explicitly forms the following argument, *if and only if (disease A or disease B or disease C) causes manifestation D, the patient has manifestation D; therefore, the patient must have (disease A or disease B or disease C). Disease A and disease B cannot be the cause of manifestation D*, as otherwise the physician would have recognized them. However, the physician knows that the disorder called psychogenicity produces weird systems. As manifestation *D* does not comport with disorders known to the physician, the manifestations are weird, thereby confirming the diagnosis of psychogenicity for the physician.

The invidious nature of the misreasoning is that *disease C*, psychogenicity, will be in every differential diagnosis. Therefore, every differential diagnosis will contain psychogenicity. The risk is that psychogenicity will be the default diagnosis for any manifestation the physician does not recognize. Further, there will be a false confidence because there is never any failure to achieve a diagnosis. There is no necessity, hence impetus, to consider another diagnosis because the process has settled on a diagnosis of psychogenicity. Further, the propensity to default to a diagnosis of psychogenicity is determined by the experience and training of the physician; in many cases the number or variety of illness scripts the physician in training is exposed to. This bodes poorly for patients with rare or unusual disorders, although epidemiologically cervical dystonia is neither rare nor unusual. Perhaps cervical dystonia is rare or unusual in the context of the physician's training. This would be most unfortunate for patients with cervical dystonia. Whether it is unjust, thus unethical, is another matter.

Educational systems failure

The critical question becomes the origin of the hypotheses, the diseases considered in the differential diagnosis. There are two general approaches, pattern recognition and the hypothetico-deductive approach; however, most procedures in everyday medicine utilize some combination (Montgomery, 2019). Pattern recognition matches the constellation of symptoms and signs with disease diagnoses. It is just the combination of

(1) tonic head twisting from muscles of the neck under constant contraction; (2) the consequent muscular hypertrophy; and (3) in many cases, the ability to temporally reduce the tonic contraction by a maneuver such as touching a certain part of the face, the *geste antagoniste*, that simply matched to the diagnosis of cervical dystonia.

Alternatively, the hypothetico-deductive approach begins with the determination of whether there is weakness (in this case, none), abnormalities in the trajectory of the movement (none), changes in reflex (none), and abnormalities of motor initiation, too little or too much (too much). These features point to a disorder of the basal ganglia. The nature of the "too much movement" is tonic contraction of the muscle around the neck, which is characterized as dystonia. These findings would not be consistent with the involvement of other components of the motor system, including the muscle, neuromuscular junction, peripheral nerve, or lower or upper motor neurons, lesions of which would all produce weakness. As the trajectory of the voluntary movement is not abnormal, such as ataxic, it is unlikely that the lesion involves the cerebellum or connections to and from the cerebellum. Further, voluntary movements are appropriate to the context and, consequently, do not suggest lesions of the motor associational cortex (Montgomery et al., 1985).

The symptoms and signs indicate the involvement of specific physiological systems, and their confluence indicates the location of the pathology. Combining the symptoms and signs to indicate the location of the pathology is not done directly, as would be the case with pattern recognition, but rather through the intermediary principles of anatomy and physiology. Thus, every patient's manifestation and diagnosis can be understood as some combination of general and fundamental principles that are subsequently used in a deductive-like approach. Without the intervening use of reasoning from general and fundamental principles, the result would be pattern matching between the manifestations and the diagnosis.

To be sure, advancing medical science is increasing the number of general and fundamental principles, posing significant challenges to the education of physicians and healthcare professionals. However, the number of principles likely is far fewer than the great variety of manifestations in individual patients. Thus, the hypothetico-deductive approach should have far greater utility when confronted with the potentially infinite variety of patients' manifestations. Indeed, it could be argued that if the education of the physician in the case presented included the approach of deduction from general and fundamental principles, it is less likely that the physician would have misdiagnosed the patient. At the very least, the physician appreciating the importance of a grounding diagnosis on fundamental anatomical and physiological principles would have acknowledged the importance of demonstrating manifestations as inconsistent with anatomical and physiological facts and would have refrained from diagnosing a psychogenic disorder.

Experience in neurology demonstrates that dystonia can be associated with specific *disorders A, B, C, ..., X* where *disorder A* could be idiopathic cervical dystonia, *B* could be Wilson's disease, *C* could be tardive dyskinesia, all the way to *X*, the last diagnosis in the reasonable differential diagnosis. The deductive logic is *if and only if dystonia implies disorders A, B, C, ..., X is true, the dystonia is true* (the patient has the physical manifestations of dystonia); *therefore, it must be true that the patient may have disorders A, B, C, …,X.* The next logical argument might be *if and only if the patient has dystonia and an abnormal serum ceruloplasmin level, the patient had an abnormal serum ceruloplasmin level; therefore, then the patient has Wilson's disease.* A laboratory test of serum ceruloplasmin is ordered and found abnormal and, consequently, out of the differential diagnosis, the final diagnosis of Wilson's disease is deduced with certainty.

The question becomes, do medical schools and universities actually teach a deductive-like approach to diagnosis based on fundamental underlying principles? Schmidt and Rikers (2007) describe the evolution of expertise in medical reasoning. Junior medical students "rapidly develop mental structures that can be described as rich, elaborate causal networks that explain the causes and consequences of disease in terms of general underlying biological or pathophysiological processes. When confronted with a clinical case in this stage of development, students focus on isolated signs and symptoms and attempt to relate each of these to the pathophysiological concepts they have learned" (Schmidt and Rikers, 2007). These students use more references to basic science than more experienced clinicians do. Next, more experienced students encapsulated pathophysiological knowledge into higher level concepts such as clinical syndromes. The final step is movement to "illness scripts," no longer invoking appeals to basic science. Others have shown that reintroduction of science and scientific reasoning may be helpful in difficult clinical cases (Patel et al., 1990).

Perhaps the greatest failure is the lack of a metacognitive skill that encourages and facilitates a self-critique on the part of the physician or healthcare professional, perhaps resulting in the physician saying to himself "Am I missing something, perhaps another opinion might be in order?" In this author's experience, it is incredibly rare for a physician to admit that he or she does not know. For example, when given the same clinical case, specialists are more likely to diagnose a condition within their specialty than outside it (Hashem et al., 2003). Note that in the study, the same case was given different diagnoses depending on the specialty of the physician. A particularly egregious case is discussed in Case 6. The large majority of the time, the physician makes a diagnosis and it only is when the patient's subsequent course proves unsatisfactory to the physician that the patient is referred elsewhere. The question is, at what cost? Those physicians perhaps overly confident are more likely to delay referral, thus delaying accurate diagnosis and treatment. Failure to diagnose is a leading cause of malpractice complaints.

In many ways, institutions of medical education and training may be complicit with the ethical lapses in the obligation to the ethical principles in the case presented, if what initially happened is considered unjust and thus unethical. However, this is not to say that the situation created by the medical educational institutions is unjust. To address that issue, one has to determine how all the stakeholders and shareholders relate to each other and which moral theory supervenes (see Chapter 5). In the relationship of the physician or healthcare professional to the patient, it would appear that a Libertarian/Contractual moral theory supervenes, as any recourse by the patient, other than trying to personally rehabilitate the physician and receiving recompense, would be through legal channels. Interestingly, there was a movement in the 1980s to encourage physicians and healthcare professionals to admit mistakes, but such noble intentions did not survive the admonitions of defense attorneys to protect against medical malpractice claims.

The patient could claim medical negligence and other forms of claims of breach of contract. In these situations, the courts or regulatory agencies would intervene as a proxy where the obligations and rights of the patient and the physician are considered. The state ultimately serves the patient's interest, as it does for the totality of patients, by requiring the physician to act according to reasonable standards, in exchange for licensure to practice medicine.

The operative presumption is that the physician is obligated to practice according to accepted standards of care. The question becomes what constitutes the standards? In many cases, the standard is what similarly situated physicians would do. Thus, the standard begins as descriptive, what similarly situated physicians do, to become normative, what any physician should do. However, in an important sense, using descriptive standards may not accomplish what would be desired ethically and morally. As Eric Fromm (1955) wrote, "The fact that millions of people share the same vices does not make these vices virtues, the fact that they share so many errors does not make the errors to be truths, and the fact that millions of people share the same mental pathology does not make these people sane." Recognizing the problem of descriptive becoming normative and thus self-reinforcing, some have looked to professional societies to establish standards of care. However, as the members of the professional societies themselves are physicians and healthcare professionals, there is the real risk of self-interest that risks unfairness to the patient (see Chapter 4). Others have refined the notion of standard care as what similarly situated *reasonable* physicians would do, recognizing the difficulty of defining reasonable.

To the extent that the descriptive standard being what similarly situated physicians would do, then the prevailing moral theory is Egalitarian. Every physician is required to treat every patient the same within the standard of practice, regardless of whether the treatment is good, bad, or indifferent. The prevalence of misdiagnosis of cervical dystonia suggests that such misdiagnosis is within the standard of care and, on this basis, it is not unjust. If the actions of the physician in this case are not unjust, then the actions of the medical education and training institutions are not unjust, even if appearing undesirable or repugnant.

Inserting the notion of *reasonable* into the standard of care invites consideration of Utilitarian and Deontological moral theories. The question becomes what constitutes reason? In Utilitarian moral theory, what provides the greater good is what is just; however, the measuring of the good is problematic. It just may be that the investment required to educate and train physicians adequately and the cost of treating once the diagnosis is made may be better spent on some other priorities that result in a greater net gain of good. The costs associated with creating the system for adequate education and training are critical. While there have been remarkable advances in pedagogy, it just may mean that educators are just better at teaching the wrong stuff (Montgomery, 2019). It may mean "radical surgery" and reconstruction of medical education for which there may be little "stomach" for—for many different reasons.

Deontological moral theories are problematic in the setting of moderate scarcity of resources. It just may be that it is impossible that every patient is seen by a physician who can make the correct diagnosis or recognize that they cannot and then refer the patient to a physician who can. Thus, some prioritization is required, which inevitably defeats a Deontological moral theory, and one then defaults to the other moral theories as described previously.

But is the situation one might expect of physicians from the Libertarian/Contractual and Egalitarian moral theories what one expects of medical schools and universities? Defaulting to a standard of care defined by the descriptive of what similarly situated physicians would do only enforces the status quo, thereby risking sustained obsolescence. Unfortunately, there are few, if any, studies that address the percentage of physicians practicing medicine in a manner considered obsolescent. It is not likely to be an ignorable small percentage. The question becomes whether those responsible for the education and training of physicians and healthcare professionals have an obligation to do more. But what then defines the obligation on the part of medical education and training institutions?

The total educational costs to teach a medical student include all activities thought necessary to train the student. These include teaching, research, scholarship, and patient care and were estimated in 1997 to be between $72,000 and $93,000 per student per year (Jones and Korn, 1997). Other more recent estimates have been much higher, even when adjusted for inflation and different currencies. For example, in Australia, the estimated cost to train a general medical graduate was $451,000 per completing student (Segal et al., 2017), which would be approximately $68,000 in the United States in 1997. Average tuition and fees for October 2019 were $60,665 for private and $37,556 for public medical schools (Association of American Medical Colleges, https://www.aamc.org/data/tuitionandstudentfees/). Consequently, the individual medical student contributes only a relatively small part of the cost to allow himself or herself to become a physician. Those excess costs must be paid by others, whether by the government or by philanthropy. The question is whether the subsidization of medical educational costs entitles those who provide the subsidization to require some return on that investment. Otherwise, the subsidies are a gift, and the question is whether those gifts are distributed unfairly. Perhaps if all the subsidized costs were provided by private philanthropy, such gifts may be not unjust, as it would be the philanthropist's discretion, although it is not clear that society as a whole would find this acceptable. However, if the government, as a proxy for its citizens, provides the subsidized costs, then it is less clear that it is not unjust given the moderate scarcity of funds, whether by fact or willingness.

The argument given suggests that the physician begins his or her practice with a debt to society that is greater than the contributions made by each physician paying his or her share of the total educational costs. The question arises as to what debt to society does medical schools and universities incur in exchange for societies' subsidization of the physicians' education and training? The total educational debt is also the goods (funds) that benefit the medical schools and universities, their faculty, and staff. Thus, the medical schools and universities acquire goods from both the graduated physician and society. The medical schools and universities provide the degree necessary for future employment to the physician in exchange for the tuition and fees paid by the graduated physician. But what about the exchange of goods to society in contractual obligation that emanates from the goods society provides the medical schools and universities? Perhaps it might be training physicians who will make correct diagnoses or, at the very least, physicians with sufficient metacognitive strengths to be self-skeptical and admit that they cannot and refer to those physicians that will. The same issues are obtained regarding healthcare professionals.

Laying blame

In 2000, a study by the US Institute of Medicine reported that up to 98,000 people die annually from hospital-based errors (Kohn et al., 2000). The report grabbed the headlines of major news outlets, stirring great public controversy (Wears and Sutcliffe, 2020). Within the academic community, where there were harsh criticisms of the Institute of Medicine report, some argued that the rate of medical errors was overestimated. Yet, even if the study overestimated by half, the rate of deaths likely would be considered too high by a consensus of reasonable persons blinded to their self-interest.

Consider the findings of a study conducted by the National Academies of Science, Engineering, and Medicine in 2015:

> A conservative estimate found that 5 percent of U.S. adults who seek outpatient care each year experience a diagnostic error.
>
> Postmortem examination research spanning decades has shown that diagnostic errors contribute to approximately 10 percent of patient deaths.
>
> Medical record reviews suggest that diagnostic errors account for 6 to 17 percent of hospital adverse events.
>
> Diagnostic errors are the leading type of paid medical malpractice claims, are almost twice as likely to have resulted in the patient's death compared to other claims, and represent the highest proportion of total payments.

Zwaan et al. (2012) studied the diagnostic efforts of 72 physicians examining dyspnea (shortness of breath) using a chart review of 247 patients over the course of 9 months. "Suboptimal cognitive acts" were found in 66% of the patients. In 34% of the patients harmed, diagnostic errors were found. The most significant factors in diagnostic errors were during the history and physical exam, where these types of errors most likely lead to patient harm (Zwaan et al., 2012). A failure to consider alternative diagnoses was identified as a cause of diagnostic error.

While there likely may be reason to question some of the data regarding the rates of medical errors, most would agree that every effort should be made to reduce them, to the degree possible given the moderate scarcity of resources. Medical errors that result in harm would be a violation of the obligation to the ethical principle of nonmaleficence. While not causing harm, but failing to provide the patient with benefit that the patient would have reasonably expected, medical errors would be a violation of the obligation to the ethical principle of beneficence. Failure to even discuss a medical error with the patient would be a violation of the obligation to the ethical principle of autonomy. Virtuous physicians and healthcare professionals, by their character, would avoid medical errors and admit them when they occur. Thus, medical errors are an important issue in the ethics of everyday medicine. Yet, there are many stakeholders and shareholders in the complex ethical ecology surrounding medical error that give rise to possibly conflicting obligations (Wears and Sutcliffe, 2020).

One then looks to a supervening moral theory to adjudicate the conflicting obligations. A Deontological moral theory would be problematic if it were to assert that no medical error can be tolerated as the effort likely required to eliminate medical errors could bankrupt the healthcare system, particularly given the extraordinary complexity of the ecology surrounding medical errors (Wears and Sutcliffe, 2020). Yet, at the same time, most reasonable persons blinded to their self-interest would find a Utilitarian moral theory unsatisfying given its inherent inclination to abandon some patients (microlevel justice) for the greater good of society (macrolevel justice). An Egalitarian moral theory likely would be moot if every patient was exposed to the same risk of medical error. A Libertarian moral theory may reduce the issues to a contract, yet it is highly unlikely that a contract would stipulate a negotiated acceptable risk of medical error. Failing to include such a term in the contract would seem to violate the notion of good faith necessary for an acceptable contract in a pluralistic modern liberal democracy. It is not clear then what effective ethical system exists for the management of medical errors. Perhaps one needs to be developed (see Chapter 3).

Some critics of the Institute of Medicine report argued that the report misrepresented the current state of medical care by its perhaps inflammatory rhetoric. The effect would be to cause patients to mistrust physicians and healthcare professionals and, interestingly enough, not the systems in which the physicians and healthcare professionals operate (see Chapter 4), on whose operations the physicians and healthcare professionals depend (see Case 12). These critics point out that relative to the total amount of healthcare delivered, the incidence of medical errors would be remarkably small (note that the definition of "remarkably small" seems to be a value judgment for which it is not clear that a group of reasonable persons blinded to their self-interest would necessarily share). Yet, it is not clear how this sentiment is much different from the response to protesters in the United States during the Vietnam war who were told "America, love it or leave it." That response seems to invite a false choice, either one has to "love it," which was taken as stop criticizing US involvement in the Vietnam war, or to leave their home in the United States. Relative to medical errors, it is not clear that countering the incidence of medical errors by the very high rate of the provision of good medical care (the last premise being arguable) is also not a false choice, that the one justifies the latter. The false choice risks foreclosing on efforts that could lead to a reduction in medical errors.

Considerable controversy regarding the report by the Institute of Medicine (Kohn et al., 2000) centers on the very use of the term "error" (Wears and Sutcliffe, 2020). Some argue that the term "error" creates a false impression of a mistake on the part of an individual such as a physician or healthcare professional. A distinction is made between error and adverse effects. For example, many feel that use of the term error leads too easily to the presumption of human error on the part of a particular individual and becomes an expediency to move on from the error. Such dismissal by attribution of the adverse effects to an error committed by an individual risks underestimating the great complexity surrounding the provision of medical care, and thus contributory or even primary causal factors, other than individual error, are ignored. Such factors ignore the risk of them being repeated in the future.

There also is a political aspect of the use of the term "error" arising from medical liability resulting from an error. There was a movement where physicians and healthcare professionals admitted to patients and their families and caregivers that mistakes were made. Perhaps the reasoning was that patients and families would be more forgiving of physicians and healthcare professionals who admitted errors, perhaps leading to a lower risk of a medical malpractice lawsuit. Perhaps it would bolster a sense of virtue in those physicians and healthcare professionals by acknowledging their errors with the presumption that such acknowledgments would reduce the risk for repeated errors. However, the movement failed almost as soon as it started (despite periodic resurrection of the idea) and medical malpractice defense attorneys nearly unanimously advise against any admission of error. It is advice that the human nature of physicians and healthcare professionals likely is predisposed to follow (Detsky et al., 2013). Further, critics were concerned that use of the term "error" would alienate the very physicians and healthcare professionals that patient safety experts need to recruit to reduce medical error. It is interesting that Wears and Sutcliffe (2020) note the paucity of use of the term "error" in professional writings on patient safety.

If one argues that error in the context of an error committed by an individual physician or healthcare professional is a misappropriation of the term "error," then perhaps error could be attributed to the system in which the physician or healthcare professional operates. Thus, systems' errors became the focus and grabbed the attention of safety experts and researchers from a variety of fields, such as engineering, psychology, sociology, and business administration. The success of involving nonphysicians in any leadership role appears to have failed to a large extent (Wears and Sutcliffe, 2020). As Wears and Sutcliffe write, medical errors were seen as a subset of medicine rather than medical errors seen as a subset of the larger science of safety that involves industry, engineering, psychology, and sociology.

If the term "error" is not to be used, then what term should be used? It appears that much of the patient safety literature focuses on the term "adverse effects." Yet, this term requires unpacking, in the philosophical tradition. The adjective "adverse" suggests harm. The adjective, however, modifies and thus takes its connotation from the term "effects." Effects imply the consequence of some action taken and, in this sense, is different from the harm due to the natural history of the disease or disorder the patient presents with. Certainly, the chest pain associated with cardiac disease is harmful. Who would welcome it? But the harm of chest pain resulting from heart disease is not likely attributable to the actions of physicians and healthcare professionals. Rather, the term "adverse effects" has an iatrogenic connotation, harm induced by the physician or healthcare professional.

The distinction between adverse effects and error offered by some is that not all adverse effects are errors while not all errors are adverse effects. The latter phrase "not all errors are adverse effects" is important. One can apply the Merriam-Webster dictionary definitions that follow (https://www.merriam-webster.com/dictionary/error):

> An act or condition of ignorant or imprudent deviation from a code of behavior
> An act involving an unintentional deviation from truth or accuracy
> An act that through ignorance, deficiency, or accident departs from or fails to achieve what should be done

The first definition suggests error as a deviation from standard behavior. Deviations need not lead to harm and the first definition does not impute harm or injury. The second definition relates to a deviation from truth or accuracy, again no harm in the sense of maleficence. This is particularly evident in the third definition where what should have resulted from an action did not happen. In the context of the case presented, what did not happen was the provision of an accurate diagnosis and appropriate treatment by other physicians. Interestingly, none of the definitions imply intention or, indeed, any blame or culpability.

Given the dictionary definitions of error, it is interesting to question where the pejorative connotations of the term "error" in medicine came from. The seriously pejorative nature of medical error is seen in the lengths to which such medical errors are kept secret, particularly in the early codes of medical ethics (see Chapter 4). The pejorative connotation of the term "error" does not follow necessarily from the dictionary definition of error. Disassociating the pejorative connotation from the definition of error, although consistent with the dictionary definition, may mute the ethical power of the term "error" to strive for justice, for example, a response to error inviting inspection and introspection. While critics of the Institute of Medicine report lamented use of the term "error," many noted the rhetorical value of the term to spur action (Wears and Sutcliffe, 2020). Perhaps the rhetorical use of error could lead to righteous indignation at the injustice resulting from medical error, with righteous indignation being an ethical virtue according to Aristotle (see Chapter 2). However, because the pejorative connotation of error has become so endemic and epidemic, nothing productive may come from using the term "error" in the vernacular.

The dictionary definitions of error suggest unintentional actions. Actions leading to adverse effects may be unintentional in the sense that the physician or healthcare professional did not know better. Perhaps more importantly, a consensus of reasonable persons blinded to their self-interest would not have expected the physician or healthcare professional to know better. The consequence would still be an error, but not the kind of error denoted by the pejorative connotation. Another sense of unintentional action follows from Thomas Aquinas' Principles of Double Effect (1485). His principles attempted to balance the obligation to nonmaleficence against the obligation to beneficence. An action may be necessary to fulfill the obligation to beneficence while at the same time violating the obligation to nonmaleficence, for example, causing bodily pain during the course of surgery important to the patient's well-being. Aquinas' principles included the following: (1) the action is either morally good or at least neutral, (2) the bad effect is not the means by which to produce the good effect, (3) the purpose must be to achieve the good effect only, and (4) the good effect must be at least important in the positive sense as is importance of the bad effect in the negative sense. The pain and injury of surgery is an adverse effect, it is harm, it is an act committed by the surgeon, and it cannot be said to be unforeseeable. Yet, Aquinas' second principle holds that the pain is unintentional in the sense that the inflicting of pain was not the means to provide benefit, such as the removal of a tumor. In this sense, pain is dissociated from the necessity of making an incision where making the incision was the intent. The pain suffered would not be an error as a consensus of reasonable persons blinded to their self-interest is likely to hold.

Perhaps the notion of adverse effects that is closest to the notion of error without having to use the term "error," and is most relevant to the case presented, is the notion of preventable adverse effects. Here, the distinction is made between adverse effects that should have been prevented and adverse effects that could have been prevented. The notion of preventable adverse effects that should have been prevented implies prior knowledge and means to prevent the adverse effects. The latter notion of preventable effects that could have been prevented is conditional, meaning that *had* there been sufficient knowledge and means, the adverse effect may have been prevented. The notion of should have been prevented, as different from could have been prevented, necessarily means that the knowledge and means were available in contrast to the condition of could have been prevented. The case presented might suggest that the misdiagnosis and mistreatment of the patient by previous physicians should have been prevented as the knowledge and means existed for the prior physicians; after all, the knowledge and means were available to the last successful physician.

One conclusion, although not the only or necessary conclusion, from the case presented is that the system for educating and supervising previous physicians failed to provide the knowledge and means that did exist and would have prevented the error. This could be taken as meaning the system for educating and supervising physicians should have—within the constraints imposed by the moderate scarcity of resources as described earlier. But note it may be that the notion of should requires more nuance. Perhaps the notion of should in adverse effects that should have been prevented should be rephrased as efforts for which there is sufficient prior knowledge and experience to prevent the adverse effects but society has deemed these efforts as not worthy of effecting.

The situation just described would be an example of justice on a macrolevel (see Chapter 1). But disputes are at the microlevel surrounding a particular individual patient. What is often unresolved is how to balance microlevel injustice in order to achieve macrolevel justice. In a pluralistic modern liberal democracy, perhaps the only means is by a consensus of those governed by the relevant ethic. For the ethic to be justified, hence ensuring a just outcome, it is necessary for the consensus to be obtained such that it is reasonable to apply the consensus to everyone (see Chapter 3).

Society seems to insist on a modicum of justice at the microlevel. An example is telishment, such as when the physician is held responsible for the failings of a charitable hospital that were beyond the physician's control (see Chapters 1 and 5 and Cases 2 and 13). The individual patient deserves justice no matter from whom the justice is extracted. In the case presented, it can be argued that the physicians who failed to properly diagnose and treat should be held liable but what about the system that educates and supervises? While ethical analysis would suggest that the system is culpable, it may mean telishment is required to protect the system and thus the sole liability falls to the physicians (see Case 12).

The issue of micro- and macrolevel justice plays out in the issue of autonomy, which may be readily apparent in the context of an individual patient (microlevel), where obligations of autonomy to the individual conflict with the presumption of autonomy to a group (a macrolevel). If autonomy is dictated at the macrolevel, then the obligation to the principle of individual autonomy on the microlevel is inherited from the autonomy to the group. In this case, a Principlist perspective with the obligation to the ethical principle of autonomy to the individual appears to conflict from the actual obligation shown and considered just at the macrolevel (McLeod and Sherwin, 2000). Whether there are concerns unique to the group, such as social or relational autonomy, that are different from particularization to the individual is an open question. Is group autonomy just an aggregate of the autonomies of individuals? Some feminists have argued that social autonomy transcends the individual, which could suggest some obligation of the individual to the group and the potential of the individual's autonomy to be subrogated to the autonomy of the group (Chapter 2). It is not clear how such a position would work in a pluralistic modern liberal democracy.

Responses from some professional medical educators

An analysis of the diagnostic errors and subsequent mistreatment in the case presented scrutinized the use of pattern recognition in diagnosis. A paper describing the exact case presented here, supporting factual data and logical analyses, was submitted to a leading journal on medical education. The concern was whether a popular form of medical education, called problem-based learning (PBL), risks reinforcing pattern recognition unless counterbalanced by an alternative by an approach of reasoning from principles. One alternative is a hypothetico-deductive approach where knowledge of the principles leads to a hypothesis of both what might be in the differential diagnosis (the possibilities other than psychogenicity) and would not be included. This would be followed by a rational scheme constructed to test the hypothesis.

To be sure, PBL is not a monolithic construct and its actual use in medical education most often is associated with a hybrid approach of both pattern recognition and an alternative such as the hypothetico-deductive approach. However, as discussed above, a great many physicians further in their training and experience default to a more likely use of pattern recognition. The paper was rejected but the reviews are telling. The first review, in part, wrote:

> The author is correct, I believe, to define PBL as he does and to suggest that it is pervasively common in medical education. He oversimplifies the respective roles of pattern recognition and the use of basic science principles in PBL pedagogy, however. And he errs in supposing that pattern recognition disproportionately contributes to diagnostic error compared to hypothesis formulation and testing using basic science. In dual process models of cognition currently au courant in education research,

pattern recognition corresponds to rapid type I apprehension of clinical problems through heuristics (including pattern recognition), while hypothetico-deductivism corresponds to the slower and deliberate type II thinking which is reflective and analytical. The author's suspicion of type 1 thinking [author – intuition which fundamentally is pattern recognition] and his preference for type II [author- analytical, perhaps logical] certainly has a venerable precedent in cognition and problem solving research and, incidentally, among medical educators. That suspicion is not borne out by recent research in medical education or elsewhere, however.

The author cites some of the earlier research on medical expertise (Schmidt and Rikers, 2007; Patel et al., 1990); he does not, however, cite recent data suggesting that both type 1 and type 2 thinking can be equally error producing and that knowledge deficits may be more important sources of diagnostic error than cognitive style [see Norman's recent review in Academic Medicine, ref. below (Norman et al., 2017)]. His argument that pattern recognition inevitably involves logical fallacies is faulty. Astute physicians do not make use of pattern recognition in the rigid fashion that the author depicts, such that failure of a case to correspond with a repertoire of patterns inevitably leads to consigning the case to a wastebasket pattern such as "psychological problems." In the case the author describes, a good physician would have recognized not only that the case did not fit usual causes of neck pain, but that it was also inconsistent with psychological causes of neck pain. Rather than premature closure on an incorrect diagnosis, the lack of fit of the case to available patterns would have led to a transition from type 1 to type 2 thinking and, perhaps, to research leading to the right diagnosis or at least to a recognition that the case was unresolved and, thus, to consultation. The author's portrayal of pattern recognizers inevitably falling prey to fallacies is a caricature. Doubtless there are physicians who do fall prey to fallacies, but the author has not shown that this is inevitable or even that pattern recognition is particularly conducive to error compared to other approaches to problem solving. Norman suggests that either conclusion would be mistaken in the present state of our knowledge. The author needs to reconsider his aversion to pattern recognition (compared to other approaches to problem solving) in light of current cognitive science and educational research.

Aside from some of the inaccuracies, the author did not single out pattern recognition as uniquely culpable. Indeed, there was no discussion of reasoning errors other than pattern recognition. However, this selectivity is not equivalent, suggesting pattern recognition is the sole problem in misdiagnosis (see Montgomery, 2019). The results of the systematic surveys were presented only to show the magnitude of the problem of diagnostic error and to the degree pattern recognition (type II errors in the reviewer's terms) may contribute to misdiagnosis is worthy of consideration in its own right. The author was not making anywhere near the sweeping judgments the reviewer seems to attribute to the author. In a sense, the reviewer makes a straw man argument by misrepresenting the argument in the paper to an extreme not actually in the paper. The more telling response is seen in the reviewer writing "a good physician would have recognized not only that the case did not fit usual causes of neck pain, but that it was also inconsistent with psychological causes of neck pain." Indeed, that is exactly what the last neurologist did and consequently made the correct diagnosis and then provided the appropriate treatment. However, the prior physicians then could not be "good physicians" by the reviewer's standard. Given the frequency of the initial misdiagnosis of cervical dystonia, the reasonable conclusion would be that there are a lot of not "good" physicians in practice. But then, is not this, in fact, an indictment of the educational system?

A second reviewer wrote:

> This article is written and presented as an interesting personal viewpoint trying to explain still quite high rate of diagnostic errors in modern day medicine by focusing on potential weaknesses of PBL-based training in contrast to hypothetico-deductive approach. Although it is well written and interesting to read, this essay is nothing more than that: a personal reflection. It cannot be regarded as a review or any other way of scientific communication. If that is the intention as far as publishing it, I find it intriguing.

The same evidentiary support written in the discussion above was provided in the paper. The question becomes, where did the attribute of "personal reflection" come from? The same issues of the logic in pattern recognition, that is the fallacy of limited alternatives was presented. It is not clear that a discussion of logic is any more a "personal reflection" than is mathematics. The more critical concern here is whether the second reviewer's dismissal of the paper as it was not a "scientific communication" was because of the ethical concerns raised. To be sure, this sentiment may merely be idiosyncratic to the second reviewer. However, if it is representative of a wider swath of medical experts, then it is hard to see how analysis and understanding the ethics of everyday medicine will gain traction in the teaching of medicine, where a form of scientism is prevalent (Montgomery, 2019).

Rawls' veil of ignorance

The kind of case presented here is not atypical in the practice of everyday medicine and therefore its questionable ethics likewise are not atypical. It is likely no shareholder or stakeholder would argue that the ethical principles of beneficence, nonmaleficence, and autonomy were not operative. Rather, it is the sense of obligation to the ethical principles that is the source of the ethical problem. The magnitude of the obligations can be estimated by the consequences of minimizing the obligations. The consequences are understood in the context of the moral theories taken as prevailing. It is the nature of the relationship between all shareholders and stakeholders that is defined by the prevailing moral theories as discussed earlier. It is the particular application of a moral theory that defines the justice in the particular case.

In the setting of moderate scarcity of resources where prioritization or rationing is necessary, a Deontological moral theory proves difficult to apply. A Deontological moral theory would seem to focus on microlevel justice; in other words, at the level of the individual patient as discussed previously. Prioritization or rationing clearly results in not all persons being treated exactly the same, which may violate both Deontological and Egalitarian moral theories, yet in terms of a Libertarian or Utilitarian moral theory, it provides a macrolevel justice. The Kantian notion of a Deontological moral theory holds that every person is an end unto him- or herself and must not be treated as a means to someone else's ends. In the case presented, the argument can be constructed that what is necessary is to provide everyone the correct diagnosis and that treatment is an inherent moral necessity. Given the possible practical impossibility, the argument turns to providing every patient with the same probability of actually being provided the correct diagnosis, an Egalitarian moral theory. The argument that any significant or meaningful probability of achieving proper diagnosis and treatment is a contractual matter follows from a Libertarian moral theory. From a Utilitarian moral theory, whatever is necessary to achieve macrolevel justice is the good to be obtained irrespective of what happens at the microlevel.

Actual experiences suggest that none of the moral theories individually seem to completely resolve what most reasonable persons blinded to their self-interests find as just. It is not likely that any patient, except those held sufficiently contemptuous, would be denied some modicum of care, even if the sole purpose is to prevent imminent death or excruciating pain. Denial of any care is certainly a logical possibility; indeed, a necessary extension of a Libertarian moral theory and such denial is a possibility under a strictly Utilitarian moral theory. Rather, such care may be justified by a Deontological moral theory as a right if that right has not been forfeited by contemptuous actions. However, the operational definitions of "contemptuous," "imminent death," or "excruciating" can be argued. Providing only a modicum of care to some persons and presumably more care to others would violate an Egalitarian moral theory. Experience suggests that most often the prevailing moral theory is some hybrid of the aforementioned moral theories. Yet, this author is unaware of any systematic attempt to create such hybrid moral theories, at least in the context of everyday medicine.

The question is how does society establish a framework for implementing the obligations to the ethical principles, particularly the principle of beneficence? A guide could be the consequences of establishing an acceptable probability threshold of obtaining a correct diagnosis where the physician, healthcare professional, and society trip from inaction, allowing the status quo, to action to create a new system of justice. The process of establishing the framework would be problematic if only those persons with cervical dystonia were to decide. It would similarly be problematic to have only those persons without cervical dystonia or at a low probability of suffering cervical dystonia, based on age and sex, be the deciding shareholder. Facing this conundrum, John Rawls calls for the notion of a veil of ignorance, writing:

> The idea of the original position is to set up a fair procedure so that any principles agreed to will be just. The aim is to use the notion of pure procedural justice as a basis of theory. Somehow we must nullify the effects of specific contingencies which put men at odds and temp them to exploit social and natural circumstances to their own advantage. Now in order to do this I assume that the parties are situated behind a veil of ignorance. They do not know how the various alternatives will affect their own particular case and are obliged to evaluate principal solely on the basis of general considerations.
>
> *Rawls (1999, p. 118)*

The question becomes, do the healthcare delivery systems in the everyday practice of medicine follow the veil of ignorance? (see Chapter 3). Who is at the center or focus of the ethical effort is a critical consideration (see Chapter 4).

References

Aquinas, T., 1485. Principles of double effect. In: Summa Theological.

Bertakis, K.D., Azari, R., Callahan, E.J., 2004. Patient pain in primary care: factors that influence physician diagnosis. Ann. Fam. Med. 2 (3), 224–230.

Bertram, K.L., Williams, D.R., 2016. Delays to the diagnosis of cervical dystonia. J. Clin. Neurosci. 25, 62–64. https://doi.org/10.1016/j.jocn.2015.05.054.

Bhatia, K.P., Schneider, S.A., 2007. Psychogenic tremor and related disorders. J. Neurol. 254 (5), 569–574.

Blum, M., Slade, M., Boden, D., Cabin, H., Caulin-Glaser, T., 2004. Examination of gender bias in the evaluation and treatment of angina pectoris by cardiologists. Am. J. Cardiol. 93 (6), 765–767.

Colameco, S., Becker, L.A., Simpson, M., 1983. Sex bias in the assessment of patient complaints. J. Fam. Pract. 16 (6), 1117–1121.

Detsky, A.S., Baerlocher, M.O., Wu, A.W., 2013. Admitting mistakes: ethics says yes, instinct says no. CMAJ 185 (5), 448. https://doi.org/10.1503/cmaj.121187.

Falk, H., Henoch, I., Ozanne, A., Öhlen, J., Ung, E.J., Fridh, I., Sarenmalm, E.K., Falk, K., 2016. Differences in symptom distress based on gender and palliative care designation among hospitalized patients. J. Nurs. Scholarsh. 48 (6), 569–576. https://doi.org/10.1111/jnu.12254.

Fromm, E., 1955. The Sane Society. Holt Paperbacks, Reissue edition (Oct. 15 1990).

Hashem, A., Chi, M.T., Friedman, C.P., 2003. Medical errors as a result of specialization. J. Biomed. Inform. 36 (1–2), 61–69.

Hermans, H.J., 1999. Self-narrative as meaning construction: the dynamics of self-investigation. J. Clin. Psychol. 55 (10), 1193–1211.

Iedema, R., Jorm, C., Lum, M., 2009. Affect is central to patient safety: the horror stories of young anaesthetists. Soc. Sci. Med. 69 (12), 1750–1756. https://doi.org/10.1016/j.socscimed.2009.09.043.

Jankovic, J., Vuong, K.D., Thomas, M., 2006. Psychogenic tremor: long-term outcome. CNS Spectr. 11 (7), 501–508.

Jankovic, J., Tsui, J., Bergeron, C., 2007. Prevalence of cervical dystonia and spasmodic torticollis in the United States general population. Parkinsonism Relat. Disord. 13 (7), 411–416.

Jones, R.F., Korn, D., 1997. On the cost of educating a medical student. Acad. Med. 72 (3), 200–210.

Kamble, N., Prashantha, D.K., Jha, M., Netravathi, M., Reddy, Y.C., Pal, P.K., 2016. Gender and age determinants of psychogenic movement disorders: a clinical profile of 73 patients. Can. J. Neurol. Sci. 43 (2), 268–277. https://doi.org/10.1017/cjn.2015.365.

Kohn, L.T., Corrigan, J.M., Donaldson, M.S. (Eds.), 2000. To Err is Human: Building a Safer Health System. Institute of Medicine (US) Committee on Quality of Health Care in America. National Academies Press, Washington, DC.

McLeod, C., Sherwin, S., 2000. Relational autonomy, self-trust, and health care for patients who are oppressed. In: Mackenzie, C., Stoljar, N. (Eds.), Relational Autonomy: Feminist Perspectives on Autonomy, Agency, and the Social Self. Oxford University Press, Oxford.

Montgomery Jr., E.B., 2019. Medical Reasoning: The Nature and Use of Medical Knowledge. Oxford University Press, Oxford.

Montgomery, E.B., Wall, M., Henderson, V., 1985. Principles of Neurologic Diagnosis. Little Brown & Co., New York.

National Academies of Sciences, Engineering, and Medicine, 2015. Improving Diagnosis in Health Care. The National Academies Press, Washington, DC.

Nestler, D.M., Gilani, W.I., Anderson, R.T., Bellolio, M.F., Branda, M.E., LeBlanc, A., Phelan, S., et al., 2017. Does gender bias in cardiac stress testing still exist? A videographic analysis nested in a randomized controlled trial. Am. J. Emerg. Med. 35 (1), 29–35. https://doi.org/10.1016/j.ajem.2016.09.054.

Norman, G.R., Monteiro, S.D., Sherbino, J., Ilgen, J.S., Schmidt, H.G., Mamede, S., 2017. The causes of errors in clinical reasoning: cognitive biases, knowledge deficits, and dual process thinking. Acad. Med. 92 (1), 23–30.

Patel, V.L., Groen, G.J., Arocha, J.F., 1990. Medical expertise as a function of task difficulty. Mem. Cognit. 18 (4), 394–406.

Quine, W.V.O., 1960. Word and Object. The MIT Press.

Rawls, J., 1999. A Theory of Justice, revised ed. The Belknap Press of Harvard University Press, Cambridge, MA118.

Razmjou, H., Lincoln, S., Macritchie, I., Richards, R.R., Medeiros, D., Elmaraghy, A., 2016. Sex and gender disparity in pathology, disability, referral pattern, and wait time for surgery in workers with shoulder injury. BMC Musculoskelet. Disord. 17 (1), 401.

Robson, C., Myers, L., Pretorius, C., Lian, O.S., Reuber, M., 2018. Health related quality of life of people with non-epileptic seizures: the role of socio-demographic characteristics and stigma. Seizure 55, 93–99. https://doi.org/10.1016/j.seizure.2018.01.001.

Ruane, L., Greenslade, J.H., Parsonage, W., Hawkins, T., Hammett, C., Lam, C.S., Knowlman, T., et al., 2017. Differences in presentation, management and outcomes in women and men presenting to an emergency department with possible cardiac chest pain. Heart Lung Circ. 26 (12), 1282–1290. https://doi.org/10.1016/j.hlc.2017.01.003.

Schmidt, H.G., Rikers, R.M., 2007. How expertise develops in medicine: knowledge encapsulation and illness script formation. Med. Educ. 41 (12), 1133–1139.

Segal, L., Marsh, C., Heyes, R., 2017. The real cost of training health professionals in Australia: it costs as much to build a dietician workforce as a dental workforce. J. Health Serv. Res. Policy 22 (2), 91–98. https://doi.org/10.1177/1355819616668202.

Smith, J.F., 1996. Communicative ethics in medicine: the physician-patient relationship. In: Wolf, S.M. (Ed.), Feminism and Bioethics: Beyond Reproduction. Oxford University Press, New York. (Chapter 7).

Stålnacke, B.M., Haukenes, I., Lehti, A., Wiklund, A.F., Wiklund, M., Hammarström, A., 2015. Is there a gender bias in recommendations for further rehabilitation in primary care of patients with chronic pain after an interdisciplinary team assessment? J. Rehabil. Med. 47 (4), 365–371. https://doi.org/10.2340/16501977-1936.

Wears, R.L., Sutcliffe, K.M., 2020. Still Not Safe: Patient Safety and the Middle-Managing of American Medicine. Oxford University Press, Oxford.

West, C., 1983. "Ask me no questions": an analysis of queries and replies in physician-patient dialogues. In: Todd, A.D., Fisher, S. (Eds.), The Social Organization of Doctor-Patient Communication. Center for Applied Linguistics, Washington, DC, pp. 75–106.

Wyatt, K.D., Branda, M.E., Inselman, J.W., Ting, H.H., Hess, E.P., Montori, V.M., LeBlanc, A., 2014a. Genders of patients and clinicians and their effect on shared decision making: a participant-level meta-analysis. BMC Med. Inform. Decis. Mak. 14, 81. https://doi.org/10.1186/1472-6947-14-81.

Wyatt, C., Laraway, A., Weatherhead, S., 2014b. The experience of adjusting to a diagnosis of non-epileptic attack disorder (NEAD) and the subsequent process of psychological therapy. Seizure 23 (9), 799–807. https://doi.org/10.1016/j.seizure.2014.06.012.

Zwaan, L., Thijs, A., Wagner, C., van der Wal, G., Timmermans, D.R., 2012. Relating faults in diagnostic reasoning with diagnostic errors and patient harm. Acad. Med. 87 (2), 149–156.

Case 10

Case 10—A 32-year-old female was admitted following a motor vehicle accident. She asked for a consultation by a neurologist as she was complaining of weakness in her right upper extremity since the accident. There were no sensory complaints except for headache. The examination was remarkable for "give way" weakness in the right upper extremity, meaning the patient would generate an inconsistent muscle force, such as involved in pushing away the examiner's hand when asked to make a sustained maximum force. When the patient was asked to make a maximal tight handgrip with the wrist flexed (bent down), she failed to dorsiflex her wrist (bend upward), suggesting that the patient was not actually making a maximal effort. The conclusion was that the weakness likely was a conversion symptom or psychogenic and not evidence of physical injury to the nervous system. The neurologist suggested that keeping the patient in the hospital overnight as is routinely done in patients who may have sustained a serious trauma. In the presence of medical students and residents, the attending physician responded that he was going to "kick" the patient out. No arrangements for follow-up care were made.[a]

Medical facts of the matter

The nature of the patient's symptoms and signs of weakness was inconsistent with known anatomical and physiological facts. In other words, based on a very long history of neurological research and clinical experience, giving rise to fundamental principles and facts, there is no location in the nervous system that, when damaged, would produce the patient's symptoms and signs. Then what is causing the patient's symptoms and signs? The presumption was that if the weakness was not physical, then what nonphysical entity could cause the patient's symptoms and signs? Whatever the cause, the presumption is that it is not physical, hence the necessity of inventing something that is not physical, yet still, it must affect the physical body. Is it the mind, spirit, or soul? If it is a disorder of the mind, spirit, or soul, how are physicians and healthcare professionals to deal with it?

A full discussion of mental illness is far beyond the scope of this text. However, certain themes seem to characterize the many and varied conceptions of mental illness. Throughout history, supernatural, somatic (a physical cause), and psychogenic (a nonphysical cause) have been considered as possible causes. The first two are clearly different from the last. A supernatural cause, such as demonic possession, has a cause outside the person that comes to "infect" the person and, generally, the

---

[a] The case presented is hypothetical, although modeled on experience. No judgment is made relative to the correctness, appropriateness, or value of any actions described on the part of any of the participants. The purpose here is to describe and explore the variety of ethical considerations and consequences that depend on the perspective of different relevant ethical systems.

The Ethics of Everyday Medicine. https://doi.org/10.1016/B978-0-12-822829-6.00010-2

patient has no control. The somatic notion holds that these types of illness are caused by abnormal physiological processes within the person, generally of the type over which the patient has no control. Increasingly, the population appreciates that many psychiatric disorders, such as schizophrenia and major depression, follow the medical model of some as a yet unidentified physical process. A corollary to the medical model is that the symptoms and signs manifest by the patient may be inconsistent with known fundamental neurological facts and principles, yet the faith is that the issue will be resolved with further scientific research. In the case of the patient presented, the presumption is there is no physical cause and any amount of research will not demonstrate a physical cause. If the possibility of a physical cause was considered, how would anyone countenance the behavior of the physician in the case presented? Those holding a phenomenological, care, or virtue ethic might expect the physician to act with grace and compassion toward the patient as any other patient according to what a patient-centered ethic might require (see Chapter 2). Yet, it cannot be presumed that such ethics are necessarily the controlling force in the ethics of everyday medicine (see Chapter 4).

Psychogenic places the cause in the sociological and psychological conditions of the patient, for example, how the patient was raised or what they have experienced. Perhaps most importantly, many may hold that it is up to the patient to "rise above" those conditions. Patients fail to do so because of some intrinsic weakness or imperfection, and thus the "blame" appears to revert to the individual patient despite the sociological or psychological situations or circumstances. With respect to psychogenic disorders, attitudes vary as to the control the patient has in the development of the psychogenic disorder. To the degree that patients are seen as in control or that their condition is a consequence of their voluntary choice, such as drug addiction, many persons hold stigmatizing attitudes toward the patient (Barry et al., 2014). As discussed later, feigning illness in the context of seeking some benefit is likely to be viewed as unjust (Chapter 1).

The clinical examination provided strong support for a presumptive diagnosis of a disorder commonly thought to be psychological in origin. Yet, there is a conundrum. If the patient's disorder was thought to be the result of neurological dysfunction, then the neurological facts and principles would have to be modified to accommodate those of the patient. This issue is addressed later.

In the case presented, the discussion addressed the syndrome of the patient presenting with neurological symptoms and signs and was diagnosed as having a functional neurological disorder. The term functional neurological disorder can be understood by the alternative diagnosis, which is an organic neurological disorder. The terminology itself presupposes that it was a functional neurological disorder, meaning without evidence of histopathology and not explainable by neurophysiology or pathophysiology, and more largely not physical, real, or organic. What remains is artificial. The wider issues of mental, psychological, and psychiatric disorders are not discussed but the issues related to functional neurological disorders are illuminated.

There could be multiple causes or disorders associated with the patient's symptoms and signs. In view of the examination, consideration of organic or physical causes, in a sense of physical injury to the nervous system, does not seem necessary. Indeed, pursuing them inappropriately could lead to maleficence, such as exposing the patient to the risks of unnecessary testing. However, there is a differential diagnosis related to a functional neurological disorder. Some of the possible diagnoses can be associated with a severe prognosis and most are treatable (Espay et al., 2019; Zeuner and Sidiropoulos, 2019). Some of the elements of differential diagnoses are listed here from Michael B. First in "DSM-5 Handbook of Differential Diagnosis" (American Psychiatric Association Publishing; https://doi-org.ezproxy.library.wisc.edu/10.1176/appi.books.9781585629992):

Differential Diagnosis for Conversion Disorder (Functional Neurological Symptom Disorder)

1. Occult neurological or other medical conditions, or substance/medication-induced disorders
2. Somatic Symptom Disorder—is characterized by distressing somatic symptoms accompanied by excessive thoughts, feelings, or behaviors related to the somatic symptoms or associated health concerns without regard to whether the somatic symptoms are adequately explained by a medical condition. In contrast, in Conversion Disorder (typically functional neurological disorder), clinical and/or laboratory findings must provide evidence that the neurological symptoms are incompatible with recognized neurological or general medical conditions (author—the notion of what is recognizable poses serious problems as discussed in Cases 6 and 9)
3. Illness Anxiety Disorder—is characterized by a focus on the "serious disease" underlying the pseudoneurological symptoms
4. Depressive Disorders
5. Factitious Disorder or Malingering
6. Dissociative Disorders—involve neurological functions (e.g., memory and consciousness) other than voluntary motor or sensory functioning

The differential diagnoses of dissociative disorders, which must be considered in the differential diagnosis of the patient in the case presented, are listed here (from First, "DSM-5 Handbook of Differential Diagnosis," American Psychiatric Association Publishing; https://doi-org.ezproxy.library.wisc.edu/10.1176/appi.books.9781585629992):

1. Depersonalization/derealization disorder experiences of depersonalization
2. Dissociative symptoms due to a general medical condition
3. Substance intoxication or substance withdrawal
4. Depersonalization/derealization symptoms attributable to the physiological effects of substances during acute intoxication or withdrawal
5. Dissociative identity disorder
6. Panic attacks
7. Posttraumatic stress disorder or acute stress disorder
8. Psychotic disorders (e.g., Schizophrenia)
9. Major depressive disorder
10. "Normal" symptoms of depersonalization or derealization

As can be readily appreciated, some of the elements in differential diagnoses to be considered require the expertise of a psychiatrist, who was not made available to the patient in the case presented, and some of the disorders that are reasonable to consider are associated with a significant risk of harm. The disorders in the differential diagnoses are treatable, at least in the sense that the associated disabilities can be managed and risks anticipated and prevented. However, this first would require a diagnosis to be established, which would require evaluation by a psychiatrist as well. In terms of functional neurological disorders, diagnostic criteria are quite sound, with an estimated misdiagnosis rate of approximately 4% (Gelauff and Stone, 2016). Thus, the ability to make the correct diagnosis is not the issue; the critical question is once the diagnosis is made, how is the differential diagnosis pursued and treated?

Psychological versus "real" medical disorders

It is likely that the attending physician did not think the disorder to be psychiatric in the sense of requiring a psychiatrist. For example, if the patient was thought to have schizophrenia or manic-depressive illness, the attending physician likely would have involved a psychiatrist. Remarkable advances in medical treatments and insights into neurobiological mechanisms of disorders such as schizophrenia and manic-depressive illness may lead physicians to consider these types of psychiatric disorders more akin to medical or neurological illness and thus more "real." However, even if schizophrenia and manic-depressive illness were not considered "physical," the possibility of their existence likely would have generated a consultation referral to a psychiatrist. For whatever reason, the patient's problems were not considered to be analogous to these other psychiatric disorders. But then, what are these psychological problems and what is so different from them so as to generate the response from the attending physician?

Conflicting notions regarding the nature of psychological and psychiatric disorders can be seen in the recent controversies related to the fifth edition of the "Diagnostic and Statistical Manual (DSM-5)," published by the American Psychiatric Association (2013). The manual was intended as a rubric of possible psychiatric diagnoses for purposes of coding, documentation, and billing, and more. Merely making a diagnosis presupposes a reality (ontology) that, in the modern Rationalist/Allopathic medical tradition, presupposes a pathoetiology and pathophysiology. The DSM-5 has a list of diagnoses and their criteria. Historically, classifications were based phenomenologically on observational inferences delimiting similarities and differences (Montgomery, 2019). Such a process need not address or posit any underlying pathoetiological or pathophysiological mechanisms, causes, or principles. In doing so, the classification risks increasing divisions with an increasing resolution of similarities and differences and, by its logical extension, the Solipsism of the Present Moment, where each patient is taken de novo without relation to the past or other patients in the present. In contrast, in a Rationalist/Allopathic physician, such a strictly phenomenology-based system represents uncertainty. What is a disease one day is not a disease the next day, if not rooted in specific neuroscientific principles. An allopath seeks to understand the variety of psychological and psychiatric disorders as variations on a fundamental economical set of principles. The great challenge for psychological and psychiatric disorders was the lack of demonstrable principles, particularly related to histopathological (observable changes in the structure of the body from what is considered normal) correlations. The histopathological correlations with the manifestations of disease provided a more compact notion of disease and, most importantly, provided plausibility that the disease manifestations were real (Montgomery, 2019).

To be sure, pharmacological models of psychiatric disorders, familial, and genetic studies do provide some basis for viewing psychiatric disorders as physically real. However, advances in technology with the advent of neurometabolic imaging, such as positron emission tomography and functional magnetic resonance imaging, have stoked the allopaths' ambitions. Neurometabolic imaging provided what traditional histopathology failed to provide for psychiatric disorders. An example of allopathic ambitions, and hence dissatisfaction with the DSM-5, is seen in the response by Thomas Insel, former director of the National Institute for Mental Health, who wrote:

> The goal of this new manual [DSM-5], as with all previous editions, is to provide a common language for describing psychopathology. While DSM has been described as a "Bible" for the field, it is, at best, a dictionary, creating a set of labels and defining each. The strength of each of the editions of DSM has been "reliability": Each edition has ensured that clinicians use the same terms in the same ways. The weakness is its lack of validity. Unlike our definitions of ischemic heart disease, lymphoma, or AIDS, the DSM diagnoses are based on a consensus about clusters of clinical symptoms, not any objective laboratory measure. In the rest of medicine, this would be equivalent to creating diagnostic systems based on the nature of chest pain or the quality of fever. Indeed, symptom-based diagnosis, once common in other areas of medicine, has been largely replaced in the past half century as we have understood that symptoms alone rarely indicate the best choice of treatment. ... Patients with mental disorders deserve better.
> *(http://www.nimh.nih.gov/about/director/2013/transforming-diagnosis.shtml)*

However, promising neuroscientific methods such as neurometabolic imaging, may be, practical utility in the management of these diseases requires a disorder to be diagnosed today, and diagnoses must have some currency among patients, physicians, healthcare professionals, insurers, and regulatory agencies. The outcome of the debate over the DSM-5 was a compromise seen in the joint communication of Insel and Jeffrey A. Lieberman, MD, then president-elect, American Psychiatric Association:

> Patients, families, and insurers can be confident that effective treatments are available and that the DSM [DSM-5] is the key resource for delivering the best available care. The National Institute of Mental Health (NIMH) has not changed its position on DSM-5. As NIMH's Research Domain Criteria (RDoC) project website states, "The diagnostic categories represented in the DSM-IV and the International Classification of Diseases-10 (ICD-10, containing virtually identical disorder codes) remain the contemporary consensus standard for how mental disorders are diagnosed and treated.... Yet, what may be realistically feasible today for practitioners is no longer sufficient for researchers.
> *(http://www.nimh.nih.gov/news/science-news/2013/dsm-5-and-rdoc-shared-interests.*
> *shtml?utm_source=rss_readers&utm_medium=rss&utm_campaign=rss_summary)*

There is a social hazard in the Scientism inherent in the statement given by Thomas Insel. What if all the neuroscientific methodologies fail to demonstrate any abnormality in a particular disorder thought to be psychological or psychiatric? Does that mean that the disorder, typically called a mental disorder or illness, is relegated to the nonreal psychological and psychiatric? Does this not perpetuate the conception that psychological disorders are just different and, in being different, that the ethics of everyday medicine do not apply to them or are at least different? As will be demonstrated, neuroscientists are increasingly demonstrating physiological abnormalities in many psychological disorders, such as those described in the case presented. Perhaps, sometime in the future, such research will allow the ethical obligations to care for psychological disorders to gain some parity with the ethical obligations in the ethics of everyday medicine associated with any "physical" disease such as heart disease, stroke, and diabetes. A critical question to be explored later is why did this distinction come about in the first place? It is that process that has to be deconstructed so that medicine can be put right, perhaps.

Are psychological disorders treatable and, if so, should they be?

In a pluralistic modern liberal democracy, reasonable persons blinded to their self-interest would hold that every person, including patients with psychological problems, still has a right to the ethical principles of beneficence, nonmaleficence, and autonomy. The obligation by the rest of persons to the patient's right to these ethical principles defines justice in that society. No reasonable person would deny the patient's rights; rather, the debate lies in the obligations of others to respect and fulfill those obligations. But if it is impossible to fulfill those obligations, then justice becomes a moot issue. It is likely that most reasonable persons blinded to their self-interest would find this unacceptable.

Ethical analysis of the case presented must define what is the benefit derived from the ethical principles of beneficence, nonmaleficence, and autonomy to be afforded to the patient. In this particular case, the beneficence may be the restoration of normal function. No reasonable persons blinded to their self-interest would argue that continued weakness, with its attendant disabilities, would be the preferred state and that restoration of ability is not beneficence worth pursuing. However, the obligation to beneficence in the context of some other psychological problems is not so clear. In the complex ecology of healthcare, there are obligations to the ethical principles on behalf of others beyond the individual patient, physician, or healthcare professional. Perhaps expenditures on treatments of patients with psychogenic disorders reduce the resources to relieve the sufferings of other disorders or other societal priorities.

As will be seen, psychogenic disorders are treatable. A Deontological moral theory would hold that these patients should be treated as it is their inalienable inherent right, or it is merely everyone's inherent obligation to offer treatment to these patients. An Egalitarian moral theory might only require that what is done for one patient with psychogenic disorders should be done for all patients with psychogenic disorders. Note that this does not imply that what is done for patients with diabetes necessarily should be done for patients with psychogenicity. The key is to what degree is diabetes comparable to a psychogenic disorder such that the obligation to treatment (beneficence) on behalf of patients with psychogenic disorders is at parity with that for patients with diabetes? If parity is not asserted, then there must be something different about psychogenic disorders compared to diabetes, as will be discussed later.

A Utilitarian moral theory argument could be made for treatment in that a treated person is less likely to demand healthcare resources and likely not to "doctor shop" until some unnecessary test or treatment results in an iatrogenic (physician or healthcare professional induced) physical disease that would demand treatment. Unfortunately, data on the healthcare costs associated with iatrogenic disease in patients with psychogenic disorders are lacking, although there is some evidence of utilitarian efficacy (Russell et al., 2016). Yet, it may well be that despite all the positive evidence of utilitarian benefit, patients with psychogenic disorders likely are not offered treatment that would be offered patients with clearly physical illnesses. Interestingly, patients with borderline personality disorders were subject to highly negative stereotyping but this appears to be improving with a greater understanding of the disease and the potential for effective therapies (Day et al., 2018). A key to many of the negative stereotypes was the presumption that patients with borderline personalities were attention-seeking and manipulative. Yet, attitudes appear to be shifting, perhaps with a greater appreciation of potential treatments. For example, if the disability associated with a putative psychological problem is induced by society, in terms of discrimination, for example, then there is a sense where obligations to beneficence, nonmaleficence, and autonomy on the part of the patient would warrant treatment, perhaps if only as a means of compensation, or mitigation of continued harm to the patient and other patients.

Considering beneficence as improvement or remission of functional symptoms and disabilities, Gelauff and Stone (2016) conducted a systematic review of functional neurological disorders, limited to motor symptoms, such as in the case presented, and reported:

> An analysis of all studies weighted according to the size of the study found an overall figure of 40% of patients with the same or worse outcome at follow-up with only 20% of patients in the whole cohort showing complete remission. In four studies with 135 patients, 66–100% of patients had the same or worse symptoms at follow-up. In 14 studies with 533 patients, 33–66% of patients had the same or worse symptoms at follow-up and in only five studies with 464 patients, 33% of patients or fewer had symptoms the same or worse at follow-up.

As can be appreciated, the outcomes were mixed, which should not be surprising given the heterogeneity of functional symptoms included in the analysis. In a subgroup analysis of those patients presenting with weakness in four studies, the percentages that had remission were 63%, 50%, and 77% (one study result was not described), while improvement percentages were an additional 27%, 44%, 20%, and 4%. Combining outcomes, the percentages that were at least improved or in remission were 100%, 50% (this percentage was only improved as the percentage in remission was not described), 70%, and 82% (Gelauff and Stone, 2016).

As Gelauff and Stone (2016) point out, such systematic studies are fraught with methodological uncertainty, as evidenced by the great variety of observations. It is unclear whether the variety is a variance intrinsic to the population. It just may be that patients with functional neurological disorders merely are variable. The variance may be in the sample—it just may be due to the nature of those who came under observations. The variance may also be in the nature of the interventions, including no active treatments. Nevertheless, clearly many patients improved, particularly those whose chief complaint was a weakness. What is unknown is whether the interventions involved contributed to the prognosis, that is, future problems, and, if so, which were effective? For any number of reasons, prospective randomized controlled blinded trials would be very difficult in this population. At the very least, there is the potential to fulfill the obligation to beneficence to the patient described in the case presented. Further, earlier intervention is most predictive of beneficent outcomes. But it is clear that in the context of the case presented, this was not afforded to the patient. However, whether the lack of fulfilling the obligation to beneficence was unethical, and thus unjust, requires further analysis, particularly under the informing moral theories.

Even in the face of being unable to fulfill the obligation to beneficence, other than perhaps the beneficence of commiseration and condolences, the obligation to nonmaleficence remains. Note that a failure to provide beneficence is not the same as maleficence, as the latter implies the actual production or significant risk of harm. Certainly, patients with psychogenetic disorders are at significant risk of iatrogenic (harm induced by the physician or healthcare professional) harm. While mentioned frequently in the literature, this author has found it difficult to find a reasonable quantitative description of the risk.

The patient as suspect

In the typical circumstance of everyday medicine, this author has never met a physician or healthcare professional who was suspicious of a patient at the onset of the relationship, at least when the situation was clear. As seen in Case 9, a patient presenting with unfamiliar symptoms and signs may well be misdiagnosed as having a functional neurological disorder. This is not to say that the physician or healthcare professional did so based on suspicion on the veracity of the patient, although, as discussed later, this is not inconceivable. More likely, the inability of the physician to "make sense" of the patient's weakness, as in the case presented, occasioned suspicion (also see Case 9).

In the differential diagnosis of a functional neurological disorder, such as affected the patient in the case presented, is factitious disorder and malingering. Both of these imply an intention on the part of the patient to deceive. Suspicion of a factitious disorder or malingering would appear to disqualify the patient from beneficence, although maleficence to the patient likely would not be condoned. The problem is that many patients, such as the patient in the case presented, likely were not malingering or consciously manufacturing factitious complaints. The problem is that these patients are liable to be "painted with the same brush" as those with the factitious disorder or malingering. The key is to develop a differential diagnosis that suggests possibilities other than a factitious disorder or malingering (Bass and Halligan, 2016). Typically, a factitious disorder or malingering is suspected when there is a possibility of primary or secondary gains. Sympathy and/or comfort on the part of others may be a source of primary gain. Dispensation from obligations or possible compensation separate from treatment, such as disability payments or financial gains through legal actions, often immediately places the patient under suspicion. In the case presented, the potential for financial reward from the motor vehicle accident is a possibility, although it is not clear that presuming secondary gain is not a violation of the ethical principle of autonomy.

There may be the presumption that the patient is unwilling to make the effort to resolve the problem of psychogenic disabilities. Such a presumption would seem to follow the consideration of factious disorder or malingering. However, it is not clear that this is also an indictment of patients with other psychogenic disorders. The notion of equity, rewards commensurate with effort, may be in play. Interestingly, very young children appear to have a concept of equity (see Wang and Henderson, 2018). Whether the children's behaviors represent innate causative mechanisms, innate mechanisms for rapid learning, or merely learning are unclear. Nevertheless, an innate causative mechanism cannot be excluded (see Chapter 1).

Neurometabolic imaging studies and psychological studies reveal possible neurobiological mechanisms that may underlie equity (Tabibnia and Lieberman, 2007). In the situation of "anomalous" giving, that is, altruistic giving independent of any external positive or negative reinforcement, different regions of the brain demonstrate increased neurometabolic activity (assumed to reflect neuronal activity, although not necessarily) when making equitable and inequitable decisions. Regions activated with equitable decisions are similar to those associated with rewards, suggesting an intrinsic value to the person providing the equity. Regions activated with inequitable decisions include the anterior insula, which is activated with aversive emotional states such as disgust and pain (Zaki and Mitchell, 2011). Another study demonstrated neurometabolic changes unique to the experience of inequality to others different from oneself (Tricomi et al., 2010). Accepting the aforementioned caveats regarding causal inferences from neurometabolic imaging described earlier, the observations are consistent with the notion of neurophysiological mechanisms uniquely related to the perception of inequality.

In a sense, humans may be hardwired to react negatively to someone obtaining goods by deception (see Chapter 1). There are implications of this "hardwired" response in the education, training, and supervision of physicians and healthcare professionals, which are discussed later. It is important to note that such a possibility of an innate neurobiological mechanism does not imply some unique and specific mechanism within the human brain, conscious or unconscious, in some detached manner, such as a miniature humanoid, which would risk the homunculus argument fallacy. The miniature humanoid within a person's brain (or mind) itself would require a mini-miniature humanoid, and so on without ending. Rather, the argument is that neurophysiological mechanisms produce a specific behavior that is rationalized post hoc into the notion of equity or inequity (see Chapter 1). However, as discussed in Chapter 1, an inborn aversion to others that may be a factor in racism can be mitigated by education and social expectations. Perhaps remedial education for physicians, such as the one in the case presented, maybe of ethical utility.

Escaping the brutality of war often raised suspicions of factitious disorders and malingering in soldiers complaining of functional neurological disorders. By extension, some physicians began to argue that development of the welfare state became a source of secondary gain that was thought to increase the rate of factitious disorders and malingering. In a letter to *The Lancet*, Fillik (1972) wrote:

> ...laws of social welfare and work insurance were made mostly for law-abiding people who really are in need. Therefore it is not the individual who causes the problem of simulation and malingering but the society which created the legal framework for exploitation.

In the setting of the moderate scarcity of resources, provision of care or largesse from the public social safety net to a person with a factitious disorder or malingering causing a patient with a "real" disease to suffer for care not afforded likely would be seen as unjust. Perhaps it is this concern that might mitigate the use of deceptions, sometimes cruel, by some physicians and healthcare professionals to demonstrate lack of a demonstrable physical basis. Unfortunately, the patient with a functional neurological disorder or simple embellishment of symptoms and signs from a "real" disease, rather than a factitious disorder or malingering, is likely to be harmed as well.

A form of suspicion was placed on patients with psychogenic disorders, that of the patients' symptoms, signs, and disabilities being suggestible and thus not "real." Following the work on hysteria by Charcot (CE 1825–1893), Stone (2016) maintained that there was a backlash against Charcot, arguing that the demonstrations of hysteria were induced by suggestion and autosuggestion. The result was to emphasize a "psychic" etiology as opposed to a somatic physical disorder or of supernatural causes. However, leading neurologists, such as William Gowers (CE 1845–1915), continued to argue for a physiological explication of the psychogenic symptoms.

Particularly, the risk of countertransference is difficult because the physician or healthcare professional unconsciously projects his or her own feelings onto the patient, as discussed later (Adler, 1981). Such countertransference, particularly if negative, may interfere with the establishment of an effective therapeutic relationship. Negative countertransference may lead patients to reject the physician or healthcare professional and to seek help from less scrupulous physicians and healthcare professionals, increasing the risk for iatrogenic disease. Thus, one aspect of an effective therapeutic relationship is to enter into a "contract" that delimits the reasonable expectations on the part of both the patient and the physician or healthcare professional; the patient is encouraged to let the physician or healthcare professional to be an honest broker of the patient's healthcare; and the physician or healthcare professional must be responsive. However, such a contract, entirely reasonable under a Libertarian moral theory, implies some surrender of autonomy on the part of the patient, perhaps in the form of a Ulysses pact (see Cases 2 and 5 and Chapter 4). Such a contract might be acceptable under an Egalitarian moral theory as long as sufficient numbers of reasonable persons blinded to their self-interest would condone such a pact. It is not clear that such a relationship would be acceptable under a Deontological moral theory. It is not clear whether the Anti-Principlist ethical system would agree or disagree with such contracts.

The problem is that currently there is no way to assess the patient's motivation, directly or indirectly, so as to differentiate a fictitious disorder or malingering from a functional neurological disorder or typical embellishment of symptoms and signs associated with a physically demonstrable disease. Bass and Halligan (2007) address this dilemma, writing:

> Moreover, given the difficulty in establishing a meaningful and reliable diagnosis based on motivation (and not deception), patients whose behaviour and presentation warrant psychiatric help should not have to endure the additional stigma of having their behaviour questioned in terms of motivations other than the sick role and causes other than their illness. In such circumstances, it seems sensible to abandon the traditional "medically specialized" use of deception and confine its standard non-psychiatric use for subjects who have the capacity (or at least the perceived capacity) to exercise choice and decision making.

However, the problem is unanswered if one holds that factitious disorders and malingering should not be medicalized and not rewarded, while those with functional neurological disorders or embellished symptoms and signs should not be penalized or abandoned. The risk is if a patient was seen by a physician or healthcare professional with a bias toward suspicion of intentional deceit and another was seen by a physician less suspicious, then the subsequent care will be a lottery depending on which physician was seen as a matter of chance. This would be unacceptable under an Egalitarian moral theory unless every citizen agreed prior that everyone is subjected to the lottery and thus everyone is exposed to the same risk of being suspected or believed. It is not clear that pluralistic modern liberal democracies would find this acceptable.

A Utilitarian moral theory may hold that the physician's actions, while perhaps unfavorable in appearance, may not be unethical and therefore not unjust. If the overall cost to all stakeholders resulting from a more solicitous action on the part of the surgeon resulted in a reduced good to other patients in the setting of moderate scarcity, then being so solicitous may actually be unethical and therefore unjust. An interesting example was discussed by Bass and Halligan (2016):

> Given that the military and the governments at the time were ill prepared to accept the large number of psychiatric casualties, "psychiatrists were often viewed as a useless burden" (Crocq and Crocq, 2000). This was well illustrated in a memorandum addressed by Winston Churchill to the Lord President of the Council in December 1942, where he wrote:

I am sure it would be sensible to restrict as much as possible the work of these gentlemen [psychologists and psychiatrists]…it is very wrong to disturb large numbers of healthy, normal men and women by asking the kind of odd questions in which the psychiatrists specialize (Ahrenfeldt, 1958 [Psychiatry in the British army in the Second World War. Columbia University Press, New York]).

Quoted in Bass and Halligan (2016)

Birth of the immaterial mind and nonphysical mental illness

Trimble and Reynolds (2016) provided a brief review of the history of hysteria, taken here as a representative of psychogenic disorders more generally as the precise definition and clinical syndromes of hysteria have been unsettled. Pointing out that descriptions of hysteria date back at least to the ancient Babylonians in second millennium BCE, these early descriptions were more phenomenological than causal. Over the subsequent millennia, the putative causes ranged widely, from effects by the uterus to changes in Galen's four humors, to abnormal blood flow, and many others. With the exception of the attribution of these disorders to witchcraft in the 1400s, most psychogenic disorders still had some causal relation to the physical body, even if those effects were initiated or influenced by the mind. However, the notion of the mind as it relates to psychogenic disorders was not explored. For example, what is the nature of the mind that it could bear some responsibility for psychogenic disorders? The question would be moot if there were not such a thing as a mind. The beta cells of the islets of Langerhans in the pancreas do not have a mind (presumably) and therefore diabetes mellitus type I cannot be a mental illness if mental illness has something to do with the mind.

To be sure, there always has been a notion of something like a mind as somehow detached or detachable from the body. The point here is that it can be rational (philosophically sound) to hold that all the functions that one might attribute to the mind need not be so attributed. If there is a complete explication of every attribute of the mind to a bodily or mechanical physical action, then there is no need to invent a mind and, by extension, a mental illness, and there would not be functional neurological disorders. Certainly, modern neuroscience is heading in that direction. Studies suggest that the motor system, from the motor cortex through to the muscles, is normal in functional neurological disorders, as is the sensory systems into the brain (Hallett, 2016). However, studies of "higher" centers and the limbic lobe demonstrate changes in patients with functional neurological disorders not seen in normal subjects, such as those mimicking a functional neurological disorder (Aybek and Vuilleumier, 2016). Such studies to date are essentially correlational from which it is difficult to confirm causality; in other words, physical changes are seen but are they physical causes of the functional neurological disorder? However, the observations suggest the possibility.

To be sure, a demonstration of psychogenic disorders as physical such as diabetes mellitus type I would certainly have an impact on the ethics of the case presented, although not in an uncomplicated manner. Regardless of the causal mechanisms, the issue of ethical obligations to treatment would remain. It merely may be that the prevailing moral theory does not hold the actions, in the case presented, as unethical and thus the actions are not unjust as described earlier. The critical issue in the case presented is whether there is a bias against patients with psychogenic disorders that reasonable persons blinded to their self-interest would find unacceptable. Holding that disease is mental and somehow related to the supernatural, autosuggestion, deceit, or moral weakness when it truly is physical likely would seem unacceptable (see Case 9). If there is such a bias, then the critical question becomes as to its origin and it is hoped, once understood, that remedial actions can be taken similarly to the remediation of omission bias discussed in Case 9.

Aristotle, perhaps a philosopher and scientist who had and continues to have the most profound impact on human intellectual history even today, was noncommittal whether there was a mind, as held separate from the body. Aristotle writes:

> A further problem presented by the affections of the soul [*mind*] is this: are they all affections of the complex of body and soul [*mind*], or is there any one among them peculiar to the soul [*mind*] by itself? To determine this is indispensable but difficult. If we consider the majority of them, there seems to be no case in which the soul [mind] can act or be acted upon without involving the body; e. g. anger, courage, appetite and sensation generally. Thinking seems the most probable exception; but if this too proves to be a form of imagination [*a form of physical perception*] or to be impossible without imagination, it too requires a body as a condition of its existence. If there is any way of acting of being acted upon proper to soul [*mind*], soul [*mind*] will be capable of separate existence; if there is none its separate existence is impossible.
>
> *Miller (2018, pp. 536–537)*

It is important to note that Aristotle suggested that only thinking might be separate from the body, but clearly left open the option that it was not separate. Perhaps the greatest schism introduced between the brain and the mind came from Descartes (CE 1596–1650). However, Descartes (1662) still attributed many attributes to the brain that persons today would attribute to the mind, writing:

> And note that I have explicitly distinguished between the two pores in order to alert you to the fact that two kinds of movement almost always follow every action: namely, external movements that serve either in the pursuit of desirable things or in the avoidance of injurious ones, and internal movements that are commonly termed passions, which serve to dispose the heart, the liver, and all the other organs on which the temperament of the blood—and as a result, that of the spirits—depends, so that the spirits produced at a particular time are those suited to causing the external movements that must follow. For assuming that the various qualities of these spirits are one of the circumstances that serve to change their course, as I shall explain in a moment, we may readily appreciate that if, for example, it is a question of avoiding some evil by force, by surmounting it or chasing it away, as the passion of anger inclines us to do, then the spirits must be more unevenly agitated and stronger than usual. And on the other hand, when one must avoid it by hiding or bearing

it with patience, as the passion of fear inclines us to do, it must be less abundant and weaker. To achieve this, the heart must be constricted, and must spare and save the blood for when it needs it. And you can judge the other passions proportionately.

As for other external movements which serve neither to avoid evil nor to pursue the good but merely bear witness to the passions, such as those consisting in laughing or crying, these occur only by chance because the nerves through which the spirits enter in order to produce them originate very close to those through which spirits enter to give rise to the passions, as anatomy will show you.

(p. 162)

I shall not pause to tell you how noise and heat, and other actions which very forcefully move the internal parts of the brain through the mediation of the sense organs, or how joy and anger and the other passions that greatly agitate the spirits, or how the dryness of the air, which renders the blood more subtle, or similar circumstances, can prevent it sleeping: nor on the other hand how silence, sadness, the humidity of the air and similar things can invite it to sleep. Nor how a great loss of blood, too much fasting or drinking, and other excesses which have something which increases or diminishes the strength of the spirits, depending on their different temperaments, make the machine either wake or sleep too much. Nor how through excessive waking its brain can be weakened, and by an excess of sleeping grow heavy like one who is senseless or stupid. Nor innumerable other things: since it seems to me that they can all be deduced easily enough from what I have already explained.

(pp. 167–168)

Certainly, Descartes was not the first to hold a Dualist view that the mind and body were separate entities and that the mind was of a different kind, immaterial (Substance Dualism). This dualism had a perverse effect on the treatment of mental illness, now attributed to diseases of the mind rather than the body (Ventriglio and Bhugra, 2015). It is striking that Descartes would create such a schism given his fascination for automata (robots and other complex mechanical devices) and his willingness to attribute so many psychological activities to reflex-like behavior in the brain. However, it may be understood in the context of the battles between theology and the beginnings of modern science. Giordano Bruno was burned at the stake in 1600 for multiple transgressions against the Church, not the least of which was his advocating a Copernican model of cosmology. Descartes fled to Amsterdam to escape the Jesuits in Paris.

The Church at that time did not unconditionally oppose science, particularly cosmology and physics. The critical question for the Church was whether God was bound by the laws he created for the universe? Many scientists at the time, such as Galileo, thought God was so bound; however, the Church viewed such a position as a threat. Perhaps Descartes was trying to give the theologians some space to retain a notion of humans being made in God's image and hence some part of humans had to be immaterial. Descartes wrote:

These men will be composed, as we are, of a soul and a body, and I must describe for you first the body; and then, also separately, the soul; and finally I must show you how these two natures would have to be joined and united so as to constitute men resembling us.

I assume the body to be just a statue, an earthen machine formed intentionally by God to be as much as possible like us. Thus He not only gives its externally the shapes and colors of all the parts of our body; He also places inside it all the pieces required to make it walk, eat, breathe, and imitate whichever of our own functions can be imagined to proceed from mere matter and to depend entirely on the arrangement of our organs.

We see clocks, artificial fountains, mills, and other similar machines which, though made entirely by man, lack not the power to move of themselves, in various ways. And I think you will agree that the present machine could have even more sorts of movement than I have imagined and more ingenuity than I have assigned, or our supposition is that it was created by God.

Descartes (2003)

Perhaps intended or not, Descartes' dualism provided room for neuroscientists to continue their work with less fear of pushback from the theologians. Yet, it has been argued that Descartes, in a philosophical sleight of hand, also undid theology based on an absolute rationalism that enabled strict empiricism; God would not deceive a human's "clear and distinct ideas." Thus, Descartes established a strong rational science because otherwise the scientist's "clear and distinct" ideas would be a deception and God would not deceive (Caton, 1970). Neuroscientists could continue their work without having to invoke any notion of God. The success of Descartes' actions is seen in a quote from G.E. Berrios (2018):

Probably, this [issue of interaction between the material brain and the immaterial mind or soul] has less to do with the possibility that practitioners have found a philosophical "solution" than with the fact that Cartesian dualism does not seem to interfere with the nature and statistical analysis of the variables chosen to map the putative relationship.

Descartes' dualism, whether or not political expediency, continues to influence. In one survey, psychiatrists and psychologist were asked to evaluate and rate clinical vignettes according to "level of intentionality, controllability, responsibility, and blame attributable to the patients, as well as the importance of neurobiological, psychological, and social factors in explaining the patients' symptoms" (Miresco and Kirmayer, 2006). Miresco and Kirmayer (2006) reported, "Factor analysis revealed a single dimension of responsibility regarding the patients' illnesses that correlated positively with ratings of psychological etiology and negatively with ratings of neurobiological etiology. Psychological and neurobiological ratings were inversely correlated." They concluded:

Mental health professionals continue to employ a mind-brain dichotomy when reasoning about clinical cases. The more a behavioral problem is seen as originating in "psychological" processes, the more a patient tends to be viewed as responsible and blameworthy for his or her symptoms; conversely, the more behaviors are attributed to neurobiological causes, the less likely patients are to be viewed as responsible and blameworthy.

Miresco and Kirmayer (2006)

In a survey of university students, likely representative of those who became physicians and healthcare professionals, there was a predominance of students holding to a form of dualism (Demertzi et al., 2009). In a metasynthesis, roughly a qualitative equivalent to a metaanalysis, the most common causes attributed to mental problems among laypersons were sociological in nature. The next most common were spiritual or supernatural causes. A biological cause was third (Choudhry et al., 2016).

While perhaps coincidental, the possibility that Descartes' dualism was an attempt to "Render unto Caesar the things that are Caesar's, and unto God the things that are God's" (Matthew 22:21, New Revised Standard Version Bible, copyright © 1989 the Division of Christian Education of the National Council of the Churches of Christ in the United States of America) to placate the theologians, the consequence may have been to render the mind as a religious entity. In a survey of university students, the majority endorsed dualism, as well as a soul, that is separate and transcends death (Demertzi et al., 2009).

A diagnosis of mental illness, such as a psychogenetic disorder, has serious health-care repercussions. Patients with mental illness are frequently discriminated against, less likely to be offered mammograms, be admitted for a diabetic crisis, or be offered cardiac catheterization in the appropriate settings (Jones et al., 2008; Corrigan et al., 2014; Henderson et al., 2014). Patients with psychogenic disorders have a higher mortality rate (Coryell and House, 1984). In many countries, there has been a long history of disparity in the ability to obtain mental health services compared to those of diseases with identifiable physical pathologies (Barry et al., 2010), which continues in many circumstances (Bartlett and Manderscheid, 2016; Zhu et al., 2017).

Ethical remediation and education

Dualism is still a pervasive perspective, even if only implicit, and such dualism may affect how physicians and healthcare professionals care for patients with mental illnesses, particularly in the case of psychogenic disorders. It may have the effect of separating the notion of patients with illnesses with demonstrable pathology, as a fellow victim deserving of help, from those without demonstrable pathology who are held morally culpable, in the case of psychogenic disorders, guilty of deceit by self-inflicted wounds. Advancing neuroscientific evidence suggests that mental illness, even psychogenic disorders, may be a consequence of abnormal neurophysiology and that the patient is no more responsible than for developing type I diabetes mellitus. Even in the absence of any demonstration of a "physical" cause for mental illness, particularly psychogenic disorders as in the case presented, it is not clear that disparities in care can be justified under at least some moral theories, such as Deontological, particularly Kantian, Utilitarian, or Egalitarian.

As demonstrated repeatedly throughout this text, the traditional dyadic approach, one patient to one clinician, fails to address the ethics of everyday medicine in meaningful ways. Merely the issue of enforcement involves insurers, courts, governmental agencies, and ultimately the citizenry, the shareholders in a pluralistic modern liberal democracy. Further, those charged with the education and training of physicians and healthcare professionals bear some responsibility or, at the very least, a case could be made (see Case 9). In cases where attitudes and perspectives that potentially risk unethical or unjust care provision and are brought by students and trainees to the institutions, it cannot be just assumed that these attitudes and perspectives will change by some undefined process during the course of medical education and training. It is an open question whether educational and training institutions are effective, whether by acts of omission or commission, in changing the attitudes and perceptions of future physicians and healthcare professionals. As discussed in greater detail in Introduction, many educational experiences in ethics focus on "headline" ethics and thus are unlikely to affect the practice of the ethics of everyday medicine. Indeed, many ethicists were unsure whether there is such a thing as the ethics of everyday medicine. However, it may be possible and perhaps imperative to make a change (Crapanzano and Vath, 2015), depending on the moral theory that empowers the obligations to the ethical principles.

References

Adler, G., 1981. The physician and the hypochondriacal patient. N. Engl. J. Med. 304 (23), 1394–1396.

American Psychiatric Association, 2013. Diagnostic and Statistical Manual (DSM-5). American Psychiatric Association.

Aybek, S., Vuilleumier, P., 2016. Imaging studies of functional neurologic disorders. Handb. Clin. Neurol. 139, 73–84. https://doi.org/10.1016/B978-0-12-801772-2.00007-2.

Barry, C.L., Huskamp, H.A., Goldman, H.H., 2010. A political history of federal mental health and addiction insurance parity. Milbank Q. 88 (3), 404–433. https://doi.org/10.1111/j.1468 0009.2010.00605.x.

Barry, C.L., Mcginty, E.E., Pescosolido, B.A., Goldman, H.H., 2014. Stigma, discrimination, treatment effectiveness, and policy: public views about drug addiction and mental illness. Psychiatr. Serv. 65 (10), 1269–1272.

Bartlett, J., Manderscheid, R., 2016. What does mental health parity really mean for the care of people with serious mental illness? Psychiatr. Clin. North Am. 39 (2), 331–342. https://doi.org/10.1016/j.psc.2016.01.010.

Bass, C., Halligan, P.W., 2007. Illness related deception: social or psychiatric problem? J. R. Soc. Med. 100 (2), 81–84.

Bass, C., Halligan, P., 2016. Factitious disorders and malingering in relation to functional neurologic disorders. Handb. Clin. Neurol. 139, 509–520. https://doi.org/10.1016/B978-0-12-801772-2.00042-4.

Berrios, G.E., 2018. Historical epistemology of the body-mind interaction in psychiatry. Dialogues Clin. Neurosci. 20 (1), 5–13.

Caton, H., 1970. The theological import of cartesian doubt. Int. J. Philos. Relig. 1 (4), 220–232. (Winter).

Choudhry, F.R., Mani, V., Ming, L.C., Khan, T.M., 2016. Beliefs and perception about mental health issues: a meta-synthesis. Neuropsychiatr. Dis. Treat. 12, 2807–2818.

Corrigan, P.W., Mittal, D., Reaves, C.M., Haynes, T.F., Han, X., Morris, S., Sullivan, G., 2014. Mental health stigma and primary health care decisions. Psychiatry Res. 218 (1–2), 35–38.

Coryell, W., House, D., 1984. The validity of broadly defined hysteria and DSM-III conversion disorder: outcome, family history, and mortality. J. Clin. Psychiatry 45 (6), 252–256.

Crapanzano, K., Vath, R.J., 2015. Observations: confronting physician attitudes toward the mentally ill: a challenge to medical educators. J. Grad. Med. Educ. 7 (4), 686. https://doi.org/10.4300/JGME-D-15-00256.1.

Crocq, M.-A., Crocq, L., 2000. From shell shock and war neurosis to posttraumatic stress disorder: a history of psychotraumatology. Dialogues Clin. Neurosci. 2, 47–55.

Day, N.J.S., Hunt, A., Cortis-Jones, L., Grenyer, B.F.S., 2018. Clinician attitudes towards borderline personality disorder: a 15-year comparison. Personal Ment. Health 12 (4), 309–320. https://doi.org/10.1002/pmh.1429.

Demertzi, A., Liew, C., Ledoux, D., Bruno, M.A., Sharpe, M., Laureys, S., Zeman, A., 2009. Dualism persists in the science of mind. Ann. N. Y. Acad. Sci. 1157, 1–9. https://doi.org/10.1111/j.1749-6632.2008.04117.x.

Descartes, R., 1662. Treatise on man. In: Gaukroger, S. (Ed.), Descartes: The World and Other Writings. Cambridge University Press, Cambridge. Online publication 2009.

Descartes, R., 2003. Treatise on Man, Translation and Commentary by Thomas Steele Hall, Great Mind Series. Prometheus Books, Amherst, NY1–4.

Espay, A.J., Ries, S., Maloney, T., Vannest, J., Neefus, E., Dwivedi, A.K., Allendorfer, J.B., et al., 2019. Clinical and neural responses to cognitive behavioral therapy for functional tremor. Neurology 5 (93), e1787–e1798.

Fillik, A., 1972. Simulation and malingering after injuries to the brain and spinal cord. Lancet 1 (7760), 1126.

Gelauff, J., Stone, J., 2016. Prognosis of functional neurologic disorders. Handb. Clin. Neurol. 139, 523–541. https://doi.org/10.1016/B978-0-12-801772-2.00043-6.

Hallett, M., 2016. Neurophysiologic studies of functional neurologic disorders. Handb. Clin. Neurol. 139, 61–71. https://doi.org/10.1016/B978-0-12-801772-2.00006-0.

Henderson, C., Noblett, J., Parke, H., Clement, S., Caffrey, A., Gale-Grant, O., Schulze, B., et al., 2014. Mental health-related stigma in health care and mental health-care settings. Lancet Psychiatry 1 (6), 467–482.

Jones, S., Howard, L., Thornicroft, G., 2008. "Diagnostic overshadowing": worse physical health care for people with mental illness. Acta Psychiatr. Scand. 118 (3), 169–171.

Miller, F.D., 2018. Aristotle: On the Soul and Other Psychological Works. Oxford University Press, Oxford.

Miresco, M.J., Kirmayer, L.J., 2006. The persistence of mind-brain dualism in psychiatric reasoning about clinical scenarios. Am. J. Psychiatry 163 (5), 913–918.

Montgomery Jr., E.B., 2019. Medical Reasoning: The Nature and Use of Medical Knowledge. Oxford University Press, Oxford. (Chapter 10).

Russell, L.A., Abbass, A.A., Allder, S.J., Kisely, S., Pohlmann-Eden, B., Town, J.M., 2016. A pilot study of reduction in healthcare costs following the application of intensive short-term dynamic psychotherapy for psychogenic nonepileptic seizures. Epilepsy Behav. 63, 17–19. https://doi.org/10.1016/j.yebeh.2016.07.017.

Stone, J., 2016. Neurologic approaches to hysteria, psychogenic and functional disorders from the late 19th century onwards. Handb. Clin. Neurol. 139, 25–36. https://doi.org/10.1016/B978-0-12-801772-2.00003-5.

Tabibnia, G., Lieberman, M.D., 2007. Fairness and cooperation are rewarding: evidence from social cognitive neuroscience. Ann. N. Y. Acad. Sci. 1118, 90–101.

Tricomi, E., Rangel, A., Camerer, C.F., O'Doherty, J.P., 2010. Neural evidence for inequality-averse social preferences. Nature 463 (7284), 1089–1091. https://doi.org/10.1038/nature08785.

Trimble, M., Reynolds, E.H., 2016. A brief history of hysteria: from the ancient to the modern. Handb. Clin. Neurol. 139, 3–10. https://doi.org/10.1016/B978-0-12-801772-2.00001-1.

Ventriglio, A., Bhugra, D., 2015. Descartes' dogma and damage to Western psychiatry. Epidemiol. Psychiatr. Sci. 24 (5), 368–370. https://doi.org/10.1017/S2045796015000608.

Wang, Y., Henderson, A.M.E., 2018. Just rewards: 17-month-old infants expect agents to take resources according to the principles of distributive justice. J. Exp. Child Psychol. 172, 25–40. https://doi.org/10.1016/j.jecp.2018.02.008.

Zaki, J., Mitchell, J.P., 2011. Equitable decision making is associated with neural markers of intrinsic value. Proc. Natl. Acad. Sci. U. S. A. 108 (49), 19761–19766. https://doi.org/10.1073/pnas.1112324108.

Zeuner, K.E., Sidiropoulos, C., 2019. Cognitive behavioral therapy in functional tremor: a promising treatment approach. Neurology 93 (19), 825–826. https://doi.org/10.1212/WNL.0000000000008438.

Zhu, J.M., Zhang, Y., Polsky, D., 2017. Networks in ACA marketplaces are narrower for mental health care than for primary care. Health Affairs (Millwood) 36 (9), 1624–1631. https://doi.org/10.1377/hlthaff.2017.0325.

Case 11

Case 11—A 72-year-old male had advanced Parkinson's disease, but was otherwise in good health. The patient responded to medications, such as levodopa. However, the response was complicated by the wearing-off effects where the symptoms, slowness and tremor, appeared before the next dose of medications was due and uncontrollable involuntary movements (dyskinesia) would appear approximately 1–2 h after taking a dose of levodopa. Both the wearing-off effects and the dyskinesias reduced the patient's quality of life greatly. The patient's neurologist said the patient was ineligible for deep brain stimulation because he was over 70 years old. The neurologist held that age 70 was the cutoff from publications he had read.[a]

Medical facts of the matter

Among experts, there is relatively little doubt that clinical studies have demonstrated the relative safety and efficacy of placing electrical stimulating electrodes into the brains of patients with Parkinson's disease, referred to as deep brain stimulation (DBS) (Montgomery, 2015). In situations where medications fail to render symptoms and disabilities tolerable, there is no alternative to DBS other than continued symptoms and disabilities. Yet, it would appear in the case presented that the obligation to the ethical principle of beneficence, affected by successful DBS therapy, was denied on the basis of the patient's age. The question is whether such denial was just.

The reason for the denial is not due to insufficient symptoms or disability as this was not a reason offered for the denial (see Case 8). It is not that the patient was "too old" in the sense that being over 70 years of age would decrease the potential efficacy and increase risk unacceptably. If so, the neurologist simply could have stated such. Indeed, in the process treating the patient by whatever means, informed consent requires discussions of all reasonable options, which include reasonable discussions of potential benefit and risk. This is true whether the informed consent is explicit or implicit. In addition to the question of whether the benefit-to-risk ratio was such that DBS would not be wise, the other question is the ethics by which the decision was reached.

A study examined any correlation between age and benefit and risk and found little correlation of age to benefit, although increasing age appeared to correlate with increased adverse effects (Ory-Magne et al., 2007). In a study by Weaver et al. (2009), in the group of patients over the age of 70 treated with DBS there was an increase in the time of effective control of 3.8 h per day compared to those patients who received

---

[a] The case presented is hypothetical, although modeled on experience. No judgment is made relative to the correctness, appropriateness, or value of any actions described on the part of any of the participants. The purpose here is to describe and explore the variety of ethical considerations and consequences that depend on the perspective of different relevant ethical systems.

The Ethics of Everyday Medicine. https://doi.org/10.1016/B978-0-12-822829-6.00011-4

best medical therapy whose time of effective control was increased by 0.5 h per day, which was a statistically significant difference and likely a clinically meaningful difference. Motor assessments demonstrated that 61% of patients over age 70 had clinically meaningful improvement with DBS compared to 27% of those patients over 70 years of age having clinically meaningful improvement with best medical therapies (Weaver et al., 2009).

It appears that the physician merely said "no," which could be an example of hard paternalism, which may not be unethical or unjust at face value (see Case 5). Indeed, it would seem that the decision was a matter of general policy not to offer DBS to any patient with Parkinson's disease who is over 70 years of age. It would have been interesting if the patient was 70 years and 1 day of age. Would the neurologist continue to deny DBS? If the neurologist would have offered DBS thinking that 1 day over the age limit was acceptable, what about 6, 12, or 18 months? What if the patient was 69 years and 360 days old?

Ethical analysis

Establishing a hard cutoff for DBS of age 70 is not necessarily unethical and, therefore, not necessarily unjust. It is clear that the patient in the case presented likely would have experienced a clinically meaningful improvement and thus likely to have received beneficence. The issue is the degree of obligation to the ethical principle of beneficence toward the patient by offering DBS in the context of perhaps competing interests. The critical issue is whether there are conflicting obligations to ethical principles of competing parties in the case presented. If there are competing interests, the question becomes which party stands to lose beneficence, experience maleficence, or be denied autonomy should the patient in the case presented be offered DBS? In virtue ethics, the question is to whom and to what degree should the physician act virtuously? In considering this question, it is important not to view the question in the dyadic dynamics solely between the patient and physician or healthcare professional (see Introduction) but in a much larger context, indeed in the entirety of the ethics of everyday medicine. Perhaps the place to start, although not end, is with the physician or healthcare professional as he or she is the focal point for the delivery of healthcare.

With respect to the neurologist in the case presented, what is the beneficence the physician may lose in recommending DBS? It is not reimbursement in the sense that the physician already has or will be paid for the patient's visit regardless of whether DBS was offered or denied. However, if the physician is in a capitated care program, the net reimbursement is the block grant of funds made available to the physician less the cost of care imposed by providing DBS, such as referring the patient to a center that specializes in DBS. Clearly, such a situation places the physician in a conflict of interest relative to the obligation to beneficence to the patient and to the beneficence of the physician's business. If the cost of providing beneficence to the patients exceeds the capitated reimbursements, the physician may be subjected to harm (financial bankruptcy) and the obligation to nonmaleficence to the physician would be violated. It may well be that capitated care, with its inherent conflict of interest for physicians and healthcare professionals, is merely a means for prioritizing societies' resources in the context of other societal priorities. In this case, a microlevel injustice resulting from the failure of obligation to the patient by not providing DBS is offset by a macrolevel justice of obligation to nonmaleficence to the society.

Adjudicating between micro- and macrolevel justices by weighing the obligations to each requires the application of a supervening moral theory. A Deontological moral theory would hold that the individual patient has an inherent unalienable right to have DBS, in which case microlevel justice would prevail. However, this may not be acceptable by a consensus of reasonable persons blinded to their self-interest if macrolevel injustice results. An Egalitarian moral theory may hold that everyone is held equally to macrolevel justice even in the face of microlevel injustice. However, one implication permissible by an Egalitarian moral theory would be that no one receives DBS regardless of age. Alternatively, a Utilitarian moral theory may hold that the good achieved relative to the risk for someone under 70 years of age provides a greater good to the society than what would be achieved by offering DBS to someone over age 70. However, a Utilitarian moral theory would be difficult to apply at the microlevel as some patients over age 70 years may do much better than some patients younger than 70 years. The only way out is not to apply an age limit but to assess the relative degree of good (beneficence) on an individual basis that likely would require wisdom even King Solomon of the bible did not have. Would the greater good achieved by offering DBS to a 71-year-old genius, statesperson, or artist be sufficient while a 71-year-old poor and uneducated person would not? It is not clear whether the "wisdom" of a consensus of reasonable persons blinded to their self-interest would be capable of achieving a consensus sufficient to be imposed on all patients, both today and in the future in the context of a Utilitarian moral theory.

A Libertarian moral theory would require a contract, without which libertarianism becomes anarchy. It may be that physicians and healthcare professionals are not contractually obligated to provide DBS to patients over 70 years of age. However, in most pluralistic modern liberal democracies, valid contracts require six elements: (1) an offer; (2) exchange of goods; (3) capacity to enter into a contract; (4) ability to give uncoerced consent; (5) good faith such that contractual obligations cannot be avoided on insignificant grounds, such as typographical errors; and (6) lawful purpose. It is the last element, lawful purpose that prevents arbitrary contracts in situations where the shortage of physicians and healthcare professionals (or their governing healthcare delivery systems) can exert a monopolistic power. In most pluralistic modern liberal democracies, physicians and healthcare professionals are required to provide the standard of care, however, that is defined. For example, such a requirement is part of licensing, statutory law, or case law. In a sense, a Deontological moral theory is the "guardrail" for unbridled libertarianism.

A standard often used is what other similarly situated physicians and healthcare professionals would do. As a type of ethical induction, the notion is that a consensus of similarly situated physicians and healthcare professionals would be sufficient to generate a binding principle (see Chapter 3). Who constitutes the group making the consensus is critical. Most ethicists would hold that persons making the consensus must be reasonable and blinded to their self-interest (Chapter 3). Increasingly, courts have gone beyond a consensus from similarly situated physicians and healthcare professionals to one of what a reasonable physician or healthcare professional would do. In the past, the standard of care what was customary among similarly situated physicians (Moffett and Moore, 2011). In the case presented, if it were customary among similarly situated physicians not to offer DBS to anyone over the age of 70, then the physician in the case presented is on firmer ground, although explanation of the potential for DBS may be an element of informed consent even if the DBS would not be offered.

Subsequently, it was realized that what was customary may also be self-serving because those involved in creating what is customary may not have been blinded to their self-interest. It would seem that some sense of blinding those deciding what is customary would be necessary to establish justice in a pluralistic modern liberal democracy, for example, John Rawls' "veil of ignorance" (Rawls, 1999). A 1932 ruling on a liability case regarding tugboat operations has served as a precedent for medical liability. Justice Learned Hand of the US Court of Appeals for the Second Circuit stated.

> In most cases reasonable prudence is in fact common prudence; but strictly it is never its measure; *a whole calling may have unduly lagged in the adoption of new and available devices. It never may set its own tests, however persuasive be its usages* [italics added]. Courts must in the end say what is required; there are precautions so imperative that even their universal disregard will not excuse their omission.
> *[The T.J. Hooper, 60 F:2d 737 (2d Cir.), cert. Denied, 287 U.S. 662 (1932)]*

In 1903, Justice Oliver Wendell Holmes stated "what usually is done may be evidence of what ought to be done, but what ought to be done is fixed by a standard of reasonable prudence, whether it usually is complied with or not" [*Texas & P. Ry. v. Behymer*, 189 U.S. 468, 470, 47 L. Ed. 905, 23 S. Ct. 622 (1903)].

The court cases just described highlight complacency at the minimum and duplicity at the worst, in deciding what constitutes the standard of care, one can refer to what reasonable persons blinded to their self-interest would reject as unjust. However, some governments and other court decisions sought to limit the otherwise very open-ended standard as to what is reasonable. In other words, attempts were made to avoid "making the perfect the enemy of the good." In large part, limitations have been placed based on what is practical or workable. Further, differences of professional opinion were held as not grounds for what is standard of care (Moffett and Moore, 2011). In the case presented, it is an open question whether the arbitrary denial of DBS solely on age is a reasonable professional opinion.

In a true Libertarian moral theory, the issue of age would only come into play as the contract that binds parties. Persons with access to sufficient wealth could simply "buy" DBS, perhaps by creating their own type of private health insurance. However, not all governments would allow this. For example, in Canada, 6 of the 10 provinces specifically exclude private insurance for medical services otherwise covered by provincial insurance plans. For the remaining four provinces, while not strictly prohibited, a regulatory burden essentially prevents their operation (Flood and Archibald, 2001). In this case, any contract deriving from a Libertarian moral theory, for the sake of avoiding anarchy, is limited by an Egalitarian moral theory. Even if a person over the age of 70 could "buy" DBS, there is the question of the cost of follow-up care, particularly in the event of severe complications. This issue plays out in "medical tourism" as discussed in Case 5.

Denial of DBS for patients with Parkinson's disease over age 70 years could be just under an Egalitarian moral theory where every patient with Parkinson's disease, regardless of age and at risk for Parkinson's disease (which is essentially everyone who does not have Parkinson's disease), is subject to the same limitation. The latter group is important, as patients with Parkinson's disease are otherwise singled out, which could be contrary to Egalitarianism. The problem with the Egalitarian moral theory is that there is no limit to the denial of care if everyone receives the same consideration, even if it is no care. It is not clear that most would find this acceptable, suggesting that there is a minimum. Yet, what is there to limit the minimum? It would seem that any justification for a minimum beneficence would require a Deontological moral theory. This would be problematic if any rights, in a Deontological sense, were dependent on age, inviting the accusation of agism. Further, a Deontological moral theory is problematic in the setting of moderate scarcity of resources and thus must be qualified by consideration of other moral theories and enforced by contract.

The aforementioned discussion again demonstrates the complexity of the ecology in which the ethics of everyday medicine plays out. As can be seen, when taking each individual moral theory alone as the prevailing theory by which to arrive what a consensus as to justice, the result likely would be unsatisfactory. Resolution of the situation requires limiting or mitigating each individual moral theory by the others. In practice, the most robust adjudication of justice arises from a consensus, whose participants operate from different moral theories. There comes to be some hybrid that attempts to integrate the different moral theories into a consensus. However, such a consensus becomes problematic when implemented in a single case by a single judge (see Application—Judging in Chapter 3).

One reason the justice consensus may not be captured in any single moral theory, or in any recognizable combination of moral theories, relates to the fact that the adjudication of justice is primary and the subsequent formulations of ethical principles, moral theories, laws, and precedents are post hoc derivative rationalizations. In coming to such post hoc derivative rationalizations, information is lost (see Chapters 1 and 2). Thus, construction of justice from ethical principles, moral theories, laws, and precedents results in Information-theoretic Incompleteness. In the case of ethics, the result is consensuses of justice that are not provable in terms of the ethical principles (taken as axioms in a logical system) and the rules of inference (taken as moral theories, laws, and precedents). As an aside, Ethical Incompleteness is this same sense of Information-theoretic Incompleteness that is at the heart of Gödel's Number-theoretic Incompleteness, Measurement-theoretic Incompleteness, and Computational-theoretic Incompleteness.

Origin of candidacy criteria and scientist-centric ethics

It is unlikely that most insurers will "drill down" to the level of dictating candidacy requirements for DBS. In the United States, the Food and Drug Administration (FDA) allows considerable flexibility of the FDA-approved pharmaceuticals, biologics, and devices in situations that are outside the standard indications, for example, for "off-label" use. Note that the issue is not whether physicians and healthcare professionals are practicing good medicine when they follow indications and guidance established by the appropriate experts, particularly when it is the experts defining what is good medical practice [arguably at risk for a type of solipsism (Montgomery, 2019)]. Indeed, great many treatments that are "off-label" are considered standard and accepted, and the failure to use these treatments in an "off-label" manner could be considered malpractice in some circumstances. Physicians are expected to use their best judgments fortified by some evidentiary support.

In the case provided, the neurologist was unwilling to offer DBS outside of indications that he believed established age 70 years as the cutoff. A highly influential paper set an age limit of 70 years (Krack et al., 2003), as did others (Schüpbach et al., 2005) (note that other studies had different cutoffs; for a review, see Moro and Lang, 2006). Yet, these papers did not offer any justification for the age cutoff. What is the justification for the neurologist in the case presented to not offer the patient DBS because he was over age 70 years old? Unless the neurologist knows something that the experts in the papers cited do not, there is no justification other than that is what these particular experts did. One might think that the obligation to the ethical principle of autonomy would require the neurologist to provide some justification to the patient beyond merely "this is what has been done," a type of solipsism. This is particularly true in the case presented because the neurologist appears to be denying the patient the beneficence of DBS.

The position of the neurologist in the case presented is interesting. On the one hand, it corresponds to the Dogmatic school of Medicine since the ancient Greeks where physicians followed the prescribed texts without critique. Yet, the authors of the American Medical Association's Code of Ethics of 1847 railed against the Dogmatics. One the other hand, these papers were published by respected scientist/physicians and would appear to comport with the admonition to follow scientific medicine. The issue becomes the scientific nature of the age criteria. Moro and Lang (2006) reviewed the history of the criteria. It appears that the basis for most criteria follows from the Core assessment program for surgical interventional therapies in Parkinson's disease (CAPSIT-PD) (Defer et al., 1999), which in turn was derived from the Core assessment program for intracerebral transplantations (CAPIT) (Langston et al., 1992). In neither of these criteria was there any mention of age.

Most of the various criteria schemes were part of research studies. Indeed, the purpose of the CAPIT was to minimize the risk of including patients with atypical parkinsonism as these patients were less likely to respond to the treatment. Had those subjects been included, the effect size of the intervention, transplant, or DBS, would have been less and likely there would have been greater variance in the results. The consequence would be to reduce the statistical power of the studies, thereby requiring more research subjects with the attendant additional costs. Note that this rationale was not mentioned in the studies reported. But what else could the rationale be? The investigators could not say that the experimental therapies at that time would not have worked for patients over 70 years of age as there were no data. Perhaps the concern over lack of therapeutic efficacy in patients over the age of 70 years was "borrowed" from experiences from other therapies. However, how is such an extrapolation justified? From a Libertarian perspective, there may be no compulsion to take any chance of enrolling a patient with atypical parkinsonism. However, this likely would not be consistent with biomedical research ethics, as what would be the compulsion to include minorities and children? It is not clear how enforcing a limit for enrollees to be 70 years of age or younger would have sufficiently reduced the risk of enrolling patients with atypical parkinsonism (Montgomery, 2015).

Historically, women, children, and minorities have been underrepresented as subjects in biomedical research. The US National Institutes of Health (NIH) requires the inclusion of women, children, and minorities but, interestingly, from the perspective of being able to generalize the results to the relevant population rather than any inherent right of the potential participant. The NIH policy states:

> Additionally, for NIH-defined Phase III clinical trials, applicants must also consider whether the study can be expected to identify potential differences by sex/gender, race, and/or ethnicity. Unless there is clear evidence that such differences are unlikely to be seen, they must include plans for valid analysis, describing how potential group differences will be evaluated.
> *(https://grants.nih.gov/grants/peer/guidelines_general/Review_Human_subjects_Inclusion.pdf)*

There is no statement in the policy that indicates inclusion of women, children, or minorities from the perspective of offering these groups any potential benefit from participation. Extrapolating to the case presented, an argument could be that patients over age 70 years would never be offered DBS and thus inclusion of patient over 70 years of age would not be relevant. However, such reasoning smacks of Argumentum ad Ignorantiam (arguing from ignorance), as how could it have been known that all patients over age 70 years would never be offered DBS?

It is important to note that, to the best knowledge of this author, no study explicitly stated that the criteria for enrollment into the research studies of DBS were intended to become the criteria for standard and accepted care. To be sure, many of these criteria would have reduced the possibility of patients with atypical Parkinson's from undergoing the risks associated with DBS but at the cost or excluding perhaps an even greater number of patients with idiopathic Parkinson's disease who would have expected benefits (Montgomery, 2015). Rather, wholesale transfer of research to clinical criteria appears to be a default based on a form of Scientism and, as an ethic, it is scientist-centric (see Chapter 4). It may be that a sufficient percentage of citizen (meaning those entitled as shareholders, for example, by the power of their vote) uphold such a scientist-centric ethic as long as the percentage is sufficient to enforce the criteria universally.

Should the neurologist have known better?

Clearly if the neurologist should have known better that enforcing an age limit of 70 years is arbitrary (meaning not based on the scientific fact as discussed previously) then such actions could be a violation of the obligation to beneficence and autonomy. However, the critical question becomes whether the patient should expect better? It might be that the neurologist is not an expert on DBS and cannot be expected to understand the intricacies of candidate criteria for DBS. Yet, the neurologist in the case presented made the candidacy determination. Perhaps the expenditure of resources in the setting of moderate scarcity of referring the patient to an expert in DBS would deprive other worthy efforts of the necessary resources, in which case a Utilitarian moral theory may prevail. However, studies have demonstrated that successful DBS more than makes up for its initial costs by reduced expenditures such as medications, visits to physicians and healthcare professionals, and use of long-term care facilities. Further, given that the average life expectancy is on the order of 85 years, DBS in the patient in the case presented would have, on average, 13 years to realize a significant return on the investment in DBS. Perhaps more importantly, how and when did physicians and healthcare professionals be granted such unfettered discretion regarding utilitarian values?

Even if the argument that a neurologist should not be expected to understand the intricacies of candidacy for DBS, shouldn't the neurologist recognized his limitation and refer the patient to an expert in DBS? What would be the costs of educating every neurologist sufficiently to understand the nuances along with supervising the neurologist's compliance? Perhaps the patient over age 70 should be sophisticated enough to demand consideration. But if the burden is considered too great for the neurologist, how is it reasonable to expect that of the patient? Perhaps there should be public advocacy efforts, for example, available on the Internet or sponsored by DBS device manufacturers. Yet, how likely is information brought to the neurologist from the Internet going to sway the neurologist? What would be the recourse to the neurologist not responding in a constructive way (see Case 1)? Such efforts on the part of the patient are likely to counter the prevailing physician-centricity of the ethics of everyday medicine.

Clearly, the difficulties outlined are not easily resolvable in terms of the ethical principles of beneficence, nonmaleficence, and autonomy or in the moral theories of libertarianism, utilitarianism, egalitarianism, or deontology. Perhaps new principles or moral theories are needed. Alternatively, the ethical principles and moral theories are retained by some new method or ethical system developed to resolve these issues in a manner that reasonable persons blinded to their self-interest would consider just. Perhaps applications of phenomenological, care, virtue, and feminist ethics may advance justice (see Chapter 2).

References

Defer, G.L., Widner, H., Marié, R.M., Rémy, P., Levivier, M., 1999. Core assessment program for surgical interventional therapies in Parkinson's disease (CAPSIT-PD). Mov. Disord. 14 (4), 572–584.

Flood, C.M., Archibald, T., 2001. The illegality of private health care in Canada. CMAJ 164 (6), 825–830.

Krack, P., Batir, A., Van Blercom, N., Chabardes, S., Fraix, V., Ardouin, C., Koudsie, A., et al., 2003. Five-year follow-up of bilateral stimulation of the subthalamic nucleus in advanced Parkinson's disease. N. Engl. J. Med. 349 (20), 1925–1934.

Langston, J.W., Widner, H., Goetz, C.G., Brooks, D., Fahn, S., Freeman, T., Watts, R., 1992. Core assessment program for intracerebral transplantations (CAPIT). Mov. Disord. 7 (1), 2–13.

Moffett, P., Moore, G., 2011. The standard of care: legal history and definitions: the bad and good news. West. J. Emerg. Med. 12 (1), 109–112.

Montgomery Jr., E.B., 2015. Twenty Things to Know About Deep Brain Stimulation. Oxford University Press, Oxford.

Montgomery Jr., E.B., 2019. Medical Reasoning: The Nature and Use of Medical Knowledge. Oxford University Press, Oxford.

Moro, E., Lang, A.E., 2006. Criteria for deep-brain stimulation in Parkinson's disease: review and analysis. Expert. Rev. Neurother. 6 (11), 1695–1705. https://doi.org/10.1586/14737175.6.11.1695.

Ory-Magne, F., Brefel-Courbon, C., Simonetta-Moreau, M., Fabre, N., Lotterie, J.A., Chaynes, P., Berry, I., et al., 2007. Does ageing influence deep brain stimulation outcomes in Parkinson's disease? Mov. Disord. 22 (10), 1457–1463.

Rawls, J., 1999. A Theory of Justice, revised ed. The Belknap Press of Harvard University Press, Cambridge, MA.

Schüpbach, W.M., Chastan, N., Welter, M.L., Houeto, J.L., Mesnage, V., Bonnet, A.M., Czernecki, V., et al., 2005. Stimulation of the subthalamic nucleus in Parkinson's disease: a 5 year follow up. J. Neurol. Neurosurg. Psychiatry 76 (12), 1640–1644.

Weaver, F.M., Follett, K., Stern, M., Hur, K., Harris, C., Marks Jr., W.J., Rothlind, J., et al.CSP 468 Study Group, 2009. Bilateral deep brain stimulation vs best medical therapy for patients with advanced Parkinson disease: a randomized controlled trial. JAMA 301 (1), 63–73. https://doi.org/10.1001/jama.2008.929.

Case 12

Case 12—A patient presented to the emergency room with suicidal thoughts. Assessing that the patient was at risk of suicide, the physician admitted the patient to the psychiatry service. The patient's admission was voluntary. The following day, the physician was informed by the patient's insurer that the insurer would not cover the cost of hospitalization. The physician asked on what basis was the coverage denied? The insurer stated that the information and methods were proprietary and would not share them. However, the insurer suggested that the decision was based on an artificial intelligence (AI) algorithm. The insurer specifically stated that the insurer was not directing the physician in the practice of medicine but only that the issue for the insurer centered on the insurer's contractual obligations. The patient, fearful of incurring significant medical financial costs, asked to be discharged. As the physician continued to be concerned about the risk for suicide, the physician and the hospital required the patient to sign a waiver indicating that the physician advised continued hospitalization. Such waivers are termed "discharge against medical advice." The patient left the hospital after signing the waiver.[a]

Medical facts of the matter

The physician assessed that the patient had a sufficient risk for suicide that warranted hospitalization as a medical necessity. The insurer stated that the patient did not meet the insurer's criteria of medical necessity. Note that this was not to say that the insurer did not believe the patient to be at a significant risk of suicide but rather did not meet the insurer's criteria for a covered service. One might argue that the dispute is solely a contractual issue and the insurer was not advising the physician as to the practice of medicine (see Chapter 4). From the patient's perspective, the net effect was that the patient could not be hospitalized to treat the patient's suicide risk because of the financial risks to which the patient was exposed.

The risk for suicide in the United States is 15.3/100,000 population and 12.5/100,000 in Canada (http://worldpopulationreview.com/countries/suicide-rate-by-country/). This is not to suggest that it is too high, too low, or somewhere in between. But it is not to say it is insignificant. It is highly likely that nearly every physician and healthcare professional have experienced someone committing suicide, and the toll on the physician and healthcare professional likely is not nil; not to mention the toll taken on the family and friends of the person who ended his or her life. The fact is that in the setting of moderate scarcity of resources (see Introduction), there are costs associated with any effort to reduce the risk of suicide. Whether the incidence of suicide is sufficient to

[a] The case presented is hypothetical, although modeled on experience. No judgment is made relative to the correctness, appropriateness, or value of any actions described on the part of any of the participants. The purpose here is to describe and explore the variety of ethical considerations and consequences that depend on the perspective of different relevant ethical systems.

The Ethics of Everyday Medicine. https://doi.org/10.1016/B978-0-12-822829-6.00012-6

warrant any changes in the allocation of resources is another issue. It may be that the system that determined the outcome of the case presented is not unjust and therefore not unethical. These issues are addressed here.

The case centers on the insurer's ability to determine medical necessity in the setting of risk of suicide. It is likely that if the risk of suicide was sufficiently large, the insurer would have provided coverage for the admission, although this is not assured. Assuming that hospitalization would have been authorized if the risk of suicide was sufficiently large, the question becomes how to estimate the risk and what is the confidence, the certainty, of the risk assessment. A number of risk assessment screening tools have been published, yet studies suggest that these may be inadequate (Mullinax et al., 2018). However, note that these tools are designed to determine the relative risk of suicide and not necessarily determine the necessity of hospitalization. The latter is an ethical question, not one of medical research, and involves allocation of resources.

The published suicide risk-screening tools at least are transparent. An informed physician or healthcare professional can review them and integrate them, in whole or in part, into the physician's or healthcare professional's reasoning from which patient's informed consent can be obtained. No such transparency existed in the case presented. There was no way for the physician to assess the validity of the method the insurer used to determine that the risk of suicide was sufficiently low that hospitalization was not medically necessary, in contrast to the physician's assessment. In a real sense, the insurer directed responsibility to the physician to determine the subsequent course of action, but from a patient-centric ethical perspective, the net effect was the insurer acted in such a manner that hospitalization was not tenable, despite the recommendations of the physician.

It may well be that the insurer was not legally or contractually obligated to divulge the methods by which the insurer determined the medical necessity of hospitalization to address the risk of suicide. As is the focus of the discussion of the case presented, it is also possible that the insurer could not explain the methodology derived from machine learning AI; thus, the insurer could not share the "reasoning" with the physician even if the insurer wanted to. As machine learning AI becomes ever more dominant in the practice of everyday medicine, how will physicians and healthcare professionals be judged and held liable for the use of machine learning AI? How would physicians and healthcare professionals be able to explain the recommendations sufficiently to the patient or the patient's legal representative in such a way to allow the patient or the patient's representative to give truly informed consent? In a very real sense, machine learning AI is asking everyone to take its word for it. To be sure, the machine learning AI application could be judged by its outcomes, but this is very problematic. Judgment by outcomes is judgment by retrospect only after the patient has been exposed to the risks of harm.

Ethical issues

The case presented, as well as other cases in part 1, clearly demonstrates that there are limits to the care physicians and healthcare professionals can offer. In the past, the limits of care were due to the limits of knowledge. In the case presented and the others in part 1, today the limits of care most often are not imposed by a lack of knowledge but by a lack of will to provide them. This is not to say that such limitations are unjust and therefore unethical.

From a Utilitarian moral theory, it may be that the physician's criteria regarding the risk of suicide is too weak or liberal in that such criteria would result in hospitalization of those unlikely to commit suicide. Their psychiatric problems could be treated less expensively in an outpatient clinic. Thus, hospitalization for which the insurer would be held financially responsible represents a failure in the obligation of nonmaleficence to the insurer, who is harmed by having to pay for the hospitalization. Further, society as a whole is harmed by the increased cost of hospitalization, as the insurer likely would increase the cost of insurance premiums or taxes to cover the insurer's financial loss. Alternatively, there is the obligation of beneficence to the patient by reducing the risk of death from suicide as well as the suffering related to those issues that put the patient at risk of suicide.

In the setting of moderate scarcity of resources, those resources expended to meet the obligation to beneficence to the patient in the case presented cannot be spent on other obligations to other patients. The question is the effect on the overall good provided to the society by hospitalization of patients at risk for suicide, as in the case presented, versus the overall good that could be achieved by diverting the resources to other patients. In may be that in the ledgering of costs, the maximal good may be obtained by denying hospitalization to the patient in the case presented so that those resources can be allocated to other priorities. However, the "goods" that are the currency for utilitarian deliberations are more than financial, they have a deontological element. No reasonable persons blinded to their self-interests would not expend resources to help a person standing on a bridge and about to jump off. Further, how one treats others defines what one is; the gift of beneficence "blesses" both those that receive and those that give. If all persons are driven by this innate sense of justice to everyone of their kind, then does not such good take on a deontological status? Thus, a Utilitarian moral theory cannot help but involve a deontological value.

From a Libertarian moral theory, the issue resolves to contractual obligations. To be sure, the patient presumably has met her obligations to the insurer in the form of payment to the insurer, whether it was the patient, her employer, or society in general. Yet, most would not allow a libertarianism that extends *caveat emptor* (let the buyer beware) to its logical extreme. Laws and traditions in pluralistic modern liberal democracies limit the degree of libertarianism by requiring the fundamental principles of contracts: (1) an exchange of goods between all parties, (2) freedom to accept or not, (3) capacity on the part of the parties to enter a contract with full knowledge, and (4) a pledge of good faith. Thus, a Libertarian moral theory cannot escape, to some degree, the obligations imposed by an Egalitarian moral theory, as all agreements among all persons are subject to the requirements of a valid contract.

Yet, there appears to be something in the case presented that presents an exception or an exclusion. Perhaps what strikes as the coldness of the insurer is some sense of lack of sympathy for the patient, the extent of which generates any egalitarian or deontological commitment to the patient. The basis of such sympathy is the extent to which such sympathy is empathetic, the latter is the ability to put oneself in the "shoes" of the person in need of beneficence. Perhaps many persons would not find themselves empathetic to a person willing to take his or her own life. Similar concerns attend most mental illnesses (see Cases 9 and 10).

The ethical ecology surrounding the case presented is complex. In addition to the obligations to beneficence, autonomy, and nonmaleficence between the patient and the insurer; there are ethical concerns centered on the relationship between the patient and the physician or healthcare professional and between the insurer and the physician or healthcare professional (see Chapter 4). As alluded to earlier, in the case presented, how can the physician or healthcare professional fulfill the obligation to autonomy for the patient mediated through informed consent? If the physician or healthcare professional simply defaulted to the insurer, then the obligation to autonomy to the patient may be violated. For example, if the physician or healthcare professional simply deferred in such a manner as to convey a sense of superiority of the insurer's decisions over the initial recommendations of the physician, then it is not clear whether there was true informed consent.

There was the instance of informed consent associated with the patient's voluntary admission to the hospital. However, there was another situation of informed consent that may not be obvious. After denial of insurance coverage and the patient electing to be discharged, the physician, likely with the support of the hospital's administration, requested the patient to sign a waiver—discharge against medical advice. In a real sense, although not obvious, the physician was asking the patient's informed consent in signing the waiver. The intent was to release the physician and the hospital from any liabilities that may follow from the patient's discharge from the hospital, for example, the successful suicide by the patient in the case presented and claims by the patient's estate against the physician and the hospital.

Signing the waiver depends on the patient being fully informed, which gets back to the reason for denying hospitalization. To be sure, there are the financial risks for the patient should the patient continue hospitalization. The patient must balance those risks against the risk of suicide. If the physician conveyed a small risk, as would be the case by defaulting to the insurer's estimation of medical necessity, it is not clear that the patient would have been fully informed. Yet, there are risks to the physician if the physician were to indicate that he did not agree with the insurer's estimation of risk. In the past, such actions may have had serious implications, such as nonrenewal of contracts between the insurer and the physician (see Chapter 4). But note that it would be very problematic for the physician to make a case for his disagreement when the physician is deprived of any rationale for the insurer's decision. As will be discussed, there may be societal means, effected through governmental regulation, that deal with the complications around the increasing role of AI in the everyday practice of medicine.

The new physician and healthcare professional—Meet Dr. AI

Machine learning artificial intelligence already has a profound impact on medicine, particularly in those disciplines whose practitioners rely heavily on imaging in their decision-making. These disciplines include radiology, pathology, and dermatology (Jalalian et al., 2013). However, machine learning AI has and will expand to fields where data are less quantifiable or vary on dimensions more difficult to combine into a single dimension necessary for a yes-no decision that is fundamental to nearly every medical decision (Montgomery, 2019a). A treatment is either given or withheld, which necessitates a diagnosis to be affirmed or denied. This is not the same as saying that a physician or healthcare professional cannot be less than 100% certain, but one does not apply only 60% of a treatment because the physician or healthcare professional is only 60% certain.

To be sure, machine learning AI in medicine has attracted incredible attention recently with the extraordinary increase in computational power. Watson, the computer which beat the best human contestants in the game based on knowledge called Jeopardy in 2011, has turned its attention to healthcare, partnering with health insurer WellPoint and the Memorial Sloan-Kettering Cancer Center (https://www.techrepublic.com/article/ibm-watson-the-inside-story-of-how-the-jeopardy-winning-supercomputer-was-born-and-what-it-wants-to-do-next/). It is not that machine learning AI is some new form of informing physicians and healthcare professionals on how to act. Indeed, physicians have used checklists and algorithms perhaps as long as there has been written communication and have been continually urged to use them. Perhaps what is so remarkable are the range and depth of possible applications made possible by the incredible increase in computational power. However, what is far more remarkable is the ability to generate "algorithms" completely independent of physicians or even scientific knowledge of causation of disease using machine learning (Chapters 1 and 3). In other words, machine learning AI can achieve remarkable success in creating conclusions without the necessity of having to or even being able to explain itself and as such allows no vouchsafing by physicians and healthcare professionals. Clearly, a powerful challenge to a physician-centric ethos in medicine (Chapter 4).

The idea that an algorithm or information processing device, whether based on hardware, wetware, or software, generating solutions to computational problems without prior knowledge of the intervening steps between question and answer is not new. Indeed, engineering long employed the notion of a "black box" where inputs and outputs are known, but hypotheses as to the translation methods between inputs and outputs are achieved by reverse engineering. The development of machine learning has placed the "black box" concept on "steroids," and its significance on the practice of medicine is now particularly salient (Holm, 2019). Unappreciated is the fact that fundamentally, evidence-based medicine, as held synonymous to randomized controlled trials, makes no claims other than being a "black box." There is no requirement in evidence-based medicine that a trial be explained in some sufficiently detailed causal way. If randomized controlled trials demonstrated that injections of ground-up kitchen sink produced statistically significant improvement in a disease, there is nothing inherent in evidence-based medicine that would prevent the administration of ground-up kitchen sink; indeed, evidence-based medicine would demand its use (Montgomery, 2019a).

Hostility to the rationalist/allopathic physician's sense of self

The ability to diagnose has always been central to physicians and, indeed, the privilege to do so has been jealously guarded for centuries (Montgomery, 2019a). Treatments that follow from the diagnosis can be delegated to others, such as the barber/surgeon or the apothecarist/pharmacist, but physician professional societies have fought to keep the ability to diagnose from others. For example, speech-language pathologists are prevented from making a diagnosis of aphasia and informing the patient. The College of Audiologists and Speech Language Pathologists of Ontario wrote:

> According to the Regulated Health Professions Act (RHPA), communicating a diagnosis is a controlled act that speech-language pathologists (SLPs) and audiologists are not allowed to perform. Communicating a diagnosis is described as: "Communicating to the individual or his or her personal representative a diagnosis identifying a disease or disorder as the cause of symptoms of the individual in circumstances in which it is reasonably foreseeable that the individual or his or her personal representative will rely on the diagnosis." (RHPA 27 (2) 1.)
> *(http://www.caslpo.com/sites/default/uploads/files/PA_EN_Communicating_
> a_Diagnosis.pdf)*

What the speech-language pathologist may be able to do is inform the patient or his representative that the patient has a *language disorder*. Now it is highly likely that the speech-language pathologist may tell the referring physician that the patient has aphasia, perhaps when asked, and it also is highly likely the physician will defer to the speech-language pathologist. It is reasonable to ask what is to be gained by requiring the physician to be the intermediary? Certainly, the speech-language pathologist conveying to the patient the diagnosis of aphasia is consistent with the ethical principle of beneficence to the patient and, arguably, it is consistent with an obligation to the ethical principle of autonomy on behalf of the patient by the speech-language pathologist. It is not likely that the speech-language pathologist violates the ethical principle of nonmaleficence in the sense of causing harm to the patient, as the referring physician is likely to provide the same diagnosis based on the speech-language pathologist's findings; indeed, it is arguable that the speech-language pathologist is better trained and experienced in making the diagnosis than the large majority of physicians.

It could be argued that retaining the physician's exclusive right to make the diagnosis of aphasia ensures the physician's paramount role in assuring the best care for all patients, present and future (Case 1). Sidestepping the physician in making the diagnosis of aphasia could have a domino effect on all other manners of diagnosis. Perhaps restricting the right to make the diagnosis of aphasia to physicians allows healthcare regulators a simplification of their responsibilities, which may be acceptable from a Utilitarian moral theory as a means to the greater good. From an Egalitarian moral theory, such restrictions applied to all patients, present and past, as long as the particular group of citizens with or at risk for aphasia were not discriminated against even if such policies benefit patients with other diagnoses. Perhaps the restriction is that of a contract consistent with a Libertarian moral theory, it just happens for physicians to come to some contract with health regulators, that one "chip" in the bargaining in the physician's hegemony is the ability to make diagnoses. Yet, in a pluralistic modern liberal democracy, these would be acceptable only if a sufficient number of citizens agreed in a meaningful way and in sufficient number to justify its application to all citizens (Rawls, 1971) as discussed in Case 2.

An interesting question is whether the Ontario government could reverse its position and allow speech-language pathologists the privilege of diagnosing aphasia and other diagnoses relevant to its profession. Consider the question by analogy to the privileges allowed to nurse practitioners. In 23 states in the United States, a nurse practitioner can practice independently, which presumably allows the nurse practitioner to diagnosis (https://online.maryville.edu/nursing-degrees/np/states-granting-np-full-practice-authority/). Interestingly, most of those states that allow nurse practitioners to practice independently are most underserved by physicians.

The case of speech-language pathologists and nurse practitioners making diagnoses may be extended to the use of machine learning AI. Given the moderate scarcity of specialists, on the presumption that specialists may be better at diagnosing than primary-care or family physicians, providing the primary-care or family physician access to optimal diagnoses through the widespread use of machine learning AI, rather than the specialist, may address any inequalities in access to the best diagnosis. To some extent, family and primary physicians already do so by obtaining neuroimaging with the radiologist supplanting the neurologist (see Case 4). However, it is a relatively short step to the ability of machine learning AI to lessen the need for primary care and family physicians when used by nurses, nurse practitioners, physician assistants, and others in the guise of reducing healthcare costs or addressing the maldistribution of physicians. However, the increased use of machine learning AI only increases the frequency of the ethical issues arising in the everyday practice of medicine but does not resolve the ethical issues, as described later.

The question becomes whether there will be some requirement for physicians to not only incorporate but use machine learning AI in their practice. If uncoerced, it is not clear that sufficiently large numbers of physicians would agree to such requirements; for example, attempts at persuasion do not appear to be entirely successful, as demonstrated by the failure to extensively adopt evidence-based medicine.

The interesting questions are why the reluctance to fully embrace evidence-based medicine and whether the issues in evidence-based medicine will affect physicians' and healthcare professionals' acceptance and adoption of machine learning AI? It does not seem that improved outcomes are the primary driving force. One possibility is that the history of medicine, going back to the ancient Greeks, has divided itself into three schools, although in actual practice the distinction blurs. The developmental history of medicine has been driven by the necessity of having to deal with the potential infinite variety of human manifestations of health and disease (for a discussion on how this conundrum affected the development of ethics, see Chapter 2). For centuries, Rationalist/Allopathic physicians, the predominant school currently, viewed the variety of manifestations as some combination on an economical set of fundamental principles (Montgomery, 2019a). The Empirics, of which homeopaths are an example, held that variety is diversity where each case must be considered de novo.

Both Rationalist/Allopathic and Empiric physicians held that knowledge of diseases could be gained by experience—scientific experience for the former and phenomenological experience for the latter. In both cases, knowledge was justified by the experiences of the physician. Both disciplines invoked the notion of causality, for the Empiric it was that "like treats like"; treatments were selected based on the patient's symptoms and signs where the agents used also produced similar symptoms and signs. The Empiric notion of causation was that the patient's symptoms and signs were manifestations of the body's attempt to rid itself of disease; consequently, increasing those symptoms and signs by the use of agents supported the body's natural defense mechanisms. However, the agents had to be given in a very dilute manner so as not to overwhelm the patient and the concentration of the agents could be increased as the patient tolerated the higher dose. The Rationalist/Allopathic physician saw the symptoms and signs as evidence of dysfunction, not bodily defense mechanisms. The key then was to interpret the symptoms and signs into internal causal mechanisms, such as imbalances in the four humors of Galenic medicine, or in terms of infectious agents and pathology. Treatments were directed at the internal mechanisms. Nevertheless, there was an implicit requirement of notions of causation. It is easy to see how machine learning AI algorithms in medicine do not necessarily invoke notions of causation.

Among the Hellenic schools of medicine, there was another, the Methodic or Dogmatic school. That school did not hold that any notion of causation was necessary, there were no appeals to inner workings. Treatments were inferred easily from the symptoms and signs without interpretation as to the inner mechanisms. Interestingly, physicians of the Methodic school were a small minority and did not persist. Perhaps this is evidence of the powerful appeal of causal explanations. Yet, it is not clear that machine learning AI is not largely the same as the Methodic school. To be sure, laboratory tests and imaging likely will be important in machine learning AI but it does not appear that any causal theory is necessary. Perhaps that might be one reason for resistance on the part of physicians and healthcare professionals to the more widespread incursion of machine learning AI.

The apparent reticence of many physicians and healthcare professionals to use algorithms not firmly rooted in causal explications is at odds with the very way many physicians and healthcare professionals practice medicine. It is fair to say that a great many physicians and healthcare professionals practice medicine by pattern recognition, not explication by causal mechanisms (Case 9). This seems little different from just following the dictates of evidence-based medicine and machine learning AI. Yet, hegemony is still claimed. At the very least, it is the physician who decides.

Relationship of machine learning artificial intelligence to physicians and healthcare professionals

Given the aforementioned discussion, there is a somewhat disingenuous situation where one "can have their cake and eat it too." Many advocates of machine learning AI in medicine seem to always attach a disclaimer that all conclusions of machine learning AI are to be "supervised" by physicians, with the presumption that if the physician disagrees, the machine learning AI conclusion is rejected. Yet, this is problematic on a number of fronts. First, the machine learning AI conclusion may be correct while the physician or healthcare professional diagnosis is incorrect. Clearly, in this case, any advantage or benefit from using machine learning AI would be obviated. Further, if every machine learning AI conclusion has to be checked by the physician or healthcare professional, then any benefit of using machine learning AI to increase access to care is lessened. Finally, there is the question of whether a physician or healthcare professional has sufficient knowledge and experience to judge the conclusions from machine learning AI, particularly when it is not possible to see the principled rationale executed by the device.

Further, it may be that requiring a machine learning AI device to be explainable in terms and principles understandable to physicians and healthcare professionals may defeat many of the advantages of machine learning AI. In a real sense, explainable machine learning AI may merely handcuff computer engineers and deny many patients of beneficence and nonmaleficence that they may experience from the use of machine learning AI. However, it is not clear that society as a whole is willing to accept unexplainable AI or, at the very least, if there is a decision counter to the desires of a citizen, that citizen has the right to an explanation. It may be that an explanation cannot be given.

Even if it were possible after an examination to explain a specific machine learning AI conclusion, it is not clear that a physician or healthcare professional would be allowed to do so. With changes in the ability to patent computer software, many companies are framing algorithms as "trade secrets," thereby preventing others from examining the software. The case of *Loomis v. Wisconsin* is an example. Loomis challenged the use of proprietary software used to determine sentencing for a crime. Loomis argued that the use of proprietary software deprived him of due process as Loomis could not rebut the sentencing recommendations. Courts in Wisconsin upheld the sentencing and the US Supreme Court denied an appeal. There are many such situations (https://www.nytimes.com/2017/06/13/opinion/how-computers-are-harming-criminal-justice.html). Interestingly, the European Union passed the General Data Protection Regulation that requires AI to explain decisions to persons subject to the use of machine learning AI.

US Food and Drug Administration (FDA) response

The US FDA attempts to respond to the potential of machine learning AI for great benefit but also at risk to patients. On April 2, 2019, the FDA issued a discussion paper and requested feedback on the regulation of machine learning AI [Proposed Regulatory Framework for Modifications to Artificial Intelligence/Machine Learning (AI/ML)-Based Software as a Medical Device (SaMD); https://www.regulations.gov/document?D=FDA-2019-N-1185-0001]. The FDA has had a policy for software used in medicine referred to these systems as "software as a medical device (SaMD)." The FDA vets these SaMDs for marketing approval. Recognizing that the technology can evolve or that user experience may necessitate changes in the software, the FDA provided guides to reporting of any anticipated change in the software. The problem is that many machine learning AI devices, the FDA terms these as A/ML SaMD, are constantly changing as additional experiences constitute new training sets by which to modify the processing of information (see the discussion on machine learning in Chapter 3). The FDA addressed this distinction by classifications of "locked" algorithms, those whose operations do not simultaneously modify the software, and adaptive or continuously learning algorithms. The latter is most applicable to machine learning AI. To date, only two continuously learning algorithms have been approved by the FDA for detecting diabetic retinopathy and for alerting providers of a potential stroke in patients (statement from FDA Commissioner Scott Gottlieb, M.D., on steps toward a new, tailored review framework for artificial intelligence-based medical devices, https://www.fda.gov/NewsEvents/Newsroom/PressAnnouncements/ucm635083.htm).

To deal with the purposeful continuing modification of software inherent in A/ML SaMDs, the FDA proposes a principle of a "predetermined change control plan." Thus, not only would the current operational software system be vetted, the algorithms by which the operational software operate would also be modified would be vetted as well. It is interesting to note that that there does not seem to be a clear call that the operational software be explainable, only that the systems for ongoing modifications be explainable. As discussed earlier and in greater detail in Chapter 3, the actual operational algorithms may not be explainable.

The FDA recognizes that it is not just the algorithm that makes changes in the operational software, the modifying algorithm requires appropriate and sufficient data for training. It is likely that the training sets used for the initial construction of the operational software will be reviewed as well as the methods for collecting subsequent training sets. As discussed in Chapter 3, the choice of training sets is critical. The training set, the sample, must be sufficiently representative of the persons for whom the device is intended, the population. To the degree that the sample does not reflect the population, the device will be at risk for the Fallacy of Four Terms, and not generalizable, and hence applicable to the population of patients for which the machine learning AI was intended. Note that the modifying algorithms may successfully create an operational software that currently characterizes the training set and yet will operate poorly on subsequent applications. For example, an A/ML SaMD for assessing stroke risk whose training set consists only of middle-aged Caucasian men without diabetes mellitus, hypertension, hypercholesterolemia, and no family history of stroke is not likely to be predictive of the population that is in need of such a predictive device. To be sure, the lack of variability (variance) in the training set increases the ability to come to an operational software, perhaps more so than a training set that involves persons other than middle-aged Caucasian men without diabetes mellitus, hypertension, hypercholesterolemia, and no family history of stroke. Thus, there may be a temptation to use only middle-aged Caucasian men without diabetes mellitus, hypertension, hypercholesterolemia, and no family history of stroke. However, the FDA likely (hopefully) would recognize the problem and prevent it.

Recognizing that the actual operations of the operative software may be inscrutable, the FDA proposes a careful examination of performance metrics. In the discussion paper, the FDA specifically mentions specificity and sensitivity of the A/ML SaMD. However, specificity and sensitivity are poor metrics to judge the diagnostic utility of a device. Rather, the positive and negative predictive values are critical. Specificity and sensitivity must be combined with prior probabilities to produce positive and negative predictive values. The prior probability, for example, would be what is the risk of the patient to have condition A? Consider a test that is 97% specific and 97% sensitive for the diagnosis of condition A. Sensitivity means that 97% of those individuals with condition A will be identified by a positive test. Specificity meant that 97% of those individuals without condition A will be identified by a negative test. If the prevalence rate of condition A in the population of concern is 3%, then essentially every individual with condition A would be identified as having condition A. However, 3 persons out of 100 in the population of concern diagnosed with condition A would not have condition A. The relationship between these factors is expressed numerically in the following equations:

$$\text{Positive predictive value} = \frac{\text{sensitivity} \times \text{prior probability}}{\begin{array}{c}\text{sensitivity} \times \text{prior probability} + \\ (1 - \text{specificity}) \times (1 - \text{prior probability})\end{array}}$$

$$\text{Negative predictive value} = \frac{\text{specificity} \times (1 - \text{prior probability})}{\begin{array}{c}(1 - \text{sensitivity}) \times \text{prior probability} + \\ (\text{specificity}) \times (1 - \text{prior probability})\end{array}}$$

It is likely that the probability of an elderly person with hypertension, diabetes mellitus, hypercholesterolemia, and a family history is at a higher risk than an otherwise healthy young woman. Thus, the positive and negative predictive values of the A/ML SaMD device for predicting risk will be very different in the two samples. The prior probability, that is, critical to the diagnostic utility, may not be known; indeed, given the uniqueness of every patient, it may not be explicitly or analytically determinable. Yet, it is critical that there be some notion of the prior probabilities and it is not likely that the A/ML SaMD will be able to provide the information.

There is a fundamental reason why the A/ML SaMD cannot determine the exact risk of stroke, for example, in an individual person. The A/ML SaMD extracts an abstraction from the training set, similar to calculating the mean (average) of a data set. In the process of abstraction, irreversible steps are taken that result in a loss of information that cannot be retrieved within the system. For example, a mean of heights may be 10, but the data set contains the following heights, 5, 5, 5, 15, 15, and 15. Alternatively, the heights could be 3, 17, 4, 16, 5, and 15. From knowledge only of the mean, one cannot know which data set determined the mean. Such an irretrievable loss of information follows from the Second Law of Thermodynamics as applied to Information (Chapter 3) (Montgomery, 2019b).

Another fundamental reason why specificity and sensitivity cannot be used to make clinical decisions lies in the fact that a threshold—yes, the person has condition A or no, the person does not have condition A—is dichotomous and based on some combination of specificity and sensitivity. Specificity and sensitivity are related reciprocally, although usually not linearly. A threshold that increases sensitivity will likely decrease specificity. What threshold to use depends on the consequences of a false positive and those of a false negative (Montgomery, 2019a). For example, treatment for condition A is benign in every connotation of benign, while condition A is severe in the suffering it causes. Thus, one might use a much lower threshold, realizing that treating persons falsely held to have condition A is far less of a concern than perhaps not providing a person falsely thought not to have condition A. Thus, the consequences are ethical and the decision is an ethical one. It is likely that there is no way for an A/ML SaMD to adjudicate the ethical concerns, at least as yet. Rather, it is up to the patient and the physician or healthcare professional within the context of the ethical considerations of all shareholders and stakeholders (see Chapter 5).

Physician and healthcare professional responsibility in the era of machine learning artificial intelligence

Given the limitations of any A/ML SaMD system, the question becomes how can the physician or healthcare professional utilize the A/ML SaMD system without being able to "look under the hood?" If it is not possible to "look under the hood," what is the physician's or healthcare professional's ethical responsibility? Is it desirable for the physician or healthcare professional to have the responsibility rather than the computer scientists who produced the machine learning AI application? There clearly appears to be a desire for the physician bearing the ultimate responsibility. Holm discusses "black box" algorithms, of which machine learning AI is an example, "Of course, letting AIs read mammograms is not controversial, in part because a human doctor checks the answer" (Holm, 2019). However, the ability of physicians and healthcare professionals to always be able to "check the answer" seems problematic as discussed earlier. Nonetheless, it appears that having physicians and healthcare professionals "check the answer" provides some solace.

Having the physician or healthcare professional "check the answer" appears to provide solace, but for whom? Certainly, that solace is not for the patient and the physician or healthcare professional as it is empty in cases where the physician or healthcare professional cannot meaningfully "check the answer." One could argue that assuring the physician or healthcare professional can and will check the answer fulfills the obligation to nonmaleficence for the patient because stating that the physician or healthcare professional will not or cannot check the answer likely will cause emotional harm, at the very least. Yet, such an approach risks violating the obligation to autonomy toward the patient and could be considered soft paternalism by merely not mentioning that the physician or healthcare professional will not or cannot check the answer. However, it might be that enough citizens agree, directly or through their representatives, that they do not want to be told that the physician or healthcare professional will not or cannot check the answer. Governmental agencies could adopt policies and laws limiting physicians and healthcare professionals in their obligations to ensure complete autonomy and nonmaleficence to the person when using machine learning AI.

Once the physicians and healthcare professionals demonstrate acceptance of the conclusions reached by the machine learning AI by signing off, the argument could be made that the manufacturer's liability goes from a medical liability to a product liability. The manufacturer may hold itself only liable for manufacturing errors, such as design and implementation, and for marketing errors (making false claims and not providing sufficient information) but not for the medical consequences of its use. Importantly, there is the learned intermediary doctrine. According to Sullivan and Schweikart (2019),

> ...a key difference arises when the products liability doctrine is applied to cases involving medicine and health care, in that such cases are typically subject to the learned intermediary doctrine. The learned intermediary doctrine addresses how patient-focused liability doctrines apply to the use of pharmaceuticals and medical devices, wherein physicians intervene between the manufacturer and the ultimate consumer. Essentially, the *learned intermediary doctrine "prevents plaintiffs from suing medical device manufacturers directly,"* [italics added] as the manufacturer has no duty to the patient directly. Under this doctrine, the "physician, rather than the patient, is considered the end consumer of medical devices because the health care provider is in the best position to weigh the risks against the possible benefits of using the device." The physician as end consumer means that manufacturers may fulfill their duty to warn about the potential dangers of their products by providing warnings to the physicians who will be using them. If a physician subsequently fails to properly warn a patient and adequately disclose the risks and benefits associated with the product, it is the physician who will face liability.

Under product liability laws, a manufacturer may be liable for negligence by the manufacturer and seller in exercising proper care in selecting those that would use it, which has been an argument in legal complaints against gun manufacturers (https://www.press.umich.edu/pdf/0472115103-intro.pdf). Thus, attaching the appropriate disclaimer, such as "the user acknowledges that they are competent to use the software," the manufacturer can claim that due diligence was used to assure that the software was made available to appropriate users. In certain circumstances, manufacturers can be required to train the users prior to the technology being provided.

There may be other considerations, particularly the best interests of society as a whole as affected by government regulations. It just may be that public interest in the advancement of machine learning AI is such that developers must be protected, much in the manner of vaccine manufacturers (http://blog.petrieflom.law.harvard.edu/2017/02/10/artificial-intelligence-and-medical-liability-part-ii/). In the case of vaccine manufacturers, the liability was covered by the government. There are other examples where one entity was deemed sufficiently important to the public interest to shift the liability to another thought more likely to withstand the findings of liability. For example, the Captain of the Ship concept held that the surgeon, for example, was responsible for everything that happened to the patient by anyone involved in the care, regardless of whether the surgeon could have changed the outcome. It originated as a way to compensate a patient who experienced an injury in a charitable hospital under the concept of charitable immunity (Murphy, 1990). The notion was that the physician was better able to afford the compensation to the patient than the charitable hospital as there was a societal interest in protecting charitable hospitals. While this notion increasingly fell out of favor, its legal enforcement in the past is evidence of "ethical social engineering" to advance the interest of the citizens. Perhaps requiring the physician or the healthcare professional to sign off on checking the answer provided by machine learning AI, however ingenuous, serves a public interest.

References

Holm, E.A., 2019. In defense of the black box. Science 364 (6435), 26–27.

Jalalian, A., Mashohor, S.B., Mahmud, H.R., Saripan, M.I., Ramli, A.R., Karasfi, B., 2013. Computer-aided detection/diagnosis of breast cancer in mammography and ultrasound: a review. Clin. Imaging 37 (3), 420–426. https://doi.org/10.1016/j.clinimag.2012.09.024.

Montgomery Jr., E.B., 2019a. Medical Reasoning: The Nature and Use of Medical Knowledge. Oxford University Press, Oxford.

Montgomery Jr., E.B., 2019b. Reproducibility in Biomedical Research: Epistemological and Statistical Problems. Academic Press, New York.

Mullinax, S., Chalmers, C.E., Brennan, J., Vilke, G.M., Nordstrom, K., Wilson, M.P., 2018. Suicide screening scales may not adequately predict disposition of suicidal patients from the emergency department. Am. J. Emerg. Med. 36 (10), 1779–1783. https://doi.org/10.1016/j.ajem.2018.01.087.

Murphy, E.K., 1990. Applications of the 'captain of the ship' doctrine. AORN J. 52 (4), 863–866.

Rawls, J., 1971. A Theory of Justice, revised ed. Harvard University Press.

Sullivan, H.R., Schweikart, S.J., 2019. Are current tort liability doctrines adequate for addressing injury caused by AI? AMA J. Ethics 21 (2), E160–E166. https://doi.org/10.1001/amajethics.2019.160.

Case 13

Case 13—A 10-year-old boy was brought to see a neurologist for consideration of deep brain stimulation (DBS) for treatment of the child's dystonia. After completing the assessment, the neurologist began to explain the nature of DBS and the potential benefits and risks to the child and the parents. The parents interrupted the neurologist. One parent then took the child out of the examination room, while the remaining parent stated that he did not want the child to hear about the discomfort associated with the surgeries and the possible risks. The neurologist explained that she was very experienced in explaining medical and surgical procedures to children and that the hospital had a comprehensive program to prepare children for such surgeries. The program had been very successful in coaching children through such procedures. Further, for optimal outcomes, the child should be awake during some parts of the procedure in order to assist with proper placement of the DBS electrodes. The neurologist also stated to the parent that it is considered standard of care to obtain the child's assessment prior to surgery in addition to the parent's consent. Nevertheless, the parents insisted that these issues not be discussed with the child and that the child remains under anesthesia during the entire procedure.[a]

Medical facts of the matter

Deep brain stimulation involves placement of a permanent set of electrodes deep into the brain. The electrodes are connected to an electronic device that provides continuous electrical stimulation; much in the manner of what a cardiac pacemaker does for the heart. The surgery is remarkably effective and reasonably safe for children with dystonia. Dystonia is characterized by severe and ongoing muscle contractions that pull the body into contorted postures. The US Food and Drug Administration has approved DBS for children over the age of 7 years.

The success of DBS depends on placing the electrodes in exactly the right location in the brain. Typically, the target is visualized on a magnetic resonance imaging (MRI) scan; however, these scans do not provide sufficient accuracy (Montgomery, 2014). To refine the target location, a temporary electrode is placed first into the brain to record electrical pulses generated by individual neurons. The optimal target location can be determined from the pattern of electrical discharges from neurons, a type of brain cell, showing how this pattern changes with movement of a limb, for example. In addition, electrical stimulation is conducted through the temporary electrodes and the patient's response as well as reports of elicited sensations are important for target localization.

---

[a] The case presented is hypothetical, although modeled on experience. No judgment is made relative to the correctness, appropriateness, or value of any actions described on the part of any of the participants. The purpose here is to describe and explore the variety of ethical considerations and consequences that depend on the perspective of different ethical systems.

The Ethics of Everyday Medicine. https://doi.org/10.1016/B978-0-12-822829-6.00013-8

Interestingly, the brain does appears not to be pain sensitive. These reasons are why the surgery typically is done with the patient awake. It is not necessary for the patient to be awake; however, without the added information of the patient's reports, there is an increase risk that the permanent electrodes will not be placed properly and the patient will have reduced benefits and possible adverse effects. This neurosurgical procedure performed in awake children even as young as 8 years of age is well tolerated, even in very severe neurological disorders (Lohkamp et al., 2019). Indeed, the risk of complications directly due to the child being awake is relatively rare.

Deep brain stimulation for dystonia is remarkably effective; some children who were wheelchair confined were able to walk independently following DBS. In clinical trials, a mean (average) improvement of over 50% in movement and disability was reported (Vidailhet et al., 2007). With any surgery, there are risks, and while generally the rate of lasting adverse effects is on the order of 2%, the adverse effects can be quite severe, such as paralysis. There is a 0.1% risk of death. There is consensus that DBS for dystonia be done early to avoid the development of irreversible muscle and skeletal changes (Krauss, 2010). However, the difficulties in the case presented are not whether the child should have DBS, as all parties in the case presented agreed with proceeding with the DBS surgery. In an online survey of 279 parents responding to vignettes of serious childhood neurological and psychological conditions with respect to possible DBS (Storch et al., 2019), a large majority of parents agreed or strongly agreed with the statement "If I were assured that my child movement control would improve greatly, I would consider DPS." Parents acceptance increased when the child participated in the decision-making.

The nature of autonomy

The uncritical default position may be just to defer to the parent's wishes that the child not be informed, effectively presuming the parent's wishes trump all other considerations. However, many experiences suggest such need not be the case, such as courts overruling the parents, as will be discussed in detail later. The ethical question appears to be an issue of autonomy, but in this case a notion of autonomy narrowly focused on self-determinations. However, as will be argued, the notion of autonomy is far more broad and relates to preserving a type of dignity or respect even in the absence of any ability to exert self-determination. It is likely that a consensus of reasonable persons blinded to their self-interest (Chapter 3) would agree that every person, including every child, has a right to dignity regardless of how it is manifested.

The discussion becomes problematic in special cases of autonomy involving self-determination. One position may be that no child has the ultimate right to self-determination until they reach the age of majority. Until such time, the parents have the right of determination for the child. Not infrequently, the parent's absolute ability to determine what happens to the child is considered an extension of the parent's right to self-determination. If it is held that the parents have an absolute decision-making capacity, absolute in the sense that the child has none, then in the case presented, the neurologist would have no choice than to accept the parent's instructions. An alternative consideration is that the parents are custodians of the autonomy due the child with custody dissipating as the child's decision-making capacity increases, as will be discussed in detail later. If the latter consideration prevails, then the child's decision-making capacity, to the degree it exists, justice would require obligation to the child's autonomy in decision-making.

Perhaps unfortunately, many discussions of the ethics of these kinds of cases are couched in a dyadic form, parents versus child. However, this would be an over-simplification of the true ecology of the ethical problem. Society, through its proxy, the government, has an interest, as evidenced by the number and variety of cases where the courts have intervened against parents' wishes, for example, a parent's refusal to have the child vaccinated against serious communicable infectious diseases. While parents may have the right to not have the child vaccinated in narrow circumstances that right may be suspended when there is an outbreak or epidemic disease. Often this is not necessarily done out of consideration of the child subject to the vaccination, but perhaps more importantly to protect others by maintaining "herd" immunity.

Perhaps the first issue to be clarified is what is meant by autonomy for a child, even an infant? In medicine, autonomy is often solely considered in terms of decision-making or the locus of control over the experiences of the child (Spear and Kulbok, 2004). However, this definition may be too narrow, as is discussed later. Awareness of autonomy is often taken from the issues of informed consent, particularly in the context of research. Autonomy was central to the Belmont report (Belmont Report: Ethical Principles and Guidelines for the Protection of Human Subjects of Research, 1978). This perspective focuses on the right of persons to self-determination. The right to self-determination may derive from a more basic principle of the right to body integrity and freedom of thought. Battery is the touching of someone without his or her consent and assault is the threat to do so.

Autonomy often has a larger connotation implying respect for decisions people make about themselves. Mill (1859, p. 84) provided a justification for this sense of autonomy, writing:

> ...neither one person, nor any number of persons, is warranted in saying to another human creature of ripe years, that he shall not do with his life for his own benefit what he chooses to do with it. He is the person most interested in his own well-being: the interest which any other person...can have in it is trifling.

Justification for Mill is that only the person knows what is best for herself or himself. When all persons are allowed to do what is good for themselves, there will be maximum good, which, under the Utilitarian moral theory, is the most just (Aveyard, 2000). Immanuel Kant created perhaps a more ideal sense of autonomy, that is, persons are ends unto themselves and to means to the ends of others, writing "Act in such a way that you treat humanity, whether in your own person or in the person of another, always at the same time as an end and never simply as a means" (Kant, 2001). Kant's notion would be an example of a Deontological moral theory.

Interestingly, in the context of medical ethics, the notions of autonomy just described would seem patient-centric. However, historically, patient-centric ethics has rarely been a driving force for medical practice (Chapter 4). Further, a person's autonomy considered as the right to self-determination has never been unqualified. Mill wrote of a man approaching an unsafe bridge "...they [others] might seize him...without any real infringement of his liberty; for liberty consists in doing what one desires, and he does not desire to fall into the river" (Mill, 1859, p. 107). Kant also would limit autonomy by holding that any concept of autonomy to be valid must be universally applicable; however, in the context of extreme scarcity of resources, the requirement of universality may be so difficult as to become moot; it just may come down to every person for themselves. Interestingly, this conundrum imposed by requiring universality of ethics is a direct response to the problem of how to deal with the infinite variety of individual ethical experiences (Chapter 2). This conundrum led Kant to impose a number of rules to limit what would appear to be unlimited autonomy (Varelius, 2006).

Consider the situation of a patient in an unresponsive state who clearly has no decisional capacity. In that case, a surrogate, such as a spouse, adult child, or parent, can make decisions on behalf of the patient in many jurisdictions. However, the prevailing notion is that the surrogate is not acting based on the surrogate's values but rather on what the surrogate believes would be the values of the patient (Beauchamp and Childress, 2013), although this is problematic in children because children are less likely to foreshadow what their preferences might be (Diekema, 2004). Thus, this is evidence of an implicit obligation to the autonomy of the patient even though the patient has no decisional capacity in that parents may be expected to act as the child would have acted even in the hypothetical. At times, what might be hypothetically the child's interest may result from a consensus of reasonable persons blinded to their own self-interest.

Narrative ethics emphasizes that people generally carry with them a sense of personal history, much like a story, whose past helps understand the present and projects into the future. Thus, every child is not merely as he or she exists now or remembered from the past but with the expectation of a future that is largely unspent. It may be like Aristotle's notion of potentiality (the possible future) and actuality (what is now) where potentiality drives what is actual. Extending to the ethics of everyday medicine in this case, it may be that the autonomy accorded the child is more driven by the child's potentiality than the child's actuality. In cases of severe disease in a very young child, often it is the child's potentiality that drives the ethical considerations more so than the child's actuality at the time (see the examples used by Kwek, 2017). At times, there may be a great disconnect between what persons, particularly parents, understand about the potentiality, influenced by unrestrained hope, compared to the actuality. Perhaps it is autonomy derived from respect to a person's potentiality that appears to make the ethics of the very young so different from the ethics of the very old.

Potentiality and actuality exert themselves in a complex interwoven relationship that constitutes a person's narrative, or story, about themselves. Importantly, the interaction between potentiality and actuality in a cohesive narrative implies a trajectory that honors its past as it anticipates the future. It is possible that each person sees the trajectory of others as an extension or, at least, validation of their own. Perhaps that may be a reason that autonomy, in the sense of respect, embedded in the trajectory, is shown even to the dead.

Autonomy as respect

Perhaps an even more telling situation demonstrating that decisional capacity may not be a necessary condition for obligation to the principle of autonomy as there are laws against desecrating the body of dead humans. Rather, individual humans, alive or dead, are accorded respect. As such, respect is perhaps a better synonym for autonomy than self-determination. Canadian law holds:

182. Every one who

(a) neglects, without lawful excuse, to perform any duty that is imposed on him by law or that he undertakes with reference to the burial of a dead human body or human remains, or

(b) improperly or indecently interferes with or *offers any indignity to a dead human body or human remains* [italics added], whether buried or not, is guilty of an indictable offence and liable to imprisonment for a term not exceeding five years.

> R.S., c. C-34, s. 178 (https://laws-lois.justice.gc.ca/eng/acts/C-46/section-182. html?fbclid=IwAR0u8DmOurJbTVCfZZ6wQGmcrZkFZie4LwpqS8EfHBnKTC_ IgMBoWoExgXo)

In England, the Human Tissue Act 2004 forbids unauthorized retention of organ or medical students "practicing" surgical skills on cadavers without consent (Jones, 2017), even though it could be argued on Utilitarian moral theory that the actions are just as they lead to a greater knowledge of disease and more skillful physicians.

Jones (2017) points out that the revulsion against desecrating a corpse is nearly a universal experience (assuming reasonable persons blinded to their self-interest) and, as such, constitutes a sufficient consensus as to codify the revulsion into an ethical principle and law. Jones notes that the revulsion is not just disrespect for the memory of the person, it is much more. Jones (2017) wrote "Moreover, the actual corpse is the physical remains of that person, as opposed to, say, pictures of the now-deceased person. It is them rather than something that simply reminds us of them." Clearly, there is a sense of autonomy, as evident by the demands for justice for dead bodies that goes well beyond individual informed consent, as dead bodies cannot give consent. Interesting, it is very difficult to codify the perhaps universal revulsion at desecrating corpses into a set of principles translated into law (Jones, 2017), again demonstrating the difficulty of abstraction of generalities from the potential infinity of experiences (Chapters 2 and 3).

These same issues will roil beneath the surface in what to do about children. Reasoning by analogy, children would seem to enjoy the right to autonomy independent of their ability to give consent. The issue is whether and to what degree do physicians and healthcare professionals have an obligation to the child's right to autonomy in the sense that transcends self-determination? This also suggests that the obligation to beneficence and nonmaleficence entertained by physicians and healthcare professionals also has to be considered in combination with the obligation to the child's autonomy, whether the obligation to the child's autonomy is limited or not.

Limits of parental paternalism

One question confronting the physician in the case presented is whether the parent's requests comport with a consensus of reasonable persons blinded to their self-interest holds as in the child's best interest. It may well be that the consensus of reasonable persons blinded to their self-interest may conclude that the parents have sole jurisdiction in any matter concerning their minor children. Historically, that has not been the case when the question has been called, for example, by recognizing disputes and petitioning courts of law for resolution. Yet, the question like the one in the case presented, as well as all the others, rarely rises to the level of recognition in everyday practice and thus rarely is the question called (see Introduction). This does not mean that there are no answers to uncalled questions, only that the answers are not recognized as well. Instead, a default ethic is employed. Given the complex evolution of the unspoken ethics, it is not clear that the legacy exerting itself as an implicit force, even as a default, is what reasonable persons blinded to their self-interest would find just today. It may well be this same legacy that enforces an implied default position that the physician should act in a manner the physician finds justifiable (see Chapter 4). Yet, it is unlikely that a consensus of reasonable persons blinded to their self-interest would just default to the physician as they would not merely default to the parents.

There are generally two approaches to the power parents can exert over children. Either children are properties of parents where parents can do with their property except as it affects others around them (exclusive of the effects on their children) or specifically enjoined by law. Alternatively, parents are ethical agents acting on behalf of their children in justifiable paternalism. While maintaining notions of beneficence, nonmaleficence, and autonomy, the obligations to these principles, particularly autonomy, are fluid (Archard, 1993). As ethical agents, parents are privileged over others by their "blood ties" to the children or by specific declarations of law, such as adoption. There is the presumption that parents always (or sufficiently often) are primarily motivated as to the best interests of their children (Diekema, 2004) and, therefore, are more likely than the child to act in the child's best interest.

While the notion of children as property with which parents can do as they please, within limits, may be not openly expressed or asserted, it still is an open question whether these notions still influence. The notion of children as chattel has long and deep historical roots and likely is reinforced when physicians and healthcare professionals uncritically default to the parent's wishes, which do not comport with the best medical practices established by a consensus of reasonable persons blinded to their self-interest. Indeed, the notion of children as chattel can be seen in one objection to interference in parents' decisions is based on the autonomy of the parents, for them to live and act without interference (Archard, 1993).

The question becomes what is the basis for parent ownership of their children? Perhaps a widely held sentiment given voice by the philosopher John Locke in 1698 is every person owns themselves and thereby owns what they created by their labor (see Archard, 1993). The idea is that the child would not have been born or survived if not for the labors of his or her parents. This sentiment is the basis of what is often said to a child by a parent "So long as you live in my house and eat my food, you will do as I say." An older child is no longer subject to his or her parents when he or she is emancipated. While the mechanisms of emancipation are highly varied, a fundamental issue is that children can meet their own needs themselves, for example, by marriage or joining the armed services. However, it is important to note that abandonment by parents generally does not result in emancipation of the child.

Emancipation can be limited to medical issues. In addition, the concept of mature minors allows older minor children the ability to decide medical care for themselves even if they are not fully emancipated (Coleman and Rosoff, 2013). Yet, the rights a mature minor can execute are limited and commentators have held that adults trump any concerns of the child. The origin of this approach comes from a long history of political theory (Coleman and Rosoff, 2013). As will be noted, this differs from some approaches by medical ethicists who take a neurobiological perspective that emphasizes the gradual and progressive increased cognitive abilities that should be a controlling influence on the notion of a mature minor. For example, the United Nations Convention on the Rights of the Child holds

> States Parties shall respect the responsibilities, rights and duties of parents or, where applicable, the members of the extended family or community as provided for by local custom, legal guardians or other persons legally responsible for the child, to provide, in a manner consistent with *the evolving capacities of the child, appropriate direction and guidance in the exercise by the child of the rights* [italics added] recognized in the present Convention.
>
> *United Nations (1990)*

The notion that children, as fruits of their parent's labor, and thus owned by the parents could be considered as a contract. The labor is expended by the parents as a good to the child for which the goods given in exchange by the child is obedience to the parent. Indeed, emancipation of a minor suggests that the child is no longer in need of the parent's labor, thereby ending the contract. This sense of the relationship between the parents and the child can be viewed from the perspective of the requirements for a valid contract. In Anglo-American common law, a contract requires (1) an offer, (2) consideration, (3) acceptance, (4) a mutual intent to be bound by the contract, (5) the parties must be capable of entering into the contract, and (6) lawful purpose. It would seem that the contract between the child and the parent does not meet many of the requirements of a valid contract. First, there is no offer to exchange goods. For the child, the contract is a *fait accompli*, essentially binding the child 9 months before the child's birth. Second, a young child is in no position to enter into a contract. Older children can enter certain contracts as their knowledge and experiences make them capable. Interestingly, for older children who break a contract, the consequences often are not enforceable until the child reaches the age of majority. Thus, even if the parents threatened withholding of support of a minor in the face of the child's disobedience, it likely could not be done until the child reached the age of majority.

Perhaps a more productive route to explore these issues would to rephrase the question of whether ownership is a question of justifiable paternalism. The question becomes whether the actions of the parents in the case presented are justifiable. Beauchamp and Childress (2013) provide a number of situations of justifiable paternalism, including justification (1) by best interests, (2) by consent, and (3) by prospective benefit. In the case of justification by best interests, the situation is when patients cannot make their preferences known. The situation in the case presented is whether the 10-year-old child could make his preference known, not merely a simple preference but rather an informed consent. More discussion of this point anon. The second situation is by consent, where patients doubt their ability to always act in their best interests and consent to others to override their autonomy (e.g., a Ulysses pact or contract discussed later). With respect to the third situation, Beauchamp and Childress (2013) wrote:

> *Paternalism justified by prospective benefit* [italics in the original]. Accordingly, the justification of paternalistic actions that we recommend places benefit on a scale with autonomy interests and balances both: As a person's interests in autonomy increase and the benefits for that person decrease, the justification of paternalistic action becomes less plausible; conversely, as the benefits for a person increase and that person's autonomy interests decrease, the justification of paternalistic action becomes more plausible. Preventing minor harms or providing minor benefits while deeply disrespecting autonomy lacks plausible justification, but actions that prevent major harms or provide major benefits while only trivially disrespecting autonomy have a plausible paternalistic rationale.

Beauchamp and Childress (2013) suggest that assessing the justification of paternalism is proportional to "a person's interests in autonomy." One might take this as meaning that the obligation to a person's autonomy depends on how much the person presses for it (also see Case 14). If the person does not requisite the obligation to his autonomy be fully executed, then paternalism is justified. However, this may strike some as equivalent to "it is not a crime if one does not get caught." A great deal of mischief occurs when one does not think he or she will be caught.

A problem with the formulation of Beauchamp and Childress (2013) is that it places the obligation to beneficence and autonomy on a sliding (relative) scale and comparing the two obligations. However, there is no common metric based on the same anchoring (reference) points with the same linear increments of value. It is not that 2 units of obligation to beneficence minus 1 unit of obligation to autonomy provides a greater justification for paternalism than the case of 3 units of obligation to beneficence and 2 units of obligation to autonomy. Beauchamp and Childress (2013) are asking one to compare apple and oranges. Interestingly, humans do make those kinds of calculations (Chapter 3).

There are at least two general approaches to resolving the problem. The first is to construct a consensus of reasonable persons blinded to their self-interest by rigorous induction from a sufficient number of similar cases (Chapter 3). This does not seem likely in this case. Alternatively, one could look to case precedents; in other words, what did reasonable persons blinded to their self-interest do in similar cases? It may well be that there are no prior cases of sufficient similarity to provide a case precedent to guide resolution; perhaps that is not even possible (see Chapter 2). However, it may be possible to find cases that define the upper and lower boundaries for resolution of the case. For example, there are considerable case precedents where parental choice places the child at immediate risk for severe disability or loss of life. One then could look to less urgent or dangerous cases in which the obligation to child's autonomy trumped the obligation to parents' autonomy.

Age of license versus age of majority and medical consent

Experiences of issues related to children decision-making suggest an important distinction between age of license and age of majority. The latter is a legal definition of what age is considered an adult. The former is quite different and represents the age at which children can exercise certain rights. For example, in many states in the United States, the age of majority to be considered an adult is 18 years. Yet, a person under age 21 is unable to purchase alcoholic beverages. In other words, the default age for independent decisional capacity is not the age of majority, for example, the mature minor concept described earlier. An example of policies regarding minors giving or withholding consent in Canada was described by a statement from the Canadian Medical Protective Association, which provides medical liability assistance. An extended quote summarizes the issues, although it must be noted that this statement does not constitute legal advice and was provided for general educational purposes:

> A patient need not reach the age of majority to give consent to treatment. In all Canadian provinces and territories the determining factor in a child's ability to provide or refuse consent is whether the young person's physical, mental, and emotional development allows for a full appreciation of the nature and consequences of the proposed treatment or lack of treatment— whether or not the patient has attained the age of majority.
>
> In Québec, however, the Civil Code generally establishes the age of consent at 14 years, below which the consent of the parent or guardian, or of the court, is required. If the medical treatment requires a hospital stay of more than 12 hours, parental notification of the stay is required if the child is over 14 years of age.
>
> Physicians usually determine whether a child has the mental capacity (competence) to provide consent on a case-by-case basis. When a child is found incapable of consenting to treatment, the parents or legal guardians are authorized to provide consent on the minor's behalf. However, when the physician determines that the child has the capacity, parental consent is not required. In such circumstances, the physician must obtain consent from the child, even when the child is accompanied by a parent or other delegated adult.
>
> How does a physician determine whether or not a child has the capacity to consent? By discussing with the child, the physician should be reasonably confident that the child understands the nature of the proposed treatment and its anticipated

effect. The child should also understand the consequences of refusing treatment. One way to gauge this capacity is to use the teach-back technique: ask the child to re-phrase what they have just been told and invite the child to ask questions. More complex medical situations may require more rigor in determining whether the child understands. It is prudent for physicians to also encourage the child to invite a family member to attend the discussion.

Physicians must use their judgment concerning a child's capacity to consent in many different circumstances, such as when a teen requests a prescription for birth control without her parents' knowledge or consent. If the physician can be reasonably confident that the patient has the capacity to consent and documents the relevant details of the consent discussion in the medical record, it is likely that a College would support the physician in the event of a complaint from a parent. Meanwhile, parental involvement is recommended when the treatment entails serious risks and may have serious and permanent effects on the patient.

(https://www.cmpa-acpm.ca/en/advice-publications/browse-articles/2014/
can-a-child-provide-consent)

The situation in Canada is illustrative; however, actual policies will vary depending on the legal jurisdiction, many of which are independent and binding only on the location holding the jurisdiction (Coleman and Rosoff, 2013). The point here is whether mature minors are deserving of an obligation to their autonomy, for example, in obtaining consent for medical treatment. However, the notion of a mature minor is a fluid one depending on the nature of the treatment and the child's maturity, knowledge, and experience.

The aforementioned statement suggests that the physician has an obligation to assess the child's maturity, knowledge, and experience independently. If so, then the case presented is particularly problematic as the parents objected to the neurologist even discussing aspects of the proposed treatment that would be necessary for valid informed consent. It is possible that if the child in this case is a mature minor relative to the particulars of the case and the physician proceeded with DBS surgery without fully discussing the issues directly with the child, the physician may be liable for technical battery by failure to obtain the child's consent prior to the surgery. Whether the physician would even be prosecuted is a different question. This is not to suggest that the physician need not obtain consent from the parents. This would be important for a number of reasons, as will be discussed later. Again, the point here is that a child is deserving of having an obligation to the child's autonomy respected. Interestingly, some have argued that legal limitations to mature minor status also serve to protect physicians and healthcare professionals by limiting their liability by simplifying lines of responsibility (Coleman and Rosoff, 2013).

As discussed in Introduction, law can be considered as an important commentator on ethics and it is in that sense that the legal issues are considered. However, there often is not a direct correspondence between what is upheld as law and what many ethicists hold as truly just (Coleman and Rosoff, 2013). Thus, the issue of a mature minor may be seen very differently depending on one's point of view. To be sure, parents, physicians, and healthcare professionals are bound to lawful conduct but that does not relieve parents, physicians, and healthcare professionals of considering the more general ethical concerns.

Assent (dissent) and consent

An alternative consideration is the notion of assent and its denial, dissent, on the part of the child. As described earlier, commentators have noted that other than emancipated minors and very specific and limited circumstances where the mature minor doctrine can be held, parental rights generally trump the decision-making capacity of children. In many ways, it amounted to "Render to Caesar the things that are Caesar's; and to God [taken here as ethicists] the things that are God's" (Matthew 22:21). Thus, the intent of child assent to treatment provides some means of holding to the obligation to the child's autonomy (in this case, in tune with God) without getting caught up in the legalities and challenging the rights of the parents (in this case, Caesar's). Consider the requirements for child assent by the Committee on Bioethics, American Academy of Pediatrics:

Assent should include at least the following elements:

1. Helping the patient achieve a developmentally appropriate awareness of the nature of his or her condition.
2. Telling the patient what he or she can expect with tests and treatment(s).
3. Making a clinical assessment of the patient's understanding of the situation and the factors influencing how he or she is responding (including whether there is inappropriate pressure to accept testing or therapy).
4. Soliciting an expression of the patient's willingness to accept the proposed care. Regarding this final point, we note that no one should solicit a patient's views without intending to weigh them seriously. In situations in which the patient will have to receive medical care despite his or her objection, the patient should be told that fact and should not be deceived.

Committee on Bioethics (1995)

It is hard to see how this formulation of assent just described is any different than informed consent with the exception that assent does not appear to have a mechanism of enforcement other than the "no one should solicit a patient's views without intending to weigh them seriously." But what does "weigh them seriously" mean? How serious does the child's views have to be before they are compelling? Some have argued that a child's assent should not be a requirement but an ideal (Lang and Paquette, 2018) but it is not clear that such a position is anything more than a reason not to try in every case. Considering assuring an adult's civil rights, assuring every citizen's civil rights should be a requirement, as most would find it unacceptable that 90% of citizens had their civil rights respected. To be sure, there may be practicalities that might limit the number of persons whose civil rights actually are respected, but one does not surrender the obligation. In this sense, civil rights are an unalienable right in the Deontological moral theory sense and it is not clear that a Utilitarian moral theory would be held as justifying failure to make the effort. If a child's right to assent is considered equivalent to an adult's civil rights, it could be argued that the notion of assent as an ideal rather than a requirement is more for the ethical comfort of the physician or healthcare professional than the child patient. This is consistent with the long history of medical and clinical ethics (Chapter 4).

Similar notions of child assent in research have been proposed (Tait and Geisser, 2017). An extended quote is interesting as it stratifies children to those 11 years and younger from those aged 12–17. Also, research ethics is frequently extended to medical and clinical ethics (see Introduction):

Information for younger children (7–11 yrs)
If assent is deemed appropriate, younger children should, at minimum, be told what procedures will be done and how the child might experience them, the purpose of the study, that there may be no expectation of personal benefit but that their participation may help other children, that the study is voluntary, and that they can withdraw at anytime.

Information for older children/adolescents (12–17 yrs)
At minimum, older children and adolescents should be told what procedures will be done and how they might experience them, the purpose of the study, that there maybe no expectation of personal benefit but that their participation may help other children, that the study is voluntary, and that they can withdraw at any time. In some cases, it may appropriate for the investigator to speak with the child or adolescent absent from his or her parent(s) or legal guardian(s). For some therapeutic trials it is important that older children and adolescents be told that the research may be different from standard clinical care in their situation.

Interestingly, the only significant difference between the two age groups is "In some cases, it may appropriate for the investigator to speak with the child or adolescent absent from his or her parent(s) or legal guardian(s)." For some therapeutic trials, it is important that older children and adolescents be told that the research may be different from standard clinical care in their "situation" in the 12- to 17-year-old group. It would appear that those 11 years of age and younger were considered not to be able to appreciate what standard clinical care is and how it differs from research and that these younger potential research subjects cannot function appropriately without their parents. Those over age 12 can assent without involvement of the parent. In the situation for assent to medical treatment, some have argued that the child need not be informed of alternative treatment options. Quoting Lang and Paquette (2018),

Importantly, the level of understanding and appreciation necessary to accept or refuse a treatment plan, which has already been decided on by one's parents, is lower than that necessary to consider all of the alternatives as required by informed consent.

This is a curious qualification on the notion of assent and is contrary to the usual notion of informed consent that requires those giving consent to be fully informed of all reasonable alternatives; the definition of reasonable is not what the physician or healthcare professional deems reasonable. Rather what is reasonable is information, for example, alternative treatments, that had a reasonable patient been informed of could have affected the patient's decision. Patients generally do not have to consent to a specific type of suture in surgical cases but surely should be informed of nonsurgical alternatives, should those exist.

It would appear that the notion of assent, as qualified earlier, rather than consent is simpler and likely reduces the risk of ethical discomfort on the part of the physician or healthcare professional and perhaps on the parents. It would seem that child's assent is optional and qualified in a way that a parent's informed consent would not be. Note that this distinction is not based on the child's decisional capacity, as this is already accounted for by the process of obtaining assent. Thus, assent would not be an issue if the child is incapable of understanding the issues surrounding the proposed treatment. However, if the child can understand to a minimum sufficient degree, then does it matter that the child has less understanding than the parent? With the decision-making capacity set aside, it would appear that the obligation to the child's autonomy would be less than the obligation to adult (parent) autonomy.

The necessity of "black or white" when data are shades of gray

Justice is an outcome from an action or failure to act. The outcome is either just or unjust, the outcome is said to be dichotomous; black or white, yes or no, true or false. However, the factors entering into taking an action (and not taking an action is an action in itself) are continuous—shades of gray. Thus, justice turns on what is taken as the cutoff, for example, which shade of gray such that anything more gray is taken as black and anything less gray is taken as white. What is the minimal decision-making capacity must a child have to be considered a mature minor? One option is to use age as a cutoff when it can be applied to a population, analogous to a macrolevel justice. An alternative is to measure the individual child's decision-making capacity, thus implementing justice at a microlevel.

The distinctions in the notion of assent between those children less than 12 years of age and those older described earlier is an example of a cutoff based on age. As such, the distinction raises the question of what was the basis for deciding on 11 years of age as the cutoff? The authors wrote:

> While we do not necessarily agree with age-cut-offs per se, our rationale for this "developmental" cut-off was that it corresponds well with established beliefs regarding when children transition from concrete to operational thought [Crain W. Piaget's cognitive developmental theory. Theories of developmental concepts and applications. Englewood Cliffs, NJ: Prentice-Hall; 1980]. Adolescents begin to understand and develop the ability to act from moral motivations at 11-12 years of age [Eisenberg N, Miller P, Shell R, McNalley S, Shea C. Prosocial development in adolescence: a longitudinal study. Dev Psychol. 1991;27:849-57] and some may also be capable of providing their own consent [Hein I, DeVries M, Troost P, Meynen G, Van Goudoever J, Lindauer R. Informed consent instead of assent is appropriate in children from the age of twelve: policy implications of new findings on children's competence to consent to clinical research. BMC Med Ethics. 2015;16:76]. Furthermore, studies have shown that children's understanding of research information improves significantly after this age [Tait AR, Voepel-Lewis T, Malviya S. Do they understand? (part II): assent of children participating in clinical anesthesia and surgery research. Anesthesiology. 2003;98:609-14; Dorn LD, Susman EJ, Fletcher JC. Informed consent in children and adolescents: age, maturation and psychological state. J Adolesc Health. 1995;16(3):185-90.

From Tait and Geisser (2017)

The problem with the rationale just described becomes what is the "gold standard" by which to assess the child's competency? Tait and Geisser (2017) appear to take the appearance of adult-like moral motivations, which they hold occurs at age 12. But is this too strong? Does it exclude children under age 12 who, although not achieving adult-like competencies, nonetheless may achieve a "sufficient" competency? For example, in a study of children and adults on tests intended to assess competency to make informed treatment decisions, Weithorn and Campbell (1982) demonstrated group differences where children 9 years old did worse than 14-year-old children who performed similar to adults. Weithorn and Campbell (1982) wrote:

> Younger minors aged 9, however, appeared less competent than adults according to the standards of competency requiring understanding and a rational reasonable process. Yet, according to the standards of evidence of choice and reasonable outcome, even these younger minors appeared competent. Children as young as 9 appear to be capable of comprehending the basics of what is required of them when they are asked to state a preference regarding a treatment dilemma. And, despite poorer understanding and failure to consider fully many of the critical elements of disclosed information, the 9-year-olds tended to express clear and sensible treatment preferences similar to those of adults.

The child in the case presented here was age 10. Applying an age-determined notion of assent rather than the capabilities determined for each child would seem contrary to the United Nations Convention on the Rights of the Child (United Nations, 1990) and the polices of the Canadian government, as described previously. Methods for determining a particular child's decision-making capacity have been offered (Larcher and Hutchinson, 2010).

The issue of deciding when a particular child has reached an age entitling the child to assent or dissent in a meaningful way can be approached as any issue of diagnosis. In this case, the diagnosis is whether the child is of an age or circumstance such that the right to meaningful assent or dissent engenders an obligation to the child's autonomy that supervenes over that of the parents. The same issues of specificity and sensitivity apply. Sensitivity can be considered the probability of children over a specific age, for example, 9 years old, are sufficiently competent when in fact they are. Specificity can be considered as the probability that a child of the specific age is not competent when, in fact, they are not competent. If the age cutoff is insufficiently sensitive, then there will be children held incompetent who are not. If the cutoff is insufficiently specific, then there will be children held competent who are not. The consequence can include false positives, meaning children are deemed competent when they are not, and false negatives where they are deemed incompetent when they are not. The critical issue is the consequences of a false positive compared to the consequences of a false negative. This is an empiric question that should be determinable by the appropriate research.

The complex ethical ecology

The ethical ecology in which the issue of obligation to the child's autonomy is embedded is complex. It is made complex because of the child's dependence on the parents, the obligations of the parents to the child, and the great variability among children in their decision-making capacities that makes stratification of the obligation to the child's autonomy very difficult. Each of these have been addressed in some degree previously. However, there is another consideration when a child is involved, that is, the unique status of the family unit.

Parents and children operate in a unique local ecology that being a family unit, typically made up of first-degree relatives, sons, daughters, mothers, fathers, brothers, and sisters. In some situations and cultures, the family unit is more extensive, for example, intimately involving grandparents. Within the local ecology, the actions of any individual within the family unit affect others in the family unit. This is particularly at issue if the actions of any individual affect the resources available to meet the obligations to beneficence and nonmaleficence to others in the family unit. In most circumstances, a family does not look to others to provide the resources and thus, within the family distribution of resources, is a zero-sum proposition. In that context, perhaps it is interesting to apply John Rawls' notion of moderate scarcity and how that may dictate notions of justice (Rawls, 1999). If there is no scarcity of resources, there is less need for rules to govern individuals for the distribution of resources. If there is an absolute scarcity of resources, the members of the family unit are at jeopardy for serious harm and the society may intervene. Moderate scarcity creates the need for ethical rules allocating resources.

Ethical resources within the family unit are obligations to beneficence, nonmaleficence, and autonomy in each and every interaction between members within the family unit, which, in a sense, becomes a matter of obligation to the ethical principles of the family unit (Diekema, 2004). It is as though the family unit takes on the ethical ontology equivalent to an individual, not unlike the notion of rights of corporations (see Chapter 4). Indeed, some ethical theories object to Principlism, such as that advocated by Beauchamp and Childress (2013), as undue emphasis on the autonomy of the individual (Chapter 2).

Given the centrality of the family unit to the raising of children, logistic necessities perhaps require some special consideration afforded to the family unit over its members. This conundrum of balancing the obligations to rights with the necessities of a collective has been fundamental whether the collective is a family or society in general through governmental proxies. Perhaps it is not only a coincidence that the state is most likely to intervene between parents and children when the harm or risk of harm to the child is sufficiently egregious (Diekema, 2004).

As it would appear, most children are at the tender mercies of their parents and indeed most children can be confident in that. However, rupturing the trusting relationship between child and parent could have severe consequences. The risks of pursuing the issues in the case presented must include the risks associated with alienation between the child and the parent, as well as alienation between the physician or healthcare professional and the parents and the child. Perhaps it is in recognition of risks of alienation that courts have proceeded with such caution.

Some commentators have argued that the standard of "best interests" appears to dominate discussions of the ethics of child care (Diekema, 2004). They point out the problematic nature of the best interest standard, similar to the arguments against

Utilitarian moral theories that argue that what achieves the greatest good is what is just. The problem always has been in situations or circumstances of conflicting interests, whose good and over what duration of time counts as the maximal good.

Diekema (2004) argues that a modified version of the Harm Principle be employed, that the objections to and interventions against parental decisions should be based on the degree of harm or risk of harm. Diekema (2004) holds that this has been the standard employed most often by the courts. Yet, the notion of harm is as slippery as is the notion of best interests. Harm and best interests are continuous variables admitting of a potentially infinite number of degrees. Again, justice demands some means of converting the continuous measures to some "yes or no" decision. If the cutoff is some degree of physical harm, such as severe pain, disfigurement, or death, then the emphasis becomes one of obligation to nonmaleficence, with nonmaleficence couched in the starkest of terms. There is little room to consider the obligations to the beneficence and autonomy on the part of the child.

The difference between the best interest standard and the Harm Principle echoes the oft confusion between beneficence and nonmaleficence, assuming one can equate best interest with beneficence and harm as maleficence. Does the denial of beneficence constitute maleficence? Interestingly, in the original formulations of the Belmont report (Belmont Report: Ethical Principles and Guidelines for the Protection of Human Subjects of Research, 1978), nonmaleficence was subsumed by beneficence (Jonsen, 2003) and later became a different concept. For example, running into a burning building to save someone could be considered as an act of beneficence. Reasonable persons blinded to their self-interest would expect a firefighter to enter the burning building, where a failure to do so would be considered causing harm, maleficence. However, most would not expect a nonfirefighter to do so; in that case, the nonfirefighter is not committing maleficence by not providing beneficence.

Perhaps one way to consider the relatively different obligations to enter a burning building between a firefighter and a nonfirefighter is on the basis of contract. The firefighter is contracted to enter burning buildings—she receives payment in exchange for her willingness to enter burning buildings. By accepting the contract and being paid, the firefighter has a contractual obligation where failure to enter a burning building under circumstances where a consensus of reasonable persons (such as firefighters) blinded to their self-interest would do so would constitute maleficence. By maintaining a distinction between failure to provide beneficence and maleficence, the issue of obligations can be addressed more readily. While a parent would be obligated analogous to a firefighter, the question is whether a physician or healthcare professional is more analogous to a firefighter or a nonfirefighter?

References

Archard, D., 1993. Do parents own their children? Int. J. Child. Rights 1 (3–4), 293–301.
Aveyard, H., 2000. Is there a concept of autonomy that can usefully inform nursing practice? J. Adv. Nurs. 32 (2), 352–358.
Beauchamp, T., Childress, J.F., 2013. Principles of Bioethics, seventh ed. Oxford University Press, Oxford.

Belmont Report: Ethical Principles and Guidelines for the Protection of Human Subjects of Research, National Commission for the Protection of Human Subjects of Biomedical and Behavioral Research, Department of Health, Education and Welfare. 1978. United States Government Printing Office, Washington, DC.

Coleman, D.L., Rosoff, P.M., 2013. The legal authority of mature minors to consent to general medical treatment. Pediatrics 131 (4), 786–793. https://doi.org/10.1542/peds.2012-2470.

Committee on Bioethics, 1995. Informed consent, parental permission, and assent in pediatric practice. Pediatrics 95 (2), 314–317.

Diekema, D.S., 2004. Parental refusals of medical treatment: the harm principle as threshold for state intervention. Theor. Med. Bioeth. 25 (4), 243–264.

Jones, I., 2017. A grave offence: corpse desecration and the criminal law. Leg. Stud. 37 (4), 599–620.

Jonsen, A.R., 2003. The Birth of Bioethics. Oxford University Press, Oxford.

Kant, I., 2001. Fundamental principles of the metaphysics of morals. In: Wood, A.W. (Ed.), Basic Writings of Kant. The Modern Library, New York.

Krauss, J.K., 2010. Surgical treatment of dystonia. Eur. J. Neurol. 17 (Suppl. 1), 97–101. https://doi.org/10.1111/j.1468-1331.2010.03059.x.

Kwek, A., 2017. The indispensability of labelled groups to vulnerability in bioethics. Bioethics 31 (9), 674–682. https://doi.org/10.1111/bioe.12379.

Lang, A., Paquette, E.T., 2018. Involving minors in medical decision making: understanding ethical issues in assent and refusal of care by minors. Semin. Neurol. 38 (5), 533–538. https://doi.org/10.1055/s-0038-1668078.

Larcher, V., Hutchinson, A., 2010. How should paediatricians assess Gillick competence? Arch. Dis. Child. 95 (4), 307–311. https://doi.org/10.1136/adc.2008.148676.

Lohkamp, L.N., Mottolese, C., Szathmari, A., Huguet, L., Beuriat, P.A., Christofori, I., Desmurget, M., et al., 2019. Awake brain surgery in children-review of the literature and state-of-the-art. Childs Nerv. Syst. 35 (11), 2071–2077. https://doi.org/10.1007/s00381-019-04279-w.

Mill, J.S., 1859. On liberty. In: Gray, J. (Ed.), On Liberty and Other Essays (1991). Oxford University Press, Oxford.

Montgomery Jr., E.B., 2014. Intraoperative Neurophysiological Monitoring for Deep Brain Stimulation: Principles and Practice. Oxford University Press, Oxford.

Rawls, J., 1999. A Theory of Justice, rev. ed. The Belknap Press of Harvard University Press, Cambridge, MA.

Spear, H.J., Kulbok, P., 2004. Autonomy and adolescence: a concept analysis. Public Health Nurs. 21 (2), 144–152.

Storch, E.A., Cepeda, S.L., Lee, E., Goodman, S.L.V., Robinson, A.D., De Nadai, A.S., Schneider, S.C., et al., 2019. Parental attitudes toward deep brain stimulation in adolescents with treatment-resistant conditions. J. Child Adolesc. Psychopharmacol. https://doi.org/10.1089/cap.2019.0134.

Tait, A.R., Geisser, M.E., 2017. Development of a consensus operational definition of child assent for research. BMC Med. Ethics 18 (1), 41. https://doi.org/10.1186/s12910-017-0199-4.

United Nations, 1990. Convention on the Rights of the Child. United Nations, New York.

Varelius, J., 2006. The value of autonomy in medical ethics. Med. Health Care Philos. 9 (3), 377–388. https://doi.org/10.1007/s11019-006-9000-z.

Vidailhet, M., Vercueil, L., Houeto, J.L., Krystkowiak, P., Lagrange, C., Yelnik, J., Bardinet, E., Benabid, A.L., Navarro, S., Dormont, D., Grand, S., Blond, S., Ardouin, C., Pillon, B., Dujardin, K., Hahn-Barma, V., Agid, Y., Destée, A., Pollak, P., French SPIDY Study Group, 2007. Bilateral, pallidal, deep-brain stimulation in primary generalized dystonia: a prospective 3 year follow-up study. Lancet Neurol. 6 (3), 223–229.

Weithorn, L.A., Campbell, S.B., 1982. The competency of children and adolescents to make informed treatment decisions. Child Dev. 53 (6), 1589–1598.

Case 14

Case 14—The patient is a 68-year-old-male member of the Mohawk Nation of Canada, who was diagnosed with Parkinson's disease by a nonindigenous physician. When asked by the physician how he was doing regarding his symptoms and disabilities from Parkinson's disease, the patient would respond after a minute with it was "OK." The physician pressed the patient to give more details. The patient responded after a delay that he was "OK." When the physician asked the patient to rate the severity of his Parkinson's disease, with 10 being almost normal and 0 being very disabled, the patient said "5." It was said in such a manner that the physician, from his perspective, suspected the accuracy or veracity of the answer. Exasperated, the physician recommended having the patient spend an entire day in the clinic where his symptoms and disabilities could be rated on an hourly basis. The physician's hope was that a series of objective evaluations by the clinic nurse would provide sufficient information about the patient's response to medications for the physician to recommend an improved medication regimen. When the physician asked the patient if he could spend the day in the clinic, the patient said nothing in reply. The physician asked the patient several more times, and the patient said that it sounded like a good idea and he would consider it. The physician then turned to the patient's wife and asked her whether she thought her husband was depressed or had a problem with apathy. The wife denied either was the case.[a]

Medical facts of the matter

Levodopa is a remarkably effective medication for the treatment of Parkinson's disease; however, the time course of action of levodopa in the body is complicated. Early in the disease, the benefit from levodopa may last many hours, allowing the patient to take the medication only a few times during the day. The duration of the levodopa benefit becomes progressively shorter as the disease progresses. Some patients find that they must take the regular formulation of levodopa on an hourly basis to maintain steady control. Some patients require a constant infusion of levodopa-like medication through a tube placed into the stomach.

Often, it is very difficult to determine how long the benefit from levodopa lasts for a given patient, and that information is critical in choosing the optimal dosing of levodopa. It would be of little use to prescribe levodopa to be taken twice a day when its beneficial effects only last 2 h. Many times, appropriate questioning of the patient can reveal the duration of benefit. At other times, it may be necessary to observe the

---

[a] The case presented is hypothetical, although modeled on experience. No judgment is made relative to the correctness, appropriateness, or value of any actions described on the part of any of the participants. The purpose here is to describe and explore the variety of ethical considerations and consequences that depend on the perspective of different relevant ethical systems.

patient over time to determine the duration of benefit and then prescribe the appropriate dosing interval. In the case presented, the physician was unable to gain a sense of the degree of benefit from the patient's levodopa dosing and the duration of the benefit. Hence, the physician recommended the patient spend the entire day in the clinic with hourly assessments.

Ethical dissonance

It is difficult to discuss this case without considering the historical and continuing discrimination against indigenous persons that affects healthcare. Canada, as a whole, ranks third among 177 countries on the Human Development Index survey, a measure of education, life expectancy, and income. When only Canada's First Nations (indigenous) communities are considered, Canada ranks 68th (https://uwaterloo. ca/canadian-index-wellbeing/sites/ca.canadian-index-wellbeing/files/uploads/files/ ACloserLookAtSelectGroups_FullReport.sflb_.pdf). Adverse stereotypes of indigenous persons affect all aspects of medical care. In a landmark case, the symptoms of an indigenous patient's urinary bladder infection were dismissed as due to alcoholism and the patient died from the infection (Brian Sinclair Working Group, Out of Sight, 2013). There is a striking failure to consider the unique situations and circumstances of indigenous patients. The State of Ohio issued a guide to implementing cultural and linguistic competence in the care of patients (State Medical Board of Ohio, Cultural & Linguistic Competency for Improved Health Outcomes in Ohio, https://www.med. ohio.gov/Portals/0/Resources/SMBO%20Cultural%20Competency%20Guide%20 for%20Providers.pdf?utm_source=BenchmarkEmail&utm_campaign=Licensees%3a_ New_Cultural_Competency_Video&utm_medium=email). While Caucasians, African Americans, and Latino/Hispanics were specifically discussed, indigenous people (Native Americans) were not. It may be argued that while indigenous people make up only 0.3% of the population of Ohio, which represents more than 35,000 indigenous people.

The case presented here does not appear to involve overt stereotyping or discrimination, yet there is something not quite right about the interactions in the case. It would seem that both the patient and the physician were unsatisfied by the encounter. As a background to discussing this case, it is important to note that the author is a nonindigenous person who grew up in a traditional Western culture and pursed a career as a neuroscientist and allopathic physician within the traditional Western reductionist empiricist ethos. Yet, having trained in an ethos where the ethical/moral imperative was never turning a patient away, and realizing what that imperative truly means (perhaps late in the game), prompted the author to examine his own presumptions and to recognize his limitations, including cultural presuppositions that were absorbed by something akin to osmosis. Yet, try as one might, it is not possible to place oneself totally in the experiences of others (empathy), but it may be possible to feel things that resonate with what others feel (sympathy), recognizing that such resonance will always be incomplete. Accepting the privilege and responsibility of being a physician requires sympathy for all patients who seek help. It necessitates Two-Eyed Seeing, an approach that recognizes the differences and common grounds between the indigenous person's perspective and that of the nonindigenous physician trained in the traditional Western ethos, which is discussed later. In a real sense, evolution of the physician's perspective is like the development of a child from the perspective (1) that all of those around are just extensions of themselves; (2) having the sense of other minds; and (3) perhaps, most importantly, that these other minds may think differently (for relevance to the ethics of everyday medicine, see Chapter 4).

In writing this case, the author looked for instruction and insight in the words of indigenous writers. Those words, understandably, are only a translation for the nonindigenous reader. Nevertheless, they shed new light on the experiences of caring for an indigenous patient. The benefit in this author describing the case presented is that what is described here may resonate with other nonindigenous physicians and healthcare professionals and may possibly help them better understand their own feelings and treatment of indigenous patients. To be clear, the author does not speak for the indigenous patient but only speaks to his own understanding, however incomplete, with the hope that it might be of some help.

To begin discussing this case, it is first necessary to consider how persons are identified or named, as many terms used to describe indigenous people are fraught with legal or pejorative connotations (http://indigenousfoundations.arts.ubc.ca/terminology/). Terms used in Canada include *Aboriginal*, a government label; and *indigenous*, meaning person who originally held and cared for the land. The term *indigenous person* will be used here except in reference to a specific Nation.

Among indigenous persons are First Nations persons, as well as persons who are ethnically Métis or Inuit. There are 634 First Nations in Canada. In Southern Ontario, where this patient received medical care, the indigenous community is made up primarily of the Six Nations of the Grand River, which include the Mohawk, Cayuga, Onondaga, Oneida, Seneca, and Tuscarora. There is no single "indigenous ethic," as in many ways the indigenous culture is more pluralistic and liberal than many traditional Western societies (discussed in more detail later). Thus, one cannot generalize across indigenous people. There are, however, commonalities among indigenous Nations that provide the basis for conversation and sharing.

In the case presented here, the physician seemed to believe that the patient was unconcerned regarding his health and perhaps dismissive of the physician's recommendations. It would appear that the physician's impression was based on the dissonance between what the physician expected and the response of the patient. There are a number of grounds on which this dissonance could be based. First, there may be different understandings of the nature of the patient's medical problems, the urgency of these problems, and the effort required for optimal management. If the physician believed that the patient understood the nature and urgency of the medical problems, then the dissonance would be increased because the patient's response would seem irrational to the physician. Perhaps the patient has a mental illness, such as depression or schizophrenia, although an assessment of such mental illnesses in the indigenous people is problematic. Frequently, mental illness in the indigenous population is overdiagnosed (Brant, 1990). Perhaps the patient has what is called an organic brain syndrome secondary to repeated neurological insults. Perhaps there is a sense of fatalism on the part of the patient given his particular history and circumstances.

It may be that the problem lies in the nature of communication and the presumptions implied in the different communication styles of the patient and the physician. These presumptions include methods of conveying information and the metaphysical or worldview presumptions that inform the manner of communication. These presumptions provide context that gives knowledge of the origin or derivation of the communicative terms used and also the direction or intended use of the words. This is evident by the fact that words in one language have no exact translation or no equivalent in another. Willard Van Orman Quine (1960) pointed to the universal problem of meaning in translation and pointed to a concept of "inscrutability of reference," meaning that two persons pointing to the same phenomenon may be pointing to two different aspects of the same phenomenon with very different meanings as a consequence.

In the case presented, the patient and the physician occupy the same space or piece of the world at the same time: a clinical encounter. The problem of such encounters, however, is that the indigenous person must navigate a medical environment created and dominated by Western traditions and values. This necessity has led to the concept of Two-Eyed Seeing (Iwama et al., 2009). By its very title, Two-Eyed Seeing suggests that there are two distinct ways of seeing the world. The concept of Two-Eyed Seeing is not to integrate or blend the two views, but rather to look at their overlap, their commonalities, at the same time, reaffirming the perspective of each eye. Citing R.D. Herring (1989), Betty Duran provided the following list that contrasts Western traditions (Anglo-American) with traditions of indigenous people (Native American) (American Indian Belief Systems and Traditional Practices, http://www.wellness-courts.org/files/Duran%20-%20American%20Indian%20Belief%20Systems.pdf):

Anglo American	American Indian
Success and happiness	Ownership sharing
"Number One"	Tribe and extended family first, before self
Youth oriented	Honor your elders
Learning is found in school	Learning is through legends
Look to the future	Look to traditions
Work for retirement	Work for purpose
Be structured and aware of time	Time is only relative
Oriented to house, job, etc.	Oriented to land
Look ahead, not to the past	Cherish the memories of youth
A critic is a good analyst	Don't criticize your people
"What are you—some kind of animal?"	Live like the animals; they are your brothers and sisters
This is America, speak English	Cherish your language
I'll raise my own; you do the same	Children are gifts of the Great Spirit to be shared with others
The law is the law!	Consider the relative nature of a crime, the personality of the individual, and the conditions of the offense.
Have a rule for every contingency	Few rules are best—loose and flexible
Religion is for the individual	Religion is the universe

The language used is important to the Two-Eyed Seeing concept. Also important is the means of conveying meaning. Indigenous people may convey meaning through stories that are metaphors instead of explicit definitional terms based largely on a reduction to more abstract concepts. For example, when describing the interconnectedness of all people, Mi'kmaw Spiritual Leader, Healer, and Chief Charles Labrador of Acadia First Nation, Nova Scotia said, "Go into the forest, you see the birch, maple, pine. Look underground and all those trees are holding hands. We as people have to do the same" (Kierans, 2003). When describing pain, indigenous persons may find the use of the Visual Analog Scale and other Western means of assessing pain problematic (Latimer et al., 2014). An indigenous person may tell a story around the pain and the length of the story may approximate the severity of the pain (Latimer et al., 2014). When asked about memory loss, an elderly lady said, "I'm like an old cedar tree; the top is turning silver and grey and the wood is no good, but the bottom is still green" (Kelly and Brown, 2002).

Just as indeterminacy of translation, such as described by Quine, affects verbal or written expressions, it also affects nonverbal communication. In a study of successful nonindigenous physicians working with indigenous patients, Kelly and Brown (2002) described a nonindigenous physician interacting with a young indigenous woman, writing of the physician:

> I would be seeing a teenage girl, who is maybe in some kind of trouble…. My approach in that situation would be to sit down, sometimes on the footstool on the floor, so I don't appear to be this large, white guy looming over a small Native girl, and not say anything for a long time. So that silent thing and getting out of someone's face and avoiding direct eye contact and avoiding pressuring someone for the answers.

Kelly and Brown described (2002) an indigenous elderly person saying, "An elder told me the story that we were born with two ears, two eyes, and one mouth and should use them accordingly. We should use the mouth just half as much as the ears and the eyes." In a study of medical residents, the average time a patient spoke before the physician intervened was approximately 12 s, with the physician interrupting the patient frequently (Rhoades et al., 2001). The risk in interrupting a patient's story risks the physician's understanding being incomplete or incorrect.

For many indigenous people, silence is a means of conveying information and meaning (Dr. Clare Brant, Native Ethics & Principles, https://www.cbu.ca/indigenous-affairs/mikmaq-resource-centre/mikmaq-resource-guide/essays/native-ethics-principles/; Knapp, 2013). Indigenous children are taught not to complain. What appears as a lack of communicativeness on the part of the patient may be a stoicism unique to his or her character or to traditional cultural values. If the patient is being stoic, this could conflict with the physician's notion of autonomy. The physician may believe that the patient has a right to self-direction, and following from that right is the responsibility to self-direct in an active manner. Indeed, the prohibition against paternalism would seem to demand active self-advocacy on the part of the patient or, at the very least, that the patient not undermine his own rights. However, the physician's view of autonomy may be too narrow (see Case 13). The notion of autonomy for the patient may be quite different than that of the physician, and the patient may deny and oppose any suggestion of a lack of self-agency. For the patient, self-agency may not be synonymous with self-advocacy.

Michael Meyer (1987) argues for two types of autonomy. There is a negative sense of autonomy holding that a person cannot be directed to do something. This notion of autonomy is evidenced in informed consent where compelling someone to act against his or her wishes would be the criminal offense of battery, and harassing the patient to do so would be assault. Positive autonomy holds that a person is actively self-directing. The notion of autonomy can be further characterized as rights-sensitive autonomy in which a right is asserted. For example, a person has the right against illegal bodily contact or threat of such. It is a right for something not to happen, hence a kind of negative autonomy as opposed to demanding an action, such as demanding a specific treatment. Stoic autonomy holds the locus of control within the individual rather than some external action. The individual chooses whether and how to assert his or her autonomy as a right.

Stoicism

Among the virtues advocated by First Nations persons is emotional restraint (Brant, 1990). Some have interpreted such emotional restraint as a prohibition against complaining. Certainly, the latter would be consistent with some interpretations of stoicism. However, emotional restraint may be closer to the notion of stoicism. Further, complaining would be a violation of the ethical value of noninterference, as the complaint could be seen as a means to coerce the person about whom the complaint was lodged. Thus, it is reasonable that stoicism is a feature of ethics among indigenous people.

There is utility in both rights-sensitive autonomy and stoic autonomy. An assertion of positive autonomy as the consequence of a right reinforces that right and reduces the harm that may come from the right being abridged—the old adage, "give an inch, they will take a yard." The emotional consequences largely depend on the degree of control of those things that threaten autonomy. In an interesting experiment, pairs of nonhuman primates were restrained and subjected to repeated electrical shocks. One of the pairs could sometimes block the electrical shocks given to both nonhuman primates. The rate of gastric ulcers was higher in the nonhuman primate that had partial control of the pain (the executive monkey) compared to the passive nonhuman primate (Brady, 1958). Perhaps the executive nonhuman primate could be attempting to assert rights-sensitive autonomy while the passive nonhuman primate was asserting a form of stoic autonomy. But stoic autonomy is much more.

It may well be that a culture established on what nature, the land, brings to them, rather than what can be taken from the land, engenders a type of stoicism. Indeed, a common theme among indigenous people is the notion of a nurturing nature that provides its bounty for which nature should be respected in return. This contrasts sharply with the exploitation of nature often inherent in Western traditions—nature is to be subjugated, it is not a collaboration. Yet, stoicism is not the same as surrender.

Stoicism is the person controlling his or her own feelings, choosing not to be hurt in the presence of pain. As such, stoicism is a positive virtue to be cultivated. It is a virtue imposed by the Principle of Harmony, as discussed later. As Meyer (1987) wrote:

> Attainment of stoic autonomy requires no social assertiveness because a fundamental realization of the stoic outlook is that the attempt of attaining outward recognition and success—in the world beyond the scope of one's will—cannot always be realized. Given this assumption, the stoic's recommendation is clear: one should not set one's heart on attaining things that are in this way not strictly identifiable with one's true self—i.e., that individual *will* [italic in the original] which should control one's life.

Stoicism may be an effective tool to deal with anxiety. Indigenous persons are more likely to become quieter under stress, whereas others tend to talk more and louder (Dr. Clare Brant, Native Ethics & Principles, https://www.cbu.ca/indigenous-affairs/mikmaq-resource-centre/mikmaq-resource-guide/essays/native-ethics-principles/). Such quietness could be mistaken as disinterest and lead to misunderstanding.

Noninterference

A key ethical constraint is noninterference (Brant, 1990; Dr. Clare Brant, Native Ethics & Principles, https://www.cbu.ca/indigenous-affairs/mikmaq-resource-centre/mikmaq-resource-guide/essays/native-ethics-principles/, Duran, B.E.S. American Indian Belief Systems and Traditional Practices, http://www.wellnesscourts.org/files/Duran%20-%20American%20Indian%20Belief%20Systems.pdf). This ethical constraint likely is prudential rather than deontological. Noninterference is derivative of the Principle of Harmony. To be sure, social groups in a setting of moderate scarcity of resources where the response is cooperative interactions to assure survival, harmony takes on a prudential virtue (Miller, 2019). This stands in contrast to "every person for themselves and the devil take the hindmost" attitude in the laissez-faire capitalistic political organization of many traditional Western societies (note that this statement is not pejorative or normative but rather descriptive). The degree to which noninterference governs ethics among some indigenous people is quite striking in contrast to more Western traditions. For example, indigenous children as young as 6 years old can decide for themselves on whether to do their homework (Brant, 1990). This is a remarkable contrast with issues of consent and assent in medical procedures by children and adolescents typical of Western traditions (see Case 13).

Noninterference extends to teaching. Many indigenous persons who teach do so by modeling or displaying the appropriate ethical behavior. If a specific ethical behavior on the part of some individual is discussed, the discussion typically involves stories and metaphors rather than any deductive approach based on principles where the appeal to rationality is the motivating force. Taking such a deductive approach risks interfering by directly telling the offending person how to behave. Considering the physician or healthcare professional as a "teacher," it may be difficult for the physician or healthcare professional to "model" what he or she wants to instill in the patient. It is not as though the physician can model what it is like to have poorly treated Parkinson's disease and then model what good control looks like.

Typically, the physician or healthcare professional provides recommendations, but the asymmetry of power, particularly in the absence of full informed consent, gives the recommendations significant force. Giving the recommendations once might be bad enough given the notion of noninterference. Repeating the recommendations under the suspicion that the patient might not be compliant only compounds the problem. Perhaps one solution would be telling a story of another patient who followed the recommendations and did well. Alternatively, a metaphor that is indirect enough so as not to be taken as interference but close enough to suggest a particular behavior may be used. In the case presented, perhaps the patient's situation where the medications are not lasting long enough and the dosing interval is too long could be likened to putting out a fire, but where the fire bucket has a hole in it. The firefighter has to go back to the well more frequently and quickly, at least sufficiently to minimize the water loss before the water is thrown on the fire.

No privileged knowledge

The notion of noninterference for many indigenous people is premised on the notion that no one person's ideas or knowledge is privileged with the exception of the elders. However, it is important to note that being an elder is not related directly to age, but that the perceived wisdom in the elder is established over years of involvement with those who elect the elder (although the election is not something that is formal, as might be the case in traditional Western societies). To allow others to have a privileged knowledge or opinion would be to invite competitiveness, whose knowledge or opinion is more privileged, thus disrupting harmony. Thus, it is not a forgone conclusion that the opinion of the physician or healthcare professional is any more privileged than the patient's own and perhaps less than that of the elders. In other words, the patient may see a symmetry in the relationship between himself and the physician or healthcare professional. This would conflict with a physician-centric ethics based on a presumed asymmetry between the physician and the patient, with the physician being in a superior position (see Chapter 4).

Changing the balance of power between the patient and the physician or healthcare professional to be more symmetrical may be difficult. It clearly would require a shift from physician- to patient-centric ethics. But to this day, there is a paucity of truly patient-centric ethics (Chapter 4). Perhaps adopting the Two-Eyed Seeing approach would be helpful, but this first requires the physician to cover (bracket) his traditional Western eye and attempt to see with an indigenous eye or, at the very least, accept the indigenous person's view. It may not be possible for the physician to see with the perspective of the indigenous person and he or she may have to ask the patient or other indigenous persons for help. This approach is central to the Two-Eyed Seeing approach to research in the indigenous community where members of the community are full, perhaps even controlling, partners in the conduct of the research (Assembly of First Nations, First Nations Ethics Guide on Research and Aboriginal Traditional Knowledge, https://www.afn.ca/uploads/files/fn_ethics_guide_on_research_and_atk. pdf). Themes often stressed are of immediate relevance and benefit to the indigenous community. Yet, this is little different than more typical research informed consent where there is some modicum of direct benefit to the subject even if it cannot be guaranteed. In the case of the indigenous people, the people hold the same position or status as the individual research subject. Sharing of knowledge is on the terms set by the indigenous people.

Elasticity of time

The notion of time may be relatively elastic in the Indigenous community (Dr. Clare Brant, Native Ethics & Principles, https://www.cbu.ca/indigenous-affairs/mikmaq-resource-centre/mikmaq-resource-guide/essays/native-ethics-principles/; Brant, 1990). For many indigenous persons, time is measured by purpose, not by the passage of minutes or hours. Things get done when the time is ripe for them to get done; importantly, time is when it is ripe for things to get done. Some may perceive this as procrastination or laziness, but this would be to misunderstand the nature of time of the indigenous people (Dr. Clare Brant, Native Ethics & Principles, https://www.cbu. ca/indigenous-affairs/mikmaq-resource-centre/mikmaq-resource-guide/essays/native-ethics-principles/; Brant, 1990). Perhaps, in the case presented, the time it took for the patient to say he would consider the recommendations was because the time did not become ripe until the physician was being seen as coercive. Note that the patient likely responded for the physician's benefit lest the physician's continued entreats would embarrass the physician. It is not likely that the physician saw it that way.

Ethics as moral-ought or prudential-ought

Like nearly every civilization, indigenous people have a creation story consonant with their ethics. It is an open question whether the relationship between the creation story and the ethics is cause and effect so much as an association. Ethics, as normative, can have the force of a moral-ought or a prudential-ought (Miller, 2019). The force that drives ethical behavior can be on the basis of some moral good, such as in received knowledge from God—moral-ought. Alternatively, it can be an abstraction or post hoc rationalization of whatever behaviors facilitate group survival—a prudential-ought. An example would be consensus among those subject to the ethics in pluralistic modern liberal democracies that penalizes those who do not conform to the consensus. It could be argued that the indigenous culture is more liberal and democratic than traditional Western cultures, thus even the notion of forming a consensus is different between the cultures. Nevertheless, the creation story, along with its ontology of nature, provides, at the very least, a metaphor to help understand the indigenous ethics.

Among the Muscogee (Creek) Nation, the natural ontology describes an energy that flows through all things, thus unifying all (Miller, 2019). The energy, in all its forms, constitutes a single entity, Ibofanga (Epohfvnkv), and humans interact through assistants to Ibofanga. The interactions between things, particularly living things, are based on an exchange of energy; there is no taking without giving. All relationships imply reciprocity, hence the appeal to the Principle of Harmony.

As Miller points out, reciprocity is a balancing act, and the manner in which balance is achieved can appear quite different from traditional Western ethics. Miller (2019) wrote:

> Since reciprocity is a way of balancing energy, then reciprocating could conceivably sometimes take the form of actions that would otherwise, or most of the time, seem bad. Actions that may be bad most of the time may be justified if that action balances energy. For example, if a transgression (e.g., an act of war, assault, the taking of hostages, etc.) has occurred, then reciprocation may require similar actions (that would otherwise be bad) as a way of balancing energy. In other words, seemingly bad actions may be justified if these actions balance an exchange of energy.

A particularly interesting example is the notion of punishment. Miller, quoting Sarah Deer (2015), writes:

> The word "force" (used twice) is an important clue that this passage describes a physical attack and the law clearly refers to women as victims (although it does not indicate the gender of perpetrators). There is a clear reference to corporal punishment ("whip or pay")—which is consistent with observed Mvskoke law in practice in the early nineteenth century. Perhaps the most remarkable component of this law is the last six words: "what she say it be law." This phrase, suggesting a… victim had legal standing to participate in sentencing decisions, is fundamentally inconsistent with Anglo-American… law in the same time period.

In a sense, the most direct balancing of energy occurs between the offender and the victim without the offloading of reciprocity affected by the punishment to a judge, magistrate, or administrator, as in traditional Western traditions.

As all things are tied to each other by the same energy that flows through everyone and the nature of an exchange of energy necessitates reciprocity, ultimately there is no escape from transgressions. This may underly the indigenous notion of taking only what is needed when hunting, so as to preserve and perpetuate future food sources. Prayers are offered to the prey killed. Miller, quoting Bear Heart, wrote:

> Never kill out of anger, nor for sport to see how many animals you can kill. Take just enough for survival and always be respectful of the four-leggeds. *If you must kill, present an offering and talk to the animal, explaining, "I need you for my family"* [italic in the original].
>
> *Heart and Larkin (1996)*

The indigenous person's view of the relationship to animals is quite different from that of the King James Bible where in Genesis 1:26 it is written "And God said, let us make man in our image, after our likeness: and let them have dominion over the fish of the sea, and over the fowl of the air, and over the cattle, and over all the earth, and over every creeping thing that creepeth upon the earth." At the other extreme are certain advocates that all animals have certain inalienable rights and another that asserts no rights for animals. Note that both are deontological in a sense that the one excluding any inherent ethical or moral standing for animals or, at least can be interpreted as such, a negative deontological sense. The one that asserts an animal's right is a positive deontological sense. It would appear than neither reflect the view of the Muscogee (Creek) Nation's people. The Muscogee (Creek) Nation's ethic may be described better as transactional rather than deontologically principled; this should not be seen as pejorative, as will be discussed later. Indeed, a transactional ethic may be virtue ethics when the transactions are virtuous (see Chapter 2), which may be better than a Principlist ethic.

Ethical system

It would seem that indigenous ethics are not principled in the sense of explicit, abstracted, generalized principles that are transcendent over the particulars (Chapters 1 and 2). This is evident in the storytelling and metaphors as a means to convey ethical knowledge. Specific admonitions as to the proper ethical conduct in particular situations and circumstances are not explicit as a set of formal abstracted rules, but intimated by story and metaphor (Dr. Clare Brant, Native Ethics & Principles, https:// www.cbu.ca/indigenous-affairs/mikmaq-resource-centre/mikmaq-resource-guide/ essays/native-ethics-principles/). Ethics are ways of behaving that lead to harmony among the people (see later). As such, the system of virtue among indigenous persons reflects an Anti-Principlist account, as it may be seen as transactional rather than principled, even though it conforms to the world ontology, as described earlier (see Chapter 2).

In many ways, the ethics are virtue ethics analogous to Aristotelian ethics. Aristotle wrote "We are not studying in order to know what virtue is, but to become good, for otherwise there would be no profit in it" (Aristotle, Nicomachean Ethics II.2). In many ways, indigenous ethics are similar to Kant's notion, where Kant (2001) wrote:

> Now in a being which has reason and a will, if the proper object of nature were its conservation, its welfare, in a word, its happiness, then nature would have hit upon a very bad arrangement in selecting the reason of the creature to carry out this purpose. For all the actions which the creature has to perform with a view to this purpose, and the whole rule of its conduct, would be far more surely prescribed to it by instinct, and that end would have been attained thereby much more certainly than it ever can be by reason.

In many respects, the ethics of Aristotle, Kant (in this reading), and indigenous persons focus on how to behave (virtue ethics) instead of why to behave (Principlist ethics); the former may be a better insurer of good behavior and harmony.

Implications for physicians and healthcare professionals

If a physician and healthcare professional hold for a patient-centric ethic (see Chapter 4), the currency of exchange with the indigenous patient should be on the basis of the ethics that the patient can relate to, i.e., what motivates the patient. Otherwise, any exchange likely will benefit neither, although the physician and healthcare professional will still receive their reimbursements. The communication may not be a matter for Two-Eyed Seeing. There may be insufficient overlap between the ethics of the indigenous person and that of the physician or healthcare professional holding to traditional Western medical ethics. The choice then is who to respect?

As the everyday practice of medicine ultimately is an exercise in ethics (see Introduction), then optimal medicine requires following the ethics that are optimal in the care of the particular patient. The first step is for the physician or healthcare professional to recognize that he or she is operating from an ethic even if it is at the "gut" level and unrecognized (see Chapter 5). Once the physician or healthcare professional recognizes that he or she is operating from a specific ethical system, it is hoped that this will generate an awareness that other ethical systems exist. Then, perhaps the physician or healthcare professional will consider that the patient may be operating from a particular ethical system that is not the same as that of the physician or healthcare professional.

The indigenous ethic is holistic, if for no other reason than the recognition that the same energy courses through all things and binds all things [as described earlier for the Muscogee (Creek) people]. Little separation exists among the mental, emotional, physical, and spiritual dimensions by which the indigenous people consider their life (https://www.fnha.ca/wellness/wellness-and-the-first-nations-health-authority/first-nations-perspective-on-wellness). Such holism would be more consistent with the Empiric school of medicine (of which homeopathy would be an example), not with the Rationalist/Allopathic school of medicine, to which medical doctors generally belong and which dominates traditional Western medicine (Montgomery, 2019). Indeed, Rationalist/Allopathic physicians railed against the Empirics as demonstrated by the American Medical Association Code of Ethics of 1847. Thus, it should not be surprising that today's Rationalist/Allopathic physicians would find such a holistic approach unsettling (see Case 7). Just the same, pursuing a Rationalist/Allopathic medicine tradition may seem unsettling to an indigenous patient.

Accepting indigenous ethics would be to embrace the Two-Eyed Seeing approach. As such, the indigenous patient may seek the advice of elders whose advice would be on par with that of physicians and healthcare professionals. It would be a violation of the noninterference ethics for the physician or healthcare professional to try and directly dissuade the patient from the guidance the patient might receive from family, friends, and elders.

References

Brady, J.V., 1958. Ulcers in "executive" monkeys. Sci. Am. 199 (3), 95–104.

Brant, C.C., 1990. Native ethics and rules of behaviour. Can. J. Psychiatry 35 (6), 534–539.

Brian Sinclair Working Group, Out of Sight, 2013. A Summary of the Events Leading Up to Brian Sinclair's Death, The Inquest That Examined It and Leading the Interim Recommendations of the Brian Sinclair Working Group; https://media.winnipegfreepress.com/documents/Out_of_Sight_Final.pdf; Nurse Thought Brian Sinclair was Intoxicated, Inquest Told, CBC News. Posted: Oct 10, 2013 6:38 PM CT|Last Updated: October 10, 2013, https://www.cbc.ca/news/canada/manitoba/nurse-thought-brian-sinclair-was-intoxicated-inquest-told-1.1959602.

Deer, S., 2015. The Beginning and End of Rape: Confronting Sexual Violence in Native America. University of Minnesota Press, Minneapolis, MN.

Heart, B., Larkin, M., 1996. The Wind Is My Mother: The Life and Teachings of a Native American Shaman. The Berkley Publishing Group, New York.

Herring, R.D., 1989. The American Native family: dissolution by coercion. J. Multicult. Couns. Dev. 17 (1), 4–13.

Iwama, M., Marshall, M., Marshall, A., Bartlett, C., 2009. Two-eyed seeing and the language of healing in community-based research. Can. J. Nativ. Educ. 32 (2), 3. Ethnic NewsWatch.

Kant, I., 2001. Fundamental principles of the metaphysics of morals. In: Wood, A.W. (Ed.), Basic Writings of Kant. The Modern Library, New York, p. 153.

Kelly, L., Brown, J.B., 2002. Listening to native patients. Changes in physicians' understanding and behaviour. Can. Fam. Physician 48, 1645–1652.

Kierans, K., 2003, May 18. Mi'kmaq Craftsman Preserves 'Old Ways'. The Halifax Sunday Herald, p. C4.

Knapp, C.E., 2013. Two-eyed seeing as a way of knowing: blending Native and Western science in outdoor learning. Green Teach. 99, 31–34.

Latimer, M., Simandl, D., Finley, A., Rudderham, S., Harman, K., Young, S., MacLeod, E., et al., 2014. Understanding the impact of the pain experience on aboriginal children's well-being: viewing through a two-eyed seeing lens. First Peoples Child Fam. Rev. 9, 22–37.

Meyer, M., 1987. Stoics, rights, and autonomy. Am. Philos. Q. 24 (3), 267–271.

Miller, J.L., 2019. Etemeyaske vpokat (living together peacefully): how the Muscogee concept of harmony can provide a structure to morality. In: Marshall, C. (Ed.), Comparative Metaethics: Neglected Perspectives on the Foundations of Morality. Routledge, New York.

Montgomery Jr., E.B., 2019. Medical Reasoning: The Nature and Use of Medical Knowledge. Oxford University Press, Oxford.

Quine, W.V.O., 1960. Word and Object. MIT Press.

Rhoades, D.R., McFarland, K.F., Finch, W.H., Johnson, A.O., 2001. Speaking and interruptions during primary care office visits. Fam. Med. 33 (7), 528–532.

Other cases

There are many more cases to consider; if one but first has the desire, then the honesty and finally the courage to look. Consider a very partial list of cases described below. It is important to note that none of these cases are considered unjust, unethical, or unprofessional. Rather, they present ethical challenges and indeed, after due consideration, may be considered just. However, it through such considerations that a greater understanding of the ethics of everyday medicine can be achieved.

An emergency room has a policy that patients brought in by Emergency Medical Technicians (EMTs, paramedics) are not considered the responsibility of the emergency room physicians and nurses until the patient is placed in a bed. However, getting a bed could take many hours. The patient is left in the care of the EMT. These actions represent a form of triage (a rational form of rationing) not based on medical necessity and urgency as assessed by a physician or a nurse. Unless it is expected that the EMTs are capable and are authorized to triage based on medical necessity and urgency, both in principle and in practice, then the triage is based on first come, first served.

Emergency room care often may only deal with the urgent and threatening symptoms and disabilities and perhaps not that causes. Yet, often no follow-up is arranged or even mentioned. Perhaps, the default presumption on the part of emergency room physicians and healthcare professionals is that the patient or the patient's representative will know to contact the patient's family physician and that follow-up care is readily available.

A physician or surgeon dismissed from one hospital because of poor performance finds a position at another hospital. The new hospital does not contact the prior hospital or its staff with questions as to why the physician or surgeon left. Perhaps the subsequent hospital's vetting process was sufficient.

A physician raised concerns about the practice of another physician through the process requested by the administration. The physician's concerns were substantiated by a subsequent enquiry. The accused physician refuses to work with the concerned physician. Consequently, the concerned physician is subjected to constructive dismissal where the terms of the employment of the concerned physician are altered as to make continued employment unacceptable.

A physician provides an exemption to vaccinations under medical necessity for a child in order to accommodate the concerns of the parents. The risk is reducing the overall immunity of the community (herd immunity) placing suspectable persons in the community at greater risk. Perhaps the physician's action represents a greater responsibility to the child by supporting the parents given that the vaccination is for an illness that is generally well tolerated by children.

The Ethics of Everyday Medicine. https://doi.org/10.1016/B978-0-12-822829-6.00015-1

A nurse practitioner, recognizing a viral upper respiratory tract infection, nonetheless, prescribes antibacterial antibiotics for a patient who otherwise is not clearly at risk of a bacterial complication of a viral illness. The patient expects to be treated by an antibacterial antibiotic, and failure to do so may adversely affect the relationship between the nurse practitioner and the patient, thus jeopardizing future care.

A patient is treated with an agent that reaches its maximum efficacy in 2 weeks and, thus, can be adjusted every 2 weeks until the treatment is optimized. However, follow-up care is scheduled for 3 months, primarily determined by the convenience of the physician and healthcare professionals. As a consequence, the time required for the patient to reach optimal therapy is greatly prolonged.

A patient calls a physician's office. The office staff asked about the patient's concerns. The office staff refuses to put the patient through to the physician and instead offers suggestions as to what the patient should do with some reassurance that the physician will call the patient at a later time. No such phone call by the physician occurs despite repeated phone calls by the patient.

A patient waited in the queue for the first available appointment several weeks later to see the physician. The physician canceled the appointment a week before the appointment. The patient was scheduled at the next available adding several more weeks to the wait.

A neurologist in solo practice in a rural area closes his office for a long weekend without any provision of coverage for continuing care. The neurologist is the only neurologist in a large geographic area and otherwise the neurologist devotes many long hours to the care of his patients. Physicians and healthcare professionals at another institution many miles away receive calls from the neurologist's patients. The neurologist on call does not have access to the patient's medical record and has never seen the patient before.

A 41-year-old man with a long history of an acquired lower spine deformity presents for consideration of surgery. The patient exhausted all previous alternative therapies. The patient is obese. The surgeon refuses to offer spine surgery stating that their policy is not to offer surgery to any patient that exceeds a certain body mass index (BMI). When asked why, the surgeon stated that his institution was able to significantly reduce the incidence of perioperative infections by instituting the policy. The patient stated he was willing to accept the risk; however, the surgeon continued to decline offering surgery.

Part Two

Explorations into the nature of ethics in everyday medicine

Justice

1

To be argued...

To understand the experience of ethical situations, circumstances, or others, it is necessary to dissect the experiences into digestible parts. Otherwise, experience remains as William James described for the infant "assailed by eyes, ears, nose, skin, and entrails at once, feels it all as one great blooming, buzzing confusion" (James, 1890). The question is how to carve the totality of experience into digestible parts. Plato offered that it is necessary "to crave nature at its joints" (Plato, Phaedrus 265e). In an instrumental (epistemological) sense, it is easier to cut through a joint rather than bone. However, in an ontological sense, the joint is a reality that separates separable components of reality. Thus, it is the nature of reality that it is separable and the existence of joints is offered as proof (Campbell et al., 2011).

The question becomes how to separate the totality of ethical experiences, what is the first joint by which ethical experiences can be cut from which all other "cuts" derive. The argument to be advanced is that the first cut is a neurophysiological affective (emotive) response to the experiences that allows separations of experiences to which the justice or injustice can be ascribed. The term emotive is used instead of emotion as the latter is held to be a noun, suggesting something static and with independent ontological status. The intent here is to use emotive as a verb, implying that it leads to actions, hopefully just actions. The ontological implication is that emotive, and thus attribution of justice, is a process. All other notions of ethical mechanisms, such as ethical principles, moral theories, laws, case precedents, and virtues, follow from that first cut. They are inductions from experiences that are first assembled on the basis of the neurophysiological affective response.

The nature of justice as described places the sense of justice on a different level than ethical mechanisms, such as the ethical principles of autonomy and beneficence (the latter subsequently divided into beneficence and nonmaleficence), moral theories, laws, case precedents, and virtues. All these are derived from justice and thus serve justice. Justice is paramount and primary, the latter from several perspectives—ethically, logically, and neurophysiologically. Humans are "wired" to act in a manner to which justice or injustice can be ascribed, and the principles, theories, laws, case precedents, and virtues are post hoc rationalizations.

To be sure, others have made the same claim as to the biological origin of ethical conceptions, but these claims to biology are mediated by psychological constructs or abstractions, which are not the same thing as neurophysiological mechanisms that are real, although they are described by abstractions. Reasoning from the psychological rather than the neurophysiological may well be the cause of many ethical conundrums, discrepancies, and paradoxes. While an exact neurophysiological explication of ethics is not at hand, and may never be, nonetheless, a great deal is known about neurophysiological systems, particularly those displaying Chaos and Complexity. Those insights can be very helpful in understanding ethics.

The Ethics of Everyday Medicine. https://doi.org/10.1016/B978-0-12-822829-6.00026-6

Justice is paramount

As will be argued later, the sense of justice is primary, and thus are the beginnings of an ethical inquiry. Perhaps the best sense of justice follows from the notion of fairness as used by John Rawls (1999). Embedded within the sense of fairness is the notion of due process, the legal analogy being the Rule of Law. Reproducibility is also critical for fairness, with different persons in the same situation and at different times being treated the same. Fairness must be credible, which typically relates to the certainty of the means to arrive at the fair decision. All of these issues are found in the sense of justice.

As will be discussed in more detail later, certainty derives from the origins of the procedures to create fairness, meaning the epistemology of the procedures for fairness [epistemology is the branch of inquiry (philosophy) that deals with the theory of knowledge's nature, origins, and discovery]. Some procedures for fairness derive from what is called received knowledge—humans receive their knowledge of fair procedures from either the word of God or the dictates of absolute sovereigns, although often presented as in the best interests of the people. In a pluralistic modern liberal democracy, where the word of God or the dictates of absolute sovereigns do not count, some other source of fairness procedure is required. Since Thomas Hobbes, governance to ensure fairness has been based on a social contract originating in the collective will of those to be governed, although Hobbes' view was a bit extreme (Hobbes, 1651). The necessary certainty derives from their origin in and the reproducibility of the consensus of the governed. Sufficient consensus among the governed allows enforcement on all and ultimately justice (Chapter 3). It is the imposition of justice on all governed that makes justice paramount.

The term fairness procedure was used as it is most general, but can be surmised as justice. There are any number of fairness procedures. For example, Principlism presumes a deductive-like derivation from ethical principles and moral theories sometimes within the context of laws. Another approach is to substitute personal characteristics, virtues, of the adjudicators, as in virtue ethics. Other approaches match the case under consideration to a case determined previously as just or unjust. All these different approaches are understood to the extent they serve justice. Yet, in a pluralistic modern liberal democracy, all these approaches have the same fundamental epistemology of ethical induction from sets or collections of prior relevant experiences defined in terms of just or unjust (see Chapters 2 and 3). All approaches are confronted with the same ontological conundrum, whether the variety of ethical experiences shows diversity such that each is unique and de novo or variations on a common economical set of principles, theories, laws, case precedents, or virtues. Significant and severe challenges confront one whichever epistemic choice is made.

Any fairness procedure is ultimately an emotive response to a particular situation or circumstance analogous to US Supreme Justice Potter Stewart's notion of pornography when the justice wrote "I shall not today attempt further to define the kinds of material I understand to be embraced within that shorthand description ["hard-core pornography"], and perhaps I could never succeed in intelligibly doing so. But I know it when I see it, and the motion picture involved in this case is not that" [*Jacobellis v. Ohio*, 378 U.S. 184 (1964)]. The implication is that any approach based on principles, moral theories, laws, case precedents, or virtue is likely to fail in some situation or circumstance, yet justice must be done. Perhaps Justice Stewart should be credited for not attempting some post hoc rationalization as his response may well represent the most honest response. The problem is that Justice Stewart's approach provides virtually no guidelines for others in future legal disputes. As such, how can anyone have certainty that such judgments would be reproducible in future applications? Nonetheless, justice is paramount, even if the methods used to achieve justice are imperfect, perhaps a charitable description.

Regardless of the fairness procedure invoked, fairness of the outcome is paramount. In Principlism, where obligations to the ethical principles, such as beneficence, nonmaleficence, and autonomy, are weighed by the supervening moral theory, thereby assigning fairness or justice, the obligations to the principles themselves can be mitigated or even denied. Particular moral theories, such as Egalitarian, Libertarian, Deontological, and Utilitarian, can be denied. What cannot be denied is justice. In Anti-Principlist ethics, consideration of a particular current ethical problem can overturn and abandon past precedents or virtues but what cannot be abandoned is justice.

Sometimes the route to justice can seem circuitous and counterintuitive, yet the drive to justice is still paramount. For example, the Captain of the Ship theory applied to hospital-based medical maleficence holds that the physician or the surgeon, not the charitable hospital, is held financially accountable, even if the physician or the surgeon was not in control of the situation or circumstance that led to the maleficence. It would seem that holding to the Captain of the Ship theory would be unjust to the physician but it is held to hold a higher justice, that is, allowing charitable hospitals to continue and not be bankrupted, what Rawls terms "telishment" (Rawls, 1955). Such telishment of liability is currently being considered in the use of machine learning artificial intelligence (AI) in medical decision-making (see Case 12). Despite being perhaps unfair to the physician or healthcare professional, some ethicists support transfer of liability from the AI programmers and manufacturers to the physician or healthcare professional so as not to stifle the emerging technology that they believe has great potential for society. In this case, justice is the final outcome to serve society by making charitable hospitals and AI technology available, yet at the same time compensating patients who may have suffered from their use.

Justice is primary—Biologically

The primary theme presented here is that justice is the consequence of neurophysiological mechanisms that create an affective (feeling or emotive) response in the observer. An analogous concept may be reactive attitudes discussed by Strawson (1962); however, the latter are psychological constructs, which are problematic, as discussed later.

Prior considerations

The argument made previously holds that the first sense of justice is an emotive from which deliberative and declarative principles, theories, laws, case precedents, and virtues are derived. The term "emotive" is emphasized rather than emotion where the former implies an action or potential action, whereas the latter implies some static state of being. As such, emotive better characterizes the sense of justice that begs action. The purpose of this section is to review what other ethicists and philosophers have said about ethics and morality as first being an emotion or emotive.

Nancy Sherman and Heath White hold one value of emotion as recognizing the moral salience of a given situation or circumstance and relate this position to Aristotle. Some have argued that such perception is independent of actions but, as demonstrated later, this is not entirely true. For example, imagining that does not result in an action nevertheless causes activations in brain areas associated with action (see Oatley and Jenkins, 1992). Emotions are held as intentional awareness, suggesting consciousness and deliberative action, and thus are different than physical sensations such as pain and hunger. Yet they need not be, as discussed later in the neurophysiology of empathy for pain in others. Perhaps a problem is that emotions have been studied by self-reports of introspection in articulate adults who have had a lifetime of inculturation by adults using conscious and deliberative methods. One wonders what Aristotle might have thought of the studies of infants described later. Nonetheless, virtuous emotive processes are the cornerstone for virtuous ethics (Chapter 2).

The affective neurophysiological mechanisms become the first organizing principle to assign justice or injustice. This sense of justice need not be articulated; indeed, nonhuman animals and human infants demonstrate a sense of justice by their actions. This is not to say that a sense of justice cannot be articulated later when language is gained. Further articulation of the sense of justice becomes a generator of one's own subsequent ethical behavior and an assessment of the ethical behavior of others. Even the latter is based on a neurophysiological infrastructure. The neurophysiology that generates an internal sense of justice also generates a sense of justice in others. This may be the basis of the human predisposition to attribute one's own feeling and motivations to others (Theory of Mind). Thus, the sense of justice is held relevant and common to all even when not articulated. This attribution of a sense of justice to all individuals in a community likely is the origin of the notion of common morality in ethics (Beauchamp and Childress, 2013).

The terms justice and injustice are those used to signify the emoted action and affective response. Arguments supportive of this theme are (1) demonstrations of neurophysiological changes [typically in the form of neurometabolic changes in neuroimaging such as functional magnetic resonance imaging (fMRI)]; (2) behaviors indicative of a sense of justice in animals, infants, and young children; and (3) neurophysiology of common morality, the mechanisms that result in transference of one's internal concerns to others, creating a sense of community now founded on a common sense of justice.

To be sure, many others offered similar perspectives suggesting a neurobiological basis for morality, and hence ethics (see Giner-Sorolla, 2014). However, the neurobiological inferences have been based on psychological constructs, not neurophysiology, with the latter reflecting the reality of how the nervous system works. As will be discussed in detail later, it is often counterproductive to conflate psychology and neurophysiology. To the degree that psychology is a poor fit to neurophysiology and psychology is thought to be central to ethics, explication of ethics in terms of neurophysiology first requires suspension of the presuppositions based on psychology. To do so, psychology, particularly as applied to ethics, is discussed first.

There is a long history that continues to hold psychology as the proper domain to explore ethics (Appiah, 2008). Indeed, experimental psychological studies have been used to investigate ethics. Thus, it may take a pause for many to accept that psychology, while illuminating human interactions, does not explicate human behaviors at the fundamental level. To be sure, psychology can be highly systemized, demonstrating high internal consistency (epistemology), but this does not mean that psychology is the true (ontological) description of the operations of the human nervous system that leads to behaviors to which justice and injustice can be attributed. Indeed, the system of Freudian psychology can be constructed so as to be entirely self-consistent, yet few would accept it as ontology.

Ultimately, an explication of ethics at the same causal level that reflects the fundamental operations of the nervous system, if possible, likely will be the most explanatory and consistent. Such an approach reduces the distance between the explanandum (that to be explained, such as the genesis of behaviors to which justice and injustice are attributed) and the explanan (the causal mechanisms responsible). The Ancient Greek philosophers recognized this in their Principle of Causational Synonymy where whatever is in the cause that produces an effect likewise must be in the effect. The reason one's hand is able to stir the water is not that the hand is solid while the water is liquid. Rather, it is due to the electrons in the outer orbit of the atoms that make up the surface of the hand that repel the electrons that make up the outer orbits of the atoms of the molecule of water. This would be the same explanation of why one's hand can move a block of ice, which is still water, although solid, or that a strong stream of liquid water can move the hand. The fundamental level of operation of all behaviors is changes in the electrical states of excitable cells, typically neurons, in the complex network of interactions that comprise the nervous system. The dynamics of those interactions determine the subsequent behaviors and thereby the behaviors that lead to the induction of ethical principles, moral theories, laws, case precedents, and virtues. This will be discussed in detail later.

To be sure, several phenomena are offered to support the ontological credibility of psychological constructs beyond internal consistency. Psychological studies can be done contrasting normal to abnormal disordered humans with the differences attributed to the physical bases of the abnormality or disability, as will be discussed later in the case of Phineas Gage. Further, psychological constructs can be correlated with neuroanatomy where imaging the brain based on changes in the metabolic activities in specific brain regions, such as the utilization of oxygen, has been taken as synonymous with the actual operations of the nervous system, with the latter being neurophysiology. Such approaches risk a categorical error, as the synonymy suggested is a poor fit to the actual operations of the nervous system, which is truly based on the changes in the electrical state of excitable cells (typically neurons) interconnected in a vast and complicated network of electrical interactions and electrically mediated neurochemical communications between excitable cells. This is not to say that psychology and neurometabolic imaging are unimportant or unhelpful; indeed, the opposite is true in that they are critical in a specific role. However, it is crucial that these approaches are always viewed with the caveat that the inferences drawn are at best an approximation of what actually happens, thus acknowledging their unique risks for misunderstanding.

The greatest risk of taking psychological constructs and neuroanatomical inferences as explications of neurophysiology is in the dynamics, that is, the change in states over time. The dynamics of many psychological-neuroanatomical concepts are relatively static where the dynamics are simple one-dimensional push-pull systems of simple excitation and inhibition, which is so oversimplified as to be counterproductive (see Montgomery, 2019a, pp. 167–177). To be sure, explication of the operations of the nervous system, particularly as it relates to just and unjust behaviors, likely will never be completely known. Nonetheless, there are general principles and robust models to begin to understand the dynamics, and these dynamics can be a lens through which to better understand ethics and morality.

The notion of a neurophysiological basis for justice is not in the philosophical trivial sense where every behavior and every thought are fundamentally neurophysiological in origin and action, although implicitly suggesting there may be something else in addition that is more important. The conundrum was recognized by Aristotle, who, being an empiricist and a biologist as well, was willing to attribute a great deal of animal and human behavior to a mechanical system, but there was one aspect that provided difficult, *nous* or reasoning. Aristotle was left with a difficult cognitive contortion saying that the mind (*nous*) was not in the body but of the body.

There is deeper meaning to the claim of a neurophysiological basis. The hypothesis is offered that a general understanding of the actual neurophysiological mechanisms provides a clearer and more consistent understanding of behaviors to which justice and injustice are attributed. This topic will be addressed at the end of this chapter. Holding all phenomena as real and independent of the mind leaves no alternative other than the changes in the electrical states of neurons (neurotransmitters, chemicals that mediate communication between neurons and are the basis for neuropharmacology, are just the messengers).

It is critically important to point out that the theory just posited does not reduce ethics and morality to "reflexes" responding in a subconscious manner. As will be demonstrated, conscious deliberative declarative analyses include those exhibited in a verbal discussion or debate in ethics. Conscious deliberative declarative analyses in terms of stated ethical principles, moral theories, laws, case precedents, or virtues are critical. Indeed, it is likely that the neurophysiological states associated with declarative terms likely affect those neurophysiological systems that generate the affective response and subsequently drive behavior. The powerful emotive force of language is a prime example. Thus, perhaps conscious debate, even of abstract notions, is associated with real neurophysiological properties and consequences. However, such a position does not require that there be "debate teams" within the head, within the neurophysiology, even though psychological theorizing often presupposes such debates, for example, the case of Phineas Gage, as discussed later. Rather, conscious deliberative declarative considerations can bias the neurophysiological mechanisms toward one behavior rather than another. The analogy would be to the phenomenon of mentally imaging a physical behavior, such as gymnastics, that improves the actual performance. Imagery activates many of the brain regions associated with the actual act. Thus, conscious and deliberative imaging, at the very least, can be considered a "dress rehearsal" for the subconscious neurophysiologic mechanisms for the act. Further, the improvement in performance argues that the actual neurophysiological mechanisms generating the act are affected by the mechanisms that underlie the conscious deliberative imagining, even if the actual conscious awareness is epiphenomenal.

Psychology versus neurophysiology

Despite or in view of the considerable concern whether the neurophysiology will ever be known, humans would seem loath to say "I don't know how to explain behaviors." Some story is necessary. Humans turn to a ready vocabulary, which are those motivations felt internally and attributed to others (Theory of Other Minds). This is the substrate of psychology, ordinary language philosophy, and the parallels of ordinary language philosophy to ethics (Lambek, 2010). Also, at the very least, psychological theories often are the necessary *a priori* supposition for subsequent empirical experimentation, for example, cognitive neuroscience. The implications for ethics are discussed in Chapter 2.

The distinction between psychology and neurophysiology can be viewed in terms of the distinction between macrolevel descriptors and microlevel mechanisms. Many psychological constructs are inferences from observed behaviors and correlations with neurophysiologically related (but not synonymous) macrolevel changes in neurometabolic imaging or behaviors subsequent to lesions, deliberate or accidental. Interestingly, the conundrum of accepting a psychological explication where a mechanical (neurophysiological) explication is not available was recognized by Aristotle. In the "De Anima" (On the Soul), Aristotle wrote:

> But the natural scientist [author – taken as the neurophysiologist] and the dialectician [author – taken as the psychologist] would define each affection [behavior or property] in a different way, for instance, in saying what anger is: the dialectician would define it as the desire for retaliation or some such thing, and the natural scientist as the boiling of blood and hot stuff around the heart.
>
> *Miller (2018)*

As the means of observing boiling of blood or measuring the heat of the stuff around the heart were not available, one settles on the notion of retaliation, a concept derived from introspection and applied to others. Yet, in Aristotle's metaphysics, life is defined by movement in response to a vital heat (Freudenthal, 1995). The boiling of blood would be the true physics of anger, not a psychological desire for retaliation.

There are many reasons why psychology is inefficient and often misleading. Cause is often conflated with purpose. Consider the operations of spell check software on a computer. It would seem reasonable to say that the operations of the software are to ensure properly spelled words, but this would not be accurate. Rather, the purpose of the computer is to establish a specific electrical state in the transistors, resistors, and capacitors in the computer circuitry such that the specific pattern of electrical states of the word typed does or does not match any specific pattern of electrical states held in the computer and, when not, that state is defined as an error.

This conundrum is well known and distills to whether psychology is a science (note that this is not the same as cognitive neuropsychology). The only point here is that psychology is not neurophysiology. An interesting example of this question is seen in Donald O. Hebb's "The Organization of Behaviour" (1949) in the first few chapters. After describing some inference to a psychological principle, Hebb would draw neuronal circuitry that might implement the psychological principle. It is as though demonstrating at least a credible hypothetical neuronal model for the psychological behavior described lent credibility to the psychological construct. Interestingly, Hebb attempted a correlation between psychological constructs and neurophysiology, but only early in his book, apparently abandoning the effort later.

Another fundamental reason of why psychology is not synonymous with nervous system function is the fact that the nervous system operates on information in its most elemental form, where information is nonrandom state changes. Psychology necessarily invokes a loss of information and thus is inherently incomplete; the fundamental significance of such loss stems from the Second Law of Thermodynamics as applied to Information, as will be discussed later. Abstraction or induction from experiences to ethical principles and moral theories can be viewed as though they were "averages" across multiple and varied experiences. The "averages" often are couched in psychological terms and their epistemological and ontological status are discussed in detail in Chapter 2 and in Montgomery (2019a,b).

Having described the limits of a psychology of ethics, psychological constructs are necessary nonetheless. Hypothesized psychological mechanisms are important generalizations that play an important role in constructing ethical induction. For example, when selecting a set of ethical experiences, some *a priori* notion is necessary to include some experiences and exclude others (see Chapter 4). As psychological theories have great value, epistemically as *a priori* hypotheses, the ethical principles, moral theories, and virtues essentially largely become psychological theories; interestingly, this was also the position of David Hume in the 1700s (see Appiah, 2008). This is not to say that psychological constructs should not be used, only that the user must be keenly aware of their limitations and the possibility of misrepresentation.

A brief example suffices here. Joshua Greene argues for subconscious mechanisms driving behaviors to which justice or injustice is attributed (Greene, 2013). The mechanisms are psychological. Justification of these psychological constructs is based on their origin in a teleological sense of evolution. In order for humans to survive, humans have to work together. Therefore, the "purpose" of psychological constructs with ethical implications is to allow humans to work together and procreate, thereby perpetuating the human along with the human's psychological mechanisms. While the theory of evolution has vast evidentiary support, whereas countertheories essentially have none, and adaptation of the species (not the individual) is a fundamental component, it cannot be said that evolution has some purpose, unlike theological apologists for evolution (Theistic evolution) holding evolution as God's means for perfecting humans. As pointed out by Steven J. Gould, evolution is far more likely to produce a new species of bacteria than a new human (Gould, 1996).

Psychological constructs need not have unique representation in the human DNA—it may be a social evolution where brain plasticity plays a role analogous to genetic mutations to achieve a different behavior. Consider Greene's description of when an individual sacrifices his or her personal interests in the interests of family, for example, advancing the progeny of kin instead of having progeny of their own. Greene, as have others, talks about the genes the individual sacrifices to advance the genes of the family or kin. Greene writes, "Why do brothers care about each other? Brotherly love (and familial love more generally) is explained by the well-known theory of kin's selection, which takes a gene's-eye view of behavior" (Greene, 2013). Perhaps it is only a figure of speech not meant to be taken literally, yet there is nothing in the discussion leading up to the quote to warn against a literal interpretation and, indeed, much to suggest a literal interpretation was intended. Yet, how can a gene have a view, what would it be a view of? To say that a gene has a view would be like saying a light switch has a view. Strangely, the quote anthropomorphizes at a group of chemicals in the cell.

A teleological evolutionary explication only postpones the question and it suggests that there is some social representation of the psychological construct in the nervous system, a product of gene expression, which is a predisposition to learning to conduct what is just. However, modern understanding of neurophysiology argues that a neuronal machine can generate behaviors that are just without invoking the psychological construct. Further, evidence of behaviors to which a sense of justice can be attributed in animals, infants, and young children suggests that learning the psychological constructs as a prerequisite to just behavior is not necessary, as discussed later.

Neurophysiological representations of just and unjust behaviors

The central point here is that humans behave in a manner that can be viewed as degrees of justice or injustice. Humans seem to be just "wired" to do so. As these human behaviors are generated at the subconscious neurophysiological level, declarative and deliberative descriptions, forming principles and theories, are post hoc attempts at rationalizations, at least initially. Once formed, the rationalizations may lead to further and different organizations of experience from which new and different inductions are made whose neurophysiological infrastructure now leads to behaviors to which the rationalizations can be attributed. For example, scientists use fMRI to examine regions of the brain that change metabolic activity, a surrogate marker for changes in electrical activities of neurons. The hypotheses informing the experiment rated the moral wrongness of harmful consequences generated by action (acts of commission) or by allowing a harm to occur (acts of omission). Two hypotheses were studied. If subjects used a rule to determine moral wrongness (rule hypothesis), then regions of the brain thought related to deliberative action should change their metabolic activity.

If the subject was making the determination "automatically," for example, not requiring deliberative action, then those regions associated with the rule following would not change their metabolic activity (Cushman et al., 2012). The study compared the same scenarios framed as an act of commission and as an act of omission, where most subjects would hold the act of commission worse than an act of omission. Differences in the changes in neurometabolic activities on the fMRI would signal a detection of differences in justice.

Subjects were asked to evaluate the degree of immorality representing a deliberative action. The response to detection of the justice of the act did not involve regions of the brain typically associated with deliberative acts, but those areas became active only on the post hoc deliberation as to the degree of immorality. These findings are consistent, although not definitive proof, with an initial nondeliberative response to an injustice (the automaticity hypothesis), and the ethical and moral assessments were likely a post hoc commentary on the actions.

Assuming that personal moral dilemmas are analogous to microlevel justice, whereas impersonal moral dilemmas are more analogous to macrolevel justice, neural mechanisms activated during personal moral dilemmas are different than those activated during impersonal moral dilemmas, as measured by changes in neurometabolic imaging (Greene et al., 2001, 2004). But it is important to note that activations associated with impersonal moral dilemmas were after the fact. Greene et al. (2004) defined personal moral dilemmas as:

> Personal moral dilemmas and judgments concern the appropriateness of personal moral violations, and we consider a moral violation to be personal if it meets three criteria: First, the violation must be likely to cause serious bodily harm. *Second, this harm must befall a particular person or set of persons* [italics added]. Third, the harm must not result from the deflection of an existing threat onto a different party.

The neurometabolic changes in brain regions in response to personal moral dilemmas were those typically associated with emotive processing, while the impersonal moral dilemmas were in regions associated with more cognitive processing involved in different regions of the brain. The argument here is the particularity of the condition for a personal dilemma that lends itself to a microlevel sense of justice, which is mediated directly by neurophysiological mechanisms. An extended quotation from Greene et al. (2004) is helpful:

> Second, we found that reaction times (RTs) were, on average, considerably longer for trials in which participants judged personal moral violations to be appropriate [*author – it is important to note the condition just described involves a dissonance, a violation with negative valence nevertheless must be viewed positively to be appropriate*], as compared to trials in which participants judged personal moral violations to be inappropriate [*author – in this case described involves a dissonance, a violation with negative valence also is viewed negatively to be inappropriate*]. No comparable effect was observed for impersonal moral judgment. We compare this effect on RT to the Stroop effect (MacLeod, 1991; Stroop, 1935), in which people are slow to name the color of the ink in which an incongruent word appears (e.g., "red" written in green ink).

According to our theory, personal moral violations elicit prepotent, negative social-emotional responses that drive people to deem such actions inappropriate. Therefore, in order to judge a personal moral violation to be appropriate, one must overcome a prepotent response, just as one faced with the color-naming Stroop task must overcome the temptation to read the word "red" when it is written in green ink. The sort of mental discipline required by the Stroop task is known as "cognitive control," the ability to guide attention, thought, and action in accordance with goals or intentions, particularly in the face of competing behavioral pressures (Cohen et al., 1990; Posner and Snyder, 1975; Shiffrin and Schneider, 1977). We interpreted the behavioral results of our previous study as evidence that when participants responded in a utilitarian manner (judging personal moral violations to be acceptable when they serve a greater good), such responses not only reflected the involvement of abstract reasoning but also the engagement of cognitive control in order to overcome prepotent social-emotional responses elicited by these dilemmas.

Inborn mechanisms generate behaviors to which a sense of justice can be attributed

Perhaps one approach is to study the neurophysiology of justice in an *a priori* condition, such as newborns. While difficult, studies have been conducted in infants 2 years of age and younger. For example, a 15-month-old child will spend greater time viewing videos representing an unequal sharing of goods, such as milk or crackers, than videos of equal sharing on average (Schmidt and Sommerville, 2011). Three-year-old children are less likely to help an adult actor portraying a harmful act than an actor portraying a neutral act (Vaish et al., 2010). These types of studies are discussed in greater detail later.

This does not exclude the possibility that just actions in these very young infants are learned but it would seem less likely. As a metaphor, similar issues have been raised with a perception of three-dimensional objects as detected by visual illusions on two-dimensional images. It is thought that illusions of different length segments on a two-dimensional image, which actually are the same, reflect the context of the other lines interpreted as generating a three-dimensional perception (Fig. 1.1). Note that there is nothing in the two-dimensional images that presuppose a three-dimensional perspective, such as images falling on slightly different locations on the retina (retinal disparity). The only suggestion would be the orientation of the other lines being what one would expect of the real world with perspective. But as will be discussed, the sensitivity to these illusions immediately on gaining sight suggests that some knowledge of the visual perspective of the three-dimensional world is present before the first visual experience.

Studies of the Ponzo and Müller-Lyer illusions very quickly after restoration of vision in adults, following removal of congenital cataract obscuring vision from birth, suggest an innate capacity to recognize three-dimensional structures (Gandhi et al., 2015). As can be seen in Fig. 1.1B, the newly sighted performed the same as those with normal sight, suggesting an innate predisposition to the illusions, suggesting an innate three-dimensional perspective. The question is whether the neurophysiological mechanisms of just actions likewise are innate. Note that it is not the visual context, such as the railroad tracks or the building as shown in Fig. 1.1A, that is the source of the three-dimensional perspective as the same illusion is seen in abstract images (Fig. 1.2).

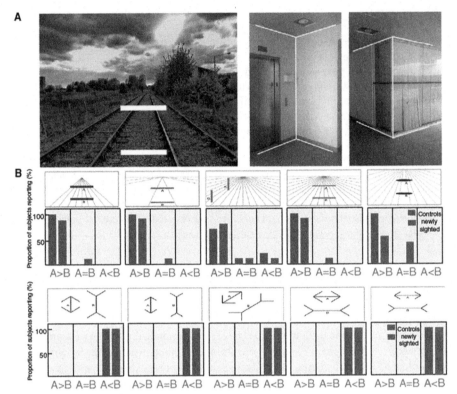

Fig. 1.1 The susceptibility of newly sighted individuals to visual illusions. (A) Ponzo and Müller-Lyer illusions superimposed on real images indicate how learned perspective cues, as proxies for distance, may be the source of the effects. (B) Results from normally sighted and newly sighted subjects on multiple displays. In each of these displays, the two lines being compared (denoted by "A" and "B") are actually of identical length. Data are represented as the proportion of subjects (%) reporting each type of response, for example, whether line A was longer than line B. (A) From Gregory, R.L., 1966. Eye and Brain. McGraw-Hill, New York; the railroad tracks image is by Darren Lewis and is in the public domain. (B) From Gandhi, T., Kalia, A., Ganesh, S., Sinha, P., 2015. Immediate susceptibility to visual illusions after sight onset. Curr. Biol. 25(9), R358–R359. doi:10.1016/j.cub.2015.03.005.

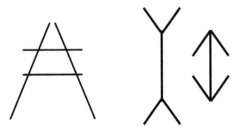

Fig. 1.2 Abstract version of the same illusion demonstrated in Fig. 1.1 demonstrating the same illusionary effect. This argues that it is not the context, as shown in Fig. 1.1, that is responsible for the illusion.

Even if it were the case that the 2-year-old learned to act justly, it is possible and indeed likely that there nonetheless existed an *a priori* neurophysiological predisposition to learn these particular behaviors. An example is imprinting of rapid and fixed recognition and attachment nearly immediately after birth. The phenomena across a great many vertebrate species, including humans, suggest a neurophysiological substate that, although not establishing an *a priori* target, nonetheless argues for an *a priori* disposition (for a review, see Mobbs et al., 2016).

Considerable research and scientific knowledge exist regarding the neurophysiological mechanisms underlying social and prosocial behaviors in laboratory animals and humans (for reviews, see Chang et al., 2013; Stanley and Adolphs, 2013; Giner-Sorolla, 2014; Chen and Hong, 2018). An explicit representation of the conceptualization of social behaviors driven by neurophysiological mechanisms is suggested by neurometabolic imaging (see Kennedy and Adolphs, 2012). In neurophysiological studies of pain empathy, observations of pain in others activate many of the brain structures that mediate the perception of pain in the observer (Benuzzi et al., 2018). In a real sense, pain in others causes pain in observers, as many humans have a physical sensation of pain while observing pain in others (Grice-Jackson et al., 2017). Thus, all perceivers have a stake in relieving pain in others, in this case reducing maleficence, if only to reduce their own pain. Such behavior can be translated into an obligation to reduce and hopefully eliminate maleficence in others. Note that even if an observer is not a shareholder or in a position to reduce the pain in others, that person would still expect those that are shareholders to reduce maleficence, thus reducing empathic pain in all stakeholders.

There are studies of empathy that may have some correspondence to beneficence and nonmaleficence. Therefore, the neurophysiology of empathy may translate to a neurophysiology of beneficence and nonmaleficence. Other studies examined the neurophysiology of the Doctrine of Doing and Allowing (harm from acts of commission and omission) and the Doctrine (Principle) of Double Effect (Aquinas, 1485; Schaich Borg et al., 2006).

It would be a mistake to infer that the neurophysiological mechanisms engaged by the observations of one's own and others' behaviors, often described at the macrolevel and generating a microlevel sense of justice, proceed by first translating to principles thought to be reflected in the macrolevel descriptions. This temptation has a long history in medical reasoning (Montgomery, 2019a). It may well be that inferences of an unjust or just behavior proceed directly from neurophysiological mechanisms not requiring translation from a transcendental notion of obligation to ethical principle, moral theory, law, case precedent, or virtue.

It is not necessary that there be a deliberative algorithm as reflected in psychological constructs that drive the sense of justice or injustice. Examples are found in machine learning AI. Consider a computer algorithm to simulate backing up a tractor-trailer to a loading dock (Fig. 1.3). Humans are able to do so only after extensive practice, and even that is often difficult or impossible for the expert to talk a novice through the process. A neural network computer algorithm based on many small computational units, taken as analogous to neurons in a human nervous system, was able to "practice" and eventually could back up the simulated tractor-trailer to any loading dock that was

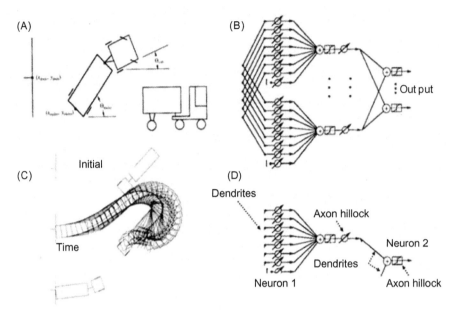

Fig. 1.3 Schematic representation of a neural network solution to the computational problem of backing a tractor-trailer (A) to a loading dock (C). The neural network consists of 25 processing units, analogous to neurons, arranged in two layers (B). Calculated values are registered in the output layer that drives a virtual tractor-trailer, the position of which is sent to the input neurons. Information is passed between each neuron, analogous to synaptic connections between neurons (D). The strength of each input is the value from the prior neuron adjusted by a synaptic weight. The difference between the position determined by the output neurons and the actual position of the tractor-trailer during training sessions is fed back into the network to modify the synaptic weights. After training, the synaptic weights in the modified neural network are able to back the tractor-trailer without error. From Nguyen, D.H., Widrow, B., 1990. Neural networks for self-learning control systems. IEEE Contr. Syst. Mag. 18–23.

physically possible. Further, analysis of the trained neural network could not reveal any kind of principle extracted from the trained network (Nguyen and Widrow, 1990).

A variety of animals, including nonhuman primates and human infants, demonstrate behaviors termed inequality aversion (for reviews, see Brosnan, 2013; Decety and Yoder, 2017). For example, nonhuman primates will not accept a reward that is a lesser value, such as a cucumber compared to a grape, than received by a nonhuman primate in an adjoining cage (van Wolkenten et al., 2007; for an entertaining demonstration, see https://www.youtube.com/watch?v=meiU6TxysCg). Indeed, evidence shows that prosocial decisions engage intrinsic neurophysiological reward mechanisms (Buckholtz and Marois, 2012; Jensen et al., 2014; Ruff and Fehr, 2014). For example, neurometabolic imaging in adults demonstrates an engagement of neurophysiological mechanisms associated with rewards in tasks where behaviors exhibit fairness (Tabibnia et al., 2008) and activation of the amygdala in behaviors thought unfair (Haruno and Frith, 2010).

Findings in adults make it difficult to establish when the changes in neurometabolic imaging develop. In other words, are the changes in the neurophysiological mechanisms reflected in the neurometabolic imaging, cause or effect? Observations in very young children suggest a neurophysiological *a priori* or, at the very least, an *a priori* predisposition enabling the behavior, such as in imprinting as discussed earlier. Observations in preverbal or young children suggest that these behaviors are unlikely driven by conscious deliberative principles (see Schmidt and Sommerville, 2011). A 15-month-old child will spend greater time viewing videos representing an unequal sharing of goods, such as milk or crackers, than videos of equal sharing on average (Schmidt and Sommerville, 2011). Further, those infants who viewed videos of unequal distribution longer than equal distribution were more likely to share a preferred toy ("altruistic" sharing) than a nonpreferred toy ("selfish" sharing). The correlation between greater awareness or concern about an unequal distribution of goods and "altruistic" sharing suggests an organization of responses consistent with an imperative, or at least a predisposition, to a sense of justice. Three-year-old children are less likely to help an adult actor portraying a harmful act than an actor portraying a neutral act. Further, these children were less helpful to the adult attempting but failing a harmful act, suggesting some understanding of intention, an important component of the obligation to the ethical principle of nonmaleficence (Vaish et al., 2010). Behaviors change with development and experience; however, such changes do not imply that the neurophysiological mechanisms are no longer valid any more than the variety of snowflakes invalidates the laws of physics (see "Ethics as an exercise in chaos and complexity" section later in this chapter).

Neurophysiology of common morality

The notion of common morality is central to ethical induction, which leads to ethical principles, moral theories, laws, case precedents, and virtues. However, this is only to say some commonality of behaviors occurs to which the term common morality can be applied unless the notion can be made rigorous, for example, as suggested in Chapter 3. However, the notion of common morality is problematic in that observation of common behaviors, thought moral, can give rise to numerous different ethical systems (see Chapter 2). It is unlikely there is one ethical universe that is different for each different ethical system. Consequently, the ontological status, its status in reality, is questionable, thereby reducing the notion of common morality from an existing entity, a solid touchstone, to an epistemic device. However, should there be a unification of ethical systems, as suggested in Chapter 2, an ontological status of the common morality may be less difficult.

One way to establish a commonality with ontological status is if the common morality can be demonstrated to be derivative of neurophysiological mechanisms. There must be some correspondence between one's behavior and the behaviors of others, in neurophysiological terms. Mirror neurons in the brain become active when a particular behavior is observed in another individual. As such, mirror neurons are thought to be fundamental to a person's behavior that is initiated or modified by the observations of others, particularly social or prosocial behaviors in the observer. Many have argued that these mirror neuron systems are fundamental to the Theory of Mind, where the intentions of the observer are attributed by the observer to those observed. From the intentions attributed to others, the observer can anticipate any future behaviors of those observed. The interactions among persons based on each prediction of the behaviors of others become a basis for social interaction, to which justice and injustice can be attributed.

A study of nonhuman primate amygdala neuronal activities demonstrated encoding of the value of objects based on the subsequent reward delivered when the animal chose the correct object over alternatives (Grabenhorst et al., 2019). Interestingly, these same neurons also changed their neuronal activities while observing a social partner learning to choose the object that gave the greatest reward. Other neurons encoded the animal's decision choice among the objects, indicating a translation of the value assessment to an act, that is, choosing the appropriate object. Interestingly, other neurons in the same animal encoded the observed animal's decision choice before the observed animal made a choice. The investigators concluded "Amygdala neurons seemed to construct predictions of partner's choices by the same mechanisms used for the recorded monkey's own choices, as indicated by the encoding of three key decision-making signatures." The interesting question is whether ethical induction via consensus by shared and observed experiences might operate by a similar mechanism. The neurophysiological studies of pain empathy described earlier are another example.

Mechanisms that relate to identification with groups may affect an empathy for pain, which, by extension, may relate to the obligation to nonmaleficence. In other words, the neurophysiological mechanisms allowing one to feel the pain of others may be the driving force of behaviors to which the principle of nonmaleficence may be attributed. For example, changes in neurometabolic activities using fMRI are seen in human subjects viewing painful events in others. In one study, videos were segregated and labeled with an identifier indicating a group in which the observer may be a member (a religious affiliation), the ingroup, or not a member, the outgroup (Vaughn et al., 2018). While similar changes were noted in the region changes of neurometabolic activities, there was a difference in magnitude between the in- and the outgroup videos. Perhaps this may be applied to the notion of the ethical principle of nonmaleficence being common to all; in other words, similar changes were seen in viewing both ingroup and outgroup painful events. However, the obligation, paralleling the neurometabolic correlates of maleficent events, differed by group, not in kind but in degree. In this sense, while the reaction to maleficence was qualitatively similar, demonstrating the neurophysiological response to the ethical principle of nonmaleficence, there was a qualitative difference that might be analogous to the obligation to the ethical principle reflected in the neuronal activities.

Similar phenomena are seen in studies of altruism, which can perhaps be seen as a form of beneficence. Studies of extreme altruism, such as kidney donors to anonymous recipients, demonstrate a larger right amygdala volume and enhanced responsiveness to fearful facial expressions in the right amygdala on fMRI (Marsh et al., 2014). Interestingly, the converse is seen in a psychopathic individual. In another study, changes in regional neurometabolic activity measured by fMRI in soccer fans who played games to win money for themselves, anonymous soccer fans, and nonfans were similar qualitatively across the groups but differed in magnitude (Bortolini et al.,

2017). The results suggest that beneficence to oneself likely has neurophysiological mechanisms. It is likely that neurophysiological mechanisms entail obligations to one's own beneficence, which likely has survival value. Similarly, there likely are obligations to others and neurophysiologically are the same qualitatively but differ quantitatively depending on the degree of personal associations. Thus, the mechanisms underlying beneficence to oneself also generate just actions to meet the obligation to beneficence to others. However, the situations or circumstances result in different degrees of obligations to the same principle of beneficence.

Evidence of the same neurophysiological mechanisms, although varying in degree, stands in contrast to the psychological dichotomy of Us versus Them invoked by some ethicists as informing morality (Greene, 2013). It may be that Us versus Them represents a continuum that is a balance between the obligation of beneficence to all and the threat posed by some. If the balance is fluid, then the key wide variations in degree in the relationship between the individual and others of differing personal distances can be understood on an economical set of underlying neurophysiological mechanisms.

The following quote by Greene (2013, p. 25) is illustrative:

> The second strange thing about morality as a device for beating Them is that it makes morality sound *amoral* or even *immoral* [italics in the original text]. But how could this be? The paradox is resolved when we realize that morality can do things that it did not evolve (biologically) to do. As moral beings, we may have values that are opposed to the forces that give rise to morality. To borrow Wittgenstein's famous metaphor, "morality can climb the ladder of evolution and then kick it away."

So, what is Greene to do? It appears that Greene invokes something that is not biological, an immaterial mind (psychological) that somehow transcends the biological. A realist or physicalist would find such an alternative nonsensical but humans are not compelled to be realists and physicalists in their rationalizations to themselves or others.

Types of justice are then described to accommodate the demonstrated differences in the obligation to beneficence. For example, if the subject worked just as hard to provide money to the anonymous nonfan group, this would be described as following from Egalitarian or Deontological (such as Kantian) moral theories. If the efforts and contributions were maximum, implying an absolute, then it could be construed that the subject perhaps recognized the inherent (deontological) right of the fan and nonfan groups to beneficence. If the efforts and contributions were not maximal, then a right could not be construed. However, if the fan and nonfan groups were apportioned the same effort and moneys, then an Egalitarian moral theory may be ascribed to the actions. It is important to note that the ascription of a theory or mode of justice is post hoc and that consideration of the abstraction, the moral theory, likely did not drive the just acts.

Challenges exist in the variety and apparent inconsistency of behaviors to which justice is attributed

The aforementioned discussion argues that a neurophysiology of justice is at least plausible. The argument advanced here is that various acts seen as justice are driven by a neurophysiological mechanism. However, there may be great differences in acts of justice among what appears to be similar situations and circumstances as well as very similar acts of justice resulting from greatly differing situations and circumstances as discussed previously. Such inconsistency is a significant challenge when deriving psychological theories, but as will be discussed, not for neurophysiological mechanisms displaying Chaos and Complexity. The lack of any simple discrete calculus with reproducible results does not mean that no economical set of neurophysiological mechanisms exists. As yet, behaviors in ethical problems are not dealt with on the level of neurophysiological mechanisms, such as changes in the electrical potentials across the cell membrane of excitable cells in the nervous system. Perhaps the inconsistencies may resolve when they are considered at the neurophysiological level, as will be discussed later in "Ethics as an exercise in chaos and complexity" section.

Injustice and biology

The discussion on the neurophysiology of the sense of justice naturally invites the question of the origins of injustice. Explorations of the issue can proceed by examining the development of young children, necessarily in psychological terms, with the presumption that the earliest findings may reflect innate mechanisms. However, one cannot discount the possibility that features found in the psychology of very young children were not learned rapidly. Even if true, rapid learning suggests, at the very least, a preexisting predisposition to the neurophysiological mechanisms underlying the psychological feature, such as imprinting, as discussed earlier.

Conceptually, it is important to recognize that the demonstration of unjust behaviors does not mitigate against innate and inherent neurophysiological mechanisms that provide the first and fundamental basis for just behaviors. Rather, it is likely that there are multiple competing neurophysiological mechanisms that underlie behaviors competing with the sense of justice. The resulting behaviors can be considered unjust. It is also possible that disorders affect the neurophysiological mechanisms that ordinarily would facilitate just behavior. For example, research into antisocial behavior and psychopathy suggests genetic and epigenetic factors (Cummings, 2015; Raine, 2018). Similarly, genetics, as well as the environment, appears to be a contributing factor in criminal behavior (Kendler et al., 2015a,b; Sabatello and Appelbaum, 2017). Also, correlations appear to exist between antisocial behaviors and differences in neurometabolic imaging (Raine and Yang, 2006). An important caveat is that correlations do not mean cause and effect. Further, in the absence of known causal mechanisms, it is difficult to assess the significance of those correlations. Nonetheless, the potential for genetic and acquired abnormalities predisposing to unjust behaviors cannot be dismissed and thus cannot be used to undermine the notion that an innate sense of justice exists.

The development of racial and gender preferences and biases might help in understanding possible inherited mechanisms that can give rise to unjust racist behaviors under particular circumstances. One area of research possibly relevant is facial recognition. There are two general aspects, facial detection and facial recognition (Simion and Giorgio, 2015). Detection need not mean recognition of a particular face but rather the general attributes typically associated with faces. Considerable evidence shows that newborns can detect images that could be considered rough approximations of the human face (Kato and Mugitani, 2015). Within 3 days, newborns can recognize their "mother's face" in contrast to other faces, suggesting a rapid learning associated with facial recognition. The speed of such learning suggests, but is not proof of, an inborn neurophysiological predisposition similar to imprinting as described previously. It may be that this early facial recognition is the underlying mechanism that differentiates Us versus Them. Note that imprinting in mammals is not limited to vision but can also include touch and smell.

The "mother's face" described is, at least partially, determined by the experience of what the newborn and infant sees. Statistically, it is likely that the earliest experiences influencing the "mother's face" are of the same race and likely female gender (Lee et al., 2017; Quinn et al., 2019). Interestingly, infant preference for female gender in the same race did not extend to gender preferences in other races (Liu et al., 2015). Gender and racial preferences in facial recognition dissipate around 9 months of age.

While teleological speculation is always dangerous, it is interesting, if only as a hypothesis awaiting refutation or tentative confirmation. One might argue that there is survival value in imprinting on the caregiver so early in life, which typically means the female gender of one's own race. Conversely, facial recognition preference may also be useful in avoiding strangers who may be more likely to inflict harm than the caregiver. Normal children often display a reaction to strangers that may be akin to an anxiety emerging between 3 and 6 months of age. Nonhuman primates also display what might be interpreted as stranger anxiety (Miller et al., 1990). The question is whether these phenomena share an underlying neurophysiology and whether the recognition of Them becomes associated with fear of Them and whether these mechanisms play a role in the development of racism.

Striking and perhaps initially disturbing is the high incidence of racial bias in young children (Doyle and Aboud, 1995). A study of Caucasian children in kindergarten and grade 3 suggested that 85% of kindergarteners were biased, which diminished to 52% of the children followed to grade 3. There was a decrease in the incidence of racial bias as the children aged. The exact cause is unclear, and the psychological development and social experiences are quite different between kindergarten and grade 3. In an interesting study, Caucasian children age 6–7 years displayed racial bias in the presence and absence of a Caucasian female adult thought to present an antiracist norm. Caucasian children age 9–10 years were less likely to demonstrate bias in the presence of a Caucasian female adult compared to the absence of an adult, suggesting that the presence of an adult served as a check on the bias.

Extending the discussion of the neurophysiological mechanisms underlying empathy for pain observed in others, Cao et al. (2015) reported neurometabolic imaging changes in Chinese students in Australia observing videos of painful stimuli applied to Chinese (same-race) and Caucasian (other-race) persons. Changes in the neurobiological imaging were qualitatively the same between same-race and other-race conditions. However, the changes were quantitatively higher for the same-race condition.

The quantitative difference between same- and other-race conditions diminished with longer duration of the Chinese research subjects' stay in Australia.

The studies just described suggest that there is an inherent predisposition that, when coupled with very early experience, generates an Us versus Them sensitivity. Further, it occurs at a time in the development of stranger fear or anxiety, although how these phenomena are connected is an open question. What is unclear is how any predisposition to recognize Us versus Them and fear of strangers should solidify into pervasive stereotyping and gain prejudicial and pejorative connotations. While perhaps of survival value early in life, fortunately their influence appears to wane with development and experience with the other.

Whether inherent mechanisms producing Us versus Them can be held as unjust is a complicated and difficult question. The situation may not be greatly removed from the question of justifiable homicide in war, for example, an ethical conundrum was addressed by Thomas Aquinas in his Summa Theologica in 1485. An argument can be made that social evolution, particularly in overcoming a severe scarcity of resources and social contracts, renders the neurophysiological mechanisms of preference and fear of the other less relevant and less acceptable. However, social evolution in modern humans does not "erase" the genetic-evolutionary endowment.

What is likely true is that humans are inherently predisposed to some form of naïve prejudice but clearly have the capacity to overcome the naïve prejudice. However, this requires experiences of the other, even if just being in the same neighborhood (Cao et al., 2015). Further, the naïve prejudice may not dissipate; indeed, it risks institution into a pernicious and relatively sophisticated prejudice unless strong social antiracist norms are enforced. Trivializing antiracist norms by saying there are "good people on both sides" of racial or ethnic conflicts likely will only reinforce a conversion from naïve prejudice to something worse and clearly unjust.

Perhaps a greater challenge to interpreting the innate sense of justice, particularly as it might be interpreted as beneficence or nonmaleficence, is the phenomenon of schadenfreude, which is taking joy at someone else's misfortune. For example, one might laugh at seeing someone slipping on a banana peel and falling. Typically, any humor dissipates if the person who fell is seen to be in pain. Perhaps it is related to the novelty and unusualness of the act. More problematic is when the observer heaps more misfortune onto the person suffering the misfortune. Very few neurophysiological studies have been done on schadenfreude, but many psychological explanations have been offered. Most have suggested that schadenfreude is a self-protecting mechanism to assert authority or greater value than the person subjected to the schadenfreude.

Often, the act is a response to some perceived slight where the phenomenon of schadenfreude seems justifiable. Indeed, a comic appears to invent the slight as a pretense to the schadenfreude. However, the response to the perceived slight is directed not at the slight but at some other aspect or character of the person subjected to the schadenfreude. For example, a politician, in response to adverse reporting, mocked a disabled reporter in front of a crowded rally. As the politician was mocking the reporting, the politician imitated the disability of the reporter's malformed arms and hands. One can see the politician's actions as revenge, however inappropriate. The question is, why did the audience appear to enjoy the act and, indeed, by their laughter reinforce the schadenfreude? It is possible that the audience shared the politician's sense of victimhood, and thus their participation in the schadenfreude was, to them, justifiable.

It is likely that most reasonable persons blinded to their self-interest would view just the act of mocking a person's disability as unjust and would explain the injustice as a violation of nonmaleficence to the person subjected to the mockery. If what the politician did was spontaneous and in the absence of any pretense of prior insult to the politician, most reasonable persons blinded to their self-interest would find the politician's behavior reprehensible. That some presumably reasonable persons blinded to their self-interest engage in schadenfreude suggests that there are factors mitigating the maleficence. For example, it may be mitigating some perceived maleficence to the person perpetrating the schadenfreude. Thus, in the context of other competing ethical claims, particularly to opposing claims of maleficence, the politician against the reporter and the presumed maleficence against the politician is predicated on justice no matter how misinformed. Again, this demonstrates that justice, in whoever's perception, is operative with the specific rationales secondary.

Justice is primary—Ethically

The subtitle of this effort is "Explorations of Justice." One motivation in how to effect justice in how patients are cared for in the settings of everyday medicine. Justice is an outcome, actual or anticipated, and is paramount. In this case, justice is a call to action in the clinic and at the bedside. Justice takes place in a complex ecology for which any and all processes for achieving justice must be considered. Most of the discussions regarding ethics involve a traditional dyadic dynamic, between the patient and the clinician, and fail to appreciate and thus consider the very extended ecology in which any dyadic dynamic is embedded (see Introduction). Justice cannot be dissociated from professional conduct, binding contracts, laws, regulations, policies, and procedures. To be sure, some authors specifically exclude law and professional codes of conduct for ethical discussions (see Introduction); however, as demonstrated in many of the cases in Part 1, justice will not be served by such narrow ethical considerations. Thus, it can be argued that a great many discussions of ethics that fail to appreciate the complex ecology are invalid and thus of limited benefit to patients, physicians and healthcare professionals, and society.

Justice is different from ethical principles, although some Principlist ethicists include justice in the set of ethical principles that also include beneficence, autonomy, and nonmaleficence. Including justice as a token of the same type that includes beneficence, autonomy, and nonmaleficence suggests that justice is negotiable, just as obligations to beneficence, autonomy, and nonmaleficence are negotiated among shareholders and stakeholders. Physicians and healthcare professionals balance the obligations to beneficence, autonomy, and nonmaleficence every time they offer a treatment that risks harm, even if it is only the pinprick of vaccinations and the attendant risks of adverse effects. Even that involves an implicit, and at times explicit, informed consent on the part of the patient and thus an obligation to autonomy. Often, the obligations to beneficence, nonmaleficence, and autonomy are constrained, even to the point of denial; yet there can be no denial of justice. It is highly unlikely that any physician or healthcare professional would willingly commit an injustice. Even the injustice to physician and healthcare professionals using the Captain of the Ship metaphor to hold physicians and healthcare professionals liable for acts they could not have effected or affected (see Case 1) or holding them liable for the use of machine learning AI when even the computer programmers do not understand how the AI algorithms work (see Case 12), what is called telishment, still asserts justice on a larger view.

The aforementioned discussion suggests that there could be multiple justices consequent to any particular ethical problem. In the case of telishment, clearly when viewed solely from the perspective of the physician, holding the physician liable for someone else's fault would strike most reasonable persons blinded to their self-interest as unjust. Yet, when viewed in the wider total ecology of the ethical problem, such telishment is taken as just. Note that it does not make sense to say that such telishment results in a "greater" justice, as the very term would suggest that there are lesser justices. If that were the case, one could not possibly agree to a lesser justice if the great justice is available. Such an act would be, in itself, unjust. Clearly, when viewed from a narrow perspective, such as the dyadic relationship between patient and physician or healthcare professional, justice is local and the local outcome can be unjust in the global ecology of everyday medicine. Unfortunately, most texts on the ethics of everyday medicine have taken this narrow perspective, either implicitly or explicitly (see Introduction).

The question becomes, why should physicians or healthcare professionals sacrifice what they may consider as justice in their particular (local) case for the global justice? One response is that they are contractually obligated to do so. As licensed physicians and healthcare professionals, they are granted a monopoly. Not just anyone can call themselves a physician, nurse, speech pathologist, physical therapist, or any other licensed healthcare professional discipline and get paid as such. Further, it is unlikely that physicians, in particular, actually paid the full cost of their education and training. Rather, society invested that cost in the physician and has a right to expect a return on the investment.

In a similar way, justice is different than law. Evidence are cases where following the letter of the law would lead to what a group of reasonable persons blinded to their self-interest would find unjust. Examples include when repeated sets of juries refuse to find a particular defendant guilty, although all would agree that the letter of the law was violated. Assuming the repeated juries represent an adequate sampling of the citizenry such that any other jury would also come to the same refusal to find the defendant guilty (see Chapter 3), then in a pluralistic modern liberal democracy, the juries are exercising justice even if contrary to the law. Other examples include prosecutorial discretion where the prosecutor may not bring charges even if the violation of law is clear. Judges may have considerable leeway in sentencing such that if the defendant is found guilty, the sentencing has a practical effect little different than had the defendant been acquitted. If law is akin to the ethical principles, as described earlier, then the actual effect of law is a derivative of justice. Justice and law are not tokens of the same type.

An alternative approach to the Principlist formalism of ethical principles and moral theories described earlier is virtue ethics. Note that most analyses of virtue hold that there are multiple virtues that in some way must be complementary as they together lead to a virtuous person who would be predisposed to just behaviors. The complexity in complex ecology of ethical problems in everyday medicine often involves competing interests that call for different obligations to the same virtues among the shareholders and stakeholders. No single virtue, *a priori*, trumps the outcome. Thus, there is not a one-to-one correspondence between the varied obligations to the virtues and the outcome that is just. Therefore, justice is different than virtues.

Consider the ethical virtues expounded by Aristotle, including (1) courage in the face of fear, (2) temperance, (3) liberality (generosity), (4) magnificence, (5) magnanimity, (6) proper ambition, (7) truthfulness, (8) wittiness in conversation, (9) friendliness, (10) modesty, and (11) righteous indignation. Consider virtues described by a group of medical students and physicians, including (1) fairness, (2) honesty, (3) kindness, (4) leadership, (5) being a good team player, and (6) good judgment (Kotzee et al., 2017). With the exception of fairness, it is not hard to imagine cases where any of the virtues described would be limited or even abandoned for the sake of justice. Thus, justice is not a token of the same type that comprises the ethical virtues (with the exception of fairness).

Ethical principles, case precedents, and virtues as induction from microlevel justice; moral theories from macrolevel justice

As discussed previously, telishment may seem a violation of justice if viewed solely from the perspective of the physician held liable; yet global considerations affirm the action as just given the interests of all reasonable persons blinded to their self-interest that find a greater good in protecting charitable hospitals and fledgling machine learning AI technology development. The situation suggests that there is a microlevel sense of justice, sometimes local, particularly in the dyadic relationship between pairs of stakeholders and shareholders. However, unless a particular shareholder is uniquely privileged, the sense of justice in the context solely of the particular shareholder is not the determining factor in the global assessment of justice. Pluralistic modern liberal democracies generally do not allow such privileged shareholders.

John stealing money from Jane because he is stronger would be considered unfair and thereby unjust by most reasonable persons. However, this is not a *prima facie* case of unfairness, for example, if Jane had more money than she needs and John's child is starving and John needs to buy food. In the specific case of a particular John and Jane, John's actions may not seem unfair and unjust or fair and just. Indeed, the situation of John, John's starving child, and Jane could be analyzed in terms of the ethical principles of beneficence, autonomy, and nonmaleficence in the context of the Principle of Double Effect following from Thomas Aquinas (1485), discussed in Case 8.

There is another notion of fairness or justice on a macrolevel that transcends the particulars of the microlevel. In the situation where Mike steals money from Mary for the exact same reason that John stole money from Jane, Mike should be treated the same as John, unless there was some reason to privilege John over Mike. In most pluralistic modern liberal democracies, most reasonable persons are likely to find that privileging John over Mike, where John is forgiven and Mike is sent to jail, would seem unfair. While there are two cases of Mike and John, they are not independent of each other. What was judged fair in one case influences how the second case is judged. Thus, there is a sense of justice in the macrolevel that may not have a one-to-one correspondence with the microlevel. Typically, at the microlevel level, the dynamics are dyadic, for example, between the patient and the physician or healthcare professional. However, at the macrolevel, the dynamics are far more complex, for example, involving insurers, regulators, legislators, and society as a whole, where each is a stakeholder and shareholder in the ethical problem. In the past, many ethicists excluded considerations at the macrolevel, which often does not lead to justice (see Introduction).

The dilemma posed by ethical problems is manifested when the ethical problem in the everyday practice of medicine is considered in the full complexity of the ethical ecology. Yet ethical problems are particulars, not hypotheticals or generalities. Justice requires a just outcome for the particular stakeholders and shareholders. Thus, ethical problems require consideration at the microlevel. Indeed, many ethicists have argued that hypotheticals and generalities are to be avoided, instead focusing on the unique and particular case, what is called specification (Beauchamp and Childress, 2013). The question that arises is, what are the tools or what are the metrics by which to characterize the microlevel issues?

Principlism would suggest the ethical principles of beneficence, nonmaleficence, and autonomy, yet when considered as all or nothing (comporting with the logical Principle of the Excluded Middle as discussed in Chapter 2), considerations of different dyadic relationships become problematic. There is no basis for making comparisons. It is unlikely that most reasonable persons blinded to their self-interests would say that one party to a dispute does not have the right to beneficence, nonmaleficence, or autonomy. If that were held, what is the sense of calling beneficence, nonmaleficence, or autonomy principles akin to rights? The same could be said of virtues in virtue ethics. Again, it would seem unreasonable to argue that a party to the dispute is entitled to generosity or kindness. What can be used as the metric are the obligations to the ethical principles or virtues. Now the different dyadic relationships operate on the same continuum of obligation and the entire ecology of all shareholders and stakeholders can be "mapped" by the degree of obligations.

A fundamental test of justice is the reproducibility of adjudications for the same token cases of the same type of case. Thus, a dispute regarding the same ethical question should result in the same judgment as every other instance. Central to the reproducibility is the notion of generalization over all instances or cases thought to be of the same ethical type. This requires that there must be some supervening system for creating the global solution that supervenes over every particular solution. One approach is that each pairwise obligation to the ethical principles or virtues can be weighted and the total weighted obligations may differentiate among the parties to the dispute, thus allowing adjudication over the parties, leading to a just global outcome. The issue becomes how to assign weights to each obligation among all stakeholders and shareholders. Moral theories can provide the weighting.

Traditionally, moral theories have been couched in terms of good, with the term "good" taken in its widest connotation. However, a good should not be considered a physical entity such as a commodity; it is more complex and often includes intangible values. An appropriate moral outcome would be that which creates the greatest good. In this sense, the obligations at the microlevel would be weighted in order to gain the maximum good, with the presumption that this would result in justice.

A Libertarian moral theory would hold that the greatest good is maximum freedom. To avoid anarchy, libertarianism often resolves to a form of contractualism where individuals trade some of their freedom for minimum regulations to avoid anarchy, thus providing a modicum of security. The obligations are then weighted by the terms of the contract. Utilitarian moral theories hold that the good to be achieved is the greatest good, a macrolevel good, considered as some sum of the goods at the microlevel. However, defining the macrolevel good is highly problematic. Deontological moral theories, particularly Kantian moral theories, hold that what is good is what is universally good for all stakeholders with the proviso that no individual person is to serve as a means to another person's end. An Egalitarian moral theory likewise holds that what is good is universally good for all stakeholders but compromises on the notion that an individual cannot be a means to others' ends so long as the ends benefit all stakeholders and, importantly, each individual has claim to equal distribution of whatever greater good. In other words, no individual has privileged claims to a good, unlike the Libertarian moral theory where claims can be privileged by contract. A Utilitarian moral theory would argue that some individual's claims are privileges so long as doing so creates the greatest aggregate good.

Each of the moral theories just described may be reasonable hypothetically. However, the limitations of moderate scarcity make the application of the logical extent of each theory very problematic. At least in healthcare, libertarianism does not operate in an unrestrained *laissez-faire* capitalistic manner. For example, there are particular elements required of a contract, for example, that all participants are free to enter the contract, which suggests a deontological notion of freedom. However, any Deontological moral theory cannot be allowed to bankrupt the healthcare system. Egalitarianism, applied narrowly, allows exploitation. For example, egalitarian healthcare would allow, in its logical extent, all patients to receive the same poor level of care. Most reasonable persons blinded to their self-interest would hold there is a minimum of healthcare that every person is entitled to, much in the manner of a Deontological moral theory. As demonstrated repeatedly in the cases presented in part 1, no single moral theory is the sole determinant of justice, at least as currently constituted in a pluralistic modern liberal democracy.

It is important to note that the specific moral theories discussed earlier do not exhaust the possible moral theories (see Chapter 2). Indeed, given the difficulties of applying each of those discussed previously, perhaps new moral theories, at least as they relate to the ethics of everyday medicine, are needed. How these might be developed is the subject of Chapter 3.

Operationalization of justice

The rule of law in modern liberal democracies would suggest that ultimately the law determines the behaviors of stakeholders and shareholders in the everyday practice of medicine. Even in the absence of statutory law, common law based on settled cases (case precedent) can extend the rule of law. In this context, the ethics of everyday medicine is ultimately subject to the rule of law; to the extent that it is, the ethics of everyday medicine is institutionalized. Yet even the rule of law is not absolute as discussed earlier in the situation or circumstance of prosecutorial discretion, jury findings contrary to law, and in discretion in sentencing. Yet the "wiggle room" provided with these discretions is precarious. Thus, the law is amendable, giving rise to the notion of the US Constitution as a living document and the interpretation of which ultimately reflects the contemporaneous will of the citizens. Often, the route to producing justice in the face of imperfect laws is circuitous. An example is the Commerce Clause of the US Constitution that was used to bar racial discrimination in restaurants [United States Supreme Court case, *Katzenbach v. McClung*, 379 U.S. 294 (1964)] and was attempted in support of the individual mandate for individuals to purchase health insurance under the Patient Protection and Affordable Care Act (otherwise known as Obamacare) but was upheld under the authority of Congress to tax. However, recently, a US appeals court ruled that because Congress reduced the liability (tax) for failure of the individual to purchase health insurance to zero, the basis for the US Supreme Court upholding the individual mandate was moot. Implications for the Patient Protection and Affordable Care Act are unclear.

The question becomes, to what degree do the obligations to ethical principles and moral theories found acceptable by reasonable persons blinded to their own interests impose an obligation to politicking so as to affect law so as to achieve justice in the clinic and at the bedside (Porter, 2007)? If so, to which of the shareholders and stakeholders does the obligation to politicking apply? To be sure, many professional medical organizations hold that physicians and healthcare professionals have obligations to ensure both patient and public good and that politics is acceptable (Earnest et al., 2010). Their argument is that physicians and healthcare professionals are in a unique and privileged position to advocate politically. The unique and privileged position stems from the expertise of physicians and healthcare professionals and the esteem in which the general public holds physicians and healthcare professionals. Whether the position of professional medical organizations is a realization of a patient-centric ethos is open to debate (see Chapter 4).

The changing nature of disease burden and the effects of "social hazard" on morality and justice

The early involvement of physicians and healthcare professionals in public policy stemmed from the shared risk of contagious diseases. Years ago, infectious diseases did not respect socioeconomic boundaries and it was to everyone's benefit to provide healthcare, in some fashion, to the ill, such as quarantine, vaccination, and treatments mandated in many cases. Hence, healthcare and microlevel justice was supervened by the global risks to society. In a way, concerns for one's self seemed to generate concern for others lest infection spread (an example of what has been called Enlightened Self-Interest). However, since 1900, death from infectious diseases decreased from 797 deaths per 100,000 population in the United States to 36 deaths per 100,000 in 1980. In the mid-1990s, there was a brief increase from new disorders and epidemics such as acquired immunity deficiency syndrome (AIDS) and influenza compounded by increased mobility of the ill with international travel (Armstrong et al., 1999). Nevertheless, the risk of infectious diseases as a driver of social and political policy has diminished and, perhaps as a corollary, the social and political role of physicians and healthcare professionals similarly has diminished.

In a very interesting article on the history of US health policy regarding the AIDS epidemic, Daniel M. Fox (2005) noted that the AIDS epidemic burst forward at a time of change in the entire ethical and political ecology—the latter termed polity to broaden consideration beyond policy—that reflected a declining importance of infectious disorders with attention turning to chronic illnesses, particularly those thought associated with lifestyle decisions. The latter then changed the discussion regarding insuring health from a collective response to help all at risk to stratifying risk and then apportioning insurance accordingly. The presumption was those with lifestyle diseases, such as smoking, obesity, drug abuse, unplanned pregnancies, and reckless behavior, were less deserving. Indeed, concerns were raised that providing support services would be a social hazard with the effect of encouraging such behavior. The "perfect storm" of the perception of decreased vulnerability to common and contagious diseases, the shift from collective to individual responsibility, isolating relative risks so as to stratify the obligations to ethical principles, and moralizing and stigmatizing of patients by identifying AIDS with homosexuality likely caused untold hardship and death that arguably did not have to be.

Experiences such as the response to the AIDS epidemic are vivid demonstrations of the complex ecology in which ethical decisions operate. The response to the AIDS epidemic involved patients, family members, caregivers, communities, physicians, healthcare professionals, insurers, governmental officials, and elected officials (Fox, 2005). As demonstrated in many of the cases in this text, the ecology in which the ethics of everyday medicine plays out is complex, at least in the range of the shareholders and stakeholders, and the complexity is not generally appreciated (see Introduction). The critical point here is there is good news and bad news about laws and acceptable ethics as they play out in everyday medicine. The good news is that laws and thus ethics are living documents that can evolve to reflect the will of contemporaneous citizens; the bad news is that laws and thus ethics are living documents that can evolve to reflect the will of contemporaneous citizens. As written by Joseph-Marie, Comte de Maistre in 1860, "Every nation gets the government it deserves" (https://en.wikiquote. org/wiki/Joseph_de_Maistre). Perhaps a sarcastic reply to the excesses of the French Revolution, it nevertheless may be true today.

Neural dynamics and implications for ethical principles, moral theories, laws, case precedents, and virtues

It likely is not readily apparent how neurophysiology can relate to ethics directly. In one sense, there must be a direct connection unless one wants to argue that there is an immaterial mind independent of the brain and body, a form of Dualism. To be sure, Dualism is rarely a topic of debate, but Dualism of varying degrees is endemic and epidemic, implicitly, in physicians and healthcare professionals (Miresco and Kirmayer, 2006; Demertzi et al., 2009). A dualist perspective clearly informs ethics and morality, as seen in the discussion of the public response to disease in terms of "social hazards" and the culpability of being in need of care (see Case 10). Disallowing Dualism as a *Deus ex machina* ("god from the machine"—a literary device where a supernatural force is used to get the author out of a corner that the author has "painted" himself into), then any behavior, whether just or unjust, must directly follow from the neurophysiology. The critical question now is whether understanding the neurophysiology can illuminate the ethics of everyday medicine.

Conceptually, neurophysiology is undergoing a remarkable revolution. In the past, much of neurophysiological reasoning was based on attempts to correlate behavior or changes in behavior to specific brain regions. In many ways this was successful but, in many ways, counterproductive. Indeed, behaviors under study had to be simplified and dichotomized to allow mapping to specific brain regions. Today, there is a shift from a notion of strict localization of function to the collective distributed over wide regions of the brain connected through a network of neuronal interconnections. Further, the interactions are dynamic such that the same set of neurons can generate different behaviors over time by shifting the patterns of interactions dynamically (Montgomery et al., 1992).

Consider the famous case of Phineas Gage, a railroad worker who had an iron rod blasted through the front part of his brain. He survived but with marked changes in his behavior. He was described as becoming antisocial and disrupted, whereas prior to the injury he had been an upstanding citizen. The theoretical explication, dating back at least to the great neurologist John Hughlings Jackson of the late 1800s, was that something resided in the frontal lobe that promoted prosocial behavior. When the frontal lobe was damaged, other areas in the brain promoting antisocial behavior became active and unbalanced, although it was never specified where those areas were. Further, the implication was that the frontal lobes inhibited other areas that, when disinhibited, manifest antisocial behavior (Macmillan, 1992). The extent of Gage's behavioral changes both initially and over time has been debated, with various features maximized or minimized as they favor presuppositions. But Gage continues to be a focal point in arguments for the neural localization of social behavior and the nature of interactions between linked structures, with the notion of inhibition being predominant. The notion of strict localization and the notion of shaping behavior by inhibition are complicated and a full discussion is beyond the scope of the present effort. Those interested can see chapters 11 and 12 in Montgomery's "Medical Reasoning: The Nature and Use of Medical Knowledge" (2019a).

Key concepts in older neurophysiology included (1) organization of the nervous system, particularly the cerebral cortex, into modules with each module serving a unique function; (2) simplification of each module to a single representative neuron (termed a macroneuron) and the physiology of the relationships between modules based on the presumed physiology of the single representative neuron; (3) simplifying and dichotomizing the actions of neurons as excitation or inhibition; and (4) disease as an absence or overactivity of the module affected by disease.

Replacing the older concepts is a neurophysiology based on distributed processing across large-scale networks of which the older modules are just nodes in the network. The actual functions of the networks are in the patterns of connection rather than in the individual neuron (Montgomery et al., 1992; Montgomery, 2004, 2017). Although not universally accepted, the notion of distributed processing is gaining experimental evidence (Parks et al., 1991; Mesulam, 1998; Prut et al., 2001; Fox and Friston, 2012). Changing behaviors are the result of changes in the physiological interconnections (a different concept from anatomical interconnections), as will be discussed later. As an interesting aside, the weakness of a modular compartmentalized system where each component serves a unique function was recognized by Aristotle, who wrote:

> Yet, to say that it is the soul [author – taken here as a part of the nervous] which is angry is as inexact as it would be to say that it is the soul that weaves webs or builds houses. It is doubtless better to avoid saying that the soul pities or learns or thinks, but rather to say that it is the man [author – taken here as the entire nervous system] who does this with his soul. What we mean is not that the movement is in the soul, but that sometimes it terminates in the soul and sometimes starts from it, sensation e.g., coming from without inwards, and reminiscence starting for the soul and terminating with the movements, actual or residual, in the sense organs.
>
> *(De Anima 408b 12–20, Aristotle, 2018)*

Parallels exist between old neurophysiology and ethical and moral theory and, as anticipated, the change from the old neurophysiology to the new will have implications for ethical and moral theories. The ethical and moral parallels to the old neurophysiology include a type of modularization of ethical principles into beneficence, nonmaleficence, and autonomy, presumably from ethical induction by reasonable persons blinded to their own self-interests from a collection of experiences. However, it is not clear whether such categorization with separate appearing principles is true or is artificial based on the methods, particularly those for assembling the experiences from which the induction was made. For example, the Belmont Report (Belmont Report: Ethical Principles and Guidelines for the Protection of Human Subjects of Research, 1978), which addressed the ethics of human experimentation initially, had three principles: beneficence, autonomy, and justice. The concept of nonmaleficence was embedded within the notion of beneficence. There may be situations and circumstances where nonmaleficence correlates with beneficence, suggesting some common element. For example, it may be that a just act requires not only beneficence but also nonmaleficence. However, there may be other situations and circumstances where the two principles do not correlate where just actions require beneficence as well as violations of nonmaleficence, such as exposing the patient to harm, surgical invasion of the body, to provide beneficence, such as the restoration of function. This was known to Thomas Aquinas in his Principle of Double Effect formulated in Aquinas' *Summa Theologica* in 1485.

An important metaphor for the new neurophysiology is machine learning artificial intelligence (AI), originally called neural network computing. The development of neural network computing derived from neurophysiology as a metaphor (Rumelhart et al., 1986). (A caveat there has been considerable controversy whether the nervous system operates in a manner exactly as suggested by the originators of parallel and distributed processing. In this author's opinion, these critiques largely are due to an overly literal reading that is related to "representation" being conflated with "localization.") An example of machine learning AI was discussed in Fig. 1.3.

The "structure" of machine learning AI as described previously is in the distribution of differing synaptic weights. Initially, the synaptic weights were random, which means there is no structure in the neural network. As one cannot point to one part of the neural network as different, there is no "localization" of function in the network. The random synaptic weights guard against bias. This notion of a random starting point is also critical in ethical induction to avoid bias in those charged with creating by induction, ethical principles, moral theories, laws, case precedents, and virtues as discussed in detail in Chapter 3. Alternatively, some ethical presumptions (bias) may be productive in ethical induction, as will be discussed later.

With training, the synaptic weights are changed in a deliberate manner and this structure is imposed on the neural network, even if it is not apparent to anyone. In other words, there is "structure" but it is not localized. The neural network cannot remain random as otherwise no information would evolve from the machine according to the Second Law of Thermodynamics as applied to Information, discussed further in Chapter 3. Importantly, the information is in the training sets and the neural network comes to represent the information in the training sets, even if one cannot explicitly identify the representation. As evidence, machine learning AI trained on experiences in one hospital may produce incorrect results when applied to data from another hospital, suggesting that the two training sets contained different information, even if unrecognized, which was internalized in the machine learning AI program (Couzin-Frankel, 2019). Thus, the selection of the training set is critical, just as the selection of cases over which to achieve some ethical consensus (ethical induction) is critical (Chapter 3).

Information in the training set may not be relevant to the computational problem and/or the initial synaptic weights may not be random; both could act as a bias. Indeed, there are significant concerns that machine learning AI used every day, such as processing loan applications or determining jail sentences, may be biased. The implication for ethics is that the ethical induction to principles, moral theories, laws, and precedents depends greatly on the collected set of experiences. For example, if the process of ethical induction is at all like machine learning AI, in a training set of experiences consisting solely of experiences of middle-aged Caucasian males, the system will arrive at a conclusion, although that conclusion may only hold for middle-aged Caucasian males. When the same neural network is confronted with a problem of deciding justice in situations and circumstances involving other than middle-aged Caucasian males, the "computational" solution may result in injustice. This is a central theme in feminist ethics (see Chapter 2). Because avoiding implicit bias is extraordinarily difficult, methods continue to be developed to control bias (Chapter 3).

Machine learning AI systems are capable of relatively extremely rapid processing of a problem once the network has been trained. Consider Fig. 1.4. Somewhere in the figure there is a dog. Typically, finding the dog requires a piecewise search of the image, which can be prolonged as the dog does not appear to jump out of the picture.

Having found the dog in Fig. 1.5, now look at Fig. 1.4. It is likely that one finds the dog quite readily. Now find the dog in Fig. 1.6; again, very often, a search of the entire image is necessary.

The point here is once the dog is found in Fig. 1.4, usually after a prolonged search, the dog is found almost instantaneously. However, when the orientation of the same image is changed, it takes longer to again find the dog. Note that it is not likely that simply knowing where to look in the image is sufficient to allow the nearly instantaneous identification of the dog on subsequent viewing of Fig. 1.4. The relative position of the dog relative to the other visual stimuli is the same in Fig. 1.6.

One need not invoke two separate mechanisms, one that involves a piecewise search strategy and a separate one that allows the image of the dog to be perceived rapidly, held as intuitive. Rather, a single neural network may be sufficient, but that prior experience biases the network to produce a more rapid solution, that is, a collection of stimuli results in a solution that rapidly fits the target dog. For example, the neural network utilizes feedback from what is perceived to some expected notion of a dog. If the latter is vague, then the error signal will be problematic and the neural network will not settle on a minimum acceptable error and thus a solution, at least rapidly. Once having seen the dog, the expectation for what constitutes the dog is settled, for example, its orientation, and the system can settle to a solution rapidly. That may be why it takes a bit longer to see the dog in Fig. 1.6—it likely is faster than when confronting

Fig. 1.4 There is a dog somewhere in this abstract figure. The challenge is to find it. Modifed from Gregory, 2005.

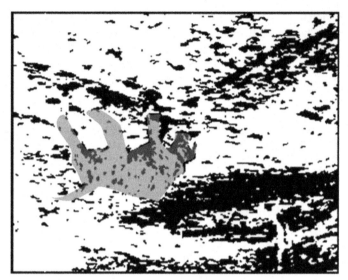

Fig. 1.5 The image is the same as in Fig. 1.4; however, the dog is outlined in *blue (light gray in print version)* and is shown upside down. Once the dog is seen, look for the dog in Fig. 1.4. Modifed from Gregory, 2005.

Fig. 1.6 Somewhere in this abstract figure is a dog. Having seen the dog before in Figs. 1.4 and 1.5, it still requires a search but it is perhaps a bit quicker. Modifed from Gregory, 2005.

Fig. 1.4 for the first time. The hypothesis described is just that—a hypothesis. But it is a reasonable one. The implication for ethics is that some have argued for intuitive ethical perceptions, fast, as separate from declarative, deliberative, and conscious processes, slow. Rather, there may only be a quantitative difference (in degree) rather than a qualitative difference (in kind). Further, the mechanisms hypothesized may help explain imprinting, which is important for creating neurophysiological mechanisms very quickly after birth and which have ethical consequences as described previously.

Not starting from random: Evolution—Biological and social

To the degree that the nervous system operates remotely like machine learning AI, any initial structure in the machine learning AI architecture would have a significant biasing effect. One example is the *a priori* that is the beginning point of ethical induction (Chapter 3). Consider the situation of the embryo frozen such that all molecular motion essentially stops, but when warmed, the embryo goes on to form a complete human. It is as though once molecular motion starts, the development has an *a priori* initial trajectory (the *a priori*). Yet it is not so simple as following a genetic blueprint. There are not enough genes in humans to dictate the detailed structure and operations of the nervous system (Edelman, 1987); instead, these systems have the capacity to self-organize, as will be discussed later in "Ethics as an exercise in chaos and complexity" section. The examples of infants' responses to just and unjust situations and circumstances argue for an *a priori* information processing system established prior to birth or rapidly after birth from an inherent predisposition.

The human nervous system has *a priori* information processing systems as a consequence of its phylogenetic legacy, such as humans' socially apt evolutionary predecessors who likely knew nothing of ethical principles, moral theories, laws, precedents, or psychological constructs. The structure of the nervous system in the evolutionary processors, in many ways, was added to and modified to become human. When applying the Second Law of Thermodynamics as applied to Information to evolution, it is less thermodynamically costly to replicate and modify information than to create it anew. What is particularly remarkable about humans, if not necessarily unique, is the capacity to comment in language on one's own actions and those of others. It may be that it is from the commentary that ethical principles, moral theories, laws, and precedents are given a declarative and deliberative sense and subsequently intervene in ethics, but perhaps one should not confuse commentary as ontology (reality).

Ethics as an exercise in Chaos and Complexity

Ethics ultimately is derived from neurophysiology unless one is willing to invoke a nonmaterial entity that can move material entities, such as Dualism as discussed earlier. Although it may sound trite, the nervous system is incredibly complex. However, it is only trite when the implications are not recognized or understood. Once they are understood, the implications are enormous and awesome. The complexity and non-linearity of neurophysiological mechanisms strongly suggest that insights of Chaos and Complexity theory may apply. Nonlinearity means that the interactions between inputs to a neuron are not simply addition but rather more complex; in other words, for the neuron, $2+2$ may not be 4. It is the nonlinearity between relatively simple interactions that gives rise to Chaos. For example, Newton's laws of motion describe beautifully the orbit of the moon about the earth or the earth about the sun. However, these laws fail to precisely predict the orbit of the moon around the earth as the earth orbits around the sun; this is referred to as the "three-body problem" (Stone and Leigh, 2019) and its chaotic nature was first alluded to by the mathematician Henri Poincaré in the late 1800s.

The argument is offered that *if the Chaos and Complexity theory implies neuro-physiology and neurophysiology implies behaviors to which justice or injustice can be attributed and if behaviors to which justice or injustice can be attributed imply ethical principles, moral theories, laws, case precedents, and virtues, then the Chaos and Complexity theory implies ethical principles, moral theories, laws, case precedents, and virtues*, by the Principle of Transitivity. Assuming, for the sake of discussion, the argument to be true, then the properties of the Chaotic and Complex systems may have implications for ethical principles, moral theories, laws, case precedents, and virtues.

Among the properties of Chaotic and Complex systems are unpredictability and dependence on initial conditions. Unpredictability means outcome of a particular instance of a particular process cannot be predicted. This is not to say that the process is random. Consider the formation of a snowflake; it is impossible to predict the particular pattern that the snowflake will take other than it will have either six points, despite the fact that the physics is well known, well-behaved, and lawlike. In fact, there are so many possibilities that it is likely that statistically no two snowflakes are exactly the same. However, the structure or regularity of the process is demonstrated by the fact that every snowflake has six points. For some systems, like the weather, there is an inherent indeterminacy (such as predicting the weather beyond 2 weeks) that likely will never be overcome, despite even more powerful computers and computational algorithms (Voosen, 2019).

A hypothetical example of a Chaotic system is shown in Fig. 1.7. Imagine a thin pillar with a weight resting atop it. At first, the pillar remains straight even as the weight is increased. At some point, the addition of a small weight will cause the pillar to buckle. As the example is confined to the two-dimensional surface of the image, the pillar will buckle to the right or left. Remaining straight, bending to the right, and bending to the left are different states of the system composed of the pillar and the weight. Thus, the system can be described as having three states—multistability. The direction of the buckling is unpredictable. Although unpredictable, the behavior is not random; it is not a matter of rolling a dice or flipping a coin. It is determinant or analytic, much like using a calculator, although in this case, a very special calculator. Indeed, an equation of the form $\dot{y} = xy - y^3$ describes this process.

The extent of the buckling can be measured as the maximal departure from the vertical (the value of y). If one were merely to plot the value y for each weight x, the

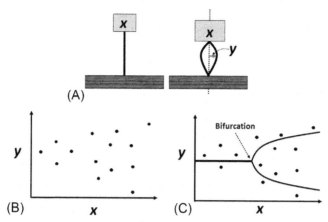

Fig. 1.7 Schematic representation of a system that demonstrates bifurcations (transitions) between states. (A) A weight is supported by a pillar. The pillar remains upright and straight as weight (x) is increased initially. At a certain weight, the pillar buckles that from the single perspective is seen as buckling to the left or right. The direction of the buckling is unpredictable. (B) The weight and corresponding distance of the buckling (y) are plotted. At first impression, it may seem as though no relationship exists between weight, x, and buckling, y. Indeed, the correlation coefficient likely is near 0. However, the physics clearly is one of a relationship between x and y given by the equation of the form $\dot{y} = xy - y^3$. From Montgomery Jr., E.B., 2019b. Reproducibility in Biomedical Research: Epistemological and Statistical Problems. New York: Academic Press.

distribution would be similar to Fig. 1.7B. It would be very difficult to infer any relationship between x and y and therefore to establish any determinant or analytic relationship, for example, as an equation. One might conclude that no relationship exists, even though the experience of an actual experiment were to argue that a relationship does exist. However, when viewed from the perspective of a Chaotic and Complex system, the relationship is recognizable. The system can be described as y does not change until a certain weight, x, is placed on the pillar. Then the pillar will bifurcate (called a pitchfork bifurcation) between bending to the right or bending to the left; the direction is unpredictable. The displacement then is the result of the amount of weight above the threshold to cause the bifurcation.

Anti-Principlists ague that ethical analyses of a set of related situations and circumstances, based on principles, moral theories, laws, case precedents, and virtues, often result in apparently inconsistent conclusions, taken as evidence against the Principlist position. It is interesting to speculate that if the neurophysiology of the ethical system underlying the analyses is Chaotic and Complex, then the variations in outcomes are still determinant, not random, but are like different solutions of the same equation, for example, bending to the right or left of the pillar in Fig. 1.7. If that were the case, then Anti-Principlism and Principlism are not incommensurate.

Chaotic and Complex systems display sensitivity or dependence on initial conditions. For example, there may be a single process but the starting point is very similar, so similar that no differences between them can be detected; nonetheless, the imperceptible difference in the starting point (initial conditions) results in outcomes that are very different. The different outcomes are not evidence of different processes but the same Chaotic and Complex process whose slight difference in the starting conditions produces the apparent discrepant outcomes. It is interesting to speculate that identifying differences in the initial conditions is the *forte* of phenomenological and narrative ethics (see Chapter 2), which would be critical if, indeed, ethical systems displayed dependence on initial conditions as the Chaotic and Complex systems.

Self-organization is a feature of many Chaotic and Complex systems. These systems tend to stabilize, even briefly, into regular patterns without outside influence. Consider the Necker cube illusion shown in Fig. 1.8. The illusion consists as a wireframe three-dimensional cube. The shaded face alternates appearing as the front or back of the cube. It is not seen as being both the front and the back and it is not seen as neither the front nor the back. Further, on prolonged gaze, the shaded face appears to switch from the front to the back and the transition is not perceived; the perception is said to bifurcate, such as the pillar suddenly bending to the right or left in Fig. 1.7. It is important to note that the visual stimulus does not change. The pattern of photons of light striking the retina of the eyes does not change with the two perceptions. Rather, the visual information self-organizes into one of two possible stable states, the shaded face seen as the front or the shaded face seen as the back. However, the perception can switch spontaneously between the two states, meaning that the states are better described as metastable. Hence, the system displays multistability. The implications of multistability for ethics will be discussed later.

Another example is the face-vase (Rubin vase) illusion shown in Fig. 1.9 where the image bifurcates between a vase and two faces facing each other. Again, the exact same visual stimulation occurs in each perception. However, one can bias the perception without changing the visual stimulation by saying the word "face." There are several sources of bias such as prior probabilities, that is, what is the probability of a particular perception? When one is on the western plains of the United States, one is likely to think of horses when hearing hoofbeats. If one is in Africa, one might think of zebras. Thus, the probability that a particular percept is associated with a phenomenon

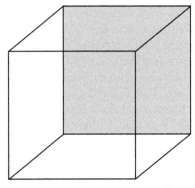

Fig. 1.8 The Necker cube illusion shows a wire-frame three-dimensional object. The shaded area may be perceived as in the front or back. The perception can alternate.

Fig. 1.9 The image is of the Rubin vase, where it can be seen alternating as two faces facing each other or a vase (Smithson, 2007).

in the environment influences whether that phenomenon is perceived. The potential consequences of perceiving a particular perception among more than one possibility influence the probability of the perception (Zhang et al., 2017). Also, the intellectual environment shaped by deliberative and declarative discussions can affect the probability of a particular perception over other possibilities. This is evidenced by demonstrations of language priming on perception. For example, the positive phrase "she thinks you are competent" influences the positive or negative estimation of a picture of a neutral face (Schwarz et al., 2013; Ryan and Turkstra, 2019).

Interestingly, social cues can also influence visual perception. For example, images can create the illusion of things floating in air and a person in the image appears to be looking at the floating objects. In the illusion the gaze of the person in the image directs our gaze to the floating object and suggests a common experience.

It is interesting to speculate whether the visual illusion shown here is analogous to "moral illusions" where persons behave in a manner inconsistent to principled ethics as suggested by Nichols and Knobe (2007). Interestingly, Nichols and Knobe (2007) explain that the moral illusion has mechanisms involved in perceiving the situation and circumstance that do not have access to the modules in the mind (or brain) that involve ethical reasoning. However, in the normal state, it is not clear why a person would not have access to modules in the mind (brain) involved in ethical reasoning. Further, assuming the argument given that ethical reasoning is post hoc, as discussed earlier, it would be immaterial to the actual behavior whether or not the person had access to the modules for ethical reasoning as the behavior, the moral illusion, would have already occurred.

The aforementioned discussion serves as a metaphor for ethics, particularly the ethics of everyday medicine. To be sure, what follows is speculative, but no more so than most theories and, even if subsequently proven wrong, then physicians, healthcare professionals, healthcare administrators, and society will have learned more. But at least one might imagine the possibility of the theory being true; from Lewis Carroll, "Through the Looking-Glass and What Alice Found There" (1871):

> Alice laughed. "There's no use trying," she said: "one *can't* believe impossible things."
> "I daresay you haven't had much practice," said the Queen. "When I was your age, I always did it for half-an hour a day. Why, sometimes I've believed as many as six impossible things before breakfast."

Many ethical problems, particularly the nonheadline type, in the everyday practice of medicine are unpredictable as to a definitive conclusion that the actions are just or unjust. As seen in the cases discussed in Part 1, complex, nonlinear interactions exist between stakeholders and shareholders with many possible conclusions. Note that the conclusions taken are direct results of the neurophysiological affective responses to the actions or potential actions and not principles, theories, laws, case precedents, or virtues that are post hoc rationalizations. However, there is not an infinite number of conclusions/actions but rather certain different types of conclusions/actions emerge, such as metastable states arising out of Chaos and Complexity. This is not to say that there is no underlying determinant (as opposed to random) set of operations (the word used in lieu of principles or laws); there are determinant operations, which are the operations subsumed in the neurophysiology. The implication for ethics is that a multiplicity of conclusions should be expected, and the presumption that the variation represents "noise," confusing the "true" ethics, should be avoided so as not to result in misdirection.

Kwame Anthony Appiah (2008) reviewed sociopsychological studies with implications for ethics. Indeed, Appiah has argued strongly for an association, although not identity, between psychology and philosophy. The studies Appiah cited demonstrated significant inconsistencies in the charitable responses in various situations and circumstances, both within a single individual and among groups of individuals, who would be considered predisposed to charitable acts. The inconsistencies raise skepticism that any coherent psychology or ethics is possible and, by extension, neurophysiology. How many exceptions does it take to disbelieve the rule? The implication is a situationist ethic, where what is ethical (just) depends on the situation and circumstance and risks the Solipsism of the Present Moment. Yet the inconsistencies seem to suggest the near impossibility of having a coherent, fully explaining theory. Appiah (2008, p. 74) writes:

> If experimental psychology shows that people cannot have the sort of character traits that the virtual therapist has identified as required for *eudaimonia* (author—human flourishing), there are only two possibilities: she has identified the wrong character traits or we cannot have worthwhile lives. Virtue theory now faces a dilemma.

This may be a false choice if the mechanisms underlying ethics display Chaos and Complexity. Recall Fig. 1.7. Just because the pillar with the weight on top bends to the right sometimes and to the left other times does not mean there are no principles or rules. Indeed, the rule for the pillar is demonstrable in the differential equations describing the dynamics. It just may be that the different responses invoking differences in charity described different metastable states of the same underlying neurophysiology that is Chaotic and Complex. The consistency is in the dynamics of the underlying neurophysiology; it only is the multistability of the outputs, behaviors, that appears inconsistent.

Appiah recounts a study by Robert Baron and Jill Thomley (1994) demonstrating that a person would most likely honor a stranger's request for change in money for a parking meter if that person was outside a fragrant bakery shop than a "neutral-smelling dry-goods store." Appiah expresses a frustration that is endemic to ethics. In considering how the smell of a bakery affects compassion, Appiah perhaps finds it incredible that such a smell should have an effect; he appears to accept the fact, but is frustrated in the attempt to understand the effect. Appiah (2008, pp. 45 and 46) writes:

> Given that we are so sensitive to circumstances and so unaware of the fact, isn't it going to be wondrously difficult to develop compassion, say, as a character trait [author – taken as ethics]? We just can't keep track of all the cues and variables that may prove crucial to our compassionate response: the presence or absence of the smell of baking is surely just one among thousands of contextual factors that will have their way with us. However, if this is so, can I make myself disposed to do or to feel the right thing [author – taken as justice]? I have no voluntary control on how aromas affect me.

Yet the ethical affect of the smell of a bakery is just what one would expect from neurophysiology, demonstrating Chaos and Complexity. First, the olfactory system is strongly involved in the limbic system, the latter thought related to emotions (Barton, 2006; Barger et al., 2014; Ghaziri et al., 2018). Thus, there is a neurophysiological substrate by which smells can affect emotion, leading to behaviors affected through emotions such as empathy (Neumann et al., 2012; Wilkinson et al., 2016).

That the smell of a bakery, presumably such a minimal factor, should have such a marked effect on empathic behavior is entirely consistent with Chaos and Complexity. Consider the example of dropping grains of sand onto a pile. With each grain of sand dropped, the size of the pile increases in a regular (linear) manner proportionate to the size of each grain until a certain number of grains drop. The next grain of sand causes the pile to collapse, yet the laws of physics are no different for the previous grain of sand that did not cause the pile of sand to collapse from the next grain of sand that did collapse the pile. One could consider the grains of sand that are taken as all the factors Appiah alluded to in determining emphatic behavior, with smell becoming that last grain of sand that collapses the pile into an empathic behavior. That a single additional grain of sand should cause such a dramatic change in the behavior of the sand pile and that the aroma of a bakery shop should have such an affect on empathic behavior may seem paradoxical, but paradoxes are not the same as dilemmas that force a choice of alternative explications. The laws of physics do not change with the addition of the grain of sand and neurophysiology does not change when smells affect behavior.

The unconscious effect of the aroma of the bakery shop on empathic behavior appears frustrating for those holding ethics to be analytical in the sense of conscious, declarative, and deliberative knowledge. Appiah (2008, p. 43) goes on to write:

> The researchers, faced with these data, may hypothesize that when you're cheerful, you are more inclined to do what you think is helpful; that is, the fact that offering change is helpful will strike you as a particularly strong reason for doing it when you are feeling particularly cheerful. But this is not something that the agent herself will notice in the ordinary course of things. So, even though she's doing what's helpful, and even though she might give that as her explanation—"He needed help," she might say—there remains another sense in which she doesn't understand why she's doing what she's doing.

The aforementioned quote presupposes that ethics is somehow the result of conscious, deliberative, and declarative processing. This need not be the case, for example, see the quote by US Supreme Justice Potter Stewart addressing pornography cited earlier. The argument here is that justice is an affective neurophysiological response to acts or potential acts. Conscious, deliberative, and declarative descriptions are post hoc. There is no paradox and hence no dilemma in the neurophysiology. The paradox is in the conscious and declarative deliberations. It is the presumption that ethics solely operates in a conscious, deliberative, and declarative manner that translates the paradox into a dilemma.

Initial conditions and Principlism and Anti-Principlism

As seen from the discussions given earlier, justice (and consequently ethics and morality) may operate as Chaotic and Complex systems if for no other reason than the sense of justice is driven by a neurophysiology that demonstrates Chaos and Complexity. Assuming that such is the case, then ethics and morality will display sensitivity to initial conditions as was discussed in the context of the effects of smell on empathetic behavior. The argument is made that many of the differences between Principlism and Anti-Principlism ethical systems derive from the issue of initial conditions in Chaotic and Complex systems.

Anti-Principlist positions operate, even if implicitly, on the basis of sensitivity to initial conditions. The Principlist attempts to maintain the notion of a unifying coherent theory operating in each and every initial condition or situation and attempts to account for the paradoxes by invoking the importance of specification. The Principlist would not want to surrender the laws of physics even if the behavior of adding grains of sand results in an unpredictable collapse of the sand pile and similarly would not want to give up ethical principles and moral theories. The Anti-Principlist will point to the unpredictability of the sand pile collapse by denying the relevance of the laws of physics, if not denying the laws of physics themselves. It is possible that the dissonance between the Principlist and the Anti-Principlist would resolve at the neurophysiological level, if that were ever known, but perhaps the neurophysiological level can be approximated. The Principlist and Anti-Principlist ethics are inextricably linked at other levels, such being the opposite sides of the same epistemic coin as discussed in Chapter 2.

Neurophysiology and moral culpability

Can a machine, such as that which might embody machine learning AI, be morally culpable? This topic is addressed in Case 12. What about a machine constituted of carbon-based materials rather than silicon, such as the human? One certainly can deny that a machine can be morally culpable, but what would be the basis for the denial? One could say that any act of maleficence instigated by a machine can be blamed on the computer programmer or the user. However, with machine learning, the computer programmer may never actually know how the machine made the decision that resulted in maleficence. Certainly, the user may not know. Yet, if the victims of the maleficence are to be made whole, someone must be culpable.

Another basis for denying moral culpability on the part of the machine is that the machine is incapable of making those kinds of decisions and therefore would not be expected to make decisions that are liable for moral culpability. Yet that is not true, as machine learning AI is used to determine sentencing for persons found guilty of a crime and to decide who receives credit cards or loans. Perhaps a machine cannot be held culpable because what would it mean to reprimand or punish a machine? One could argue that only humans are culpable, but if humans are considered a type of machine, what is it that distinguishes the culpable human from the machine that is not culpable? Such a distinction would seem to necessitate an immaterial mind or soul independent of but instantiated in the human nervous system (see Case 10). However, if there is no distinction, then what does that say about the ability to render any moral culpability?

The same conundrums associated with assigning moral culpability to a human, as a neurophysiological machine, also confront psychology and ethical considerations based on ethical induction from psychology (see Appiah, 2008). Appiah (2008, pp. 101–110) explores these issues in his discussion of folk psychology. In ways to be discussed shortly, psychology may be less able to resolve these conundrums than neurophysiological explication. Appiah describes participants in a study who judged a particular person who cheats at paying taxes as less morally reprehensible (culpable) than a person who kills a husband in order to pursue the husband's wife; yet both violated the principle against maleficence (not paying taxes, thereby forcing others to pay more than their share, and murder, respectively). Note that the issue was not the severity of the violations; it was the degree of culpability.

Appiah cites Strawson's (1962) notion of "participant reactive attitudes," such as approbation, reproach, and gratitude, as frustrating principled ethics. Appiah cites these cases as being "affect laden" so as to elicit the participant reactive attitudes. Appiah then discounts the notion of affect, suggesting that those using such explanations attribute some ontology (reality) to affect. However, the point is that "affect" does have an ontology; it is in the neurophysiological mechanisms discussed previously. Appiah (2008, p. 106) writes:

> Knobe's findings are, as he recognizes, congruent with Strawson's arguments about
> reactive attitudes—in particular, the hovering thought that such attitudes may be the
> basis of our judgments about responsibility, rather than vice versa.

It is not clear whether Appiah agrees with Knobe and Strawson. However, if one substitutes "reactive attitudes" with "affective neurophysiological responses," then the position attributed to Knobe and Strawson is exactly the point of the argument in this chapter. What is different are the properties and dynamics of an affective neurophysiological response and those of psychological reactive attitudes. Even though the current state of neurophysiological mechanisms is unknown (and perhaps will never be known), reinterpreting the accessible psychological explications in the context of what is generally known about the neurophysiology likely will nonetheless improve the utility of psychology, particularly as it forms the basis for ethical induction to become ethical decision-making systems.

The dependency or sensitivity to initial conditions and the bifurcations to multiple metastable states described earlier for Chaotic and Complex systems may be applicable. The "reactive attitudes" that seem to be at odds with ethical principles may represent the initial conditions, just as the aroma of a bakery affected empathy as discussed previously. The different multiple metastable states represent outcomes that appear to be inconsistent on analysis at the psychological level or ethical conscious and deliberative descriptive level, yet they truly are consistent when viewed from the common underlying Chaotic and Complex neurophysiological mechanisms. While the opposing psychological or ethically descriptive inferences may seem irreconcilable, they are consistent with each other at the shared origin in the neurophysiology.

What can be said of Chaotic and Complex neurophysiological mechanisms underlying moral culpability? First, the exact outcome of Chaotic and Complex neurophysiological processing of a situation or circumstance is unpredictable. The notion of free will, a necessary prerequisite for moral culpability, would be a nonsensical claim [a category error according to Gilbert Ryle (1949)] in the case of unpredictability. Even in a determinant nonrandom process such as discussed previously regarding the weight atop a pillar (Fig. 1.7), unpredictability allows for the attribution of free will even if it is a categorical error. Second, neurophysiological processing, assuming that it is in the manner of machine learning AI with prior structure as described earlier, depends on the initial structure, perhaps as determined by genetics and intrautero experiences modified by postpartum experience.

None of the discussion invalidates the notion of moral culpability but perhaps can help reframe it. To be sure, humans and everything that humans depend on or require protection as there are behaviors that are unjust and counterproductive. When necessary, humans can compel vaccination against, treatment of, or quarantine from maleficent agents. It is not necessary to rationalize those actions based on the innocence of the victims and the badness of the maleficent agents, as these terms might be used as virtues or vices. Such a view likely would not change the ethics radically, how humans treat each other, but likely would change morality, perhaps to be less vindictive.

Conflation of neuropharmacology and genetics with neurophysiology—A categorical error

This chapter has argued for the importance of understanding neurophysiological mechanisms in the development of ethical systems. The ontology of the neurophysiological mechanisms is the change in the electrical potential across the cell membrane of excitable cells, particularly neurons, in the nervous system. Unfortunately, humans' abilities to fully explicate these mechanisms are severely limited, and perhaps will remain so. As a consequence, neuroscience has used surrogate markers—in modern terms, biomarkers. The danger is when the surrogate markers "masquerade" as neurophysiological mechanisms. Neurochemistry, and its technical implications in neuropharmacology, has a long history of masquerading as neurophysiology based on a false equivalence between the actions of neurotransmitters and the changes in the electrical potential across cell membranes of excitable cells (Valenstein, 2005). Consequently, there perhaps would be great risks inferring ethics from neuropharmacology, but the temptation appears irresistible.

Social behaviors, held analogous to ethical behaviors, are a subject of neuropharmacological studies. Manipulations of oxytocin levels in a variety of animals, particularly the prairie vole, have effects on mate and offspring social bonding and behaviors. This has led to questions such as "what does oxytocin do?" posed by Patricia Churchland (2019, p. 50). The answer provided by Churchland is a great many things. Churchland is skeptical of those taking neuropharmacology too literally, particularly as it relates to conscience but nevertheless appears quite sympathetic, writing:

> Since the 1950s the has been remarkable advances in neuropharmacology. The development of Thorazine literally resulted in closure of vast state mental health hospitals. The development of replacement of the natural neurotransmitter chemical, dopamine, with levodopa dramatically changed the fate of patients with Parkinson's disease. The list of such accomplishments is extensive. Similarly, use of chemical agents, such as natural neurotransmitters or their analogs, markedly affects behavior, such as social bonding and anticipating rewards thought relevant to ethical behaviors.
>
> The observations above and others lends credence to a neurobiological basis of ethical behaviors. Unfortunately, the discussions of neurochemicals as substrate for ethical behaviors is so simplistic that undermines any argument for a neurobiological basis. There are two prototypical examples, the neurotransmitters oxytocin for social bonding and parenting, and dopamine for anticipating rewards.
>
> *Churchland (2019)*

Yet, nonetheless, Churchland asks the question without apparent reticence or caveat. The following behaviors are attributed to oxytocin (Jenkins and Nussey, 1991):

human parturition, particularly onset of labor
milk ejection
ACTH secretion
prolactin and gonadotrophin secretion
osmoregulation
sexual activity and maternal instincts

The following list regarding oxytocin is from Argiolas and Gessa (1991):

penile erection
lordosis and copulatory behavior
yawning
memory and learning
tolerance and dependence mechanisms
feeding
grooming
cardiovascular regulation
thermoregulation

As the number of things a drug can do increases, one often goes from viewing the claims as incredible to incredulous. Fig. 1.10 shows the chemical structure of oxytocin. The question becomes, which part of the oxytocin molecule "does" human parturition, particularly onset of labor; milk ejection; ACTH secretion; prolactin and gonadotrophin secretion; osmoregulation; sexual activity and maternal instincts; penile erection; lordosis and copulatory behavior; yawning; memory and learning; tolerance and dependence mechanisms; feeding; grooming; cardiovascular regulation; and thermoregulation? If there is some part of the molecule that "does" each of the things just described, how does the brain know which part to pay attention to in order to do the right thing? If it is the entire molecule, how does the brain know which thing to do when presented with the molecule?

Likely there is no information contained within the oxytocin molecule that is specific for any of the things oxytocin is purported to do. Rather, it is where in the neurophysiology that the oxytocin is applied that determines what thing occurs, as recognized by Churchland. Thus, the answer to the question "what does oxytocin do?" is a category error. The actual question is, "what is the interaction

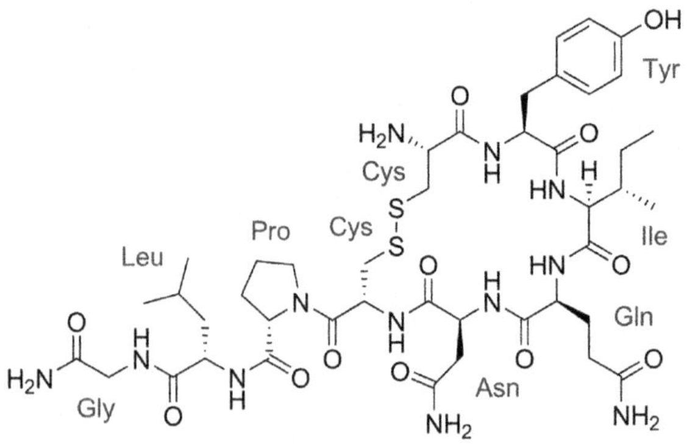

Fig. 1.10 Structure of the oxytocin molecule where the letters in *black* are atoms of hydrogen (H), oxygen (O), nitrogen (N), and sulfur (S) and those in other letters are labels for various amino acids, the specifics are unnecessary for the point to be made. By Edgar181—Own work, Public Domain, https://commons.wikimedia.org/w/index.php?curid=8482270.

between oxytocin and specific changes in the electrical potential across the cell membrane of excitable cells, particularly neurons, that comprises the neurophysiological mechanisms?"

Advances in genetics and molecular neurobiology have, to an increasing degree, supplanted neuropharmacology with genetics. Now there often is spoken "the genetics of social behavior" such as related to the genetics of oxytocin. Nevertheless, the same issue arises for the gene as for oxytocin. The gene, presumably, is in every cell in the body that contains nuclear DNA (red blood cells do not, for example). Thus, the oxytocin gene, alone, can do no more than the oxytocin molecule alone. This led Landis and Insel, former directors of the National Institute of Neurological Disease and Stroke and the National Institute of Mental Health, respectively, to write:

> Already it is clear that, for the study of behavior, genomics is not destiny. Indeed, if genomic sequence "determines" anything behaviorally, it determines diversity. It is important that we be wary about extrapolating from model organisms to humans. We must also avoid using small statistical associations to make grand claims about human nature. Obviously, we have much to discover before understanding how genes influence behavior—a discovery process that will closely involve the brain.
>
> *Landis and Insel (2008)*

Several approaches held sufficient to understand behavior without any deference to actual neurophysiology historically included first psychology, then neuropharmacology, and now neurogenetics. Yet it points to a feature of humans to value an intuitive appealing supposition even when contrary to fact (Johnson-Laird, 2008). The same caveat would seem valuable in ethical reasoning.

References

Appiah, K.A., 2008. Experiments in Ethics. Harvard University Press.

Aquinas, T., 1485. Summa theologica.

Argiolas, A., Gessa, G.L., 1991. Central functions of oxytocin. Neurosci. Biobehav. Rev. 15 (2), 217–231.

Aristotle, 2018. In: Miller, Fred D. Jr (Eds.), On the Soul and Other Psychological Works. Oxford World's Classics, Oxford.

Armstrong, G.L., Conn, L.A., Pinner, R.W., 1999. Trends in infectious disease mortality in the United States during the 20th century. JAMA 281 (1), 61–66.

Barger, N., Hanson, K.L., Teffer, K., Schenker-Ahmed, N.M., Semendeferi, K., 2014. Evidence for evolutionary specialization in human limbic structures. Front. Hum. Neurosci. 8, 277. https://doi.org/10.3389/fnhum.2014.00277.

Baron, R.A., Thomely, J., 1994. A Wiff of reality: positive affect as a potential mediator of effects of pleasant fragrances on task performance and helping. Environ. Behav. 26, 766–784.

Barton, R.A., 2006. Olfactory evolution and behavioral ecology in primates. Am. J. Primatol. 68 (6), 545–558.

Beauchamp, T., Childress, J.F., 2013. Principles of Bioethics, seventh ed. Oxford University Press, Oxford.

Belmont Report: Ethical Principles and Guidelines for the Protection of Human Subjects of Research, National Commission for the Protection of Human Subjects of Biomedical and Behavioral Research, Department of Health, Education and Welfare. 1978. United States Government Printing Office, Washington, DC.

Benuzzi, F., Lui, F., Ardizzi, M., Ambrosecchia, M., Ballotta, D., Righi, S., Pagnoni, G., et al., 2018. Pain mirrors: neural correlates of observing self or others' facial expressions of pain. Front. Psychol. 9, 1825. https://doi.org/10.3389/fpsyg.2018.01825.

Bortolini, T., Bado, P., Hoefle, S., Engel, A., Zahn, R., de Oliveira Souza, R., Dreher, J.C., Moll, J., 2017. Neural bases of ingroup altruistic motivation in soccer fans. Sci. Rep. 7 (1), 16122. https://doi.org/10.1038/s41598-017-15385-7.

Brosnan, S.F., 2013. Justice- and fairness-related behaviors in nonhuman primates. Proc. Natl. Acad. Sci. U. S. A. 110 (Suppl. 2), 10416–10423. https://doi.org/10.1073/pnas.1301194110.

Buckholtz, J.W., Marois, R., 2012. The roots of modern justice: cognitive and neural foundations of social norms and their enforcement. Nat. Neurosci. 15, 655–661.

Campbell, J.K., O'Rourke, M., Slater, M.H. (Eds.), 2011. Carving Nature at Its Joints: Natural Kinds in Metaphysics and Science. MIT Press, Cambridge, MA.

Cao, Y., Contreras-Huerta, L.S., McFadyen, J., Cunnington, R., 2015. Racial bias in neural response to others' pain is reduced with other-race contact. Cortex 70, 68–78. https://doi.org/10.1016/j.cortex.2015.02.010.

Carroll, L., 1871. Through the Looking-Glass and What Alice Found There.

Chang, S.W.C., Brent, L.J.N., Adams, G.K., Klein, J.T., Pearson, J.M., Watson, K.K., Platt, M.L., 2013. Neuroethology of primate social behavior. Proc. Natl. Acad. Sci. U. S. A. 110 (Suppl. 2), 10387–10394.

Chen, P., Hong, W., 2018. Neural circuit mechanisms of social behavior. Neuron 98 (1), 16–30. https://doi.org/10.1016/j.neuron.2018.02.026.

Churchland, P., 2019. Conscience: The Origins of Moral Intuition. W.W. Norton & Company, p. 50.

Couzin-Frankel, J., 2019. Medicine contends with how to use artificial intelligence. Science 364 (6446), 1119–1120. https://doi.org/10.1126/science.

Cummings, M.A., 2015. The neurobiology of psychopathy: recent developments and new directions in research and treatment. CNS Spectr. 20 (3), 200–206. https://doi.org/10.1017/S1092852914000741.

Cushman, F., Murray, D., Gordon-McKeon, S., Wharton, S., Greene, J.D., 2012. Judgment before principle: engagement of the frontoparietal control network in condemning harms of omission. Soc. Cogn. Affect. Neurosci. 7 (8), 888–895. https://doi.org/10.1093/scan/nsr072.

Decety, J., Yoder, K.J., 2017. The emerging social neuroscience of justice motivation. Trends Cogn. Sci. 21 (1), 6–14. https://doi.org/10.1016/j.tics.2016.10.008.

Demertzi, A., Liew, C., Ledoux, D., Bruno, M.A., Sharpe, M., Laureys, S., Zeman, A., 2009. Dualism persists in the science of mind. Ann. N.Y. Acad. Sci. 1157, 1–9. https://doi.org/10.1111/j.1749-6632.2008.04117.x.

Doyle, A.B., Aboud, F.E., 1995. A longitudinal study of white children's racial prejudice as a social-cognitive development. Merrill-Palmer Q. 41 (2), 209–228.

Earnest, M.A., Wong, S.L., Federico, S.G., 2010. Perspective: physician advocacy: what is it and how do we do it? Acad. Med. 85 (1), 63–67. https://doi.org/10.1097/ACM.0b013e3181c40d40.

Edelman, G.M., 1987. Neural Darwinism: The Theory of Neuronal Group Selection, second ed. Basic Books.

Fox, D.M., 2005. AIDS and the American health polity: the history and prospects of a crisis of authority. Milbank Q. 83 (4), https://doi.org/10.1111/j.1468-0009.2005.00432.x.

Fox, P.T., Friston, K.J., 2012. Distributed processing; distributed functions? Neuroimage 61 (2), 407–426. https://doi.org/10.1016/j.neuroimage.2011.12.051.

Freudenthal, G., 1995. Aristotle's Theory of Material Substance: Heat and Pneuma, Form and Soul. Oxford University Press, Oxford.

Gandhi, T., Kalia, A., Ganesh, S., Sinha, P., 2015. Immediate susceptibility to visual illusions after sight onset. Curr. Biol. 25 (9), R358–R359. https://doi.org/10.1016/j.cub.2015.03.005.

Ghaziri, J., Tucholka, A., Girard, G., Boucher, O., Houde, J.C., Descoteaux, M., Obaid, S., et al., 2018. Subcortical structural connectivity of insular subregions. Sci. Rep. 8 (1), 8596. https://doi.org/10.1038/s41598-018-26995-0.

Giner-Sorolla, R., 2014. Intuition in 21st-century moral psychology. In: Osbeck, L.M., Held, B.S. (Eds.), Rational Intuition, Philosophical Roots, Scientific Investigations. Cambridge University Press, pp. 338–361. (Chapter 15).

Gould, S.J., 1996. Full House: The Spread of Excellence From Plato to Darwin. Harmony Books.

Grabenhorst, F., Báez-Mendoza, R., Genest, W., Deco, G., Schultz, W., 2019. Primate amygdala neurons simulate decision processes of social partners. Cell 177 (4), 986–998. e15.

Greene, J., 2013. Moral Tribes: Emotion, Reason, and the Gap Between Us and Them. The Penguin Press.

Greene, J.D., Sommerville, R., Nystrom, L.E., Darley, J.M., Cohen, J.D., 2001. An fMRI investigation of emotive engagement in moral judgment. Science 293, 2105–2108.

Greene, J.D., Nystrom, L.E., Engell, A.D., Darley, J.M., 2004. The neural bases of cognitive conflict and control in moral judgment. Neuron 44, 389–400.

Gregory, R.L., 2005. The Medawar Lecture 2001 knowledge for vision: vision for knowledge. Philos Trans R Soc Lond B Biol Sci. 360 (1458): 1231–1251.

Grice-Jackson, T., Critchley, H.D., Banissy, M.J., Ward, J., 2017. Common and distinct neural mechanisms associated with the conscious experience of vicarious pain. Cortex 94, 152–163. https://doi.org/10.1016/j.cortex.2017.06.015.

Haruno, M., Frith, C.D., 2010. Activity in the amygdala elicited by unfair divisions predicts social value orientation. Nat. Neurosci. 13 (2), 160–161.

Hebb, D.O., 1949. The Organization of Behaviour: A Neuropsychological Theory. John Wiley and Sons, New York.

Hobbes, T., 1651. Leviathan or the matter, forme and power of a common-wealth ecclesiasticall and civil.

James, W., 1890. Discrimination and comparison. In: Principles of Psychology. New York, Henry Holt and Company (Chapter 13).

Jenkins, J.S., Nussey, S.S., 1991. The role of oxytocin: present concepts. Clin. Endocrinol. (Oxf.) 34 (6), 515–525.

Jensen, K., Vaish, A., Schmidt, M.F., 2014. The emergence of human prosociality: aligning with others through feelings, concerns, and norms. Front. Psychol. 5, 822.

Johnson-Laird, P., 2008. How We Reason. Oxford University Press, Oxford.

Kato, M., Mugitani, R., 2015. Pareidolia in infants. PLoS One 10 (2), e0118539. https://doi.org/10.1371/journal.pone.0118539.

Kendler, K.S., Lönn, S.L., Maes, H.H., Sundquist, J., Sundquist, K., 2015a. The etiologic role of genetic and environmental factors in criminal behavior as determined from full-and half-sibling pairs: an evaluation of the validity of the twin method. Psychol. Med. 45 (9), 1873–1880. https://doi.org/10.1017/S0033291714002979.

Kendler, K.S., Maes, H.H., Lönn, S.L., Morris, N.A., Lichtenstein, P., Sundquist, J., Sundquist, K., 2015b. A Swedish national twin study of criminal behavior and its violent, white-collar and property subtypes. Psychol. Med. 45 (11), 2253–2262. https://doi.org/10.1017/S0033291714002098.

Kennedy, D.P., Adolphs, R., 2012. The social brain in psychiatric and neurological disorders. Trends Cogn. Sci. 16 (11), 559–572. https://doi.org/10.1016/j.tics.2012.09.006.

Kotzee, B., Ignatowicz, A., Thomas, H., 2017. Virtue in medical practice: an exploratory study. HEC Forum 29 (1), 1–19. https://doi.org/10.1007/s10730-016-9308-x.

Lambek, M. (Ed.), 2010. Ordinary Ethics: Anthropology, Language and Action. Fordham University Press, New York.

Landis, S., Insel, T.R., 2008. The "neuro" in neurogenetics. Science 322 (5903), 821. https://doi.org/10.1126/science.1167707.

Lee, K., Quinn, P.C., Pascalis, O., 2017. Face race processing and racial bias in early development: a perceptual-social linkage. Curr. Dir. Psychol. Sci. 26 (3), 256–262. https://doi.org/10.1177/0963721417690276.

Liu, S., Xiao, N.G., Quinn, P.C., Zhu, D., Ge, L., Pascalis, O., Lee, K., 2015. Asian infants show preference for own-race but not other-race female faces: the role of infant caregiving arrangements. Front. Psychol. 6, 593. https://doi.org/10.3389/fpsyg.2015.00593.

Macmillan, M., 1992. Inhibition and the control of behavior: from Gall to Freud via Phineas Gage and the frontal lobes. Brain Cogn. 19 (1), 72–104.

Marsh, A.A., Stoycos, S.A., Brethel-Haurwitz, K.M., Robinson, P., VanMeter, J.W., Cardinale, E.M., 2014. Neural and cognitive characteristics of extraordinary altruists. Proc. Natl. Acad. Sci. U. S. A. 111 (42), 15036–15041. https://doi.org/10.1073/pnas.1408440111.

Mesulam, M.M., 1998. From sensation to cognition. Brain 121 (Pt 6), 1013–1052.

Miller, F.D., 2018. Aristotle: On the Soul and Other Psychological Works. Oxford University Press, Oxford.

Miller, L.C., Bard, K.A., Juno, C.J., Nadler, R.D., 1990. Behavioral responsiveness to strangers in young chimpanzees (Pan troglodytes). Folia Primatol. (Basel) 55 (3–4), 142–155.

Miresco, M.J., Kirmayer, L.J., 2006. The persistence of mind-brain dualism in psychiatric reasoning about clinical scenarios. Am. J. Psychiatry 163 (5), 913–918.

Mobbs, E.J., Mobbs, G.A., Mobbs, A.E., 2016. Imprinting, latchment and displacement: a mini review of early instinctual behaviour in newborn infants influencing breastfeeding success. Acta Paediatr. 105 (1), 24–30. https://doi.org/10.1111/apa.13034.

Montgomery Jr., E.B., 2004. Dynamically coupled, high-frequency reentrant, non-linear oscillators embedded in scale-free basal ganglia-thalamic-cortical networks mediating function and deep brain stimulation effects. Nonlinear Stud. 11, 385–421.

Montgomery Jr., E.B., 2017. Deep Brain Stimulation Programming: Mechanisms, Principles and Practice, second ed. Oxford University Press, Oxford.

Montgomery Jr., E.B., 2019a. Dynamic. In: Medical Reasoning: The Nature and Use of Medical Knowledge. Oxford University Press, Oxford. (Chapter 14).

Montgomery Jr., E.B., 2019b. Reproducibility in Biomedical Research: Epistemological and Statistical Problems. Academic Press, New York.

Montgomery Jr., E.B., Clare, M.H., Sahrmann, S., Buchholz, S.R., Hibbard, L.S., Landau, W.M., 1992. Neuronal multipotentiality: evidence for network representation of physiological function. Brain Res. 580 (1–2), 49–61.

Neumann, D., Zupan, B., Babbage, D.R., Radnovich, A.J., Tomita, M., Hammond, F., Willer, B., 2012. Affect recognition, empathy, and dysosmia after traumatic brain injury. Arch. Phys. Med. Rehabil. 93 (8), 1414–1420. https://doi.org/10.1016/j.apmr.2012.03.009.

Nguyen, D.H., Widrow, B., 1990. Neural networks for self-learning control systems. IEEE Contr. Syst. Mag. 18–23.

Nichols, S., Knobe, J., 2007. Moral responsibility and determinism: the cognitive science of folk intuitions. Noûs 41 (4), 663–685.

Oatley, K., Jenkins, J.M., 1992. Human emotions: function and dysfunction. Annu. Rev. Psychol. 43, 55–85.

Parks, R.W., Long, D.L., Levine, D.S., Crockett, D.J., McGeer, E.G., McGeer, P.L., Dalton, I.E., et al., 1991. Parallel distributed processing and neural networks: origins, methodology and cognitive functions. Int. J. Neurosci. 60 (3–4), 195–214.

Porter, D., 2007. Doctors, the state, and the ethics of political medical practice. Virtual Mentor 9 (12), 832–837. https://doi.org/10.1001/virtualmentor.2007.9.12.msoc1-0712.

Prut, Y., Perlmutter, S.I., Fetz, E.E., 2001. Distributed processing in the motor system: spinal cord perspective. Prog. Brain Res. 130, 267–278.

Quinn, P.C., Lee, K., Pascalis, O., 2019. Face processing in infancy and beyond: the case of social categories. Annu. Rev. Psychol. 70, 165–189. https://doi.org/10.1146/annurev-psych-010418-102753.

Raine, A., 2018. Antisocial personality as a neurodevelopmental disorder. Annu. Rev. Clin. Psychol. 14, 259–289. https://doi.org/10.1146/annurev-clinpsy-050817-084819.

Raine, A., Yang, Y., 2006. Neural foundations to moral reasoning and antisocial behavior. Soc. Cogn. Affect. Neurosci. 1 (3), 203–213. https://doi.org/10.1093/scan/nsl033.

Rawls, J., 1955. Two concepts of rules. Philos. Rev. 64 (1), 3–32.

Rawls, J., 1999. A Theory of Justice, revised ed. The Belknap Press of Harvard University Press, Cambridge, MA.

Ruff, C.C., Fehr, E., 2014. The neurophysiology of rewards and values in social decision making. Nat. Rev. Neurosci. 15, 549–562.

Rumelhart, D.E., McClelland, J.L., PDP Research Group, 1986. Parallel Distributed Processing, Explorations in the Microstructure of Cognition: Foundations. A Bradford Book. vol. 1. MIT Press, Cambridge, MA.

Ryan, C.W., Turkstra, L.S., 2019. Sex-Based Differences in Emotion Recognition in Context in Typical Adults. . (Unpublished Data).

Ryle, G., 1949. The Concept of Mind. University of Chicago Press, Chicago.

Sabatello, M., Appelbaum, P.S., 2017. Behavioral genetics in criminal and civil courts. Harv. Rev. Psychiatry 25 (6), 289–301. https://doi.org/10.1097/HRP.0000000000000141.

Schaich Borg, J., Hynes, C., Van Horn, J., Grafton, S., Sinnott-Armstrong, W., 2006. Consequences, action, and intention as factors in moral judgments: an FMRI investigation. J. Cogn. Neurosci. 18 (5), 803–817.

Schmidt, M.F., Sommerville, J.A., 2011. Fairness expectations and altruistic sharing in 15-month-old human infants. PLoS One 6 (10), e23223. https://doi.org/10.1371/journal.pone.0023223.

Schwarz, K.A., Wieser, M.J., Gerdes, A.B., Mühlberger, A., Pauli, P., 2013. Why are you looking like that? How the context influences evaluation and processing of human faces. Soc. Cogn. Affect. Neurosci. 8 (4), 438–445. https://doi.org/10.1093/scan/nss013.

Simion, F., Giorgio, E.D., 2015. Face perception and processing in early infancy: inborn predispositions and developmental changes. Front. Psychol. 6, 969. https://doi.org/10.3389/fpsyg.2015.00969.

Smithson, J., 2007. English Wikipedia, transferred from en.Wikipedia to Commons by P. Hansen.

Stanley, D.A., Adolphs, R., 2013. Toward a neural basis for social behavior. Neuron 80, 816–826.

Stone, N.C., Leigh, N.W.C., 2019. A statistical solution to the chaotic, non-hierarchical three-body problem. Nature 576, 406–410.

Strawson, P.F., 1962. Freedom and resentment (philosophical lecture). In: Proceedings of the British Academy. vol. 48, pp. 187–211.

Tabibnia, G., Satpute, A.B., Lieberman, M.D., 2008. The sunny side of fairness: preference for fairness activates reward circuitry (and disregarding unfairness activates self-control circuitry). Psychol. Sci. 19 (4), 339–347.

Vaish, A., Carpenter, M., Tomasello, M., 2010. Young children selectively avoid helping people with harmful intentions. Child Dev. 81 (6), 1661–1669. https://doi.org/10.1111/j.1467-8624.2010.01500.x.

Valenstein, E.S., 2005. The War of the Soups and the Sparks: The Discovery of Neurotransmitters and the Dispute Over How Nerves Communicate. Columbia University Press.

van Wolkenten, M., Brosnan, S.F., de Waal, F.B., 2007. Inequity responses of monkeys modified by effort. Proc. Natl. Acad. Sci. U. S. A. 104 (47), 18854–18859.

Vaughn, D.A., Savjani, R.R., Cohen, M.S., Eagleman, D.M., 2018. Empathic neural responses predict group allegiance. Front. Hum. Neurosci. 12, 302. https://doi.org/10.3389/fnhum.2018.00302.

Voosen, P., 2019. A 2-week weather forecast may be as good as it gets. Science 363 (6429), 801. https://doi.org/10.1126/science.363.6429.801.

Wilkinson, D., Moreno, S., Ang, C.S., Deravi, F., Sharma, D., Sakel, M., 2016. Emotional correlates of unirhinal odour identification. Laterality 21 (1), 85–99. https://doi.org/10.1080/1357650X.2015.1075546.

Zhang, X., Xu, Q., Jiang, Y., Wang, Y., 2017. The interaction of perceptual biases in bistable perception. Sci. Rep. 7, 42018. https://doi.org/10.1038/srep42018.

Ethical systems

2

To be argued...

On one interpretation, justice results from adjudicating the degree of ethical obligations to conflicting parties. In one system, a form of Principlism, the obligations are to the ethical principles of beneficence, nonmaleficence, and autonomy. Other systems may hold different principles, for example, the autonomy of the individual may be supplanted by the autonomy of a group, such as a family unit (see Case 13). Other systems may reject the notion of any principles and justice is achieved by the extrapolation of relevant precedents or the fulfillment of virtue. These alternatives, termed here as Anti-Principlist, include narrative, phenomenological, virtue, care, and some versions of feminist ethics. Principlism has the virtue of generalizability of its applications, whether valid or not, by combinatorics on an economical set of principles. Principlism has the lability of being seen as a forced fit. Alternatively, Anti-Principlism can be seen as more accommodating to the particular individuals but risks being reduced to the Solipsism of the Present Moment when different individuals, situations, and circumstances are encountered. To avoid solipsism, an economical set of precedents are collected to serve as a benchmark for justice. But how this is different from principles remains unclear.

The perhaps unasked question is whether these different approaches are really all that different. While they may seem very different in implementation, it will be argued that they share the same origin in the fundamental conundrum that confronts any empirically based knowledge domain. Modern ethics in a pluralistic modern liberal democracy requires derivation and validation of any ethics based on the consensus of those to be governed, hence it is empirical. This is true even if the consensus of those governed is to off-load the authorities and responsibilities to proxies. The experiences of those to be governed are the empirical substrates for the development of ethical systems. Yet, those experiences are highly variable; indeed, no two experiences are exactly alike, thereby creating a potentially infinite variety. How to deal with this variety hinges on the ontological nature of the variety. Regardless of the ontological status, an epistemic choice is required. It is the epistemic response to that conundrum that drives the specifics of the various ethical systems, defined here as the methods to adjudicate justice.

Either the potentially infinite variety of ethical experiences is taken as diversity, where each case is unique and de novo, or the variety is variations on a set of economical principles. A problem for the Anti-Principlist adjudicating resolutions to ethical conflicts is that there must be some benchmark. This may be a case precedence but this means that the case of concern cannot be unique or de novo, as otherwise how would a case precedent be applicable? The case of concern must be sufficiently similar to a previous case for the previous case to act as a precedent. But how to define

The Ethics of Everyday Medicine. https://doi.org/10.1016/B978-0-12-822829-6.00027-8

"similar enough?" Thus, the Anti-Principlist must "borrow" some means from the Principlists in order to establish case precedents.

The challenge to the Principlist approach is that the empirical substrate of particular experiences has to be organized in order to induce general principles. Relevant experiences must be separated from those irrelevant. However, the criteria for relevance must be evaluated prior to the induction from the experiences. As the criteria for relevance is an abstraction, so as to apply to the variety of experiences, Principlism risks "begging the question," presuming the very abstractions it is meant to discover. Clearly, the process is the (1) application of some principles to select relevant particular experiences, (2) induction of different principles, and (3) testing of the principles against particular experiences. The fascinating question is where did the principles used in step 1 come from? It is argued in Chapter 1 that those "first principles or first philosophy" lie in the neurophysiology.

The process contains epistemic components of both Principlism and Anti-Principlism, with the application of principles to collect relevant experiences and then induction from the particulars, each understood from an Anti-Principlist perspective. Indeed, the development of ethical systems is an iterative process between the two approaches as both are responses to the common fundamental ontological conundrum.

Variety as diversity or variation

Justice demands action (see Chapter 1) in any situations or circumstances where different parties vie for resources in the setting of moderate scarcity. How to adjudicate such disputes depends on the ethical system held to prevail. The challenge is to create an ethical system that will be appropriate, independent of the particular, and applicable to all particulars. Given that multiple ethical systems are possible, how to determine which is to prevail so that justice can be done? Assuming a pluralistic modern liberal democracy, acceptable ethical systems cannot just be based on received knowledge, such as the word of God or the dictates of an absolute sovereign, as they are "inadmissible." Interestingly, sovereigns before the work of Thomas Hobbes (1651) claimed a received knowledge subsumed under the divine right of kings. Since Hobbes, most consider governance as a form of social contract validated by the will of those governed.

The challenge is defining the consensus that will shape the ethical system. As the consensus is based on the experiences of those governed, directly or through some form of representation, establishing the consensus is an empiric process, that is, observation of the experiences of those governed. In this sense, modern ethics in the setting of a pluralistic modern liberal democracy is little different from other empirically derived knowledge domain, such as science, although perhaps different in the methodologies applied (see Chapter 3). Theories of ethics face the same fundamental ontological conundrum that requires an epistemic choice as any other empirically derived knowledge, such as medicine and biological science (Montgomery, 2019a, 2019b).

Modern ethics depends on induction. The ethical theories under consideration are those derived from experiences of human actions and are first categorized as just or unjust from which principles, theories, laws, case precedents, and virtues are subsequently derived (see Chapter 1).

If the variety of ethical experiences is diversity, where each experience is considered de novo and unique, then rendering generalizations to past, present, and future, as well as to other situations and circumstances, is problematic. Alternatively, is the variety just variations on an economical set of fundamental principles and each experience is just some combinatorics on the set? If so, then past, present, and future can be understood as some combination of the fundamental principles. How this conundrum plays out in the development of ethical systems will be explored. However, this conundrum and how to gain certainty of presumed knowledge are common to all empirically derived knowledge domains. This conundrum has driven medicine since the very beginning, with the response to the conundrum determining whether the Rationalist/Allopathic/Regular school (today's medical doctor) or the Empiric/Irregular school (today's homeopath) prevails (Montgomery, 2019a). As will be shown, the epistemic choice affects not only the nature of medical knowledge, but also the ethics of everyday medicine.

The Rationalist/Allopathic/Regular school held that the variety of patients' manifestations of health and disease is just individual variations on an economical set of fundamental principles. Thus, the Allopathic physician would reduce the particular manifestations of a particular patient to a value on limited number dimensions reflecting principles of physiology and pathophysiology. The Empiric/Irregular school held that variety is diversity, which cannot be reduced to an economical set of principles. Rather, manifestations of the individual case are just matched to a set of properties associated with a potential treatment. The process essentially is pattern matching rather than a type of deduction. In homeopathy, the principle is "like treats like," if disease A has manifestation B and agent C also produces manifestation B, the patient is treated with small and then increasing doses of agent C. As discussed in detail later, in ethics, the Allopathic approach would be analogous to the Principlist approach, while the Empiric approach would be analogous to the narrative, phenomenological, care, and feminist approach to ethics, at least in some versions, and is subsumed under the term Anti-Principlist.

Modern science, particularly since the founding of the Royal Society in 1660, has defaulted to variety as variations on economical fundamental principles, such as natural laws in physics. The dominance of this approach led to the enthusiastic embrace of scientism and the Enlightenment. In philosophy, scientism led to the development of Logical Positivism, where the theme was that all of human knowledge could be reconstructed on the basis of self-evident axioms and rules of inference to relate axioms. This theme was extended to Logical Empiricism, where all of scientific knowledge similarly could be reconstructed from fundamental irreducible observations and rules of inference to relate the observations. This same scientism can be seen in the Principlist approach to medical ethics, such as that championed by Beauchamp and Childress (2013). As Francis Bacon commented in his "Novum Organum" (1620), scientific principles were induced from sensory stimulation associated with the scientific experience (what would it mean to deny that scientific experiences were not determined sensorially), thus scientific principles are the perceptions that derive from the stimulations, just as the percept of a cat follows from but is not identical to the sensory experience of a cat. The argument here is that the specific situations and circumstances of human ethical interactions are much like sensory stimuli, while the sense of justice from observing those specific situations and circumstances is analogous to perception (Chapter 1).

Persons only experience particulars, adding to the conundrum. One does not experience generalizations, such as principles qua principle (principle as principle). Yet, generalizations, as principles, case precedents, or virtues, are necessary as they are transcendent, meaning they cover a range of experiences. As no experiences in any two cases are exactly the same, there must be something general or transcendent in the case taken as the precedent for other cases. As the past and present are not exactly the same, there must be something found in the past that is transcendent to explain the present and to predict the future, for example, a specific ethical case that must be adjudicated. There must be something that exists in some manner in the different individual experiences, yet at the same time is relevant to all the different individual experiences. Yet principles, theories, case precedents, and virtues cannot be isomorphic with any one particular experience.

When a person is asked to consider a cat, one does so without picturing a particular cat, usually. The fact is that no two cats are exactly the same, yet the concept or percept "cat" has to be applicable to all cats, no matter the variations in size, shape, and coloration. As such, the variety of cats is reduced to a single percept, the principle, theory, case precedent, or virtue of "cat." Transcendent principles, theories, case precedents, and virtues are necessary to afford the reproducibility of perceptions—knowledge extracted from sense experience. Thus, whatever particular cat is seen, the same principle, theory, case precedent, or virtue of "cat" operates, and each and every different cat is seen as belonging to the principle, theory, case precedent, or virtue "cat." In an important sense, the repeated perception of the principle, theory, case precedent, or virtue "cat," despite the infinite variety of particular cats, demonstrates the reproducibility of the percept "cat." It is no different for principles, theories, case precedents, or virtues that make up ethics and morality.

Reproducibility is necessary for survival, for example, the ability to recognize a poisonous snake, no matter the variety of shapes and perspectives of the day on which the snake was encountered. In some sense, there must be a canonical snake (snake in principle, theory, precedent, or virtue) in a manner that allows the perception of a snake whenever there are sensory stimuli of any particular snake. This is not to suggest that the canonical snake has some ontological existence, such as a Platonic Form. Nevertheless, the range of behaviors generated by encounters with a number of poisonous snakes is sufficiently similar to imply a canonical behavioral response as though there was a canonical snake. It is the response that is canonical, not the experience. The sense of justice is a canonical response due to an affective (emotional) neurophysiological response (see Chapter 1). Similarly, behaviors over a range of specific ethical situations and conditions are sufficiently similar to suggest a canonical ethical or moral response. Without the transcendent principles, theories, case precedents, or virtues, perception becomes limited by the Solipsism of the Present Moment—what is true is only what is experienced immediately. Human perception would be reduced to William James' description of the baby "assailed by eyes, ears, nose, skin, and entrails at once, feels it all as one great blooming, buzzing confusion" (James, 1890). The specifics of establishing transcendental principles, theories, case precedents, and virtues and the means to have confidence are discussed in Chapter 3.

As will be discussed later, justice requires ethical systems to establish normative claims, what people should do. Thus, any normative claim must be transcendental regardless of the ethical system considered. It is clear that Principlism readily admits of the transcendental but it is not clear that Anti-Principlism, such as narrative and phenomenological and versions of care and feminist ethics, do not also depend on the transcendental, although persons holding the Anti-Principlist perspective might object to the use of the term "principle." Thus, the different ethical systems will be seen as some aggregate of variety as diversity or variety as variation. But as will be shown, there is a common dependence on a prior principle, theory, case precedent, or virtue (the "first philosophy") to organize experiences from which transcendental principles, theories, case precedents, and virtues can be induced. Hence, the different ethical systems are not and perhaps should not be seen as antithetical. As such, the ethics of everyday medicine might be advanced by an integration of the ethical systems.

Transcendent principles, theories, case precedents, and virtues and the *A Priori* Problem

Some sort of transcendental principle is necessary just to start to make sense of the observations of ethical experiences, and thus must be prior to, *a priori*, the experiences. Considering sensory experiences may be a helpful metaphor. For example, why should a particular subset of sensations, such as certain photons falling on the retina, be grouped together and considered separate from others so as to recognize a dog? Why are not other stimuli, such as grass or leaves, considered part of the dog? The necessity of some prior organizing principle is the *A Priori* Problem for Induction. Consider Fig. 2.1 containing the figure of a dog. At first it is not at all clear which blobs of black are to be taken together to illustrate the dog. Next, look at Fig. 2.2 with the page upside down and the dog becomes perceptible shaded in *blue or gray*. Finally, when looking back again at Fig. 2.1, one almost immediately perceives the dog. Seeing Fig. 2.2 with the dog upside down shows the general location and orientation of the dog, which then serves as the *a priori* when Fig. 2.1 is looked at a second time.

To illustrate the notion of the transcendental, one can close the eyes and think of a triangle. It is likely that what is perceived as a triangle in one's consciousness is not a particular triangle, such as that shown in Fig. 2.3. Yet all the images shown in Fig. 2.3 would easily be considered a "triangle" even though the thought of a triangle is not isomorphic with any of the images in Fig. 2.3. Further, it is unlikely that the thought of a triangle would be any particular triangle but a transcendental notion of all triangles, including those physically impossible, such as that shown in Fig. 2.3B. Indeed,

Fig. 2.1 There is a dog somewhere in this abstract figure. The challenge is to find it (Gregory, 2005).

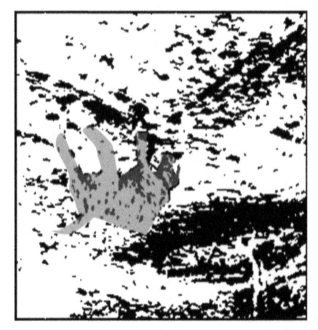

Fig. 2.2 The image is the same as in Fig. 2.1; however, the dog is outlined in *blue (light gray in print version)* and is upside down.

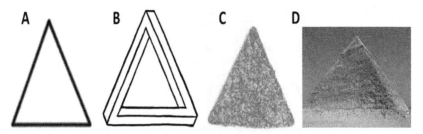

Fig. 2.3 When asked to close one's eyes and visualize a triangle, is that triangle any of the ones shown in this figure? Likely it is not; indeed, it is likely that no specific triangle is visualized but rather some general notion consistent with Plato's notion of Ideal Forms.

Plato called such figures transcendental, such as the thought triangle as Ideal Forms (sometimes referred to as Platonic Forms). Similarly, transcendental ethical forms, be they principles, theories, laws, case precedents, or virtues, inform the organization of experiences from which other principles, theories, laws, case precedents, and virtues are inducted, those new principles, theories, laws, case precedents, and virtues inform judgments of new experiences.

From the other side of the conundrum, the principles, theories, laws, case precedents, and virtues are transcendent and may not be isomorphic with any actual experience. In this case, the transcendent principles, theories, laws, case precedents, and virtues are "disconnected" from the actual experiences that constitute reality (ontology). The interesting question is whether the transcendental principles, theories, laws, case precedents, or virtues have any ontological status or are only epistemological devices. For example, the average height of a group of persons may be 5 ft 6 in. (168 cm), although no one in the group is actually 5 ft 6 in. (168 cm). The average (mean) height is an abstraction, it cannot be any kind of reality as there is no real experience of any person in the group being 5 ft 6 in. (168 cm) (see Chapter 8 in Montgomery, 2019a). The average in this case is an abstraction as no person in the group has a height of 5 ft 6 in. (168 cm); this average height is thought somehow to "represent" the group of persons whose height has been measured. In a sense, the average was inducted from the group of specific observations. Further, the abstraction of the average is thought to be relevant in some manner to all the persons in the group. It is an open question whether the abstracted average is the best representative of the group compared to just a simple enumeration of the particular observations (Gigerenzer et al., 1989).

The situation is analogous to the development of ethics in a setting where the ethics is not dictated by received knowledge, such as the word of God or the edicts of an absolute sovereign. In a modern liberal democracy, the ultimate system of ethics must somehow originate from those on whom the ethical system operates. Yet, it is likely that there would be some, perhaps considerable, variability or differences in the opinions or positions of the individuals subject to the ethical system. This does not mean that a consensus is not possible or that whatever consensus reached would be unjust. The critical question is whether the consensus reached is sufficiently representative of a sufficient number of those to be governed so that the consensus can be enforced on all. This is an empirical question, meaning it can be understood and analyzed in terms of observations of the opinions and positions of those to be governed, or their proxies. It is an empirical question of the same nature that confronts biomedical research. However, in both the ethics of everyday medicine and biomedical research, it is problematic.

Just as in the group of persons whose heights are being measured, a group of persons making ethical judgments are observed and an attempt can be made to abstract something like the average ethical judgment, such as ethical principles, moral theories, laws, case precedents, and virtues. However, just as in the case regarding the average of an enumerated list of particular observations of heights, it is an open question which provides a better understanding of ethics, principles abstracted from the group of observed ethical judgments in the manner of an average or an enumeration of the actual ethical judgments of the individuals. The Principlist position would argue that the abstracted "average" ethical judgment is the best or most useful approach, while the Anti-Principlists, represented by narrative, phenomenological, and feminist (in some cases) ethics, would hold that an enumeration is the most appropriate. It is likely that simple enumeration is not feasible in large groups of consensus makers and that too small of a group risks misrepresentation of the society to be governed (see Chapter 3).

Humans operate from some type of knowledge when dealing with ethical situations so as to render judgments of justice or injustice and, in its most rudimentary form, such knowledge is innate (Chapter 1). Yet, it is highly unlikely that any innate knowledge is sufficient given the social evolution of human interactions that are far more complex than the social situation and circumstance at the time when evolution selected out those innate mechanisms. Rather, any innate knowledge must be elaborated, whether by observational experience or by declarative, deliberative, or conscious rationalism (itself derived from previous experiences). The Principlist might argue for elaboration based on declarative, deliberative, or conscious rationalism, whereas the Anti-Principlist may argue for elaboration by observational experiences alone. The critical question is how these different ethical approaches work to resolve ethical conundrums, which is dependent on the presuppositions regarding how the knowledge relevant to each approach is obtained.

For the Principlist, principles appear to be generated and validated by consensus, a type of induction, among the appropriate persons allowed to participate, such as reasonable persons (Beauchamp and Childress, 2013) or persons operating under a veil of ignorance where the person is unaware of his or her own interest and thus is expected to act in an unbiased or nonprejudicial manner (Rawls, 1999). To be sure, such an approach that combined both generation and validation by consensus risks circularity and narrow relevance perhaps only idiosyncratic to the group of participants. As such, the logic or rationale risks being tautological. Further, as in all such inductions and demonstrations of knowledge, such as biomedical research (Montgomery, 2019b), the critical question of its utility is the ability of those inductions, such as ethical principles and moral theories, to generalize beyond the consensus group used to generate the inductions. Generalization from an originating group is necessary to apply prospectively to future ethical problems; however, such a schema is inherently at risk of the Fallacy of Four Terms as discussed in Chapter 3. The same problem exists for most scientific investigations; indeed, probability and statistics, in a sense, evolved to deal with just this sort of problem. Chapter 3 examines the inductive process in ethics from the perspective of probability and statistics in an effort to make the ethical inductive process more rigorous.

Application of the Principlist approach to resolve ethical questions investigates the relationships between all stakeholders and shareholders, which are analyzed and characterized in terms of ethical principles such as beneficence, autonomy, and nonmaleficence (in one form of Principlism). The obligations to the characterized accounts of the ethical principles are then interpreted in terms of competing moral theories. The moral theory held paramount then determines what is the just outcome (see Chapter 5). However, such a characterization based on a principle is difficult, just as it is to apply the average height of 5 ft 6 in. (168 cm) to a group of persons, none of whom are 5 ft 6 in. (168 cm). The Anti-Principlist might argue that this amounts to "pounding a square peg into a round hole." In a statistical analysis of quantifiable observations, one could relate each person whose height is observed by calculating a z score based on the average (mean) and the standard deviation, a measure of the variety of heights within the group. However, such a process would be difficult in cases of qualitative data that cannot be ordered or ranked or where calculation of the incidence of each specific observation is not possible. Even at that, an infinite number of groups whose elements or participants are very different across groups can have the same average (mean) and same measure of variability such as the standard deviation. The potential for the Fallacy of Four Terms remains and makes precise and accurate resolutions problematic (see Chapter 3).

The Anti-Principlist approach might argue that only actual experiences are relevant and that it is not possible to induce some general principle that can be applied with accuracy and precision to any particular ethical problem. In its pure form, the Anti-Principlist approach may be to enumerate a complete set of prior experiences where ethical problem A had just outcome $A*$ and so on for all experiences in the set of experienced ethical problem—experienced as justice-outcome pairs. The particular ethical problem under consideration is matched exactly to an experienced ethical problem—just-outcome pair in the set and the experienced ethical problem—just-outcome pair matching experienced ethical problem under consideration is taken as just.

The problem arises in that the match must be exact in every way. Note that the pure Anti-Principlist cannot say that it is sufficient to match only for the relevant features as this necessarily has invoked some prior abstraction as to what the relevant features are. The relevant features would have to transcend all the possible features, thus defeating the purely Anti-Principlist approach. The Anti-Principlist cannot say that the match is close enough, as what would be the measure of "close?" Again, there would be the necessity of a transcendental notion of "close." Finally, even if there were an exact match, the experienced ethical problem in the set to which the current ethical problem is being matched existed in a different time. According to Hume's Problem of Induction, to argue that the past experienced ethical problem is relevant to the problem under analysis, prospectively and normatively, requires a transcendental supposition that time does not change the relevance of the exact same experience. Hence, the pure Anti-Principlist is caught up in the Solipsism of the Present Moment.

To be sure, an Anti-Principlist ethicist can respond to an individual person's ethical actions, characterizing them as just or unjust without any appeal to a principle, theory, law, case precedent, or virtue, it is just a reaction and no further qualification is necessary. By analogy, consider the statement US Supreme Court Justice Potter Stewart, regarding pornography, wrote in *Jacobellis v. Ohio* in 1964, "I shall not today attempt further to define the kinds of material I understand to be embraced within that shorthand description, and *perhaps I could never succeed in intelligibly doing so* [italics added]. But I know it when I see it, and the motion picture involved in this case is not that" (*Jacobellis v. Ohio* 378 U.S. 184). However, this approach is not likely to be helpful except in circumstances where the position of the judge is unassailable. The necessary comparison of the competing claims requires some transcendent measure to be applied; in other words, a principle, theory, law, case precedent, or virtue.

Principlism and Anti-Principlism—Two sides of the same coin

As can be seen, the fundamental ontological conundrum that requires an epistemic choice, whether variety is diversity or variation, cannot be overcome by the pure Principlist, the Anti-Principlist, or some mixture of both. The two approaches are epistemically irreconcilable. One way to understand the problem is from Aristotle's notion of combinations [*mixis*] and mixtures [*sunthesis*]. Aristotle wrote "[I]f combination [*mixis*] has taken place, the compound must be uniform—any part of such a compound is the same as the whole, just as any part of water is water" (On Generation and Corruption, I.10, 328a10ff). This is not the case with mixture as *sunthesis*. If the Principlist and Anti-Principlist were to combine [*mixis*] rather than mix [*sunthesis*], according to Aristotle's notion, there would no longer be a Principlist or an Anti-Principlist but something altogether different. To the degree that the Principlist and Anti-Principlist are mutually exclusive, then it will not be possible to have a uniform understanding or *mixis*. Rather, the situation is that of Aristotle's mixture, *sunthesis*, where the contrasting elements do not blend into some other element but rather the contrasting elements are juxtaposed. The "composite body" would be a mixture, and, as one moved through the composite body, in this case the particular ethical problem, one would encounter each contrasting element in succession.

Principlism and Anti-Principlism are two sides of the same composite body, that being two sides of the same coin, in this case two sides of the fundamental ontological conundrum coin. The process of resolving ethical problems involves flipping between the two sides. For example, some initial transcendental notion, a first principle (or "first philosophy")—the "heads" side of the coin—is necessary even to begin the process of organization and then enumeration of experiences for subsequent consensus building. It just might be the neuromechanical mechanism by which persons, beginning in infancy, just come to respond to problems such that the response can be described as indicating justice or injustice. The response as just or unjust becomes the first cut to organize ethical experiences. As the person's experiences expand in complexity and scope, the first principles are no longer sufficient to collect and characterize a set of experiences that would be responded to be just or unjust in contrast to the prior experiences. Consequently, the neurophysiological mechanisms underlying the first principles then adapt and elaborate in response to social experiences.

The first principle is then elaborated but how it is elaborated is determined by the change in the set of experiences whose responses are characterized as just or unjust, the Anti-Principlist—"tails" side of the coin. Indeed, as discussed later, Anti-Principlist approaches, such as narrative, phenomenological, and feminist (in some cases), bring the failure or limitations of the application of the prior principle to light. For example, feminist ethics has been powerful in demonstrating past accepted ethical behaviors as unjust, thereby requiring modification, perhaps to the point of abandonment (rare) of prior principles. The next set of experiences, whose responses are characterized as just and unjust, leads to new inductions to elaborated principles, theories, laws, case precedents, and virtues, for example, by consensus, and the process of iterations continues.

An analogy may be seen in the machine learning (referred to in the past as neural networks) solution to Piaget's balance beam problem. Young children are presented with a beam that is balanced on a fulcrum. Weights of different sizes are suspended at different distances from the fulcrum. The beam is held in position and the children are asked in which direction the beam will tilt, either the left or the right end will go down, and are then shown what actually happens when the beam is released. Very young children appear to adopt a weight rule (Rule 1 in Fig. 2.4) in which the side with the greatest weight will go down regardless of the distance to the fulcrum. Later, after repeated experiences, older children see that the side with the greatest weight may not necessarily go down. At that point, they may adopt a mixed strategy of the weight rule when circumstances seem to dominate (Rule 1 in Fig. 2.4) or a distance rule when the distances are particularly salient (Rule 2 in Fig. 2.4). In unclear cases, the child "muddles through" (Rule 3 in Fig. 2.4). Older children develop a notion of torque where both weight and distance are considered (Rule 4 in Fig. 2.4).

McClelland modeled this process using neural network computing with a back propagation of error (McClelland, 1988). In this form of machine learning, many relatively simple computational units, analogous to neurons in the nervous system, are arranged into input, middle, and output layers containing the computational units. Information as to the experience is entered into the neurons of the input layer. The information is manipulated by the middle layer of units and sent to the output layer. The results of the output layer are compared to the actual results of the experiment. If there is an error, connections between neurons between the layers are adjusted and continue to be adjusted until outputs from the neural network match the actual experience.

Fig. 2.4 shows the results. The vertical axis is the probability that the children's or the computer's response would correspond to the rule-governed response in various conditions. The balance condition (BAL) is the situation where the same weights are placed at the same distances from the fulcrum on each side of the fulcrum. The weight condition (WEI) has different weight sizes at the same distances from the fulcrum in which the side with the greater weight would fall according to the weight rule. The distance condition (DIS) has the same weight sizes but at different distances from the fulcrum. Then there are conflicted conditions where weights of different sizes

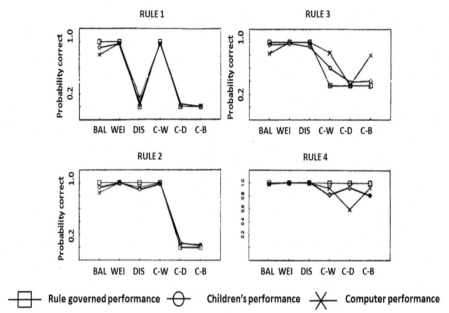

Fig. 2.4 Performance on the balance beam test under conditions where the beam was balanced (BAL) with equal weights on each side of the fulcrum; greater weights on one side but at the same distance (WEI); same weights on each side but at different distances (DIS); and conflicted conditions, where there was a mixture of different weights and distances. In the conflicted state C-W, the weights determined the torque, in the C-D state the distances determined the torque, and in C-B the torques were equal on each side of the fulcrum. Rule 1 is where the torque is judged only by the weight, regardless of distance. Rule 2 is a mixed strategy where the effect of distance on torque is unambiguous and then the distance rule dominates, as otherwise the weight rule determines the response. Rule 3 is "when in ambiguous conditions, the child muddles through." Finally, Rule 4 represents a clear application of the notion of torque. Results show what the responses should be according to the rules (rule-governed behavior) and how children and the computer performed. As can be seen, there is considerable agreement (McClelland, 1988).

are at different distances from the fulcrum. The conflicted case, where the size of the weights dominates, that is, producing greater torque on the side with the greater size in weights, is called the conflict-weight condition (C-W). Similarly, the conflicted-distance is where torque is driven by the distances of the weights from the fulcrum (C-D). Finally, the conflicted-balance condition is that where the weights and distances are all different but torques about the fulcrum are the same (C-B). As can be seen from Fig. 2.4, a good agreement exists among the rule-governed children's and computer's performance, except in the conflicted conditions. Interestingly, the adoption of various rules, migrating from Rule 1 to 2 to 3 and then to 4, during training of the computer paralleled the development typical of children.

As can be seen from Fig. 2.4, the first rule adopted by the young children and by the computer early in training focused on just the weights on each side without taking into consideration the distance of the weight from the fulcrum. As situations were encountered where the consideration of weight alone would not result in 100% accuracy as to the direction the lever would tilt (Rule 1), the children and the computer program then incorporated knowledge as to the distance of the weight from the fulcrum, leading to greater success, the adoption of Rule 2. Note that the weights were the same, although the distances to the fulcrum changed. By the time Rule 4 evolved, the children and the computer displayed responses suggesting an understanding of the concept of torque.

The physical notion of torque is weight through a distance from the point of rotation, such as the fulcrum in the balance beam problem; torque $(t) =$ weight (w) X distance (d), where d is negative when the weight is to the left of the fulcrum and positive when to the right. With a series of i number of weights at different distances, the net torque is given by the following equation:

$$t = \sum_{1}^{i} w_i * d_i.$$

If the net torque (t) is negative, the left side of the beam would rotate downward. In a sense, the equation illustrates the principle of torque. However, it is not possible to know if the child actually uses something like the equation, as it is clearly not in the case of Rules 1–3. However, as it is possible to know the exact state of the computer down to the operation of each transistor, one should be able to know exactly how the computer solved the balance beam problem when fully trained. Yet, no clear representation of the equation for torque will be found, which is not an unusual circumstance in machine learning (Reyzin, 2019). Nevertheless, observation of the computer at the final stage would give the intuition that the computer knew the principle instantiated in the equation for torque.

Interestingly, the first rule adopted is the pure weight rule, which would not be surprising as young children likely have greater experiences with weights than distances. Similarly, the training set used by the machine learning was such that the weight factor was the most common factor. For the computer, changing the training set to have a predominance of the distance factor leads to the first rule being based on distance, not weight. Nonetheless, the computer still would have learned the final rule for the correct prediction of torque.

Reproducibility as justice

Perhaps the single most important factor in favor of the Principlist approach is the issue of reproducibility. For the purposes here, the term "reproducibility" will be used instead of consistency in order to align the notion more closely to similar concerns in biomedical research. The hope is that reproducibility in ethics may then borrow from similar notions in biomedical research. The basis for the attempts to draw parallels is that ethics, in a pluralistic modern liberal democracy, is empirical in many of the same ways as biomedical research (see Chapter 3). Reproducibility is fundamental to many notions of justice as it is in biomedical research and law (indeed, any empirically derived knowledge domain). In fact, there has been a crisis in biomedical research to the extent that studies published in leading scientific journals cannot be replicated (Montgomery, 2019b). It just feels (affective neurophysiological response) unjust for the same person to be dealt with differently in repetitions of the same ethical problem; a person cannot be found innocent of one crime only later to be found guilty of the same crime or tried a second time for a crime for which the person was already found not guilty—the prohibition against double jeopardy. In law, the concept of *stare decisis* holds that a legal precedent should not be overturned, at least easily, thus ensuring legal reproducibility.

In the case of medical ethics, it would just seem unfair that a particular patient would receive a treatment 1 day only to be denied the same treatment on another day, assuming that all relevant factors are sufficiently similar on both days. If there were a material difference between the 2 days, then the subsequent denial may not be responded to as unjust. A Principlist approach would be useful in the case where the same patient is treated differently for the same problem on two different dates, as the injustice would be prima facie (on its face). However, this assumes that the actual principles do not change over time, thus negating Hume's Problem of Induction. It need not be the case that such irreproducibility is indicative of some underlying failure or reasoning or in establishing the facts, as will be discussed later in the context of Chaos and Complexity.

The analogy in biomedical research would be narrow reproducibility where the attempt to replicate the exact same experiment produces unreproducible results (Montgomery, 2019b). Even at that, it is not the nature or science of the experiments that is questioned, but rather suspicions are raised as to fraud, insufficient statistical rigor, and lack of transparency. The presumption is that in the absence of fraud, insufficient statistical rigor, and lack of transparency, the experiments just must have the same outcomes. However, no two experiments always have the exact same outcome, as there always is some variance, typically attributed to biological or instrumental variance. Nevertheless, the experiments must have sufficiently little variance as to be held invariant and thus reproducible.

Where the issue of ethical reproducibility becomes more difficult is whether it is fair for particular person *A* from group **1** be treated differently from particular person *B* in group **1**. Is it unjust for Mary Smith (person *A*) to receive treatment for her Parkinson's disease while a bit later Jane Doe (person *B*) is denied treatment for her Parkinson's disease? Mary's receiving treatment establishes a precedent that patients with Parkinson's disease should be treated, and if that precedent is to be followed (the legal concept of *stare decisis*), then Jane should receive treatment as well. Note that both Mary and Jane are said to belong to group **1**. The implication is that by virtue of both being in group **1**, they share the exact same material facts that are relevant to receiving treatment. Again, it would appear prima facie that Jane should be treated and observers would react to the case where Jane is denied as though it were unjust.

If both Mary (person *A*) and Jane (person *B*) are in group **1**, which means they have Parkinson's disease and are to receive treatment, should Jack (person *C*) in group **2** receive treatment? For the answer to be yes as inferred from the treatment of Mary and Jane, group **2** would have to include the fact that members of group **2** have Parkinson's disease.

Further, persons in group **2** should have the same relevant material facts that led persons in group **1** to be treated. Also, persons in group **2** should not have any relevant material facts that would contraindicate treatment, as would members of group **1**. Note that members of both group **1** and group **2** cannot be exactly the same in ALL facts as there would not be any group **2** that would not belong to group **1** but rather a group that has been mislabeled group **2** rather than group **1**. Because groups **1** and **2** are not identical, then the question becomes which subset of ALL facts constitutes the relevant facts such that both groups are identical for the relevant facts related to treatment and thus what is true of group **1** is true of group **2**? Note that there is nothing prima facie (on its face) that would distinguish relevant from irrelevant facts, as the decision is transcendental over the specific facts. Thus, there is, of necessity, an appeal to the transcendental, hence principles, thereby invoking a Principlist approach. Note that the necessity of the Principlist approach need not exclude an Anti-Principlist approach, as the latter is important for reproducibility in the treatment for each person and for all persons similarly situated, as will be discussed later. Indeed, persons experience particulars, not principles, and therefore some aspect of an Anti-Principlist approach is necessary.

In science and medicine, reproducibility is critical to the validity of any scientific claim or medical decision. Indeed, there has been a great crisis in biomedical research as high degrees of unproductive irreproducibility have been discovered, resulting in significant changes in the practice of science (Montgomery, 2019b). In biomedical research, there is narrow reproducibility, such as within a single experiment. Most scientists replicate their studies in their own laboratories to demonstrate reproducibility in a statistical sense. In some ways, ethics proceeds similarly, the question "what is just in case (experiment) X?" is given to a number of appropriate persons. Reproducibility in subsequent judgments is taken to vouchsafe the validity of the judgments obtained. The Principlist would then extend the judgment, perhaps as an abstraction, that can be applied to all similar cases, whereas the Anti-Principlist would only be able to apply the actual judgments, not the abstracted judgment to exactly the same cases.

In biomedical research, it is generally well established that insufficient sample size, meaning the number of trials conducted in a single experiment, can lead to type I errors, claiming that some experimental manipulation or comparison resulted in a claim of difference that actually was not true. In the case of a type II error, the conclusion is that there was no difference when there truly was a difference. Another version of reproducibility is when another scientist in a different laboratory attempts to replicate the same exact science. Reproducibility in the broad or conceptual sense is demonstrating that a general principle, for example, is true in different circumstances. For example, a claim related to glucose metabolism demonstrated in a rodent is again demonstrated in a human. Similar concerns are obtained in the ethics of everyday medicine. An analogy to narrow reproducibility is the case described earlier where Mary and Jane are of the same group and if Jane does not receive treatment while Mary does, then the result is ethical irreproducibility in the narrow sense. An analogy to broad or conceptual reproducibility is the situation described previously where Mary and Jane receive the same treatment but Jack does not, presumably because Jack is in a different group, group 2. However, if it is held that the concepts leading to treatment in those in group

1 are conceptually the same for group 2, then failure to provide Jack treatment would be an example of ethical irreproducibility in the broad or conceptual sense.

As will be shown, any ethical theory claiming to reflect a consensus among reasonable persons, as argued by Beauchamp and Childress (2013), or persons operating in a particular manner, such as under the veil of ignorance as argued by Rawls (1999), in response to a class or category of ethical situations and circumstances is going to concern reproducibility. For example, the methods of Reflective Equilibrium (Chapter 3) essentially involve induction from particulars to general principles that is little different from that done in biological sciences. Indeed, in biomedical science, statistical methods fundamentally are developed to assess and assure reproducibility. Indeed, the critical question becomes what degree of irreproducibility, such as between different trials of the same experiment, which is inevitable, is still considered reproducible? The same types of methods can and should be applied to Reflective Equilibrium used by philosophers and ethicists (see Chapter 3).

Reproducibility is fundamental to the ethical notion of equality; all persons should receive the same considerations in the same situations and circumstances. While the notion of equality often is aspirational, such as in the US Declaration of Independence (see Introduction), the test of the notion of equality is in the reproducibility of its applications. At first, it might appear that there are unequal applications of ethics and law, but the question is whether the unequal application is a matter of principle or practice in pluralistic modern liberal democracies. For example, Deontological and Egalitarian moral theories would appear to support equality and thus the same reproducible application of ethics and law. At first appearance, Libertarian and Utilitarian moral theories would not, as clearly citizens can be treated differently. However, one might expect in pluralistic modern liberal democracies that all persons, or sufficient numbers so as to justify universal application, agree to the common notion of ultimate good for a Utilitarian moral theory or the maximization of individual liberty as in a Libertarian moral theory in order to justify application of those moral theories. In the latter scenarios, equality is maintained even in unequal distributive justice (the distribution of goods and liabilities). A libertarian moral theory does not necessitate equal justice but rather equal responsibility to contracts. However, in a pluralistic modern liberal democracy, laws and ethics help ensure a degree of equality in contracts (Contractualism); a person cannot be excluded from an opportunity to participate in contracts open to the public based on factors irrelevant to the contract, such as discrimination in housing.

Reproducibility is also fundamental to the concept of equity where those disadvantaged are supported to achieve outcomes comparable to those without disadvantage. In this case, the goal is equality of the end, such as quality of life and self-actualization, even if the means to achieve those goals are not equal in their distribution. An example might be the Americans with Disabilities Act of 1990 (42 USC § 12101).

The notion of equality also appears to be a neurophysiological predisposition such that violation leads to psychophysical discomfort (Chapter 1), although such reactions are abstracted to notions of ethical or moral angst. More strikingly, humans also appear to be wired to perceive the quality of equity in social interactions. Thus, equality can be said to have a neurophysiological basis, although for reasons discussed later in the context of Chaos and Complexity, the same neurophysiological mechanisms need not lead to the exact same resulting applications and instantiations. Thus, equity leads to equality and thus reproducibility even if the means appears unequal.

Reproducibility is necessary for generalizations over space and time and has significant implications embodied in the neurophysiological mechanisms underlying behaviors held as just or unjust. With respect to time, the fact that the sun has risen every day in the past allows the generalization that the sun rises every day and the prediction that it will rise tomorrow, although it can be argued that this is not proof that it will happen, for example, some galactic catastrophe could occur. The fact that the past is not proof of the future is referred to as David Hume's Problem of Induction. More generally, Hume wrote "instances of which we have had no experience resemble those of which we have had experience" (Selby-Bigge, 1888). What if, while one's eyes are closed, the world is destroyed instantaneously and then duplicated exactly? Far-fetched, yet there is no way to prove it doesn't happen. Consider a person standing in front of a bridge such that middle section of the bridge is not seen for an observer on the ground. How does the observer know that the middle part of the bridge is there? Perhaps there is a gap. One could have the person in front of the bridge walk away, but given Hume's Problem of Induction, how would one know that it is the same bridge that the person was standing in front of only a few moments earlier?

Fortunately, the world does not operate in the ways just described, to the best of our knowledge. But more importantly, humans do not expect the world to operate in such a way. For example, infants as young as 4 months appear to understand object permanence, where the object persists even if visualization is interrupted, and rudimentary physics. In an experiment, a screen was rotated, in the manner of a drawbridge, with a box placed behind the screen so as not to impede the full rotation of 180 degrees. In one condition, the screen was rotated to 112 degrees, and then rotated back to the origin, suggesting that the screen had encountered the box and could not fully rotate (the possible condition). In a second condition, the screen continued to rotate through the full 180 degrees, giving the impression that the screen was moving through the box (the impossible condition). The infants spent more time looking at the impossible condition, suggesting an awareness of the notion that an object cannot move through the same space as another object (Baillargeon, 1987). Interestingly, essentially the same results were obtained when the scene was presented in two rather than three dimensions (Durand and Lécuyer, 2002). The latter experiments suggest that the infant's response was generalized from a three-dimensional world typically encountered to an abstracted world of two dimensions and would be expected to be an example of reproducibility as evidenced by the child's reaction.

The studies of infants just described suggest that object permanence, taken here as analogous to reproducibility of the object in different circumstances, could be inborn (pre-"wired"), learned rapidly, or there being an inborn predisposition to rapid learning (see Chapter 1). However, evidence of object permanence has been demonstrated in a wide range of animals, including mammals and birds. Further, neurons in the temporal cortex of nonhuman primates change activity differently between expected and unexpected conditions; the latter condition is a violation of object permanence.

The theory offered here (and just that, a theory) is that reproducibility is critical to generalizations, which are critical to concepts of categories and prediction. As all of these functions offer survival value, it would not seem surprising that, over time, the nervous system evolved to have neurophysiological mechanisms innate in the organism to serve such functions. Being born with those functions, either innate or the predisposition to learn those functions, could have significant survival value. It likely is no less true for the social environment shaping neurophysiological mechanisms that increase the survival value for social animals, such as humans (see Chapter 1).

The nature of abstraction

A central question is the origin and evolution of transcendental ethical principles, moral theories, laws, case precedents, and virtues, particularly if there is a first principle (first philosophy), analogous to Aristotle's Prime or Unmoved Mover, that starts or enables the first perception. Aristotle argued that the essence of life was movement and believed in the Principle of Causational Synonymy, that all moving things had to be moved, but then what moved the very first thing? The *A Priori* Problem for Induction would argue that a first principle is necessary. There must be some prior principle to organize experience, such as things that generate the experience, that leads to nonrandom behaviors, thereby signifying an organization of experience. The form of a snake must be differentiated from the surrounding leaves so that the person moves nonrandomly to avoid the snake and, in doing so, provides evidence that stimuli are organized such that the snake was responded to. There likely are many first principles. In fact, humans, as well as most living organisms, appear to start life with innate mechanisms that facilitate or act as first principles, although they are not necessarily the types of principles that can be described declaratively, consciously, and deliberately. These first principles likely are neurophysiological mechanisms, a type of neuromechanical reflex, using reflex to denote a sequence of neurophysiological actions (with due respect to Descartes, see his "Treatise on Man," 2003) (Chapter 1). These are principles in that they are transcendental, meaning the same principle operating over a variety of experiences—avoidance of snakes of many varieties, situations, and circumstances. Yet, if principles are not experienced, that is, arise spontaneously in some manner from the sensory stimulation of ethical experience, how is it possible for an organism to have a first principle? Evolution, particularly the constraints imposed by the Second Law of Thermodynamics as applied to Information, likely plays a role in the evolution of first principles. The argument will be made that the very first principle may be very simple, but through evolution, incremental additions accumulated to such a degree that the organism now displays Chaos and Complexity, obscuring the neuromechanical notion of the subsequent first principles, as discussed later.

It is important to note that the evolution of principles does not occur only by stimulus-bounded experiences. Accumulated abstract declarative, deliberative, and conscious experiences can also shape the evolution of principles. In fact, Gedanken (thought) experiments constructed from logical and mathematical principles based on hypotheticals, such as those used in physics and philosophy, are a powerful way to shape experience that then shapes the evolution of principles. Consider a knowledge of physics, where children demonstrate behaviors relative to physics that are consistent with Aristotelian physics, which, in many, persist into adulthood (DiSessa, 1982). As persons begin to encounter situations where Aristotelian physics does not hold but Newtonian physics does, persons begin to change their notions to accommodate Newtonian physics. However, it is difficult to demonstrate predispositions to Newtonian physics in common everyday experiences, as the majority of experiences comport more with Aristotelian physics. One does not try to avoid a stone thrown from a great distance because of Newton's first law—an object in motion remains forever in motion unless acted upon by an outside force. Thus, there would not seem to be any survival value to Newtonian physics predispositions—indeed, there may be disadvantages—outside of academics or outer space. Learning to accommodate Einsteinian or Relativity and Quantum physics are additional examples.

Consider the Galilean (later Newtonian) notion of inertia which states that an object in motion continues in motion in a straight line unless acted upon by an outside force. Clearly, in everyday human experience, nothing is seen to move for an infinite time. How then does one modify human thinking to comport with inertia? Galileo attempted an experimental "demonstration" of this principle but obviously the principle was never, and could never be, realized and thus demonstrated explicitly. Rather, Galileo used two inclined planes. A ball was released from a certain height on one plane, rolled down the plane, and then up the second plane (Fig. 2.5). Galileo noted that the ball would roll up the second plane equal to the height from which it was released on the first plane. Subsequently, the angle of the second plane was reduced and the ball traveled further to reach the original height. The process could be repeated with continued reduction in the angle of the second plane. By extrapolation, it could be understood that at an angle of 0 degree on the second plane, the ball would roll forever. Note that this notion of "demonstration" depends on the extrapolation that is more believed than known. The demonstration is in the process, not the result, and is referred to as the Process Metaphor. Indeed, Reductionism in science is a Process

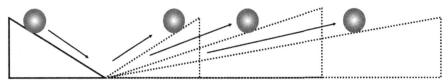

Fig. 2.5 Schematic representation of a demonstration of Galilean (Newtonian) inertia. Two inclined planes are oriented such that a ball released from a height on the first inclined plane would roll down and then up the second inclined plane to the same height from which the ball was released on the first inclined plane. In subsequent experiments, the angle of the second inclined plane is reduced systematically and the ball has to travel further on the second inclined plane to reach the height from which the ball was released on the first inclined plane. As one continues to reduce the angle of the second inclined plane, presumably the ball would roll forever as it would never reach the height from which the ball was released on the first inclined plane. The experiments become a Process Metaphor where the conclusion of inertia is never demonstrated but is extrapolated from the process.

Metaphor, but the metaphor only provides an abstraction or principle (Montgomery, 2019a). Critically, the Process Metaphor depends on reproducibility, in the case of the demonstration of inertia, of the principle that transcends each specific instance of a different angle of the second incline plane.

Similarly, much progress in ethics has been achieved by Gedanken experiments such as the runaway trolley car dilemma first authored by Philippa Foot in 1967 (Foot, 1978) and extended by Judith Thomson in 1976. The dilemma is posed by a runaway trolley car that will kill five workers unless a switch is thrown to divert the car to another track but resulting in the deaths of two workers. Most reasonable persons would throw the switch. The scenario is changed to two persons standing on a bridge over the track where if the smaller person pushes the larger person from the bridge onto the rail, the trolley car will be derailed, saving the five persons at the end of the rail. In the latter case, one life is lost to save five. Yet, most persons would not push the larger person off the bridge, whereas minutes ago when presented with the first scenario, they would pull the switch to kill two persons but save five. As it is highly unlikely that any person would actually experience the exact situation and circumstance of the runaway trolley car, dealing with this dilemma nonetheless allows one to explore and exercise their ethical reasoning. Note that the effect may be analyzed in a declarative, deliberate, and conscious manner to affect subsequent behaviors, yet the experience of the Gedanken experiment may become another training set used to affect the neurophysiological mechanisms as described previously for the balance beam problem. It would seem that asserting the necessity of consciousness, particularly self-consciousness, would be favored by Principlists as it presupposes principles. Anti-Principlists need not posit consciousness as they do not suppose principles, perhaps.

Epistemic risk

As discussed previously some abstraction to a transcendental form is critical in most ethics where adjudication of ethical disputes are based on appeals to principles, theories, laws, case precedents, and virtues. However, adjudication of ethical disputes relates to particulars to which the transcendental forms are applied. In Principlism, considerations of general principles, moral theories, and laws must be made specific to the particulars of the specific case under consideration—a process called specification of norms (Rauprich, 2008). For example, the syllogistic deduction is of a general form where the major premise is the creation of the transcendental form, the minor premise relates a particular to the antecedents to the transcendental form so that the conclusion relates to the particular in the minor premise to the transcendental form, as shown in Argument 2.1:

Argument 2.1

Major premise: *All persons* (bridging term) *hold that autonomy is to be respected* (major term)
Minor premise: *John Doe* (minor term) *is a person* (bridging term)
Conclusion: *John Doe* (minor term) *holds that autonomy is to be respected* (major term)

In this case "all persons" is the collective from which "autonomy is to be respected" the transcendental, in this case a principle, is inducted from the collective. "Autonomy is to be respected" is transcendental as it is applicable to any member of the collective "all persons." It is through the transcendental nature of "autonomy is to be respected" derived from the collective that "autonomy is to be respected" now applies to John Doe, a particular.

The syllogism is true only if "person" (bridging term) that John Doe is in the minor premise is the same as "all persons" (bridging term) in the major premises. If they are the same, there is only three terms. If the nature of persons in the major and minor premises are different, the argument is the Fallacy of Four Terms.

The similarity of the bridging terms in the major and minor premises in Argument 2.1 is relatively straight forward and there is great confidence in the argument. However, it need not be the case, and often it is not, as discussed earlier on the Fallacy of Four terms. The question is how to assess the risk of holding the bridging terms in the major and minor premises as being sufficiently the same? What is the Epistemic Risk?

One approach is to consider the Epistemic Risk as made up of the Epistemic Distance between the two forms of the bridging term and the Epistemic Degrees of Freedom, for example, how many conceptual twists and turns are necessary to get from one version to the other version of the bridging term? For example, rodent models of human diseases are often used to predict whether a particular treatment will be effective in humans. The argument comes down to the presumption that what is true of the rodent having received the treatment will be the same for humans who would receive the treatment. This depends on how similar rodents are to humans relative to the disorder being treated. The metabolism of glucose in rodents may be very similar to humans. Thus, any differences (Epistemic Distance) between rodents and humans relative to glucose metabolism may be relatively small, making the Epistemic Risk of the Fallacy of Four Terms in using the animal model likewise small. However, cognition in rodents may be of a very different kind than humans (Epistemic Degrees of Freedom), and thus predicting the effects of treatment on cognition in humans may be very difficult, leading to the Epistemic Risk of the Fallacy of Four Terms being very high (Montgomery, 2019b). Similarly, if an ethical principle, moral theory, law, or precedent derived from Caucasian males in the source domain for the metaphor by which to make inferences that involve African-American females, the Epistemic Risk will be quite high.

Ethics and consciousness

Chapter 1 attempted to argue that behaviors thought in the domain of ethics are generated by neurophysiological mechanisms. This would seem philosophically trivial in that only the staunchest substance dualist would hold otherwise. Pluralist modern liberal democracies generally drive a secularist or humanist conversation regarding ethics, perhaps to avoid conflicts resulting from enforcement of received knowledge, such as the word of God. An example would be the notion regarding the separation of church and state in the United States. However, there appears to be property dualism where the primacy of the nervous system is not disputed but only that the nervous system alone does not drive all behaviors, at least in humans. To be sure, behaviors are not possible without the nervous system, but this may be no different than the fact that an automobile is necessary to drive to a different location but it is not the car that determines the goal or the path to get there.

Ethical property dualists would not argue that the sensory nervous system is necessary if only to perceive the ethical problems and that the motor nervous system is necessary to affect a solution to the ethical problem. Yet, these ethical property dualists argue that there is something more and that something more transcends the nervous system. This position can be seen in the responses to an article by Pfaff et al. (2008). These authors argued that a motivating factor in ethical behaviors is shared feelings of fear and that mirror neurons in the nervous system are the vehicle for those fears and thus the source of ethical behaviors, not unlike Thomas Hobbes' position in his "Leviathan: Or The Matter, Forme, & Power of a Common-Wealth Ecclesiasticall and Civill" (1651) described in Chapter 1. Clearly, the position of Pfaff et al. (2008) would be thought naturalistic, as in the form of naturalized ethics. Interestingly, this is not the notion of naturalized ethics that many others would hold. da Rocha and Bergareche (2008) argued that consciousness is required, particularly self-consciousness, which supervenes on the neurophysiological mechanisms. da Rocha and Bergareche wrote:

> This is the ground "underlying" M's action [the neurophysiological mechanisms], and as such it can be present both in human and non-human primates. But Korsgaard (2006) adds that, when acting morally, humans are conscious of something else: we are conscious that we fear or desire something, and that we are inclined to act in a certain way as a result. As she puts it, we are conscious of the ground as a ground.
>
> *da Rocha and Bergareche (2008, p. 24)*

The position of ethical property dualists is that consciousness, particularly self-consciousness, is causal to the behavior, with the neurophysiological mechanisms simply being implementation of the conscious causal forces. Such a position would be sympathetic to the Principlist approach to ethical systems. This does not negate neurophysiological mechanisms as a necessary and sufficient cause of ethical behavior. Similar to beliefs, emotions, and other so-called mental states, the abstract ethical notions that require consciousness are commentaries on actions made possible by the very self-same neurophysiological mechanisms. However, there is reason to believe that consciousness and self-consciousness are epiphenomenological post hoc experiences or consequences. It may well be that consciousness and self-consciousness are not causal at all. What is left are neurophysiological mechanisms, even though no full neurophysiological explication is possible as yet (and perhaps never will be). In important ways, this position is sympathetic to the Anti-Principlists.

Discussions of consciousness are far beyond the scope of this study, but the temptation to leave at least a "teaser" for those interested in pursuing the problem is hard to avoid. Often, Gödel's Incompleteness theorems are taken as the death knell of logic. The effect of the theorem is that no sufficiently robust formal system, in Gödel's case, an arithmetic of natural numbers, can prove "truths" that can be derived from the formal system. On first appearances, Gödel's proof would suggest that formal logical systems are futile. One can readily see that the fundamental failure of logical systems would be difficult for the Principlist approach to ethics. One need only to substitute neurophysiological mechanisms for the arithmetic of natural numbers to the extend that any neurophysiological system is incapable of proving every truth related to human behavior. Neurophysiology as a complete formal explanatory theory is futile. Indeed, any formal principled system would be futile, which would lend support to the Anti-Principlist approach to ethics. Yet this is not entirely the case.

In a sense, Gödel demonstrated that certain truths, while not provable in a formal system, nonetheless can be incorporated into and acted upon by the formal system. The inability to prove these true statements is a reflection of a universal Informational Incompleteness of which Gödel's Incompleteness theorems can be described as a subtype of Number-Theoretic Incompleteness. Other subtypes include Physical-Theoretic Incompleteness—the Heisenberg Uncertainty Principle—and Computational or Algorithmic Incompleteness—the Halting Problem. Similarly, it is argued that there is a Neurophysiological-Theoretic Incompleteness.

A type of mathematical statement that is true but unprovable is the one that comments on what might be ordinary mathematical statements such as "$2 + 2 = 4$." An example of a metamathematical statement would be "the statement '$2 + 2 = 4$' is true." Gödel devised a system where both types of statements could be encoded in the same formal language and operated on by the same principles. Such a system had no difficulty in proving the mathematical statement such as "$2 + 2 = 4$" but any proof of the metamathematical statement, "the statement '$2 + 2 = 4$' is true," would result in a paradox and therefore, by *reductio ad absurdum*, cannot be proved valid [perhaps the best, although lengthy, introduction to Gödel's Incompleteness theorems is Hofstadter, 1999].

Consider the neurophysiological logical statement "I see dog" that has binary ASCII encoding as 01001001 00100000 01110011 01100101 01100101 00100000 01100100 01101111 01100111, while the coding for metastatement, implying consciousness, "I know I see dog" is 01001001 00100000 01101011 01101110 01101111 01110111 00100000 01001001 00100000 01110011 01100101 01100101 00100000 01100100 01101111 01100111. Any computer would be able to handle both the statement and the metastatement by the same formal rules underlying the operations of the computer. For example, the computer voice synthesizer would "speak" "I see a dog" in the context of being in the vicinity of a dog, as well as "I know I see a dog."

The statement "I see dog" would be relatively understandable from the perspective of a computation, as one would trace the actions of the photons of light energy as they are transduced into changes in the electrical potential across the cell membranes of excitable cells that constitute the visual system. A neurophysiological program could process both the neurophysiological ("I see dog") and the metaneurophysiological ("I know I see a dog") statements, and a neuroscientist, in principle, might find both statements expressed in the nervous system. In this sense, no difference exists in the neurophysiological computation of "I see dog" and "I know I see dog," as both have the same form of encoding and thus are amendable to the same form of computation. However, the notion of such a representation in the nervous system is limited by the incompleteness of the logical operations of the nervous system, on many levels, but this does not mean that the reality (ontology) of "I see a dog" and "I know I see a dog" is any different.

Extending this discussion to ethics, it is quite possible that reacting in an affective (emotive) manner to an unjust situation utilizes the same or at least a subset of the neurophysiological mechanisms that encode consciousness of the unjust situation. In a way, both Anti-Principlist and Principlist approaches utilize the same machinery.

Chaos and complexity

Given the fundamental ontological conundrum that requires an epistemic choice for constructing the ethics and every empirically based knowledge domain, reproducibility of justice from the application of principles, theories, laws, case precedents, and virtues is critical. Reproducibility is related to the degree of consistency between similar experiences, in terms of particulars, and the outcomes are adjudicated as just or unjust. A one-to-one correspondence provides the greatest certainty, epistemically and perhaps ontologically, and gives the greatest assurance of reproducibility. The same result obtained with the same ethical situation or circumstance provides a one-to-one correspondence from the ethical situation or circumstance to the specific implementation of the principles, theories, laws, case precedents, and virtues. A many (different attributions of justice or injustice to a single type of ethical experiences)-to-one (a specific set of principles, theories, laws, or precedents) correspondence undermines the robustness and, hence, significance of any set of principles, theories, laws, case precedents, or virtues to explain experiences to which justice or injustice has been attributed, as does a one (attribution of justice or injustice but not both to different instances of one type of experience)-to-many (principles, theories, laws, case precedents, or virtues) correspondence. Lack of a one-to-one correspondence has been used as an argument against the Principlist approach to ethics.

Yet, it is possible for a single set of principles, theories, laws, case precedents, or virtues to give rise to different attributions of justice or injustice to different instances of what would appear to be the same type of ethical experience. Thus, the apparent inconsistency of the single set of principles, theories, laws, case precedents, or virtues would not be an argument against Principlism. Indeed, as will be demonstrated elsewhere, such an apparent inconsistency is entirely consistent with neurophysiological systems that demonstrate Chaos and Complexity. Thus, inconsistencies appearing as paradoxes or dilemmas are understandable and resolvable in neurophysiological terms but are not easy in psychological terms. A full discussion is provided in Chapter 1.

Defining Chaotic and Complex Systems is the dependence on initial conditions. Slight, and perhaps imperceptible, differences in the experiential starting positions could give rise to very different judgments of justice or injustice. For example, slight or imperceptible differences in initial atmospheric conditions can give rise to a potentially infinite number of snowflakes where the only commonality is having six points. This does not negate the same laws of physics for each snowflake.

The argument to be advanced is that dependence on initial conditions plays to the arguments made by the Anti-Principlists. It is critical to understand the initial conditions constituted by the particular ethical experiences, the forte of the Anti-Principlist, in order to appreciate why Principlism can result in seemingly inconsistent outcomes of justice without rejecting Principlism. As importantly, the valid use of principles, theories, laws, case precedents, and virtues does not obviate the importance of the particular ethical experiences and thus supports the role of the Anti-Principlists. The importance of the Anti-Principlists in understanding the initial conditions on which any principled outcome is dependent is discussed in detail from this perspective.

Examples of Anti-Principlist ethical theories

What follows is a very brief survey of Anti-Principlist ethical theories, although the prior discussion demonstrated that both Principlist and Anti-Principlist approaches are two sides of the same epistemic coin in response to the fundamental ontological conundrum that requires an epistemic choice. The focus of this section is the initial organization of experiences that becomes the substrate for abstraction to principles, theories, laws, case precedents, or virtues.

On the issue of organization of the experiences to serve as a substrate for ethical induction, Principlism is relatively quiet. It often appears that the experiences relevant for ethical induction are just "given," meaning requiring little or no effort outside of the processing of the sensory stimuli that comprise the experiences. The notion of a "given" is not an unreasonable epistemic position, given that the first cut at organizing experiences is an affective (emotional) neurophysiological response that does not necessarily invoke any conscious awareness. However, this is not the case as realized, even implicitly, by the Anti-Principlists.

As will be shown, the organization of ethical experiences as initial conditions for the generation of ethical systems is very complex and the putative failure of the Principlists to accommodate such complexity is held as a severe limitation. From this perspective, several ethical systems that are Anti-Principlist, in whole or in part, and will be reviewed.

Before the review, there is another important limitation of the Principlist approach. One presumed strength of the Principlist approach is borrowing confidence from logical deduction. The principles, such as beneficence, autonomy, and nonmaleficence, are taken as premises in a logical deduction and the consequence is taken as absolute and thus only true. However, the actual use of principles, theories, laws, case precedents, and virtues as premises in an argument are not typically true forms of deduction (Chapter 3). The ultimate adjudication of justice is in the negotiations among competing claims of different parties to the obligation of the principles of beneficence, autonomy, or nonmaleficence and which moral theory allows one set of obligations to trump the others. Particularly problematic is the principle of autonomy, as noted in many of the cases presented in Part 1. It is not that respect for persons is not held as true, but rather the concept of autonomy is also extended to a collective, such as a society (local or global).

The argument raised throughout this book, particularly in Introduction and in Chapter 1, is that while the principles may be absolute, that is, not contravened by experiences of reasonable persons blinded to their self-interest, the operating factor is the obligation to those principles. The obligations are a continuous variable capable of any value or weight ranging from absolute to denied. As such, obligations do not comport with the Principle of the Excluded Middle—a statement, such as a premise or proposition, must be either true or false but not both and not either. The Principle of the Excluded is a necessary precondition for logical deduction. In this case, Principlism is not a deductive process and any confidence based on its resemblance to logical deduction is a misrepresentation. To be sure, a number of formal logical systems do not require an absolute Principle of the Excluded Middle, such as predicate, fuzzy, and modal logic, yet these logics are not as certain as a valid deductive argument would be. It would be interesting, and left to future efforts, if these alternative logical systems were explored for utility in ethics.

The Anti-Principlist argument that many situations and circumstances call the ethical principles, moral theories, laws, and case precedents into doubt. In contrast to a strict Principlist account, any appeal to principles, theories, laws, case precedents, and virtues (the latter particularly important to Anti-Principlist accounts) are contingent. But if what is susceptible to doubt are the obligations to those principles, theories, laws, case precedents, and virtues—the principles, theories, laws, case precedents, and virtues themselves need not be defeated or rejected. Many Anti-Principlist ethicists call for an analysis of context and interpretation (Greenfield and Jensen, 2010). Some Anti-Principlists then focus on values and goals, but it is not clear that values are anything different than obligations to beneficence and nonmaleficence and that goals are any different than obligations to autonomy, at least in many respects.

Some have used codes of ethics, such as those of professional organizations, by which to judge the Principlist position with the conclusion that such codes of ethics are ill-suited to the actual care of patients (Greenfield and Jensen, 2010). To be sure, these ethicists recognize the regulatory role of codes of ethics, but fail to appreciate that most codes are professional-, government-, or industry-centric (see Chapter 4). Conflating professional codes of ethics with the patient-centric ethics of everyday medicine would be a category (logical) error (Ryle, 1949), such as using apples to describe oranges.

In describing the limitations of codes, as they illustrate the Principlist ethical theory, Greenfield and Jensen (2010) wrote:

> The core principles and specific rules that inform the physical therapy Code of
> Ethics, such as autonomy, justice, beneficence and veracity, are prima facie duties
> that obligate all physical therapists. The problem is that principles that are contained
> in a code of ethics and applied to an ethical dilemma may be equally compelling
> to a physical therapist. As a result, a physical therapist is faced choosing between

conflicting duties and justifying his or her choice. But the code of ethics does not describe how physical therapists apply these ethical principles or specify rules to solve an ethical dilemma.

However, in the Principlist framework offered here, the ethical principles are not guides, rather they are stipulations allowing ethical analyses of the situation and circumstance in terms of the relationships between all shareholders and stakeholders in terms of obligations (see Chapter 5). Thus, it is not a matter of choosing between the ethical principles, but rather adjudicating the varying and, at times, conflicting obligations to the ethical principles, moral theories, laws, case precedents, and virtues among the shareholders and stakeholders.

Phenomenological ethics

When asked "who are they and what are they?" people generally do not respond, "I am a thing or object," "I am a bag of chemicals," or "I am a bag of self-actuating chemicals." Rather, they are likely to say "I am a person" or, for the more philosophically inclined, "I am a being." The ethical extension is that physicians and healthcare professionals are dealing with a being, not an object or thing, not a self-actualizing bag of chemicals. Thus, what differentiates the latter from the former, what it is to be a being, is critical to any patient-centric ethics (which, admittedly, is not the general rule, see Chapter 4). An ethic based on autonomy to beings, in the fullest sense of the word, being, is a focus of Phenomenological ethics, which originated in the philosophical discipline of Phenomenology.

Perhaps Ethical phenomenalism is at one extreme of the Anti-Principlist approach, focusing nearly exclusively on the particular experiences of those involved in the ethical problem. Ethical phenomenalism follows from the development in philosophy of the program of phenomenology, largely due to the work of Edmund Husserl in his "Ideas Pertaining to a Pure Phenomenology and to a Phenomenological Philosophy—First Book: General Introduction to a Pure Phenomenology" in 1913 (Kersten, F., trans. The Hague: Nijhoff, 1982). Phenomenology begins with what is perceived and is usually recognized as what one is conscious of. Not addressed is the manner in which the world out there (or the world within that is not accessible to conscious appreciation) ultimately ties to perceptual consciousness or conscious intentions on the part of the organism. Phenomenologists do not deny this problem, but, at least in an operational sense, great progress in understanding human behavior does not require an understanding of how, in a neurophysiological sense, the world out there connects with the conscious perception of the world out there. In a sense, it is hoped that phenomenalism represents an epistemic approach to make analysis and understanding of consciousness more tractable. This approach "craves" the experience of phenomena at one possible "joint," that is, conscious perception versus the prerequisite neurophysiological mechanism as suggested in Chapter 1. At least for Husserl, this was not a denial of the reality of the external world or the internal world not accessible to consciousness. Rather, Husserl sought to marginalize, "bracket" in his terms, the effects of the naturalist standpoint of any presumption or denial of an existence of an external or internal world inaccessible to consciousness with all their problematic implications.

The first mode of conscious effort was introspection. Intuitions were the focus and the fundamental or primitive substrate for a science of consciousness. Husserl offered his Principle of All Principles, writing "every originary presentive intuition is a legitimizing source of cognition, that everything originally (so to speak, in its 'personal' actuality) offered to us in 'intuition' is to be accepted simply as what it is presented as being, but also only within the limits in which it is presented there" (Husserl, 1913).

Application of phenomenology to ethics primarily follows epistemology rather than ontology, as the latter focuses on concepts of the "self" disconnected to any other external or internal reality and has been accused of Idealism. In general, phenomenological ethics argues that intuitions are fundamental, yet this does not deny underlying principles, particularly neurophysiological mechanisms (see Chapter 1), only that they are not necessary to achieve a workable ethical system. One value is that consideration starts with intuition, not principles and moral theories, thereby avoiding, according to some, prejudicing the intuitions by some prior presumption. Yet, as discussed frequently in this chapter, there is no escaping the necessity of some *a priori* operation, principled or not, to organize the relevant experiences that would generate the intuition. The phenomenologist can argue that whatever *a priori* is necessary, it is incorporated into the formation of an intuition and thus the intuition can "account" for the necessary *a priori*.

There is a psychological analog to Husserl's phenomenology philosophy that may help appreciate the *a priori* that may be relevant to phenomenology, that being Gestalt psychology. The main difference lies in the embrace of experimental methodologies by the Gestalt psychologists, which were rejected by Husserl as being too empirical and, in being empirical, tied to specific phenomenon that presupposed an external world in the Lockean tradition (Heinämaa, 2009). However, being empirical, Gestalt psychology brings phenomenology closer to an ontology. A major focus of Gestalt psychology is understanding perception, particularly visual, in a primary sense, that is, not a reconstruction from a reductionist sensory neurophysiological basis.

This brings up a semantic difficulty, which is differentiating phenomena in the sense of observable entities and phenomenology. Both can share the same notion of a conscious perceptual experience but the difference lies in the primary or most direct source of the perceptual experience. Husserl argued that the conscious perceptual experience as held by Gestalt psychologists arose from the phenomenon itself. Husserl sought a source that was transcendental; in other words, not tied to the sensations that gave definition to the conscious perceptual experience, much in the manner of Kant's *synthetic a priori* in his "Transcendental Aesthetic of his Critique of Pure Reason" (1781). The Berlin School of Gestalt psychology was very sympathetic to Husserl's position (Wagemans et al., 2012a,b). Perhaps a significant difference is that Gestalt psychology focused on consciousness in a broader sense, whereas phenomenology focuses primarily on self-consciousness.

Gestalt psychology has developed an extensive set of principles related to the organization of conscious perceptual experience, too numerous to discuss here; the reader is referred to Wagemans et al. (2012a,b). Just a few examples are considered to demonstrate the relevance of Gestalt psychology and its origination in phenomenological philosophy and its relevance to neuroscience.

The principle of completion can be seen in the phenomenon of illusionary contours. Consider the visual illusion shown in Fig. 2.6. Most persons would say there are two triangles in the image, one side with the vertex down and sitting on top of the other with its vertex on the top. However, the perimeter of the upside-down triangle is not given by any line that would reflect photons onto the retina. These contours are illusory and are referred to as subjective contours. The key is that the illusion demonstrates that perceptual consciousness is not synonymous or isomorphic with immediate and primary sensation.

Perhaps one of the greatest contributions of Gestalt psychologists is the exploration of visual perceptual illusions, particularly those that cannot find explication at the most fundamental levels of sensory transduction of physical stimuli such as the retina (Spillmann, 2006). A striking example is the "filling-in" effect demonstrated in Fig. 2.7 in which the image that generates the photons of light to strike the retina consists of a black background with white vertical slashes. Superimposed is a small white square. If one maintains gaze on a distant part of the image, such as the upper left corner of the image, the white square is perceived to disappear. There is no "blank" left behind, as the white square appears to be filled in by the background.

The interesting question is whether phenomena such as illusionary contours and filling-in are important for human behavior and increase survival value. One could argue that because illusionary contours and filling-in are not in the sensory experience

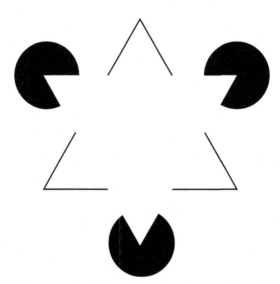

Fig. 2.6 The typical first perception is that of two triangles superimposed. Most persons perceive that one triangle has the apex pointed up and the second has its vertex pointed down (the Kanizsa triangle). Yet, there is no image, such as a line, that demarcates the Kanizsa triangle. In this case, the two triangles appear to be perceived without the actual sensation that completes the perception.

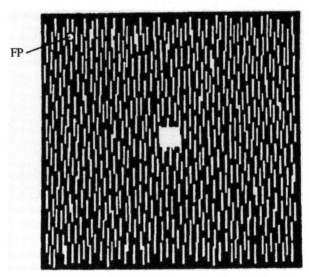

Fig. 2.7 Demonstration of the Gestalt phenomenon of "filling-in." If one maintains gaze at a distant part of the image, FP, the white square appears to disappear to be filled in by the background (De Weerd et al., 1995).

then they must be extrasensory. If one were to equate extrasensory with consciousness, then Gestalt psychology would seem to support independence and hence an independent importance of consciousness. It certainly suggests an *a priori* mechanism for perception originating in consciousness. It is important to note that modern neuroscience is "chipping away" at such presumptions to demonstrate specific underlying neurophysiological mechanisms (De Weerd et al., 1995). However, historically and perhaps currently in many quarters, this presumption continues and could be considered as fundamental to phenomenology and the ethics implied by phenomenology.

Starting from intuitions is closer to the actual experiences of a particular situation and circumstance. As argued previously, humans do not perceive principles but rather particulars, although prior principles facilitated recognition of the particulars. Phenomenologists might argue that the experiences of the particulars are conscious experiences, particularly self-conscious experiences. Humans communicate their experiences in a language of which nearly always is a self-conscious experience. As discussed previously, language—at the very least a tool of declarative, deliberative, and conscious experiences—has the power to affect human behavior. The declarative, deliberative, and conscious experiences are derived from intuition and thus the operations of any ethical system are derived from the intuitions. While intuitions brought to consciousness can be categorized and related, the process need not be a reduction to more fundamental or primitive principles, such as beneficence, nonmaleficence, and autonomy.

There is a form of reductionism in phenomenology, that being eidetic reduction; however, this consists of analyzing the perceptible features of a phenomenon which can be removed while leaving the phenomenon as a transcendental intact. What is left is called the essence of the phenomenon. Descartes used a similar approach in his skepticism to come to the fundamental *Cogito, ergo sum* (I think therefore I am) as what remains when all other features of human experience are removed (by attributing them to illusions). However, this is not a reduction in the scientific sense, which presumes that all phenomena reduced to a set of entities such that any experience can be reconstructed from those entities. Note that the terms beneficence, nonmaleficence, and autonomy can be used as descriptors in phenomenological ethics, but they are not normative. An adjudication of ethical conflicts is based on what the particular situations and circumstances "mean" to the participants, with the notion of "meaning" derived from the intuitions of the parties that are conscious. It is not clear that a Principlist approach can capture what is "meaning" and, therefore, may be ill-equipped in adjudicating ethical disputes.

It would be an understatement bordering on the inane to say that the notion of meaning is complex. Yet, the notion of meaning has such ubiquity and utility in human interactions that occasional reminders of its complexity and problematic nature are warranted. Whatever constitutes meaning, it is connected with lived experiences, in phenomenological terms. Thus, discussion of the ethical implications of phenomenology is a very large and complex discussion (Sanders and Wisnewski, 2012). However, how the notion of meaning and phenomenology plays out in the ethics of everyday medicine will be discussed briefly as examples. One place to look is where Principlism fails.

Many of the arguments against Principlism center on the nature of autonomy (see Cases 13 and 14). Typical Principlism, as least as articulated by Beauchamp and Childress (2013) and their critics, is that autonomy is paramount and individual. This notion of autonomy is particularly salient when autonomy is conflated with the notion of self-determination. Thus, the notion of autonomy is particularly exposed in situations where a patient is unable to exercise self-determination actively and directly. One then looks to surrogates for the patient but this is difficult, particularly when the patient has not left relevant specific instructions (Brudney, 2009). The surrogate is then left to reconstruct how the patient would assess and judge the situation; in other words, what would be meaningful to the patient. Certainly, self-determination is meaningful to all sentient beings but in this case it is moot. What other values then serve as a source of meaning of the patient's life that would inform the surrogate?

To be sure, meaning can derive from what are considered virtues. Aristotle offered (1) courage, (2) temperance, (3) liberality (expenditure of effort proportional to the reward), (4) magnificence (a proportionate expenditure of effort), (5) magnanimity (a proportionate expenditure of effort but amplified by the significance of the reward such as acts of honor or great beauty), (6) proper ambition, (7) truthfulness, (8) wittiness in conversation, (9) friendliness, (10) modesty in the face of shame or shamelessness, and (11) righteous indignation. Yet, these appear to be relevant to the actions of the surrogate rather than the patient (see discussion of virtue ethics). In this case, virtue ethics, as might be applied by the surrogate, would appear to be surrogate-centric instead of patient-centric [note that there is a long history of other than patient-centric ethics (see Chapter 4). While the patient also had the capacity to want to act virtuously, this is not a relevant question. Perhaps the surrogate could decide based on the patient's ability to act virtuously in the future. Unqualified, it is not clear how the last approach is self-evidently different from the Principlist notion of the principles of beneficence, nonmaleficence, and autonomy.

Another concept is authenticity, which recognizes a patient's right to be an authentic person. Typically, the latter refers to what the patient or person "naturally would evolve to" in the absence of external influences. What the person would be is indeterminant but it would be authentic to that person. The notion of authenticity is difficult for many reasons. As a matter of practice, can there be a person absolutely unaffected by external forces? Also, external forces have a way of being internalized. Once internalized, they contribute what the person holds as authentically themselves, but the person cannot claim that his or her authentic person is independent of external forces.

The notion of authenticity has two aspects. First, the person or patient has a right to self-determination, that is, a right to be authentic (Brudney, 2009). Helping a person achieve the results of self-determination would be an act of beneficence. However, the second component is what exactly is authentic for particular individual person A? Does the obligation to the principle of autonomy, which necessitates recognition of the right to self-determination, in principle, extend to an obligation relative to what the patient sees as beneficial to her authentic person? Do the obligations to autonomy and beneficence mean providing medical assistance so that a genetic (sex) male can actualize himself as a female gender? A physician or healthcare specialist may not recognize as beneficent what the patient believes to be beneficent. Indeed, if the physician or healthcare professional acts in a way to deny the patient's gaining beneficence as she sees it, then the act could be harmful and a violation of the ethical principle of nonmaleficence.

There are more common examples in the everyday practice of medicine. Most normal persons act as though they believe in the integrity of the whole mind/body (see Case 14). Disease disrupts that integrity separating mind from body. For some, disease pathology is seen as a revolt or betrayal by the body, causing hurt in the mind. Many Rationalist/Allopathic physicians see their purview as the body, not the mind, and it is to the body that the beneficence is extended, perhaps not what the patient would have in mind (see Cases 7 and 14). One might argue that attending to the totality of the mind and body as would be beneficent to the patient would be a patient-centric ethic. What the physician is willing to provide seems to be different and physician-centric ethic (see Case 6 and Chapter 4). It would seem that a phenomenological ethic would be, by its nature, patient-centric. Yet, this may prove unwieldy (see Chapter 4).

Narrative ethics

Phenomenological ethics follows from conscious deliberations based on intuitions. It is easy to see such actions as static snapshots of one's self-conscious existence, although this is not necessarily the case. Yet, self-consciousness often is the form of an ongoing story. The phenomenon has been described as the "stream of consciousness" considered as a psychological state (James, 1890), which suggests a continuity of self-consciousness even as the content of the self-consciousness changes over time. The concept found extension as a literary device. Its mode of expression has been called the interior monologue and can be ascribed the structure of a narrative or evolving story. One's own narrative is not merely a commentary on past experiences but the narrative is self-informing, the person is affected by her own narrative. Persons are conscious not only of the present, but also of the past, and a form of consciousness extends to expectations for the future. However, unlike phenomenology, there may be unconscious subtexts that affect behavior and are informative.

All of these time domains exert an effect on behaviors in the here and now. The ongoing narrative allows persons to situate themselves in the ethical situation and circumstance that are evolving and have a trajectory. The narrative then forms the substrate for personal ethical induction and the resulting transcendental abstractions form the basis of future expectations. The associated emotions then drive behavior that can be assessed as just or unjust.

In short, the narrative of a patient's life is a story with all the literary devices, for example, rhetoric. Emotive words and rhetorical arguments are often employed, typically implicitly, because this rhetoric has impact, conveys urgency, seeks understanding and commiseration, and demands attention (see Case 14). Often, this has been examined from the perspective of the use of rhetoric on the part of the physician or healthcare professional (Dubov, 2015). Indeed, some have viewed the use of rhetoric as contrary to the obligation to autonomy due the patient. Others have argued for the use of rhetoric in the obligation to beneficence by inducing patients to act in what the physician or healthcare professional sees as in the patient's best interest. However, this is not to say that patients or their surrogates do not also engage in rhetoric. The use of rhetoric by the patient is often meant to elicit a positive countertransference in the physician or healthcare professional (Lucius-Hoene et al., 2012) (see Chapter 4).

It is telling that many physicians and healthcare professionals hold the patient's narrative as suspect. Studies demonstrate that the patient's narrative is not entirely consistent with the patient's actual past experience and perhaps an encounter with a physician or healthcare profession is collaborative work; the physician's or healthcare professional's interactions can affect the subsequent course of the patient's narrative greatly. In a study of medicine residents, the average time a patient spoke before the physician intervened was approximately 12 s, with the physician interrupting the patient frequently (Rhoades et al., 2001). One might dismiss such findings, arguing that these are just resident physicians in training; however, that would presuppose that there is something in the resident's subsequent experience that would change the pattern. In the author's experience, this is unlikely.

Most importantly, the patient's narrative is held to reflect what is most important to the patient, thus clearly indicating what the patient would take as beneficence and nonmaleficence, which is the object, in part, of the physician's or healthcare professional's obligations to beneficence and nonmaleficence and, importantly, patient autonomy. In important ways, considering the patient's narrative from a literary perspective allows literary methods to help physicians and healthcare professionals understand the patient. An extended quote from Shapiro (1993) is helpful:

> Central to the plot are human predicaments and their attempted resolutions, of which disease and its cure or amelioration are merely specific subsets. The theme, or definitive message, insight, or organizing purpose of a story, provides us with a context in which to understand specific disease-oriented complaints. Finally, style, or the manner in which narrative is presented, yields significant clues about the person of the narrator. How does the patient tell his or her story? How do patients use detail, language, and dramatic climax? Where does their story become obscure, confusing, or unbelievable? These questions all help us assess potential areas of confusion and miscommunication in the doctor-patient interaction.

The self-narrative has a trajectory that projects expectations, which contrasts with the subsequent turn of events. The contrasts allow discovery of the narrative's trajectory. Its collisions with ethical realities can be illuminating. Such narratives are intensely personal and private. Even attempts at a declarative description more likely represent translations rather than transcripts. Consequently, intensely personal and private narratives are not an inviting environment for the Principlist approach. Exploring each shareholder and stakeholder in a difficult ethical problem for a narrative perspective can serve a number of important functions. First, each person's narrative can be seen as a story, and the tools of literary criticism or literary theory can be brought to bear to explicate and understand that narrative relative to the ethical problems.

Second, as the narrative involves past and present experiences and expectations for the future, it provides a fuller representation of persons' values for themselves and for others. Often, this can be a basis to inform persons' senses of obligation to the ethical principles for themselves and others and to give weight to a particularly moral theory that ultimately must prevail. Indeed, this is the basis for the Consequentialist approach to ethics. However, it is important to note that the narrative need not be dismissive of ethical principles, moral theories, laws, case precedents, or virtues.

Third, narratives are evocative, as is most good literature. It generates emotions that provide the force to justice. A good narrative likely will make any formalized logical argument appear acerbic at worst and irrelevant at best; hence the disinclination of narrative ethics toward Principlism. The power of the narrative is its ability to resonate with similar feelings in the observer, giving force to the observer's own sense of justice, even if in a neurophysiological sense (see Chapter 1).

Given the evocative power of the narrative, the important question is whether the narrative provides greater power to a particular moral theory, thereby influencing the ultimate expression of justice strongly. As the narrative on the part of the speaker evokes similar feelings on the part of the listener, there is a presumption of commonality of values. Moral theories, such as libertarianism and utilitarianism, would not necessarily recognize, indeed may be antithetical to, a common value and hence the bond between the speaker and the listener. Hence, Libertarian and Utilitarian moral theories are not likely to have much sway. An Egalitarian moral theory would recognize the relationship between the listener and the speaker but not to the exclusion of any and all other potential speakers and listeners. This would deny a special relationship between the individual and particular speaker and the listener. Finally, a Deontological moral theory would give an unassailable privilege to the relationship of the individual particular speaker with listener even if it does not deny such privilege to any other pair of speakers and listeners. Yet, it is not clear that the interactions driven by the personal narratives of all shareholders and stakeholders necessarily will comport with the moral theories given previously and perhaps new and different moral theories will emerge.

Feminist ethics

As the notion of gender, rather than biological sex, is recognized as "fluid," so too does feminist ethics admit to a variety of positions. However, there are several common features, perhaps best described as family resemblances. Central to feminism is the notion of cultural gender inequality that is historical fact. Feminism concerns itself with the origin and maintenance of such inequality and then means for rectification. Maintenance of such inequality, in part, may be related to a lack of awareness of the inequality and also to active and passive efforts to enforce the inequality. Indeed, one means of enforcement is to reinforce the lack of awareness. The latter is particularly important in liberal democracies where there are no received ethical or moral rules, such as the word of God or dictates by absolute sovereigns.

In a pluralistic modern liberal democracy, creation and maintenance of inequalities cannot be obvious, as an extreme would generate revolution, notwithstanding the revolutions that pop up from time to time. Further, it is argued that all humans would react to situations and circumstances of inequality as to suggest these are unjust. Consequently, creation and maintenance of inequalities require some mitigating factors that would offset the clear injustice of the inequalities. Offsetting factors include the unique responsibilities and hence prerogatives of the male gender. Some may argue historical tradition as though history was some impersonal power and therefore exonerates humans, particularly males, from responsibility. Another offers a privileged position for the female gender shielded from the trials and tribulations of full engagement with the world "red in tooth and claw" (Tennyson, 1849) in exchange for the unequal status.

Whatever relative advantages or disadvantages of biological sex thought to be conveyed to gender in physical survival in the past, modern society renders moot, as witnessed by women in the military and other circumstances where biological differences, once thought to warrant differential treatment, no longer exist. Yet, inequalities persist nonetheless, suggesting that some form of coercion is in play. Feminism has clearly demonstrated the subtle (and not so subtle) coercive forces that still operate in the ethics of everyday medicine. Indeed, the culture and structure of the interactions between patient and physician or healthcare professional reinforce such inequalities (see Case 5 and Chapter 4). Further, the ubiquity and often insidious inequalities become so ingrained that "to see what is in front of one's nose needs a constant struggle" (Orwell, 1946). Feminism has been a potent spotlight to illuminate these issues as it affects patients as well as physicians and healthcare professionals.

Gender differences in the expected healthcare roles blend gender and political issues, with the latter related to the different privileges and prerogatives afforded to physicians as compared to healthcare professionals (see Case 1). Until relatively recently in the history of healthcare, most physicians have been male gender while most healthcare professionals have been female gender. The interesting question is whether this was happenstance or indicative of differences in the genders. Consider the notion that even when a child of two physicians is ill, most often it is the female gender who stays home to care for the child. Some feminist ethicists hold that gender differences are ontological, perhaps biological, breeding different perspectives built of different values. There is a female gender perspective that is different from the male gender perspective (Baron-Cohen et al., 2005). Contrasts include the female gender perspective as more group focused, emphasizing shared values and commitment to shared values, whereas the male gender perspective is individualistic. As Susan Sherwin (1989) wrote, "It is clear that in doing feminist ethics, it is important to be critical of the maleness of mainstream ethical theory, given its tendencies to demand a very high degree of abstractness and to deny the relevance of concrete considerations."

As in pluralistic modern liberal democracies, ethical systems derive from ethical induction, consensus; the critical question is who makes up the group that is responsible for generating the consensus? It is those charged with developing a consensus to gather what are held as relevant experiences to be the substrate of ethical induction to consensus. Feminists would argue that the male gender, generally speaking, is not capable of reaching the same consensus as the female gender. Thus, applying consensus derived from a male gender group and applied to the female gender would be the logical Fallacy of Four Terms (Chapter 3). It is not clear that enforcing a consensus reflecting only male gender onto the female gender would be considered just.

As discussed in Chapter 4, medical or clinical ethics has been physician-centric until the late 1960s and early 1970s. During the physician-centric period, as physicians were mostly men, it would not be surprising that the ethics were, in a sense, male-centric. Subsequent to the 1960s and 1970s, ethics has come to be business- and government-centric. To the degree that business and government is dominated by men, then the critical question is whether the new and dominant business- and government-centric ethics is male-centric ethics. The argument is that proactive measures are needed to counter the historical dominance of male-centric ethics.

As principles are more difficult to apply over a particular group, given human diversity, many feminists discount the Principlist approach. However, it is unlikely that feminists would say that the ethical principles and moral theories do not apply; rather the obligations to those principles differ between the group and the individual and possibly between genders. It is not that the individual is not deserving of beneficence, autonomy, and nonmaleficence, but rather those same considerations are due to the group and often those obligations to the beneficence, nonmaleficence, and autonomy of the group prevail over those obligations to the individual, for example, obligations to the family unit in contrast to obligations to the individual child who is the patient (see Cases 13 and 14). Yet, for this to be countenanced, a sufficient number of individuals in the group would have to acquiesce to apply the commitment to the group over that of the individual in pluralistic modern liberal democracy. In that sense, justice, even in a feminist ethic, ultimately derives from the individual.

Care ethics

Another alternative to the Principlist ethical theory is what is termed care ethics, which emphasizes the dynamics between the individual person receiving care and the individual providing care. Interestingly, care ethics, in part, arose from feminist ethics, despite the emphasis of the group over the individual, at least for some feminist ethicists as described earlier (Kuhse, 1997). However, care ethics can be seen consistent with the female gender perspective of commitment to care, emphasizing the individual, rather than the principle. However, it is not as though the ethical principle of beneficence (care) obviates the other ethical principles of autonomy and nonmaleficence. Rather, the obligations to beneficence are embedded in a Deontological moral theory that emphasizes the obligation to beneficence over autonomy and maleficence (consistent with the Principle of Double Effect). However, the situation and the circumstance of moderate scarcity make such Deontological moral theories difficult, as revealed in attempts at enforcement as discussed in the cases presented in Part 1.

Virtue ethics

Similar to care ethics, virtue ethics focuses on the individual providing care but not necessarily on the individual person receiving care. Virtue ethics emphasize behaviors of the care provider while the patient is considered in a more generic manner. In some ways, emphasizing the patient in a generic manner makes the ethical interactions less contingent and variable, thus more binding. Good ethics is just what a good (virtuous) physician or healthcare professional does. A failure to follow good ethics just means that the physician or healthcare professional was not good, thereby violating the terms of the conditions by which physicians and healthcare professionals are privileged to practice medicine. Interestingly, the same claim was made by biomedical researchers opposed to the Belmont Report (Belmont Report: Ethical Principles and Guidelines for the Protection of Human Subjects of Research, 1978), which established committees on human studies and institutional review boards to govern human research. These critics argued that good human-based research is just what good researchers do. Yet, human subject research scandals subsequent to the Belmont Report obviated such criticism. The same issue then confronts the regulation of virtue ethics.

Virtue ethics focuses on the particulars, not on the principles, considered in a declarative, conscious, and analytic manner. Aristotle held that virtue was derived from practical wisdom (*phronēsis*) rather than theoretical knowledge (*sophia*). Curiously, scientific (empirical) knowledge (*epistēmē*) is generally not considered, although justice is empirical and should be scientific knowledge, as argued in Chapter 1. The implication is that virtue, known by practical wisdom, is contingent and de novo and consequent to the particulars. However, such a position invites all the problems described previously for the Anti-Principlists, such as Solipsism of the Present Moment. How is a person to judge the ethics of another's behavior if there is no normative standard? Perhaps the only recourse is to compare the present behavior to some previous occurrence of the exact same situation and circumstance and to apply the judgment achieved then, a precedent. However, if every experience is anew, there can never be a precedent and thus no ethical judgment is possible. Even Aristotle senses the problem, writing:

> But this [judgement as to the good] is no doubt difficult, and especially in individual cases; for *it is not easy to determine both how and with whom and on what provocation and how long one should be angry* [italics added]; for we too sometimes praise those who fall short and call them good-tempered, but sometimes we praise those who get angry and call them manly. The man, however, who deviates little from goodness is not blamed, whether he do so in the direction of the more or of the less, but only the man who deviates more widely; for he does not fail to be noticed. *But up to what point and to what extent a man must deviate before he becomes blameworthy it is not easy to determine by reasoning, any more than anything else that is perceived by the senses; such things depend on particular facts, and the decision rests with perception* [italics added]. So much, then, is plain, that the intermediate state is in all things to be praised, but that we must incline sometimes towards the excess, sometimes towards the deficiency; for so shall we most easily hit the intermediate and what is right.
>
> *The Nicomachean Ethics (2009)*

It seems that Aristotle is arguing for a mean (average) as the standard with the degree of departure from the mean determining the ethical culpability of the present behavior. However, in the absence of any past experiences of exactly the same situation and circumstance, how is any mean determinable? Further, how does one establish the degree of departure where judgment is dichotomized as either just or unjust as it cannot be both and cannot be neither?

Virtue has always been a touchstone in medical ethics, for example, as noted by Thomas Percival in 1803, as discussed in Chapter 4. Members of the professions are required to be of good moral character (virtuous) and standing (acting virtuously). However, the notion of virtue was on behalf of the shareholders, the physicians, for the sake of the shareholders. Nothing in this sort of virtue is driven or demanded by the stakeholders, the patients. Physicians were extolled to be virtuous by means of their special privileges, with nothing to enforce virtuous behavior other than charity to the patient or judgment by the physician's peers, with the former reflecting on the virtue of the giver of charity. Appealing to such virtue for justice, as demonstrated repeatedly in cases in Part 1, is a "dicey proposition."

There may be some value in virtue as a predisposition (instinct) to act independently of declarative, deliberative, and conscious analyses of principles and theories. Immanuel Kant recognized the importance of ethical instincts, inherently procedural, as the best way to ensure ethical behavior. Kant wrote:

> In the physical constitution of an organized being, that is, a being adapted suitably to the purposes of life, we assume it as a fundamental principle that no organ for any purpose will be found but what is also the fittest and best adapted for that purpose. Now in a being which has reason and a will, if the proper object of nature were its conservation, its welfare, in a word, its happiness, then nature would have hit

upon a very bad arrangement in selecting the reason of the creature to carry out this purpose. For all the actions which the creature has to perform with a view to this purpose, and the whole rule of its conduct, would be far more surely prescribed to it by instinct [author—taken here as procedural knowledge], and that end would have been attained thereby much more certainly than it ever can be by reason [author—taken here as declarative knowledge]. Should reason have been communicated to this favored creature over and above, it must only have served it to contemplate the happy constitution of its nature, to admire it, to congratulate itself thereon [author—taken here as examples of post-hoc declarative commentary on the consequence of procedural knowledge], and to feel thankful for it to the beneficent cause, but not that it should subject its desires to that weak and delusive guidance, and meddle bunglingly with the purpose of nature. In a word, nature would have taken care that reason should not break forth into practical exercise [author—taken as procedural knowledge], nor have the presumption, with its weak insight, to think out for itself the plan of happiness, and of the means of attaining it. Nature would not only have taken on herself the choice of the ends, but also of the means, and with wise foresight would have entrusted both to instinct.

Kant (2001)

It is very difficult to discount any of the Anti-Principlist ethical systems. Indeed, most humans would identify with the particulars in each of the ethical systems. Using a logical device to demonstrate the truth of a statement by asserting the denial of the claim, who would not want to be an authentic person as described by the phenomenologists? Who does not believe that they have a story that puts into a context, their past, and who would deny that their story has a future, as held by narrative ethicists? Who would deny that the male gender has dominated the medical professions as physicians and that most shareholders, those with the power to effect justice or injustice, are predominantly men? Who would deny that most physicians and healthcare professionals derive value and satisfaction from caring for others, which is why they are in clinics and hospitals as opposed to laboratories or board rooms? Who would hold themselves as vicious rather than virtuous? These are the real human experiences that defy easy abstractions to principles, moral theories, laws, case precedents, and virtues, although *they must*. These experiences must be fully understood and appreciated so that they form the proper substrate for induction to the necessary principles, theories, laws, case precedents, and virtues. Anti-Principlists are on the other side of the same coin occupied by Principlists.

References

Bacon, F., 1620 (republished 2010). Novum Organum: True Directions Concerning the Interpretation of Nature. Kessinger Publishing.

Baillargeon, R., 1987. Object permanence in 3 ½- and 4 ½-month-old infants. Dev. Psychol. 23 (5), 655–664.

Baron-Cohen, S., Knickmeyer, R.C., Belmonte, M.K., 2005. Sex differences in the brain: implications for explaining autism. Science 310 (5749), 819–823.

Beauchamp, T.L., Childress, J.F., 2013. Principles of Biomedical Ethics. Oxford University Press, Oxford.

Belmont Report: Ethical Principles and Guidelines for the Protection of Human Subjects of Research, National Commission for the Protection of Human Subjects of Biomedical and Behavioral Research, Department of Health, Education and Welfare. 1978. United States Government Printing Office, Washington, DC.

Brudney, D., 2009. Choosing for another: beyond autonomy and best interests. Hast. Cent. Rep. 39 (2), 31–37.

da Rocha, A.C., Bergareche, A.M., 2008. Wired for autonomy. Am. J. Bioeth. 8 (5), 23–25. (discussion W1-3) https://doi.org/10.1080/15265160802180042.

Descartes, R., 2003. Treatise of Man, Translation and Commentary, Thomas Steele Hall. Great Minds Series. Prometheus Books, Amherst, New York.

De Weerd, P., Gattass, R., Desimone, R., Ungerleider, L.G., 1995. Responses of cells in monkey visual cortex during perceptual filling-in of an artificial scotoma. Nature 377 (6551), 731–734.

DiSessa, A.A., 1982. Unlearning Aristotelian physics: a study of knowledge-based learning. Cogn. Sci. 6 (1), 37–75.

Dubov, A., 2015. Ethical persuasion: the rhetoric of communication in critical care. J. Eval. Clin. Pract. 21 (3), 496–502. https://doi.org/10.1111/jep.12356.

Durand, K., Lécuyer, R., 2002. Object permanence observed in 4-month-old infants with a 2D display. Infant Behav. Dev. 25 (3), 269–278.

Foot, P., 1978. The Problem of Abortion and the Doctrine of the Double Effect in Virtues and Vices. Basil Blackwell, Oxford. (Originally appeared in the Oxford Review, Number 5, 1967).

Gigerenzer, G., Beatty, J., Kruger, L., Daston, L., Porter, T.M., Swijtink, Z., 1989. The Empire of Chance: How Probability Changed Science and Everyday Life. vol. 12. Cambridge University Press.

Greenfield, B., Jensen, G.M., 2010. Beyond a code of ethics: phenomenological ethics for everyday practice. Physiother. Res. Int. 15 (2), 88–95. https://doi.org/10.1002/pri.481.

Gregory, R.L., 2005. The Medawar Lecture 2001 knowledge for vision: vision for knowledge. Philos. Trans. R. Soc. Lond. B: Biol. Sci. 360 (1458), 1231–1251.

Heinämaa, S., 2009. Responses to Gestalt psychology. In: Heinämaa, S., Reuter, M. (Eds.), Phenomenological in Psychology and Philosophy. Lagerlund, H., Yrjonsuuri, M. (Series Eds.), Studies in the History of Philosophy and Mind, Vol. 8. Springer, New York, pp. 263–284.

Hobbes, T., 1651. Leviathan or the matter, forme and power of a common-wealth ecclesiasticall and civil.

Hofstadter, D.R., 1999. Gödel, Escher, Bach: An Eternal Golden Braid. Basic Books.

Husserl, E., 1913. Ideas Pertaining to a Pure Phenomenology and to a Phenomenological Philosophy. F. Kersten, Trans. Nijhoff, The Hague. 1982, sec. 24, quoted in http://www.iep.utm.edu/phenom/.

James, W., 1890. Discrimination and comparison. In: Principles of Psychology. (Chapter 13).

Kant, I., 1781. Critique of Pure Reason. Translation Norman Kemp Smith, 1965. Saint Martin's Press, New York.

Kant, I., 2001. Fundamental principles of the metaphysics of morals. In: Wood, A.W. (Ed.), Basic Writings of Kant. The Modern Library, New York, pp. 153.

Kuhse, H., 1997. A history of subservience. In: Caring: Nurses, Women and Ethics. Blackwell Publishers, Oxford. (Chapter 2).

Lucius-Hoene, G., Thiele, U., Breuning, M., Haug, S., 2012. Doctors' voices in patients' narratives: coping with emotions in storytelling. Chronic Illn. 8 (3), 163–175.

McClelland, J.L., 1988. Parallel Distributed Processing: Implications for Cognition and Development (The Artificial Intelligence and Psychology Project Technical Report AIP-47). Available at: http://www.dtic.mil/dtic/tr/fulltext/u2/a219063.pdf.

Montgomery Jr., E.B., 2019a. Medical Reasoning: The Nature and Use of Medical Knowledge. Oxford University Press, Oxford.

Montgomery Jr., E.B., 2019b. Reproducibility in Biomedical Research: Epistemological and Statistical Problems. Academic Press, New York.

Orwell, G., 1946. Tribune, London, March 22.

Percival, T., 1803. Medical Ethics, or a Code of Institutes and Percepts, Adapted to the Professional Conduct of Physicians and Surgeons. Cambridge University Press.

Pfaff, D.W., Kavaliers, M., Choleris, E., 2008. Mechanisms underlying an ability to behave ethically. Am. J. Bioeth. 8 (5), 10–19. https://doi.org/10.1080/15265160802179994.

Rauprich, O., 2008. Common morality: comment on beauchamp and childress. Theor. Med. Bioeth. 29 (1), 43–71.

Rawls, J., 1999. A Theory of Justice, revised ed. The Belknap Press of Harvard University Press, Cambridge, MA.

Reyzin, L., 2019. Unprovability comes to machine learning. Nature 565 (7738), 166–167. https://doi.org/10.1038/d41586-019-00012-4.

Rhoades, D.R., McFarland, K.F., Finch, W.H., Johnson, A.O., 2001. Speaking and interruptions during primary care office visits. Fam. Med. 33 (7), 528–532.

Ryle, G., 1949. The Concept of Mind. University of Chicago Press, Chicago.

Sanders, M., Wisnewski, J.J. (Eds.), 2012. Ethics and Phenomenology. Lexington Books.

Selby-Bigge, L.A. (Ed.), 1888. Hume's Treatise of Human Nature. Clarendon Press, Oxford, pp. 89. (Originally published 1739–1740).

Shapiro, J., 1993. The use of narrative in the doctor-patient encounter. Fam. Syst. Med. 11, 47–53.

Sherwin, S., 1989. Feminist and medical ethics: two different approaches to contextual ethics. Hypatia 4 (2), 57–72.

Spillmann, L., 2006. From perceptive fields to gestalt. Prog. Brain Res. 155, 67–92.

Tennyson, A., 1849. In Memoriam A.H.H., Edward Moxon, London, 1850.

The Nicomachean Ethics, Aristotle. 2009. Oxford University Press, Oxford. D. Ross, Trans.

Thomson, J.J., 1976. Killing, letting die, and the trolley problem. Monist 59 (2), 204–217.

Wagemans, J., Elder, J.H., Kubovy, M., Palmer, S.E., Peterson, M.A., Singh, M., von der Heydt, R., 2012a. A century of Gestalt psychology in visual perception. I. Perceptual grouping and figure-ground organization. Psychol. Bull. 138 (6), 1172–1217. https://doi.org/10.1037/a0029333.

Wagemans, J., Feldman, J., Gepshtein, S., Kimchi, R., Pomerantz, J.R., van der Helm, P.A., van Leeuwen, C., 2012b. A century of Gestalt psychology in visual perception. II. Conceptual and theoretical foundations. Psychol. Bull. 138 (6), 1218–1252. https://doi.org/10.1037/a0029334.

Reflective equilibrium and ethical induction from the perspective of logic and statistics

3

To be argued...

Assuming that one lives in a pluralistic modern liberal democracy, the general principle often is held that the individual is governed in a manner determined by all (or a sufficient number) of those governed. The great challenge is the fundamental ontological conundrum of the potentially infinite variety of human experiences that forces an epistemic choice. Is variety a variation on some economical set of operating principles, theories, laws, case precedents, and virtues? Alternatively, is variety of diversity with each experience de novo and unique? The epistemic choice largely determines the ethical system that subsequently is developed and applied (Chapter 2). The argument presented here presumes the choice of variety as a variation on an economical set of principles, theories, laws, case precedents, and virtues. Importantly, these same issues confront traditional science. Thus, traditional science will be discussed in order to illuminate how these same issues may play out in the development of ethical systems.

The challenge then is to define the economical set of principles, theories, laws, case precedents, and virtues. There must be some sort of ethical induction, in its widest connotation, from the particulars of past experiences to something more general that explains the present and can be applied prospectively. The test is the reproducibility of future judgments, as irreproducibility would undermine any sense of justice. Any such set of ethical principles, moral theories, laws, case precedents, and virtues, giving rise to inconsistent applications of justice in cases of similar situations or circumstances, likely would be held suspect and unjust. Thus, any endeavor to develop any system of ethics would want to first address the means to ensure reproducibility.

This challenge is little different from any other empirically based system of knowledge, including biomedical research. All begin with a set of experiences of particulars, specific patients with disease *A* or ethically disagreeing participants. Next, an attempt is made to derive a commonality among the particular experiences, for example, by induction. Yet, the notion of commonality is at odds with the variability of the experiences. How will induction succeed? What is necessary for its success?

In biomedical research, and science in general, many tools have been developed, particularly probability and statistics, to address these questions and provide at least some modicum of confidence or certainty in the outcome (Montgomery, 2019). Indeed, traditional science often provides confidence intervals around result variables that define the variation around the Central Tendency reported in the result. These concepts also apply to scientific investigations whose variable are not quantifiable. In ethics, an example might be that in cases of the kind that is more difficult to quantify, it is more likely that the obligation to nonmaleficence will trump the other obligations because there was the consensus of a panel of judges. And one can know this because (1) the judges were determined to be competent and free from bias prior to the deliberations, (2) there were a sufficient number of judges, (3) a sufficient number of cases were reviewed, (4) the cases were sufficiently similar to constitute an ethical kind, and (5) the confidence (certainty) of the consensus was judged sufficient to allow the application to all future cases of the same ethical kind. These same types of questions are fundamental to traditional science as well. Indeed, the specific methods used in science and ethics are tokens of the same type.

It has been argued that the drive to certainty in biomedical science appeals to deductive logic following the establishment of hypotheses through induction, at least implicitly. This is the Logic of Certainty, which paradoxically does not provide new knowledge. Rather, new knowledge results from the judicious applications of deductive logical fallacies to create the Logic of Discovery (Montgomery, 2019b). This chapter pursues the same course of analyses for ethical systems, the combinations of ethical principles, moral theories, laws, case precedents, and virtues used to adjudicate ethical disputes and to prevent them.

It is necessary to refute some widely held myths in medicine, biomedical research, and ethics (Montgomery, 2019a, 2019b) that falsely generate resistance to statistics. Statistics does not measure what is "true," "what is real," or "what is out there." Statistics only measures internal consistency in the data, the observations or experiences under consideration. The jump from what is internally consistent to reality, "what is out there," is enormously problematic and perhaps intractable. To be sure, a great deal of the history of human experience would seem to vote in favor of the claim statistics cannot create, on its own, any ethical system. All statistics can say is that relative to the experiences, the judgment resulting from the ethical system has some degree of reproductivity, hence certainty. Thus, the ethicist may well benefit from knowledge of these methods.

Moral and ethical objective truth from ethical induction

New approaches to ethics are needed. The numerous and disparate current approaches (see Chapter 2) become problematic as to their use in the actual adjudication of ethical disputes as shown by the cases in Part 1. Which is correct? A unifying theory of ethics is necessary. Chapter 1 offered the innate sense of justice as a starting point for the post hoc construction of ethical systems, yet experience markedly shapes and adapts what once were inborn mechanisms. Experiences begin to model the human from the first moments of life; consequently, it is likely impossible to reconstruct from direct witness by humans. The very young cannot provide a verbal report and elders are already too changed. Rather, the effects of experience and how it shapes subsequent senses of justice have to be explored empirically, primarily by the observation of behaviors to which justice or injustice has been attributed.

The necessity to explore empirically what are the operational predispositions to the judgment of situations and circumstances also follows from the social contract. Pluralistic modern liberal democracies eschew appeals to the word of God, the divine right of kings, and the totalitarianism of absolute dictators. There is the idea that ethics reflects the will of those to be bound by the ethics, even if imperfect in the application. Any social contract given the plurality of perspectives and interests makes certainty difficult in any particular ethical system. Yet, certainty is critical to any sense of justice. The problem becomes how to gauge certainty from an empirical analysis of such a heterogeneous set of observations. This chapter explores the issue of certainty in ethical inductions to ethical systems for the purpose of just actions in a pluralistic modern liberal democracy.

In many ways, using received knowledge such as from God, some objective truth, or the dictates of an absolute sovereign in ethics is appealing. At the very least, one can have ethical and moral certainty, if not clarity. In these cases, the certainty is imposed by something external to the person caught in an ethical or moral problem rather than a self-imposition of those subjects. In the usual connotation of traditional science, the laws of physics are omnipotent as they are not contingent. Physical laws have always been and they will always be, at least since the first seconds following the Big Bang. It is not possible to violate them although there can be adverse consequences in the attempt to do so; even if the violation appears successful momentarily.

It is human and reasonable to long for moral certainty that might be obtained if there were objective ethical and moral truths. Fundamentally, and in a philosophically trivial way, objective ethical and moral truths must exist as the solutions to problems are consequences of the changes in the electrical potentials (voltages) across the cell membrane of excitable cells, particularly neurons, in the human nervous system. It could not be otherwise, even if the solutions are not explicable in readily understandable human terms (see Chapter 1).

Perhaps the notion of objective needs to be unpacked, in the philosophical sense, that is, dissected into its components and dynamics. In one sense, objective truth is external to and independent of the human. But what could be seen as objective if it is not obtained by human means? Humans only have what impinges on their senses. The only ontological reference possible is the consistency, reproducibility, of some knowledge, such as objective truth. A truth that is extracted from the senses is cross-checked by sufficient and adequate experiences beyond what could happen by chance. It is the consistency of the abstraction that seems independent of humans and thus objective.

There is another sense of objective, that is, a claim that the relevant experience could not have been anything else in the past and the future. This is not to say that every experience is the same—they may be tokens (instances) of the same type. Rather, the presumption is that every relevant experience will be sufficiently the same following the removal of the human and instrumental error. Indeed, as will be discussed, statistical analyses provide means to "remove" error or at least the effects of error.

It is at least theoretically possible to have objective ethical and moral truths. Consider the situation of taking many thousands of measurements of a phenomenon.

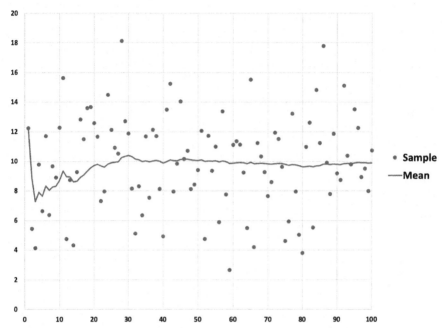

Fig. 3.1 Graph of samples created by a random number generator using normal distribution with a mean of 10 and a standard deviation of 3. Also shown is the value of the mean beginning with a set containing 1 sample and progressing to containing 100 samples.

The measurement values may converge (become asymptotic) on a single value, the Central Tendency, such as the mean or average, based on the Large Number Theorem (Fig. 3.1). (Exceptions to the Large Number Theorem are discussed later in "Chaos, Complexity, nonlinear dynamics, and far-from-equilibrium steady states" section.) Every subsequent measurement value will regress to the Central Tendency. Thus, if one measurement results in a value that is very far from the Central Tendency, the next value is likely to be closer to the Central Tendency—the principle of Regression Toward the Mean. In a very real sense, the Central Tendency is presumed to be the objective truth because every measurement in the future will move to it and every measurement of the past will be seen as having moved to it.

Similarly, if every instance of a relevant ethical problem results in a series of adjudications displaying Regression to the Mean analogous to Fig. 3.1, the system can be said to center around an objective truth. Note that every adjudication doesn't need to have a result which is same as what is held as the truth. However, if one case displays a marked difference from what is thought to be the truth, then the next case result would be more similar to what is taken as the truth. Over the total of all cases, there will be a convergence to what is taken as the truth. It is as though what is taken as the truth is a point attractor that "pulls" every relevant case to the same point or conclusion. If all cases "pull" to the same point, the point can be taken as the truth. It is an objective truth in that there is no other truth possible.

The necessity of ethical induction

If human interactions are not to be determined by received ethics, such as the word of God or the dictates of an absolute sovereign, and if anarchy is to be avoided, from where do the rules, ethics, originate to ensure harmoniously and mutually beneficial human interactions? There seems to be no other source than from some analysis of observed human interactions based on the attribution of justice or injustice (see Chapter 1). As such, the development of ethical systems is an empirical analysis. Further, sheer enumeration of all observable human interactions would not be effective for ethical rule generation as the number of rules would be equal to the number of possible observed human interactions. Short of exhaustive enumeration and in the absence of any generalization across observations, Solipsism of the Present Moment would result. Consequently, an economical set of rules is necessary where the economy is where the number of rules is sufficiently less than the number of possible human interactions, with the latter being potentially infinite. It would seem that the only method for obtaining an economical set of rules would be by induction, in this case, ethical induction to one or few ethical principles, moral theories, laws, case precedents, and virtues that are applicable to a potentially infinite number of situations and circumstances.

Reflective Equilibrium has been advocated as a method of such induction. The origin of Reflective Equilibrium, as a general epistemic procedure, has been attributed to Nelson Goodman (1983) but is more fully explicated by John Rawls (1999) as it relates to his theory of justice.

Ethical induction, in the sense used here, is no different from induction from any other empirically derived knowledge, such as medicine and traditional science. All empirically derived inductions face the same challenge—how to deal with the potentially infinite variety of experiences. The variety of experiences raises an ontological dilemma that forces an epistemic choice. Is variety actually diversity where every experience is treated as de novo and unique or is variety actually variation where all experiences are resolvable to some combination of principles, theories, laws, case precedents, or virtues? How the ontological dilemma and epistemic choice plays out in ethics is discussed in Chapter 2. As discussed in Chapters 1 and 2, the distinction between variety as diversity and variety as variation, in some ways, is artifactual, just as the separation between Anti-Principlist and Principlist ethics is artifactual. These two approaches are just two sides of the same epistemological coin reflecting a single ontology.

The challenge is to make the notion of induction robust, in the case of ethics, to create systems that will result in just outcomes consistently. Measures of robustness include certainty, which stems from reproducibility, yet reproducibility is a great challenge to the induction of all types, for example, biomedical research (Montgomery, 2019b). Thus, the question becomes which methods can be used to assure the greatest robustness of the ethical inductions achieved?

The methods may seem quite different between empirical inductions where the observations can be quantized and those appearing unquantifiable. Quantifiable data allow mathematical analyses, particularly statistics. The apparent inability to quantize observations of ethical experiences would seem to suggest that mathematics and statistics are irrelevant, at worst, or unhelpful, at best. Such pessimism is ill-founded. As will be discussed, statistics is a derivative of the Logic of Discovery, which is derivative of the Logic of Certainty as described in Montgomery (2019b). The Logic of Discovery requires the judicious use of what the Logic of Certainty would hold as fallacies. The robustness of the Logic of Discovery is how close it can approximate the Logic of Certainty; in quantifiable observations, the closeness to the Logic of Certainty can be calculated.

The first order of business is dispelling the misconception that logic is rigid and analytic and thus independent of empirical observations and independent of humans. Logic is synthetic—the term to be used here is organic. In other words, logic arises from the need of humans to be certain, in any knowledge discipline. The great logician Bertrand Russell said that if nature were different, so would be logic, clearly indicating a dependence of logic on nature. To be sure, logicians use a type of calculus, for example, propositional logic argument of the form *if a implies b is true and a is true, then b is true* (modus ponens). The logician would say that the argument is true regardless of whatever is substituted for *a* or *b*. But the argument necessarily supposes the truth of *a implies b*. Whether or not *a implies b* is true is not a matter of logic, it is empirical and observation bound. It is the purview of the Logic of Discovery.

The second order of business is dispelling the notion that statistics solely relates to mathematics and numbers. To be sure, statistics, *as explicitly practiced*, is about mathematics and numbers but this is only superficial manifestations of a much more fundamental logic. Indeed, many problems of reproducibility in biomedical research arise from the sound use of numbers and mathematics but the wrong Logic of Discovery (Montgomery, 2019b). Indeed, it is possible, daresay critical, to use the "spirit" if not the letter of mathematical statistics even in qualitative observations of the kind experienced in ethics.

Ontological issues with induction

The operations of induction, whether implicit or explicit, involve abstracting generalizations from a set of observations of particulars. In ethics, a group of reasonable persons blinded to their self-interest considers a set of particular instances of ethically relevant situations and circumstances from which a set of principles is abstracted, for example, ethical principles, moral theories, laws, case precedents, or virtues. Some Principlists would say that the induced ethical principles are beneficence, nonmaleficence, and autonomy (these need not be exclusive). The moral theories may include Deontology, Utilitarianism, Egalitarianism, and Libertarianism (these need not be exclusive). These abstractions stand in some ontological and epistemological relationship with the set of observations from which they were derived. For example, the induction *all ravens ever seen were black, therefore all ravens are black*. Note that the induction that *all ravens are black* is related to the set of observations. The induction would not be possible if the set of all ravens seen contained nonblack ravens. The induction is more than just an enumeration as an epistemic device, as evident by the likely answer to the question *what color will the next raven be?* The answer likely will be *black.* Blackness, while derived from a logical and epistemological construction, is taken as an ontological property of ravens, at least in this context.

This conundrum has a parallel in statistics where the Central Tendency of a set of quantitative observations is calculated, such as the average (mean). The Central Tendency is taken as the true reflection of the phenomenon, even if there are no actual observations that are the same as the Central Tendency. Thus, the Central Tendency transcends the set of actual observations and measurements. This issue is discussed in detail in Chapter 2.

Epistemological issues with induction

For any generalization, such as an ethical principle, moral theory, law, case precedent, or virtue derived by induction, it takes only a single counterexample to disprove the generalization. For example, it takes only a single nonblack raven to disprove the induction that *all ravens are black.* It does not matter whether 10, 100, 10,000, or 1,000,000 other ravens were examined and all were black. Further, it is virtually impossible to exclude the possibility of an unknown nonblack raven, hence the Fallacy of Induction. One could argue that the single nonblack raven is an exception, an anomaly, and that for all intents and purposes, the induction still holds and the ontological status of the induction is maintained. One might be willing to continue the induction as ontologically true if 1,000,000 ravens were examined and only 1 found was not black; however, one would be less willing if only 10 ravens were examined.

The lesson of ravens holds for ethical induction to ethical principles, moral theories, laws, case precedents, and virtues. Just because the examination of 1,000,000 cases resulted in a consensus that principles of beneficence, autonomy, and nonmaleficence exist, there is no guarantee that the 1,000,001th, 1,000,002th, ... cases will demonstrate the same principles. The most honest induction, even in science, is to a probability, for example, there is a high probability that the next ethical case will demonstrate the principles of beneficence, autonomy, and nonmaleficence. The critical question becomes what degree of probability is sufficient to hold the induction and, importantly, to use the conclusion to enforce justice? Further, what is the probability that the probability determined truly reflects the reality? The Principlist likely will hold that the probability established to date (even if it cannot be quantified) is sufficient to allow the use of the principles in all cases. Other ethical theories may hold that the probability is not sufficient or that the probability is wrongly determined (see Chapter 2).

In ethics and other disciplines, pure logical deduction often is unhelpful because relationships among premises and propositions are not that of identity but rather causal relationships, thus translations from pure logical deduction to practical syllogism are also uncertain. These terms will be explained later. Thus, ethicists are forced to use the probability syllogism even if the probability cannot be quantified. The difference from basic science, such as biology, is that most scientific phenomena can be quantified.

Interestingly, considerable progress exists in assigning more explicit probabilities to qualitative phenomena, such as in ethics. These notions can be leveraged to increase the robustness of ethical induction, as explored later in this chapter.

Induction, particularly by enumeration, for example, counting all ravens seen and noting the number that were black, has limitations. Enumeration involves an identity operator, a particular raven *is* in the set of all ravens and the raven *is* black. Not all relations are identity relations. Indeed, the Logic of Discovery is derived by changing from the identity relations in the Logic of Certainty to causality (Montgomery, 2019b). Being a raven causes the raven to be black, for example, the genetics of ravens is such that the raven will be black. This accommodates the white raven as a departure, a genetic anomaly that still preserves the principle that all ravens are black. To be sure, there is the risk of circularity by making the color black be a necessary prerequisite to being included in the set of ravens.

The notion of cause is readily appreciated in the case of genetics where the physiology and pathophysiology are sufficiently understood. Causality is more problematic in ethics, yet causality is understood as relevant. Unethical behaviors may be accepted because some abnormality of otherwise true causal mechanisms occurred, such as by disease or injury.

Justice is action, either doing something or actively not doing something. Thus, justice is a dichotomous variable, something is either acceptably just or it is unacceptably unjust. Yet, the factors that go into creating the dichotomous outcomes are continuous, reflecting a range of values or degrees. Thus, even the *sense* of justice, in terms of an affective neurophysiological response (see Chapter 1), rather than the obligation is continuous, as there are varying degrees to an affective neurophysiological response. Also, there is a continuous range of obligations from none to absolute. An interesting illustration of the conundrum of converting a continuous to a dichotomized variable or outcome is shown in Fig. 3.2. At what point along the continuum is the object a face or a woman's figure?

Central to justice in the action is some threshold or cutoff that precipitates an action. Yet, the threshold can be quite fluid depending on the situation and circumstance. In medicine and biomedical research, the cutoff of a claim as true or false is often a default to a specific statistic such as a P value, which is derived from the observations. In the search for the Higgs boson in physics, the cutoff was a measurement that was five standard deviations (a quantitative measure of the variability of the results from repeated trials) from the noise level (just chance). In these cases, the decision as to

Fig. 3.2 A series of drawings proceeding from the left, clearly a man's face, to the right where the last image is a female figure. The question is where along the transition does the perception go from that of a man's face to a female figure? Interestingly, that transition point is affected by at which end one starts. Drawings devised by Gerald Fisher in 1967.

the cutoff depends on the consequences. In the case of the Higgs boson, five standard deviations were chosen because *a priori* cutoff of three standard deviations resulted in an error (5-Sigma—What's That?, 2012). In medicine, acceptance of a treatment based on clinical trials that have little or no adverse effects for a severe and devastating disease would mandate a low cutoff (*p* value) (Montgomery and Turkstra, 2003). The cutoff in the case of treatment with a high risk and a disease that is very mild and non-threatening of life or limb should be very high. Unfortunately, very often in medicine a $p < .05$ is the de facto standard without deference to the consequences.

The situation in ethics is no different. As the word of God and the dictates of an absolute sovereign are inadmissible and justice demands action, some sort of threshold or cutoff is necessary. It is extremely difficult to see how anything other than a consequentialist approach in a pluralistic modern liberal democracy is feasible. This is discussed further in "Continuous versus dichotomous character of principles, theories, laws, case precedents, and virtues" section.

Meno's paradox and the *A Priori* Problem of Induction

Induction, ethical and otherwise, first requires the assembling of experiences as the substrate for subsequent induction. Not just any bird can be included in the set of ravens seen from which to induct that all ravens are black. Thus, assembling the experiences requires some prior knowledge; in this case, what are the properties of birds, other than being black that allow the assembling of only ravens? This conundrum was noted in the Socratic dialog in Meno's paradox, which can be paraphrased as *if one doesn't know what one is looking for, how will they know if they found it and if they already know what they are looking for, why look*? (Plato, Meno 80d-e, https://marom.net.technion.ac.il/files/2018/09/Meno.pdf). What aspects of the experiences of cases of ethical concern are taken as relevant so these ethical experiences are taken together as the substrate for subsequent ethical induction? Is the color of one's eyes relevant? Before dismissing certain aspects as *prima facie* irrelevant, recall the effect of the smell of a bakery influencing empathy described in Chapter 1. This conundrum is referred to as the *A Priori* Problem of Induction (Montgomery, 2019a). Clearly, some prior "organizing" principle is required.

The principles, theories, laws, case precedents, and virtues inducted are abstractions and thus internal to the set of observations, such as those judged unjust. It may be that no expert actually holds the inducted principle, moral theory, law, case precedent, or virtue, just as the average height of set persons can be calculated, even though none of the particular persons in the set observed is at the average height. Also, one can take an average of virtually any set of observations if there is at least a single dimension along which the elements of the set vary. For example, it is possible to determine the average weight of a set of elements that include humans and duckbill platypuses, but it is not clear what meaning or sense there would be in such an average. [As an interesting aside, the statistician Stephen M. Stigler (2016), aware of this conundrum, defended the ontological status of the average by saying no good statistician would construct such an average, leaving undefined who is a good statistician.] In a sense, a consensus can be achieved just as an average is described. Thus, a consensus from a set of experts can be obtained but what its validity outside the set of observations is problematic.

Any induction as to the general properties of ravens, such as being black, is more suspect when those making the generalizations are not ornithologists. Similar concerns hold for those generating the inductions to ethical principles, moral theories, laws, case precedents, and virtues. Beauchamp and Childress (2013) came to require reasonableness on the part of those persons who would generate a consensus. John Rawls included several requirements, including IQ, for a person to qualify to be involved in consensus as to what is just (Rawls, 1951). As some prior "organizing" principle to assemble the proper cases, the risk is that the resulting ethical inductions are predetermined (biased) because of those selecting the relevant cases; for example, men deciding ethics that affect women. Further, there likely is considerable variability in the perspectives, knowledge, and experiences among all those who could participate in the consensus generation. In that case, one would suspect that a small number of persons generating the consensus will result in ethical principles and moral theories that are insufficiently generalizable to justify their imposing them on all persons. These issues are addressed in detail later in this chapter.

As the subject matter of ethics involves individuals and their interrelations, it is striking that public/private organizations, such as corporations, are often put on par with individuals. These organizations have many of the same rights afforded individuals, for example, entering into contracts and owning property (Chapter 4). In a sense, corporations are part of the consensus to arrive at governing principles, moral theories, laws, case precedents, and virtue, either by the power of the contracts with physicians and healthcare professionals or by the laws the corporations were instrumental in enacting. This is demonstrated frequently in the cases of Part 1. It is interesting that most discussions of ethics either fail or purposefully exclude consideration of public organizations, such as corporations, in establishing ethical norms (see Introduction).

Discovery versus judgment

In a sense, the consensus must be discovered. Further, the consensus should be agreed to by a sufficient number of citizens, either directly or by their representatives, such that the consensus is enforceable on all citizens (Rawls, 1999). In many situations and circumstances, it is not possible to poll every citizen who would be subject to the consensus, the population. Consequently, a sample (a subset of citizens) from the population (every citizen) must be obtained and studied. The argument constructed is shown in Argument 3.1:

Argument 3.1

Major premise: *A sample of citizens holds principle, theory, law, case precedent, or virtue A*
Minor premise: *The population of all citizens is the sample of citizens*
Conclusion: *The population of all citizens holds principle, theory, law, case precedent, or virtue A*

For Argument 3.1 to be valid and thus held as a validation of *principle, theory, law, case precedent, or virtue A*, it must not be the Fallacy of Four Terms; specifically that *the sample of citizens* is sufficiently the same as the *population of all citizens*. Thus, sampling to obtain the sample will be critical and is discussed further later.

The discovery of the principle, theory, law, case precedent, or virtue must be "extracted" from all stakeholders and shareholders, both actual and potential. Rawls (1951) writes once "...competent judges are selected, there remains to discover [*taken here as constructive*] and formulate a satisfactory explication [*evidentiary support*] of the total range of these judgments." Subsequently, any principle, theory, law, case precedent, or virtue extracted is then used to render a judgment on future experiences of the ethical problem under consideration. As it is highly unlikely that the judgment follows from some specific calculus or logical deduction (with regards to Principlism, it cannot be a deduction in the formal logical sense, see Chapter 1), some interpretation of the principles, theories, laws, case precedents, and virtues is required. Yet such interpretations must be reproducible as applied to past, present, and future considerations of the ethical problem at hand. Further, as it is unlikely that the adjudication will involve the same judges, then the concept of reproducibility extends to other judges positioned to adjudicate the ethical problem. Most persons would react negatively to different judgments rendered by different judges for the same ethical problem. Indeed, this may be the basis for a hierarchical system of courts in legal cases. Those judges selected in a specific instantiation of the ethical problem can be considered a sample as in Argument 3.2:

Argument 3.2

Major premise: *A population of judges holds principle, theory, law, case precedent, or virtue A to prevail*
Minor premise: *The sample of all judges is the population of judges*
Conclusion: *The sample of all judges holds principle, theory, law, case precedent, or virtue A to prevail*

Argument 3.2 would be valid and its conclusion true if the population of judges is the same as the sample of judges. Thus, the critical question becomes how were these specific judges sampled from the population of judges? For example, a population of judges may include some that are very conservative and others that are liberal. If the judges selected as the sample that renders judgment were taken from a convention of conservative judges, it is suspicious whether the sample would reflect the population of all judges. This is discussed further in "The experts, judges, and officials as those who make the consensus" section.

Logic and philosophy

To be argued here is that logic, particularly the extension of the logic of a certain kind, can help understand ethical induction and make the inductions more robust, thereby assuring greater justice in their application to subsequent cases. As will be seen, this expanded version of logic extends the use of valid deductive logic to the judicious and necessary use of logical fallacies. Indeed, the judicious use of these logical fallacies is critical to the substantiation of ethical inductions in order to gain new ethical and moral knowledge. It is no different than the case in biological knowledge (Montgomery, 2019b). The robustness or reproducibility of new ethical knowledge depends, in part, on the degree of difference between certain deductive Logic of Certainty and the logical fallacy necessary to use in order to discover new knowledge, the Logic of Discovery. As will be discussed, attempting to reconcile the gap leads to probability and statistics, particularly in their conceptual sense. This is as true for ethics as it is for science.

To be sure, probability and statistics can be derived from logic and that a sense of probability and statistics is necessary in any system of ethics where the ethical principles and moral rules for relating principles, as well as laws, case precedents, and virtues, are inducted from experience and applied prospectively. The relevance of probability and statistical thinking to ethics is supported if humans were "hardwired" for probability and statistical intuition in some sense. Many psychologists have argued such is the case (Sturm, 2017); given the relevance of psychology to ethics (Appiah, 2008), the psychological studies would seem relevant to ethics.

Evolution of inherent probability and statistical thinking and logic

Working effectively in a complex and highly varied ecology enabled by the appropriate neurophysiology has survival value, which helps ensure that subsequent generations have the same neurophysiological mechanisms. Such neurophysiological mechanisms must be capable of dealing with the variety of ethical situations and circumstances to facilitate useful predictions of consequences of one's own behaviors and those of others. As no two experiences are the same, humans rely on some transcendental operations that generalize over sets or types of experiences. These transcendental operations are a form of induction. The application of these transcendental operations to future experiences or possible experiences is logic.

An inherent logic, as a means to deal with others, would have survival value and thus evolutionary force. Further, logic is derived from the experience much in the manner of induction. As such, logic must conform to the structure inherent in reality even if that structure is not apparent. Formal logic may represent an attempt to formalize the inductions and consequent deliberations. Most importantly, formal logic imposes constraints on those deliberations in a manner that maximizes certainty—the Logic of Certainty (Montgomery, 2019b). For example, the Principle of the Excluded Middle, where a statement or claim only can be true or false, not both true and false, and not neither true nor false, makes the problems of certainty more tractable, even if in the process, logic was made less useful for gaining new knowledge (Montgomery, 2019b). Nonetheless, the value of the Logic of Discovery is its approximation of the Logic of Certainty.

The problem is that humans did not evolve in an environment that followed the rules of logic, probability, and statistics, as traditionally understood, if for no reason other than the rules of logic, probability, and statistics are abstractions from what happens in reality. Kahneman and Tversky (see Sturm, 2017) describe cognitive illusions, such as the gambler's fallacy (my luck has been so bad, I am due for a good poker hand; generalizing from past experiences to future experiences, when those experiences are independent). Yet, arguably, there is survival value in many cases of overgeneralization. Indeed, the nervous system seems predisposed to overgeneralization. For example, spontaneous patterns of neuronal activities in visual areas of the brain arise in the absence of visual stimuli. A pattern specific to a previous stimulus-induced pattern is more likely to arise spontaneously following a visual stimulation that is no longer present (Han et al., 2008). A cognitive illusion is the predisposition to think that politicians are corrupt, which is not necessarily the case (Sturm, 2017). However, given that the typical person's exposure to politicians is through the public media, the predisposition to presume that all politicians are corrupt is proportionate to the media coverage of corrupt politicians. Thus, induction to politicians are corrupt is a reasonable induction and is not evidence against induction (rational intuition) as Kahneman and Tversky would suggest, and replacing induction with the term heuristic seems to confound the issues unnecessarily.

It should not be surprising that actual human behavior would not always be consistent with the rules of logic, probability, and statistics. However, this does not mean that they do not have value or some counterpart in human neurophysiology. Humans do not comport with Newtonian physics as one would not evade an object thrown at them at a very large distance as experience would demonstrate that the object is likely to fall short. The behavior is more consistent with an Aristotelian rather than Newtonian physics. This is not evidence that humans do not have an intuitive sense of physics (Sanborn et al., 2013). Further, infants and young children appear to have a grasp of certain fundamental physical properties (Hespos and Baillargeon, 2001; Hespos et al., 2016), as well as statistical inferences (Gopnik, 2012).

The centrality of certainty

Central to all knowledge, including ethics, is the element of certainty although this may not be explicitly understood or appreciated. To see the crucial nature of certainty, consider the negation, knowledge is uncertainty. To be sure, the situation is not dichotomous between absolute certainty and absolute uncertainty, as the latter would be equivalent to randomness that cannot contain information and therefore cannot be a form of knowledge. Hence, knowledge is a certainty to whatever degree of certainty achieved.

Perhaps the greatest contribution of philosophy was the consideration of logic as a first philosophy, that is, not a philosophical consequence of biology, mathematics, physics, chemistry, history, and so on, per se. Rather, the origin of logic as a first philosophy came from philosophy, particularly the early Greek philosophers such as Aristotle. Logic is a first philosophy as it is transcendental to the particular experiences that gave rise to logic. It is only since the rejection of the medieval scholastic natural philosophers by modern science and the rejection of modern science by philosophers that a schism developed between philosophy (and the humanities) and science, technology, engineering, and mathematics (the so-called STEM disciplines). The rejection of philosophy is evident in the founding of the Royal Society in 1660. The rejection of scientists by philosophers, such as Thomas Hobbes and Samuel Taylor Coleridge, was based on objections to modern scientists identifying themselves as natural philosophers. It was that rejection that caused William Whewell to coin the term "scientist" in 1834. Even then, the schism was not complete, with many identifying themselves as philosopher-mathematicians, as well as philosopher-biologists (Appiah, 2008). It is relatively recently that "philosopher" was dropped from the hyphenated descriptions. Thus, philosophers of ethics became ethicists (see Introduction).

The case for philosophy

The term "philosopher" is derived from the Greek *philos*, meaning loving, and *sophos*, meaning wisdom. The relationship between wisdom and ethics is fairly straightforward, particularly one of the forms of wisdom articulated by Aristotle, *phronesis*, that is, of practical knowledge. The goal was to use *phronesis* to lead a virtuous life, for which virtue ethics is a direct descendent (see Chapter 2). However, the relationship between wisdom and science is more problematic. In addition to *phronesis*, Aristotle held that *sophia* is another aspect of wisdom. Whereas *phronesis* was contingent, how to act virtuous in a specific situation or circumstance, *sophia* sought general principles that transcend the contingent particulars and instances. There would seem to be sympathy between *sophia* and *epistēmē*, science, as both sought to discover the transcendental. Yet, philosophy, *sophia*, seems to be very distant to science. Aristotle draws the two together in a way that *sophia*, as logic, is essential to science. To explain, an extended quote from Osbeck and Robinson (2005) is helpful but note that ethical knowledge could easily be substituted for science:

> What, then, is the distinction between scientific knowledge and theoretical wisdom [author – *epistēmē* and *sophia*, respectively]? This is clarified by examining Aristotle's conception of science itself. The immediate focus of scientific knowledge is the observation of existing things in the sensible world, but the end and purpose of this observation is knowledge of what is invariant [author – taken as transcendental]—what is impervious to changes over time or in a particular "situated" human standpoint, namely, universals and lawful relationships (first principles) governing their functioning. Induction based on observation (sensory experience) paves the way to the discovery of relationships through identification of similarities between observations, but the necessity of any relationship between observed particulars is ascertained through demonstration. By demonstration, Aristotle refers to "a scientific deduction ... . one in virtue of which, by having it, we understand something."

> Demonstration leading to understanding is accomplished through the syllogism
> [author – logic], the point of which is to make clear that some conclusion follows by
> necessity from something assumed previously. Understanding carries an assumption
> that "it is not possible for this to be otherwise" [author – related to the Principle of
> the Excluded Middle].
>
> *(Post. Anal., 70b16–19 quoted in Osbeck and Robinson, 2005)*

Demonstration, as used previously, is a type of logical argument used to explicate and prove a claim to knowledge. Indeed, any claim to scientific knowledge is defined by the arguments and demonstrations made (Montgomery, 2019b). One of the purposes of the Royal Society founded in 1660 was to provide a forum where demonstrations, a type of argument, could be made to justify a knowledge claim. The claim to knowledge without demonstration is conjecture. Consequently, science organizes a set of sense experiences (observations) to induct a general principle that becomes the hypothesis that is demonstrated to be a scientific principle. The development of ethical systems is a little different. The explorations of ethics in Part 2 of this book attempt to take a scientific approach to the ethics of everyday medicine by understanding how claims to knowledge of ethics can be demonstrated logically. Therefore, the discussion of science [*epistēmē,*] and *sophia* [logic] articulated by Aristotle and described in the quote is relevant to ethics.

Deductive syllogistic logic

If there were a system of ethics that was absolutely certain, there would be little controversy and everyone would know how to act. Clearly, this is unlikely to ever be the case. The questions here are "what can be certain" and, from starting from what is certain and its nature, how can any ethical system approximate certainty? The fact that any ethical system could never be absolutely certain does not negate the value in attempting to approximate certainty. Indeed, the notion of reproducibility in ethics, for example, different persons in the same ethical situation or circumstance are treated the same, is directly proportional to the certainty of the ethical system.

For all intents and purposes, perhaps the only claim that is certain is the Identity Principle, which is of the form *a is a*. This is certain, as its denial, *a is not a*, would not be reasonable or perhaps not even possible. One can stipulate that *a is b*. If all agree, then *b* cannot be anything else but *a*. Similarly, *b is c* can be stipulated as just stated and now the Principle of Transitivity can be established, which is of the form *a is b and b is c, therefore a is c*. If we substitute *implies* for *is*, one can construct the logical proposition that *if a implies b is true and a is true, then b is true*, which is the same as *a is b* (stipulated by acceptance of *if a implies b is true*), then *b* must be whatever *a* is. This form is called modus ponens in deductive logic. The point here is not to derive the propositions of deductive propositional logic but rather to demonstrate that logical relations are state-of-being verbs that link the variables, such as *a*, *b*, and *c*. State-of-being verbs are an identity. The utility of logic can be extended by limiting the "direction" of *implies*, for example, if *a implies b is true*, it does not follow that *b implies a* is true. From the perspective of set theory, *a implies b* can be taken that all members of set *a* are contained within set *b*, but not all members in set *b* need be in set *a*, although some are.

Ethical arguments involve relationships between individuals, and those relationships often are in terms of ethical principles, moral theories, laws, case precedents, and virtues. If those relationships were state-of-being linking verbs, the ethical arguments would be certain, assuming a valid structure to the arguments and true premises. However, it is highly unlikely that the relationships involved in ethical arguments are ever certain and, therefore, not explicable by state-of-being linking relationships. This is no different in traditional science. The greater the degree of departure from state-of-being relationships, the greater the uncertainty. The effort that follows is an attempt to understand how a departure in ethical arguments from state-of-being relationships generates uncertainty and opens the door to injustice. There are moral theories that do presuppose state-of-being relationships, for example, Deontological moral theories where being a person *is* having a certain right or expectation. However, there are other moral theories where the relationships are not the state-of-being.

For the purposes here, syllogistic deduction will be used instead of propositional logic, as described previously although syllogistic deduction can be recast as propositional logic. Nonetheless, the relationships remain state-of-being linking verbs. For example, consider the syllogism in Argument 3.3 that follows:

Argument 3.3

Major premise: *All humans are mortal*
Minor premise: *Socrates is human*
Conclusion: *Socrates is mortal*

The certainty of the conclusion is because the set of *Socrates* is contained within the set of *all humans*, which is contained in the set of *mortals*. Set membership established the state-of-being linking verbs *is* and *are*. An analogous ethical argument might be Argument 3.4:

Argument 3.4

Major premise: *All humans have a right to autonomy*
Minor premise: *Socrates is human*
Conclusion: *Socrates has a right to autonomy*

The conclusion of the argument is certain, and assuming true premises and valid argumentation, Socrates has an absolute right to autonomy, perhaps in a deontological sense. In this sense, being human *is* having the right to autonomy.

The question is what do the deductive arguments just described have to do with induction, particularly ethical induction necessary to develop ethical systems? It would appear that deduction is the process transitioning from a general statement, *all humans are mortal* and *all humans have a right to autonomy*, to the particular, in this case, *Socrates*. Induction is held as transitioning from a set of particulars, *ravens seen are black*, to a general, *all ravens are black*. It is from the observations of many persons, on par with Socrates but not including Socrates, that the general statement *all humans are mortal* originates by induction and then can be applied to Socrates in a syllogistic deduction. Similarly, it is from observations of persons, other than Socrates, in situations where their right to autonomy is in question allows generalization that *all humans have a right to autonomy*. Thus, in many ways, the syllogistic deduction presupposes the induction to the general principles utilized in the deduction.

As will be more fully discussed later, the critical point is that Socrates was not included in the set of observations that led general principle inducted to form the major premise of Argument 3.4. For the syllogism to hold, Socrates must be the same as those humans observed. This distinction is critical in several ways for the development of ethical systems. First, if Socrates is insufficiently similar to all previous humans observed, then Arguments 3.3 and 3.4 fail as the arguments would be the Fallacy of Four Terms, to be discussed later. Second, there is a large epistemic jump from the set of humans observed to the statement *all humans have a right to autonomy*, as that step requires acceptance as the Fallacy of Induction prevents certainty. The set of humans observed that was the substrate for the generalization was observed in the past, where the conclusion about Socrates is in the future. These are fundamentally different and this conundrum is called Hume's Problem of Induction. It just may be that Socrates is a new mutation and the generalizations do not apply. As will be discussed later, the issue of how to be sure that the set of observations of humans in the past is sufficiently the same as the situation or circumstance at hand in order to do work, such as resolving ethical problems, is critical to science and ethics.

In most ethical situations, it is not clear that the conclusion is true, as there could be circumstances in which Socrates is not afforded autonomy, in the sense that autonomy means Socrates gets what he wishes if that is taken as the definition of autonomy. Next, consider Argument 3.5:

Argument 3.5

Major premise: *All humans may have a right to autonomy*
Minor premise: *Socrates is human*
Conclusion: *Socrates may have a right to autonomy*

The linking verb *have* is qualified by the word *may*, which does not necessarily mean identify or subset membership. Therefore, the conclusion is true but it is contingent—Socrates may or may not have the right to autonomy respected. The conditions that determine the nature of *may* determine whether Socrates does or does not have his right to autonomy respected. As will be seen, the adverb *may* can be qualified, as to a likelihood or a probability.

The linking verb *may* is not quantifiable, yet still contains some sense of quantity as it just could become a binary like *will*. However, the syllogism constructed on "may" is of little utility. The binary nature of the linking verb *will* results in the Principle of the Excluded Middle, which holds that any claim must be true or false, not both true and false, and not neither true nor false. The linking verb *may* violates the Principle of the Excluded Middle, thus making Argument 3.5 invalid because it violates the Principle of the Excluded Middle.

The Principle of the Excluded Middle is critical to ethical decision-making although not that ethical premises follow the Principle of the Excluded Middle. Rather, it is the degree that ethical premises deviate from the Principle of the Excluded Middle that makes ethical certainty so difficult. Greater certainty may be achieved by making ethical principles comport more closely with the Principle of the Excluded Middle, perhaps by reframing the ethical principles and moral theories.

More utility can be obtained by rephrasing Argument 3.5 into Argument 3.6:

Argument 3.6

Major premise: *All humans have an 80% probability of exercising the right to autonomy*
Minor premise: *Socrates is human*
Conclusion: *Socrates has an 80% probability of exercising the right to autonomy*

One might object that is it is impossible to quantify the probability of exercising a right to autonomy; however, consider Argument 3.7:

Argument 3.7

Major premise: *All humans have a more likely chance of exercising the right to autonomy*
Minor premise: *Socrates is human*
Conclusion: *Socrates has a more likely chance of exercising the right to autonomy*

Argument 3.6 is of the type encountered more likely in science, while Argument 3.7 is more likely to be encountered in ethics. Yet, these arguments are fundamentally the same logic.

The interesting question is how is the probability of a person exercising the right to autonomy derived, such as in Argument 3.6? In any situation where there is no scarcity in exercising the right to autonomy, the probability likely would be 100%. Likewise, in a setting of an absolute absence of exercising the right to autonomy, the probability is 0%. In between, the situation of moderate scarcity, the probability likely varies between 0% and 100%. Estimating the probability is critical to the establishment of any ethical system in a pluralistic modern liberal democracy.

In a situation where a strictly Deontological moral theory prevails, the probability is limited only by the resources in the actual sense of physical possibility. If there are only sufficient resources to provide persons with 80% of the ability to exercise their right to autonomy, then the probability of exercising that right would be 80%. In a situation where a Utilitarian moral theory prevails, the resources that would limit the probability of exercising the right to autonomy would depend on the cost of doing so. The probability is not directly determined by physical opportunity but by a decision of how much opportunity to provide willingly. In situations where an Egalitarian moral theory prevails, the probability of exercising the right to autonomy does not matter as long as every person has the same probability. However, even in an Egalitarian moral theory, the probability may still be limited by the physical opportunity. In a situation or circumstance where a Libertarian moral theory prevails, the probability is negotiable between those that have and control the opportunities and those seeking the opportunities. In the pure form of Libertarianism, the probability of exercising the right is immaterial except that the probability has to be established and be greater or less than 50%.

The critical person might ask how valid is the probability that the right to autonomy has an 80% of being upheld? The answer is that was the incidence of the right in the sample studied. This incidence may not continue to be true if persons in the entire universe were studied (the population). If the incidence in the right being respected was 80%, then the probability used in Argument 3.6 would be absolutely true by tautology.

As studying the entire universe is not possible, a critical person might ask just how many persons whose rights are under discussion were studied. As the number of persons actually studied (the sample size) increases to nearly every person in the universe, the probability estimate would approach being absolutely true and valid (the Large Number Theorem in statistics). Yet, this may not be feasible. The critic then might demand that additional samples be studied. If a reasonable number of additional studies on additional samples all resulted in an incidence of 80%, the critic then has confidence that Argument 3.6 is valid and guarantees true conclusions. Note that the notion of probability is not solely of "academic" interest, it may well translate into ethical systems to affect justice or prevent injustice.

A series of samples can be studied, and each sample provides a somewhat different probability. However, the average of the probabilities reported is 80%, and 95% of the probabilities reported lie between 60% and 100% (the range of the standard deviation subtracted from and added to the average, respectively). Now, Argument 3.6 can be rephrased as Argument 3.8:

Argument 3.8

Major premise: *All humans have an 80% ± 20% probability of exercising the right to autonomy*
Minor premise: *Socrates is human*
Conclusion: *Socrates has an 80% ± 20% probability of exercising the right to autonomy*

Argument 3.8 has the same utility as Argument 3.6 but there is more confidence in Argument 3.8.

An ethical experiment could be constructed such that there is a particular ethical problem that results in interests competing with the person's right to exercise autonomy. In fact, there may be a continuum of conflicting interests that cannot be reduced by some calculus to a number. A sample of ethicists is assembled and a poll taken finds that 80% of the ethicists would require enforcement of the opportunity to exercise their right to autonomy. In other words, the panel of ethicists votes "yes" or "no" whether the right to autonomy should be upheld. Note that this is one example where a qualitative phenomenon, that is, the continuum of competing interests, may only be qualitative, yet by forcing a yes-no question onto the ethicists, a quantitative metric can be achieved (discussed in more detail later). But what if a different sample of ethicists is convened? How sure can one be that this next sample or any number of subsequent samples will still find that 80% of the ethicists would enforce the ability to exercise their right?

Alternatively, one could just ask whether the number of ethicists insisting on enforcement is greater than chance, such as just flipping a coin. The probability of an ethicist insisting on enforcement due only to chance would be 50%. All one would need to test is whether the actual incidence of insistence on enforcement was greater than chance, 50%; if so, then one could argue that the right should be enforced for all. One could perform a Chi-squared, X^2, test where the question of interest is what incidence of enforcement would be more than just random? If one requires a 90% approval for enforcement (the effect size), the minimum number of ethicists or others making the decision would be 10 in order to have an 80% chance of detecting an approval rate (the power of the test) that had only a 5% chance (the significance level) of being random (equivalent to flipping a fair coin). For an 80% required approval, 20 voters are required (sample size); for 70%, 47 voters are required; and for 60% approval, 194 voters would be needed. While it seems paradoxical, the greater the threshold approval percentage, the relatively easier it is to have confidence that the final decision is not just due to chance, which, if due to chance, most would believe the result to be unjust. All of these statistics require that the votes are independent so that one voter is not coerced by any other voter; hence the value of blinded or anonymous voting.

The sample size required to have confidence in the results depends on the variance within the sample, the degree of difference from chance expected (the effect size), the degree in confidence required (the significance level), and the probability of detecting a difference (the power). A greater effect size and a lesser variance increase the power and reduce the necessary sample size. The X^2 test can reduce the sample size, compared to the role of the die, because it makes some assumptions about the variance in the flipping of a coin, for example.

The X^2 test described earlier is used primarily to test a hypothesis. In this case, the consensus reached by a sample of ethicists or others empowered to enforce justice is not just due to chance or a flip of a coin. As an interesting contrast, the US Supreme Court requires only a 56% approval (5 of the 9 justices agreeing), which, according to the X^2 test, would require nearly 200 justices. Given that there are 9 justices, the approval rate would have to be 90% agreement—8 of the justices would have to agree to be sure the decision was not random. Near unanimity would be required to assure that any decision is not due to chance. Clearly, this would not be workable. However, there is a marked difference in the reasoning behind the X^2 test and the decision-making process of the court. The X^2 test looks only at data, the set of particular ethical or legal problems, for validation. In this sense, the attempt at validation is purely empirical and inherently based on the characteristics of the data. The validation is internal to the data set. This would comport with the Anti-Principlist approach (see Chapter 2). However, the justices are not looking at data alone to justify their decisions but rather how the data comport with principles that transcend the particulars of the ethical problem at hand, much in the manner of a Principlist approach. In a way, the Principlist rests on external validation.

One might say that all the discussions just given in attempting to interpret ethical induction and demonstration based on probability and statistical syllogisms are interesting in theory and may not actually reflect how human beings arrive at ethical inductions and then use those inductions in an effort to achieve justice. First, it is important to note that the modes of probability and statistical-like thinking need not be quantitative but rather can be qualitative. For example, the statement such as "it is more likely than not that a person in this particular situation would be accorded deference" (the principle of autonomy) is an example of probability-like thinking. A statement such as "I understand your opinion but I would like to get additional opinions of some others" is statistical-like thinking that does not require quantification. As will be seen, there are methods to at least approximate a type of quantification (see "Implications of qualitative research for ethics" section).

The Fallacy of Four Terms

Argument 3.3 would seem to be absolutely certain. But what if Socrates was a special kind of human, one who does not die? Then Argument 3.3 would be invalid. The reason is that there are four terms: all *humans* (bridging term) in the major and minor premises; the major term, *mortals* (major term) in the major premise and conclusion; and *Socrates* (minor term) in the minor premise and conclusion. As long as there are only three terms, the argument is valid; if the premises are true, the conclusion will absolutely be true. However, in the case where Socrates is a different kind of human, there are four terms: two versions of the bridging term, that is, two types of humans, those who die and those who do not. Consequently, the argument would be invalid and an example of the Fallacy of Four Terms.

Argument 3.9 is the ethical version of Argument 3.3:

Argument 3.9

Major premise: *An ethical problem A involving principle, theory, law, case precedent, or virtue B requires the just result C*
Minor premise: *John Doe is subject to ethical problem A involving principle, theory, law, case precedent, or virtue B*
Conclusion: *John Doe requires the just result C*

The question becomes whether *ethical problem A invoking principle, law, case precedent, or virtue B* (bridging term) in the major premise is the same *ethical problem A invoking principle, law, case precedent, or virtue B* in the minor premise. For example, the *ethical problem A* held as *invoking principle, law, case precedent, or virtue B* associated with the sample from which it was induced to require the *just result C* is the same as the *ethical problem A* held as *invoking principle, law, case precedent, or virtue B* in the minor premise. If the ethical problems are different, then there are four terms and the conclusion that *John Doe requires the just result C* is invalid. To the degree that they are different, there is a corresponding risk for the Fallacy of Four Terms. Note that the source of the difference also could be the *principle, theory, law, case precedent, or virtue B*.

The same issues confront medicine (Montgomery, 2019a) and biomedical research (Montgomery, 2019b). Indeed, the Fallacy of Four Terms is a major cause of irreproducibility in medicine and biomedical research. Many of the principles and methods in statistics are used to specifically avoid the Fallacy of Four Terms. Most risks for the fallacy center around creating the sample, for example, the group of persons involved in the induction so that the sample truly reflects the population of concern, all persons affected by the induction.

The critical question is just exactly how close the sample, used to create the major premise in Argument 3.9, resembles John Doe's particular situation and circumstance so that the conclusion is valid? The ethical problem is a serious concern, particularly for phenomenological ethics (see Chapter 2), as there seems no limit to the depth of detail that can be explored in a particular case. This increases the dimensions over which the sample and Jon Doe may differ, making it harder to avoid the Fallacy of Four Terms.

Methods can be used to avoid the Fallacy of Four Terms by assuring that the principles, theories, laws, case precedents, and virtues are generalizable, for example, the sample used to create the major premise in Argument 3.9 is generalizable to John Doe. The structure of the argument presumes that the major premise is generalizable. Indeed, it is held generalizable over relevant persons and hence is transcendent or universal over all particulars, of which John Doe is an example. These methods should ensure that principles, laws, case precedents, and virtues are sufficiently representative of the entire population of concern, past, present, and future.

Implications of qualitative research for ethics

One could reasonably argue that most interactions among humans, hence ethics, are qualitative, that is, any conversation that involves concepts that are relatively intangible, such as those that evoke meanings, emotions, and beliefs. How does one quantitate notions of love, loyalty, patriotism, and bravery? Interestingly, methods have been developed to attempt some quantification of these "intangibles," but sacrificing much of the complexity or "richness" of the interactions. [Interestingly, the notion of "richness" in the sample of subjects undergoing qualitative analysis, while evocative, would appear to be something quantifiable but such quantization has proved elusive (Palinkas et al., 2015).] Nonetheless, there are methods to analyze qualitative data. While often used in psychology, such approaches also lend themselves to studying ethical systems.

Remarkable advances have been made in qualitative research, both in situations and in circumstances that allow no quantification or only limited quantification, making the application of traditional statistical analysis difficult as typically practiced. Nonetheless, the conceptual basis of statistics is still applicable. The next section discusses qualitative research from the perspective of how it can be used in research into ethics. It is striking that there are a great many publications on the ethics of conducting qualitative research, but this author was unable to find any publication on the use of qualitative research to develop systems of ethics; however, it is hoped that the discussion that follows can make a considerable case for such efforts.

As will be seen, sampling of a set of observations (the sample) from all possible relevant observations (the population) is critical to avoiding the Fallacy of Four Terms. An excellent review of sampling in qualitative studies was provided by Palinkas et al.

(2015) in the context of research on the implementation of healthcare policies, specifically those used to implement evidence-based medicine. The important aspect here is reasoning to ethical analogues.

As can be readily appreciated, nearly every sampling method requires some prior assumption (deliberative, declarative, and conscious criteria) or presumption (not necessarily deliberative, declarative, and conscious criteria, but may be procedural, such as convenience sampling). All of this is necessary when dealing with the *A Priori* Problem of Induction, where the sampling may beg the question, in other words, the specific observations presuppose the induction (see Chapter 2). For example, using the criterion of requiring being black as a means to select a population of ravens so that the induction can be made that *all ravens are black* would be an error. The most explicit *a priori* assumption is seen in theory-based sampling where the constructs of a given theory are used to assemble the sample with the intent of validating an inference or prediction of the theory. The issue of the *A Priori* Problem of Induction perhaps is the least in purposefully random sampling. However, even purposefully random sampling still requires some prior criteria to determine what is relevant, for example, to exclude the inclusion of duckbill platypuses in a study of the weights of humans.

Note that in most of the criteria offered by Palinkas et al. (2015), there is an implicit tension between sampling criteria that emphasize homogeneity versus heterogeneity; this relates to variance, as will be discussed later. Typical case sampling may use as selection criteria what is thought to be typical, the norm or average in order to induct some other generalization (otherwise it would be a tautology). Even if a tautology is avoided, typical case sampling first requires knowledge of what is typical, the norm, or average. The subsequent induction becomes what is typical, the norm, or the average of the collection of typical, norms, or averages. The quantitative analysis equivalent is the mean (average) of means (averages). The Central Limit Theorem holds that the variance of the mean of the means will not be more than the variance associated with any one mean (Fig. 3.3).

The consequence is that the typical, norm, or average of the set of typical, norms, or averages may reduce the variability of the resulting typical, norm, or average compared to the variability in the actual data that was used to achieve the set of typical, norms, or averages used for induction. The reproducibility of the mean of means will be greater than the reproducibility of the individual studies where the variances are greater. Translating to the qualitative analysis, the typical case sampling of a set of cases will have more reproducibility than each set of typical cases. The question is which is more appropriate? If the typical case among the set of typical cases is more valid in an ontological sense and its reproducibility is more certain, what does it mean if the actual typical case is, in itself, less reproducible in any of the typical cases in the set? Is what actually happened, the set of typical cases, more relevant than an abstraction, the typical case from the set of typical cases of what happened? The consequence could be an increase in the risk of the Fallacy of Four Terms if the typical case is very different from each typical case in the set of typical cases. This has been a problem since the first days of statistics. Which is more "true," the Central Tendency (mean) or the distribution (variance) (Gigerenzer et al., 1989)? An important question is whether the induction of ethical principles, moral theories, laws, case precedents, or virtues as

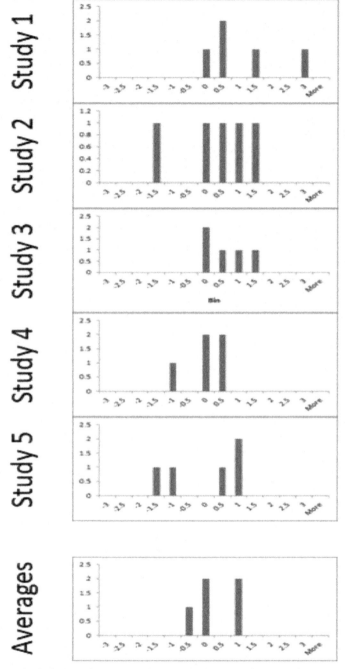

Fig. 3.3 Hypothetical example of five studies (1–5) where the histogram demonstrates the observations in each study set. Also shown is a histogram of the Central Tendencies (means or averages) from the individual studies. As can be seen, the distribution of the original observations is wider (less coherence) in the individual studies than the distribution of the averages, demonstrating the Central Limit Theorem. Note that the histograms are not to the same vertical scale.

a consensus among experts, such as ethicists, and not on the actual experiences of the citizens is analogous to typical case sampling and therefore at risk.

The Fallacy of Four Terms may be less of a risk if the variance in the sample is very large so as to be sufficiently heterogeneous, minimizing its difference from the population. Qualitative sampling methods, such as extreme or deviant case and maximum variation sampling, emphasize variance (heterogeneity) (Palinkas et al., 2015). However, even this does not exclude the potential for the Fallacy of Four Terms, as the excessive weight given to outliers could sway the consensus inappropriately. Further, greater variability means that there is less power to detect meaningful differences between different situations and circumstances, limiting the resolution of judgment (see discussion of statistical power earlier).

Confirming and disconfirming case sampling is somewhat unique and perhaps one of the strongest for establishing the ethical principle, moral theory, law, case precedent, or virtue. In this situation, the method is not hypothesis generating as is the case from inductions from sampling methods described earlier. Rather, it is hypothesis testing. The method starts with a hypothesis or set of hypotheses. Then the population of related and relevant ethical problems is sampled to find those that confirm a hypothesis and those that disconfirm the hypothesis. There are two perspectives to the confirming and disconfirming case sampling approach. The first follows from the Principle of the Excluded Middle, which holds the hypothesis under examination as true or false, not both true and false, and not neither true nor false under the instantiations of the particular ethical problem. Thus, it takes only a single counterexample (a disconfirming case) to demonstrate that the hypothesis is false. For example, it takes only a single nonblack raven to disprove the argument "because every raven seen is black, all ravens are black." Similarly, there must be at least one confirming example as otherwise the hypothesis is vacuous.

Grounded theory

Grounded theory [for example, proceduralized in the Delphi method (Brady, 2015)] is particularly interesting from the standpoint of developing ethical systems. It is a complex methodology (Murphy et al., 1998) where a full explication is beyond the scope of this effort. However, fundamental to the grounded theory approach is an explicit method in which the induction is expected to evolve over repeated sampling. It is this aspect of the grounded theory approach in a qualitative analysis that will be considered.

The evolution of consensus over time in grounded theory parallels the evolution of ethics, particularly the ethics of everyday medicine (Chapter 4). The evolution occurs as physicians and healthcare professionals continue to encounter similar ethical problems and attempt to come to a continuing modified consensus that accommodates subsequent experiences, particularly if concerns are raised. Indeed, much of the modern history of medical and clinical ethics demonstrates such evolutions, for example, from physician-centric ethics to patient-centric ethics in the context of informed consent and then to commercial-centric and governmental-centric ethics (Chapter 4). Typically, the strictly medical question does not fundamentally change although the context may, and thus it is the ethical consequences that generated the evolution of ethical consensus.

In a sense, the grounded theory approach involves discovery (constructive) and judgment (evidentiary) methods used iteratively. An example would be a panel of ethicists considering a set of ethical problems, where the judgment (induction) of one ethicist influences the provisional consensus until the reactions of other ethicists change the consensus. The process continues until no further changes in the iterative consensus are seen, and the final consensus is taken as principle, theory, law, or precedent. The presumption is that the final equilibrium after such reflection is the truest representation of the nature of the ethics involved in the problem. This same concept underlies the Large Number Theorem where the Central Tendency, the mean (average) of the sample, becomes equal to the mean of the population as the number in the sample increases. In situations and circumstances that do not involve Chaos and Complexity, as will be discussed in greater detail later, the Large Number Theorem and its qualitative analogue likely hold.

However, the key assumption is that each element in the sample, whether it is the weight of a human or the typical ethical case, is independent of all the other elements. Clearly, in grounded theory, induction follows from prior inductions and is not independent; thus, the notion of the final induction being most reflective of the ethical nature of the problem is suspect. Consider the concept of Brownian motion where, for example, a pollen grain placed in a drop of water appears to jitter and move through the liquid when observed under a microscope. The jitter is due to collisions between the water molecule and the grain of pollen. Proceeding from the reasonable assumption that all the collisions are random in strength and direction and are independent, how is it the grain of pollen moves through the water? The magnitudes and directions of the collisions should "average" out and the net movement (the consequence of all prior collisions) should be zero. Yet, the pollen grain moves (Fig. 3.4).

In the distribution of the magnitudes of the collisions, the most likely collisions are very small so the pollen grain does not move much. Subsequently, collisions likely also are small and some might be in the opposite direction to return the pollen grain to its prior position. However, on rare occurrences and by chance there may be a collision of a very large magnitude in a specific direction that will move the pollen grain to a large distance. The next collisions likely are very small (regression toward the mean) and will not move the pollen grain back to its original position. In this case, each collision is random and independent of each other. However, the specific position of the pollen grain is not independent of the entire history of the collisions, rather it is a "running total" of all the collisions and thus the effects of each subsequent collision on the net position of the pollen grain are not independent of the previous collisions and the pollen grain moves.

The operations of grounded theory are analogous to Brownian motion in that the induction to principle, theory, law, case precedent, or virtue represents the net ethical position just as the current position of the pollen grain. If the original positions of the participants in the consensus are relatively homogeneous in the direction and strength of their opinions, there will be relatively little movement of the evolving consensus. However, it takes only a relatively few participants with strong opinions to shift the

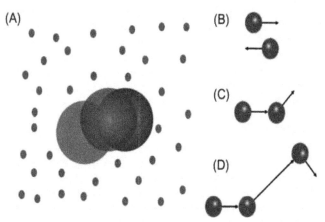

Fig. 3.4 Brownian motion is a phenomenon exemplified by the movement of a very small particle, under a microscope, that has been placed in a liquid (A). The particle appears to jitter and moves through the medium. The direction and magnitude of the particle movement are related to the direction and magnitude of the collisions between the particle and molecules of the liquid. If the collisions are equal and opposite (B), the particle will not appear to move as the collisions cancel out. However, the collisions may not be opposite, in which case the particle will be seen to move (C). As most of the collisions are very small (C), the particle will appear to jitter. Relatively rarely there will be a very big collision that will move the particle through the medium (D). The subsequent collisions are too small to bring the particle back toward the original position. From Montgomery, E.B. Jr., 2019. Reproducibility in Biomedical Research: Epistemic and Statistical Problems. Academic Press, New York.

evolving consensus greatly and the subsequent opinions of the more likely participants may not be sufficient to move the final consensus back to a position that is most reflective of the entire set of participants if each of their opinions had an independent effect. The effects of ethical Brownian motion are most likely in the extreme or deviant case- and maximum variation-sampling methods described by Palinkas et al. (2015).

An example of such a qualitative Brownian motion effect is the telephone game phenomenon. There is a sequence of persons to whom a story is told to the first person who relays the story to the second person and so on. Often the final story is very different from the original story. The difference in the final story is likely obtained even if the changes created by each person were random because the cumulative effect of the variations is not independent. Each variation starts at the point affected by the prior variation.

The final consensus principle, theory, law, case precedent, or virtue resulting from the application of grounded theory may be very different than the Central Tendency of the ethical principle, theory, law, case precedent, or virtue based on an independent assessment of the participants in the consensus building. The problem lies when the principle, theory, law, case precedent, or virtue derived by grounded theory, which is not reflective of the true Central Tendency, is now applied to a subsequent ethical problem to be adjudicated, as the process will be at risk for the Fallacy of Four Terms.

The criticism does not invalidate the use of grounded theory or undermine its utility. Rather, the criticism is to raise caution and, if attended to, could lessen the risk and impact of the Fallacy of Four Terms.

Mixed-methods research

Often, some means to create some degree of quantification are possible and thereby, traditional statistical analysis methods, in addition to concepts, can be brought to bear. Indeed, computer programs available commercially allow interesting metrics to be extracted from qualitative experiences, including ethical experiences (Banner and Albarrran, 2009; McLafferty and Farley, 2006; Weitzman, 1999; Williamson and Long, 2005). Some methods include examining words, phrases, or concept usage in conversations among ethicists, shareholders, and stakeholders to provide some rough quantification of the relations even as a distribution in some "space." Agree-disagree voting allows quantification. Pairwise comparisons of every judgment by all participants can create a rank order among the experiences that allow for the use of quantitative analyses. Finally, a quantifiable surrogate marker of what are qualitative assessments, such as the use of neurometabolic imaging, may be possible (Chapter 1). To the degree that quantification is possible, traditional statistical approaches can be employed. The subsequent "hybrid" is often referred to as "mixed-methods" research, and these methods can be brought to bear on the establishment of ethical principles, moral theories, laws, case precedents, and virtues, as well as the application of these transcendental notions to the resolution of particular ethical problems.

In line with the thesis of this text, quantification of the obligation to the ethical principles and moral theories of some sort is critical. The ethical principles are binary but the obligation to those principles is not. In a modern liberal democracy, the principles of beneficence, nonmaleficence, and autonomy must operate or there is no liberal democracy and a different ethical universe would exist. The moral theories provide the context with which the different obligations to the ethical principles on the part of all shareholders and stakeholders become a way of weighing the obligations to produce some judgment. Thus, the critical factor is the obligation to the ethical principles.

As emphasized repeatedly in this text, because ethical problems require enforcement of justice, there must be some transcendental notion, such as the law in the concept of the Rule of Law that must be used in some fashion. Such a necessity would require some type of quantifiable metric, considered here quantifiable in the largest possible sense. The metric may be just a vote or series of votes resulting from the application of ethical principles, moral theories, laws, case precedents, and virtues. However, some moral theories, in the setting of moderate scarcity, make quantification difficult, such as Deontological moral theories, where the obligation is an inalienable unqualified right or Phenomenological ethics. The result is that Deontological moral theories and Phenomenological ethics are at a disadvantage in the enforcement of justice. However, Deontological theories, as a starting point or first principle, and Phenomenological ethics for "discovering" the dynamics of the ethical problems are very important (see Chapter 2). The actual resolution of an ethical problem requires some sense of a quantifiable metric. Therefore, obligations to the ethical principles, moral theories, laws, case precedents, and virtues are subject to the same constraints. Thus, the logic inherent in traditional statistics is still critical to qualitative analyses, even if the mathematical methods are not.

Ergodicity

The best way to ensure that ethical principles, moral theories, laws, case precedents, and virtues are representative of and thus applicable to the population of concern—past, present, and future—is to construct them well. In pluralistic modern liberal democracies, what becomes principles, moral theories, laws, case precedents, and virtues is determined empirically from sampling from the population of concern, for example, voting by the citizens or their representatives. As discussed earlier, the variability (variance) within the sample and the population has a great effect on the inductions obtained and the credibility of those inductions.

Ideally, ethical inductions from multiple repeated sampling should be the same. For this to be the case, the population should demonstrate high ergodicity. This means that a sample of sufficient size, although less than the entire population, taken anywhere within the population should have the same results. Consider the hypothetical example of two fisherpersons fishing in a lake represented schematically in Fig. 3.5. In both lakes, the distribution of the fish is random. However, as can be seen, the fish are not dispersed equally in lake B. The distribution of fish in lake B would have low ergodicity, whereas lake A will demonstrate high ergodicity. In lake A, fisherpersons

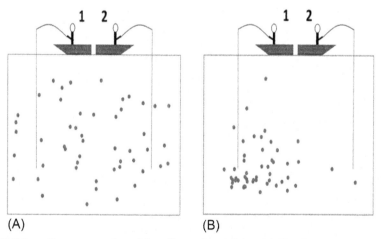

Fig. 3.5 Schematic representation of the effects of ergodicity. (A) The fish are distributed evenly in the lake—high ergodicity. (B) The fish are concentrated under fisherperson 1. The experience of the two fisherpersons in condition A will be the same and considered reproducible. Their experiences will be different under condition B and the experiments considered irreproducible.

1 and 2 will have the same sampling experience. Further, anywhere the fisherpersons move to, the sampling experience will be the same. The experiences will not be the same in lake B.

Consider moral theories along some dimension, for example, "conservative" and "liberal." If there is an even distribution of conservative and liberal ethics in the population of ethicists, say in a large ballroom, then surveying a small part of the room should result in relatively the same number of conservative and liberal ethicists. Any consensus from small groups anywhere in the room should be the same. However, if all the conservative ethicists are at one end of the room and the liberal ethicists are the other end demonstrating low ergodicity, then the consensus achieved from groups at each end of the room will be very different. The effects of ergodicity are discussed in more detail in "The experts, judges, and officials as those who make the consensus" section.

A great many factors may determine how the conservative and liberal ethicists distribute themselves throughout the ballroom. If they could not identify themselves or others as being conservative or liberal, in accordance with John Rawal's notion of persons blinded to their self-interest, they may distribute themselves with high ergodicity and any consensus of any sufficiently sized subgroup may be consistent. If their affiliations are known yet there was no need for any ethicist to work with any other ethicist regardless of affiliation, the distribution of conservative and liberal ethicists may be sharply divided (low ergodicity). If all the ethicists had to work together to find some consensus of the whole group, then the distribution of conservative and liberal ethicists may be quite different and likely more complex. These issues are taken up in more detail in "Chaos, Complexity, nonlinear dynamics, and far-from-equilibrium steady states" section.

Continuous versus dichotomous character of principles, theories, laws, case precedents, and virtues

Justice requires dichotomous judgments—one must act or not act. There is no deferral as any deferral is a decision not to act. The problem is that most of the factors that would determine the just act are not simple dichotomous, "yes or no" or "present or absent." Rather, the problem is in degrees, for example, the degree of obligation to the ethical principles, moral theories, laws, case precedents, or virtues. In a pluralistic modern liberal democracy, it is unlikely that any citizen would hold that there is no such thing as ethical principles, moral theories, laws, case precedents, or virtues, meaning that they are absent during the consideration of any ethical problem. What is continuously variable, for example, is in the degrees of obligations.

As to ethical principles, moral theories, laws, case precedents, or virtues, in distinction to the obligations, is that ethical principles, moral theories, laws, case precedents, or virtues follow the Principle of the Excluded Middle (also referred to as the Principle of Bivalence). Logical deduction, as discussed earlier, gains its strength certainly by insisting on the Principle of the Excluded Middle. The problem with many examples of Principlism that purport to be deductive, and therefore conveying a sense of certainty, is that the discussion is of the ethical principles, moral theories, laws, case precedents, or virtues rather than the obligation to the ethical principles, moral theories, laws, case precedents, or virtues. The application of such ethical principles, moral theories, laws, case precedents, or virtues without mitigating them by the obligation often is unsatisfactory (see Chapters 1 and 2).

Coherence and correlation

A central theme of this book is the ecological complexity of the ethics of everyday medicine. There are many instances or variations in each of the ethical principles, moral theories, laws, case precedents, or virtues. Further, the ethical principles, moral theories, laws, case precedents, or virtues play out over every combination of stakeholders and shareholders. Yet, there can only be one decision that is held just. (Note that one cannot be satisfied by the "second best" just decision as this would violate the sense of justice.) Thus, there must be some means to "combine" all the varying obligations to ethical principles, moral theories, laws, case precedents, or virtues. Yet, it is not as though two units of obligation to beneficence equal one unit of obligation to nonmaleficence.

There are several approaches to "combining" disparate obligations. For example, the pattern of obligations can be compared to previous case precedents of just actions. The previous case precedent most similar to the ethical problem under consideration is then used to determine what is just. This same approach is the heart of case or common law operative in many countries. The critical question becomes how similar does the ethical problem under consideration has to be to any past case precedent so that the particular case precedent can be applied? How similar is similar enough?

One can approach the issue of "similar enough" in a statistical sense. One can consider the variability (variance) around the case precedent. At first, this may not seem possible as the case precedent typically is a single case and there is no variability assessable. However, one can examine all previous cases in the past held to the case precedent and the differences between these cases and the difference between each case and the case precedent can provide a qualitative estimate of the variance around the case precedent. The question then is whether the case under consideration lies within the variance around the case precedent and, if so, then the case precedent would hold. In other words, what is the degree of coherence between the case precedent and the case under consideration?

It is important to distinguish the facts of the case precedent rather than its interpretation. Interpretation of the case precedent can evolve much in the manner of the grounded theory (or Delphi approach) as described previously, leading to a migration of the interpretation and possibly away from the actualities of the case precedent. In that case, to use the migrating interpretation of the case precedent to determine whether the case under consideration is governed by the case precedent could lead to error and injustice. It may be that the interpretation associated with the case precedent is what is actually used in adjudicating the ethical problem under consideration. However, the justification for doing so depends on the "similarity" [assessed qualitatively as the Epistemic Risk, see Chapter 2 and Montgomery (2019a, 2019b)] between the interpretation and the case precedent and whether the case under consideration is within the variance of the case precedent.

The Principlist may wish to develop a "calculus" for assessing the degree of obligation to the ethical principles and moral theories but the wish is challenged by the number of ethical principles and moral theories as they are parceled out among all combinations of shareholders and stakeholders. This is a daunting task but not insurmountable. Further, just a straightforward combination of the degrees of obligations spanning several ethical principles and moral theories risks "comparing apples to oranges" or a categorical error. However, one can normalize each obligation in the particular case under consideration by calculating the value of the obligation ranging from 0% to 100% obligation. Now each dimension is reduced to the same unitless scalar quantity, allowing a calculated combination, a type of induction, across the dimensions.

One group of methods for combining the unitless scalar values of each obligation over the ethical principles and moral theories is correlational analysis. The values of the obligations for the set of cases are the substrate for induction, for example, the set of relevant cases found to have been justly adjudicated. The actual values can be plotted or graphed onto a multidimensional space (one dimension of space for each dimension of the ethical principles or moral theories). One then creates a model, which, if successful, becomes the induction. The model is adjusted mathematically until the model that is predictive of the actual observations is "similar enough" to the actual observations. Further, the "similar enough" model can be quantitated, often referred to as the residuals and a threshold can be applied to demonstrate that the model is "similar enough" to the observations; therefore, the observations justify the use of the model. The internally validated model can be used to adjudicate future similar cases.

Most often in correlational analyses, the model is linear (like a straight line) among the observations. Thus, future cases can be judged based on the similarity of the analysis of the case under consideration to a homologous data point predicted by the model. If the model predicts what is just and the case under consideration, its combination among the obligations, falls far from the homologous point on the model, the case may not be just and thus considered unjust. This could be interpreted as the assigning of values to the obligations to the ethical principles and moral theories was sufficiently different from relevant prior cases to suggest that the assessment of the obligations in the case under consideration was in error. Proceeding from the original assessments could lead to injustice.

Another type of correlation is logistic regression, where the model to which data are fitted is sigmoidal in shape (Fig. 3.6). The advantage of logistic regression is that the outcome, the ethical decision, is dichotomous, meaning that it is just or unjust, not neither nor both, and approximates the Principle of the Excluded Middle. In this case, the model's sigmoid function classifies the observations of the past at the top of the sigmoid curve as just and injustice past observations at the bottom of the curve as unjust. To be sure, there is a transition zone at the S bend in the model and

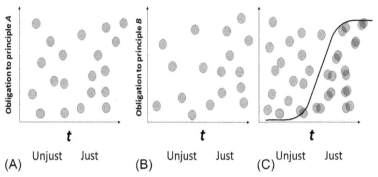

Fig. 3.6 A hypothetical situation where a series of observations, at times t, are the magnitudes of the obligations to Principle A (shown in A) and Principle B (shown in B). Note that these need not be quantitative but can be qualitative assessments that are rank ordered as described in the text. Further, each observation of an instance of ethical behavior of a certain type is determined to be just or unjust based on a consensus of shareholders and stakeholders or on a panel of ethicists. In C, the results of logistic regression are shown with the sigmoid curve demonstrating that it is possible to derive a regression equation that can predict just or unjust outcomes from the combined Principles A and B when it could not be by considering just Principle A and Principle B.

observations on the S bend would be problematic in their assessment of justice or injustice. Ideally, the S bend would be as sharp as possible, thereby minimizing the probability of a case falling on this indeterminant zone. Once the model is validated internally by it being "similar enough" to past observations, it can be applied prospectively to determine whether the assignment of the degrees of obligations in the case under consideration would result in a just or unjust outcome. If the outcome is unjust, then the assignments of the degrees of obligations would be reconsidered to ensure a just outcome.

There are several analyses that assess the robustness of the logistic regression and thus the confidence one can have of applying the model to adjudication of future cases. For example, the results of logistic regression can be analyzed in a receiver-operator characteristic (ROC) curve and the area under the curve is indicative of its "diagnostic" potential to differentiate just and unjust assignments of obligations. A system operating only by chance, for example, the flip of a coin, would have an area of 0.5 under the ROC curve, while a perfect system would have an area of 1.0.

Generally, logistic regression analysis is explicitly analytical following from the application of well-behaved mathematical functions that are well understood. The development of machine learning artificial intelligence (AI) can solve these same cases with a dichotomous outcome as just or unjust. It does not presuppose any explicitly analytical algorithm. Indeed, the great power of machine learning AI is the lack of any need for such an algorithm. Unfortunately, most humans depend on there being some understandable algorithm, which need not be the case for machine learning AI to solve these types of problems. Machine learning AI is discussed in greater detail in "Artificial intelligence and ethical induction" section that follows and the ethical implications are discussed in Case 12.

Humans do solve the types of ethical problems described earlier. It is unlikely that humans do so by mathematically analytical methods, but this does not evidence that it cannot be done so. Even if a mathematically analytic approach has not been attempted, this does not evidence that it cannot be done. The question is testable by future attempts, that is, ethical experimentation. [As an aside, there are mathematically and computationally analytic algorithms that are valid but one either cannot prove them (Gödel's Incompleteness Theorem) or cannot know ahead of time whether they will work (the Halting Problem).] Whether humans solve these ethical problems in a manner analogous to machine learning AI is an open question. Regardless of the methods used, the same fundamental limitations still apply and thus cautions and caveats must be observed. Thus, however humans solve such ethical problems, the same cautions and caveats understandable from mathematically analytic or machine learning AI are relevant. This issue of machine learning AI is discussed in greater detail later.

The experts, judges, and officials as those who make the consensus

In a pluralist modern liberal democracy, individuals are held accountable to the ethical consensus among those governed. Ideally, the consensus is generalizable fairly to all citizens, but rarely do all citizens participate directly in creating the consensus. Rather, as a matter of practicality, a group of representatives, a sample, determines the consensus. Yet, historically there is another reason for a representational form of governance to create the consensus that becomes the ethical principles, moral theories, laws, case precedents, and virtues—avoidance of "mob mentality." Thus, those forming the consensus (shareholders) are a select group and likely not true representatives of all those affected (stakeholders). Beauchamp and Childress (2013) would hold that those making the consensus be reasonable persons. Rawls (1951) would hold that consensus makers are blinded to their self-interest. The presumption is that the "mob" is not sufficiently reasonable nor can they blind themselves to their self-interest. This is the main reason why Hobbes held the necessity of an absolute sovereign (individual or collective). Perhaps the contrasts between the American and the French Revolutions make the case. However, how this plays out in ethics is difficult and risks violating the fundamental concept that ethics are the consensus of those subject to its normative consensuses. Perhaps the best that can be achieved is that the "mob" submits to a Ulysses pact, as described in Case 5, perhaps unwillingly as a consequence of political power or by acquiescing to ideals regarding who should be set to judge in their stead.

The experience selected to serve as a substrate for ethical induction is critical, just as is the training set in a neurophysiological version of machine learning AI (Chapter 1). In the usual form of machine learning AI there is little structure as the initial strengths of the connections between computational units, neurons, are random to avoid biasing, which would affect subsequent machine learning. However, the initial state of the neurophysiological mechanisms in humans is not random and therefore proceeds from a prior structure, that is, a bias that has a critical effect on ethical learning and the consequences of what is learned. Subsequent experiences can introduce additional biases. These same concerns apply to those selected to form the ethical consensus. Yet, if these persons did not have biases, then they would never have been differentiated from the mob, assuming that such differentiation from the mob is considered to be of value.

Beauchamp and Childress (2013) use the criteria of reasonableness for those selected to create consensus—the concept of the *reasonable person*. The concept of the reasonable person has a long history in law and ethics. Often, disputes are resolved based on what a reasonable person would demand, expect, or accept in any ethical situation or circumstance. To be sure, there are any number of attributes of what a *reasonable person* might look or act like. Typically, these constitute virtues, such as honesty, modesty, charity, respectfulness, and fair-mindedness (see Chapter 5). Yet, it is not clear how these virtues would differentiate those typically called upon to create consensus from those in the "mob." Clearly, there are other criteria.

John Rawls offered a set of criteria for moral judges that may be applicable to the issue of selecting those who would be responsible in creating a consensus that becomes ethical principles, moral theories, laws, case precedents, and virtues. An extended quotation is useful.

> **(i)** A competent moral judge is expected to have a certain requisite degree of intelligence, which may be thought of as that ability which intelligence tests are designed to measure. The degree of this ability required should not be set too high, on the assumption that what we call "moral insight" is the possession of the normally intelligent man as well as of the more brilliant. Therefore I am inclined to say that a competent moral judge need not be more than normally intelligent.
>
> **(ii)** A competent judge is required to know those things concerning the world about him and those consequences of frequently performed actions, which it is reasonable to expect the average intelligent man to know. Further, a competent judge is expected to know, in all cases whereupon he is called to express his opinion, the peculiar facts of those cases. It should be noted that the kind of knowledge here referred to is to be distinguished from sympathetic knowledge discussed below.
>
> **(iii)** A competent judge is required to be a *reasonable man* [author – italics added] as this characteristic is evidenced by his satisfying the following tests: First, a reasonable man shows a willingness, if not a desire, to use the criteria of inductive logic in order to determine what is proper for him to believe. Second, a reasonable man, whenever he is confronted with a moral question, shows a disposition to find reasons for and against the possible lines of conduct which are open to him. Third, a reasonable man exhibits a desire to consider questions with an open mind, and consequently, while he may already have an opinion on some issue, he is always willing to reconsider it in the light of further evidence and reasons which may be presented to him in discussion. Fourth, a reasonable man knows, or tries to know, his own emotional, intellectual, and moral predilections and makes a conscientious effort to take them into account in weighing the merits of any question. He is not unaware of the influences of prejudice and bias even in his most sincere efforts to annul them; nor is he fatalistic about their effect so that he succumbs to them as being those factors which he thinks must sooner or later determine his decision.
>
> **(iv)** Finally, a competent judge is required to have a sympathetic knowledge of those human interests which, by conflicting in particular cases, give rise to the need to make a moral decision. The presence of this characteristic is evidenced by the following: First, by the person's direct knowledge of those interests gained by experiencing, in his

own life, the goods they represent. The more interests which a person can appreciate in terms of his own direct experience, the greater the extent to which he satisfies this first test. Yet it is obvious that no man can know all interests directly, and therefore the second test is that, should a person not be directly acquainted with an interest, his competency as a judge is seen, in part, by his capacity to give that interest an appraisal by means of an imaginative experience of it. This test also requires of a competent judge that he must not consider his own *de facto* [italics in the original] preferences as the necessarily valid measure of the actual worth of those interests which come before him, but that he be both able and anxious to determine, by imaginative appreciation, what those interests mean to persons who share them, and to consider them accordingly. Third, a competent judge is required to have the capacity and the desire to lay before himself in imagination all the interests in conflict, together with the relevant facts of the case, and to bestow upon the appraisal of each the same care which he would give to it if that interest were his own. He is required to determine what he would think to be just and unjust if each of the interests were as thoroughly his own as they are in fact those of other persons, and to render his judgment on the case as he feels his sense of justice requires after he has carefully framed in his mind the issues which are to be decided.

Rawls (1951, pp. 178–180)

Rawls' notion of a reasonable person contains many of the virtues just described but it also appears to require a certain intelligence [not too small, not too large, but just right—perhaps to ensure that the samples (the judges) are representative of the population], experience that may be considerably larger and further ranging than the typical citizen, a sense of self and others that is expansive and egalitarian, and at least some training, for example, in the logic of induction, formally or informally.

The question arises as to whether these competent judges, as Rawls would describe, actually exist? Even if competent moral judges and reasonable persons do not exist, groups must be assigned to create consensus. How, then, to select persons for consensus creation so that the aggregate effect approximates what the competent moral judge who is a reasonable person might do without risking the Fallacy of Four Terms?

Biomedical research is faced with the same conundrum, and the response of biomedical researchers can provide insight in assisting ethicists. Consider testing of a drug to reduce the risk of stroke. In the case of ethics, the experiment presents a series of ethical experiences and possible ethical outcomes to a panel and tests which ethical consensus is more likely, which then may be taken as an induction to an ethical principle, moral theory, law, case precedent, or virtue. A medical clinical trial is constructed where patients at a high risk of stroke are given the medication or the best alternative therapy (placebo, or "sugar pill" if no alternative therapies are available). The patients are followed over time and the rate of actual stroke in the two groups is assessed. The problem is that not all the patients are at the same risk for stroke, which will affect the probability of having a stroke, which in turn would bias the outcomes of the clinical study. For example, some patients may have preexisting heart disease, diabetes mellitus, high blood pressure, or a genetic predisposition. If one group has more or different risk factors, any difference in the outcomes from the two groups may not be attributable to the actual treatment but rather due to the risk factors. The clinical study confounding risk factors may be analogous to certain ethical predispositions and assumptions among members of the panel.

In the case of the clinical trial, admission to either group, those receiving the experimental medication and those treated with the best alternative, is done in a randomized and blinded fashion to counterbalance the different confounding risk factors. The incidence of a particular confound, such as diabetes mellitus, is the same in the two groups. Thus, the effects of diabetes mellitus in the experimental group are counterbalanced by the effects of diabetes mellitus in the group receiving the best alternative treatment. The question then becomes how to counterbalance the different presumptions and assumptions among the members of the ethics panel to minimize bias.

The investigators in the clinical trial of the experiment treatment are blinded during their assessments—the investigators do not know who received the experimental medication or the best alternative. In many ways, this is analogous to Rawls' insistence that competent moral judges "not directly acquainted with an interest" is the best way to avoid bias (Rawls, 1951, p. 180). The difficulty is what does "directly acquainted with an interest" (bias) mean in ethics?

Counterbalancing of the risk factors in the clinical trial is an epistemic necessity to avoid biasing the outcomes. However, there is the presumption that the outcome is not just an epistemic consequence but actually reflects the nature, ontology, of the experimental medication. This is a very problematic inferential jump from the epistemological to the ontological. Each group, in a sense, has AND does not have diabetes mellitus and similarly for all the other risk factors. Thus, what can be honestly concluded is that the outcome of the experimental medication study is only true for the person who does AND does not have diabetes mellitus and, similarly, does AND does not have any of the other risk factors. Similarly, the process of Reflective Equilibrium appears to proceed from a certain ontological sense. Rawls writes a conclusion arrived at by Reflective Equilibrium that "would survive the rational consideration of all feasible moral conceptions and all reasonable arguments for them" (Rawls, 1974). The survival could simply be epistemic, that is, the result of a process. "All feasible moral conceptions and all reasonable arguments for them" would settle on some single equilibrium with differences between them effectively canceling each other out. Rawls writes "Individual predilections will tend to be *cancelled out* [italics added] once the explication has included judgments of many persons made on a wide variety of cases. Thus, the fact that the principles constitute a comprehensive explication of the considered judgments of competent judges is a reason for accepting them" (Rawls, 1951, p. 187). Yet, just as in the case of the medical clinical study, the outcome of Reflective Equilibrium is truly reflective of an ethical situation that does not exist, that is, of a group that does AND does not have certain predilections. The problems associated with inference to those that do AND do not have certain predilections are analogous to medical clinical trials where outcomes derived from the groups that do AND do not have confounding risk factors are then applied to an individual patient who cannot have AND not have a confounding risk factor. In many ways, this increases the risk of irreproducibility in biomedical research (Montgomery, 2019b) and, by analogy the same situation in ethics increases the irreproducibility of an ethical guideline and thus injustice.

With a sufficiently large sample of competent moral judges, a sufficient range of moral conceptions and reasonable arguments may be obtained and any additional moral conceptions and reasonable arguments would not meaningfully affect the final judgment rendered. This is very analogous to the statistical concept of the Large Number Theorem where the Central Tendency, taken as the final judgment based on several moral conceptions and reasonable arguments, approaches a stable limit, and taken as THE final judgment, with increasing sample size. Presumably, this is the reason for juries and panels of judges rather than a single judge.

A type of counterbalancing can be seen in questions of justice in the notion of the governmental loyal opposition that takes on an official status (the Official Opposition in Canadian government, for example) and creates a shadow cabinet paralleling the official ministers. While perhaps not explicit or intentional, there is a sense of counterbalancing in which successive US federal administrations representing different political perspectives appoint Supreme Court justices with similar perspectives. Assuming that there is a sufficient turnover of administrations with different perspectives, a plurality of perspectives comes to be represented in the court.

Epistemic conclusions convey a sense of ontological unease—the epistemic conclusions are seen as tautologies internal to the set of particular cases. In other words, a final judgment in problems of ethics or justice is true simply because the panel of judges holds it to be true. For whatever reason, most humans are uncomfortable with such a situation and look to some reality in nature that survives and transcends any particular set of particular experiences. This is reflected in the notion of natural law and Deontological moral theories. The preamble to the US Declaration of Independence is a statement of natural law, which, interestingly, did not make its way into legal law (Introduction). Similarly, the Central Tendency in medical science takes on the sense of a natural law, where the Central Tendency reflects the reality more than the particulars (Chapter 2; Montgomery, 2019a, 2019b). Thus, Reflective Equilibrium may have the connotation of "discovering" the natural law of ethics and justice. Rawls writes that particular judgments "show a capacity to hold its own (that is, to continue to be felt reasonable), against a subclass of considered judgments…" (Rawls, 1951, p. 188).

Chaos, complexity, nonlinear dynamics, and far-from-equilibrium steady states

A central theme of Reflective Equilibrium is the notion of a process that converges onto a coherent claim, consensus, or judgment that can withstand repeated testing in the future, just as the Central Tendency in quantitative analyses with increasing sample size, the Large Number Theorem, or repeated testing, the Central Limit Theorem. However, a great many mathematical and physical systems do not converge onto single values, such as the Central Tendency, from the same starting points. Consider a weight supported by a narrow column (Fig. 3.7). The column remains straight and vertical until a critical load is reached, at which point the column buckles, called here a bifurcation. For the sake of discussion, it is assumed that the column will buckle either to the right or to the left with equal probability. Replications of the experiment will result in two different outcomes, buckling either to the right or to the left. Increasing the

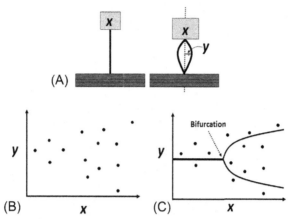

Fig. 3.7 Schematic representation of a system that demonstrates bifurcations (transitions) between states. (A) A weight is supported by a pillar. The pillar remains upright and straight as weight (x) is increased initially. At a certain weight, the pillar buckling from the single perspective is seen as buckling to the left or right. The direction of the buckling is unpredictable. (B) The weight and corresponding distance of the buckling (y) are plotted. At first impression, it may seem as though no relationship exists between weight, x, and buckling, y. Indeed, the correlation coefficient likely is near 0, little or no coherence. However, the physics is one of the relationship between x and y. Indeed, an equation of the form $\dot{y} = xy - y^3$ describes this process (C). As can be appreciated, high coherence exists between the observations, $Y(t)$, and the model shown by the *solid line* in (C).

number of replications, even to a point near infinity, will not result in a convergence to a single direction of buckling; indeed, the Central Tendency (mean or average) would be no buckling, given that the average displacement of the midpoint on the column would be zero, this certainly cannot be the case. Further, there will not be any coherence in the direction in which the column buckles with increasing weights, even when there are no describable differences in the initial or starting conditions. Importantly, this does not mean that the physical principles, laws, or equations that completely describe the system are in epistemic doubt but only that the outcome cannot be predicted. Note that in all the cases where the column buckles to the right the observations will cohere, as will all the observations in the cases where the column buckles to the left.

In the situation just described, where the dynamics can be specified in relatively simple mathematical terms, in this case $\dot{y} = xy - y^3$, the outcome is indeterminant, which is referred to as Chaos. Typically, these involve dynamics where the interactions between the predictors or the determinants are nonlinear. Complexity differs where the outcome is indeterminant because of the very large number of agents or interactions even if the equations are relatively well behaved, such as linear. For example, the laws of physics directly determine the formation of snowflakes, but the exact pattern of each snowflake (other than having six points) cannot be predicted. Indeed, there is so much unpredictability that statistically no two actual snowflakes may be the same. However, this does not mean that the laws of physics do not apply or that there is a different set of laws for each snowflake. If that were the case, the virtually infinite number of possible different snowflakes would mean an infinite set of different laws of physics, not likely a workable situation.

It would seem that the dynamics that underlie the process of Reflective Equilibrium, such as in ethics, would be highly nonlinear. It is not as though one measure of obligation to beneficence equals three measures of respect for autonomy or that a doubling of obligation to beneficence results in a sixfold increase in respect for autonomy. Thus, it would be reasonable to suspect that the same considerations of Chaos and Complexity in mathematical and physical systems may be relevant to the use of Reflective Equilibrium and ethical induction generally.

An indicator of Chaotic and Complex systems is the dependence on initial conditions. In other words, two repetitions of some instance of the operations of a Chaotic or Complex system, from the same initial conditions (meaning that no discernible differences exist at the beginning), produce very different outcomes. In other words, outcomes do not cohere. This is often taken as evidence against the Principlist position but such an argument would be tantamount to arguing that the laws of physics do not apply to the formation of snowflakes, as no two snowflakes are the same. This situation has been recognized in a variety of situations to which the Reflective Equilibrium process may be brought to bear. The argument is that experts following the process impeccably can arrive at very different, noncoherent outcomes, which becomes an argument against Reflective Equilibrium (Rawls, 1974 cited in Kelly and McGrath, 2010). This may not be a cogent argument against Reflective Equilibrium if the situation is analogous to the buckling columns described earlier. Thus, there may be coherent processes, coherent in terms of following determinately from principles and facts, yet produce what appears to be incoherent outcomes.

The notion of apparently incoherent outcomes needs clarification. For example, observations of atmospheric turbulence that form the basis of weather may look completely incoherent as evidenced by the inability to accurately predict the weather for more than a few days. Yet, when the dynamics are reconstructed in a different view, the regularity becomes apparent in the Lorenz attractor (Fig. 3.8). The question becomes whether the different outcomes from the same starting point in the discussions of the ethicists really are different and perhaps can be understood when viewed from the perspective of nonlinear dynamics in the Chaos and Complexity theory.

The Lorenz attractor (Fig. 3.8) resulted from attempts to model atmospheric convection currents, such as during weather events. The situation was modeled mathematically as two-dimensional flow through a fluid (such as air) based on variations in temperature, gravity, buoyancy, heat flow, and fluid resistance (http://mathworld. wolfram.com/LorenzAttractor.html). The complex set of equations was reducible to three related nonlinear equations. When the model was solved over multiple iterations, the result is shown in Fig. 3.8. As can be seen, as the number of iterations increases, there will not be a convergence onto a single value. Rather, the results of iterations will alternate between the two different loops in the attractor.

The notion that ethicists starting from the same set of propositions and using the process of Reflective Equilibrium impeccably should come to a highly coherent set of conclusions assumes that the specific process involved is not Chaotic or Complex. In part, this expectation may derive from the presumed notion of equilibrium in physics

Fig. 3.8 Lorenz attractor. The Lorenz attractor was first derived from a simple model of convection in the Earth's atmosphere. From Stewart, I., 2000. The Lorenz attractor exists. Nature 406, 949.

and Reflective Equilibrium in ethics and justice. Central to the notion of equilibrium is a steady state where the final outcome appears constant. Consider a chemical reaction where reactants A and B combine to form AB and AB decomposes to A and B. The rate by which A and B combine is given by a rate constant, k_1, and the concentrations of A and B, whereas the breakdown of AB to A and B is given by a rate constant, k_2, and the concentration of AB. The breakdown of AB to A and B increases the concentration and thus the rate of forming AB will increase. Opposing that reaction will be an increased rate of AB breaking down into A and B. At some point the two processes will balance and achieve equilibrium, such that the relative concentrations of A, B, and AB will be constant (cohere) over time, defining a steady state.

In the aforementioned case, the steady state recognizable by the coherence of its product is the consequence of an equilibrium resulting from opposing dynamics acting on the same reactants. The propositions of ethical experts can be taken as the reactants and the interplay between the propositions analogous to the dynamics for an equilibrium process described previously. An example is the aforementioned discussion of the group of conservative and liberal ethicists in the ballroom who were collectively charged with arriving at a consensus for the entire group. There will be forces of divergent interests and the force of the necessity of coming to an agreement. The dynamic balance of those forces will create an equilibrium whose final steady state becomes the consensus that becomes the ethical principles, moral theories, laws, case precedents, and virtues. This is seen in Rawls, who writes "Individual predilections will tend to be *cancelled out* [italics added] once the explication has included judgments of many persons made on a wide variety of cases. Thus, the fact that the principles constitute a comprehensive explication of the considered judgments of competent judges is a reason for accepting them" (Rawls, 1951, p. 188). The notion of *canceled out* presupposes opposing dynamics or mechanisms.

Many systems achieve steady state but not via an equilibrium between opposing dynamics. In the example of the conservative and liberal ethicists given earlier, in the situation where there is no requirement for a general consensus, the conservative and liberal ethicists separate themselves into different areas within the ballroom and will reach a steady state. There are no forces operational to oppose their self-organizing into their respective areas of the ballroom other than the physical constraints of the ballroom. Living things, ethicists included, often maintain steady states, which in biology is called homeostasis. Often the homeostatic steady state is achieved by regulatory negative feedback mechanisms, for example, an increase in blood sugar triggers the release of insulin to reduce the blood sugar. A drop in blood sugar reduces insulin secretion. Yet, living things change, such as growth and decay, which argues against a purely homeostatic equilibrium between opposing forces, as otherwise there would be no change. There must be some force that is not balanced by an opposing force. These systems can still come to a steady state if only because of some limit on the unopposed force, such as exhaustion. Such nonequilibrium steady-state experiences can become ethical principles, moral theories, laws, case precedents, and virtues.

The interesting question is to what degree does the development of ethical systems represent equilibrium or nonequilibrium steady states? To address that question, it may be helpful to consider the dynamics of nonequilibrium steady states in general. Perhaps the most general framework is in the thermodynamics of physical systems, particularly the notion of entropy, which is a state roughly equivalent to randomness. The Second Law of Thermodynamics suggests that in many or most systems that are closed, meaning no interactions with the outside world, there is an irreversible increase in entropy. This irreversibility means that there is an ultimate bias in the interactions in any physical system such as a change in one direction cannot be undone, although generally at a level generally unobservable or recognizable by humans. Thus, there will be a "drift" in the direction in the evolution of any physical system but, at any one time, the system may appear steady. To be sure, the effects of entropy can be counterbalanced by the application of energy, but typically that expenditure of energy comes at the cost of increased entropy elsewhere.

In the context of ethics and justice, there may be unopposed forces (energy), such as a bias that maintain a steady state in the consensus as to what is ethical and just. The risk is that such a consensus could be viewed as the natural consequence by which obligations to principles, moral theories, laws, case precedents, and virtues just balance out, resulting in the best possible world. The impression would provide a sense of certainty and, consequently, less compunction against applying the consensus to future problems. However, if the situation is a steady-state far-from-equilibrium, then the same confidence cannot be held. Perhaps it is better to speak of a Reflective Steady State rather than Reflective Equilibrium.

The critical point to be made here is that any better ethical system would be helped along by as complete an understanding of the dynamics as possible. Further, that understanding can be facilitated by metaphors to understandings achieved in other systems. Thus, science and mathematics are not irrelevant to ethics. However, the relevance goes beyond enlightening metaphors to include common fundamental physical mechanisms, such as neurophysiology as discussed in Chapter 1. Too often ignored is the sheer complexity of understanding the systems, which is true of the ethics of everyday medicine (see Introduction).

Complexity does not imply randomness and should not lead to nihilistic trepidation. The remarkable fact is that the physics of the universe allow or drive self-organization of the physical systems, which is particularly true in living things, particularly living things as complex as humans. Such self-organization likely is relevant to any attempt to develop ethical systems. An interesting example is seen in the behavior of puppies when offered milk to drink (Fig. 3.9) (Perony, 2014). At first, the behavior of the puppies appears incoherent, except for their attention on the human providing the milk. Once the milk is poured into the saucer, the puppies push forward toward the milk.

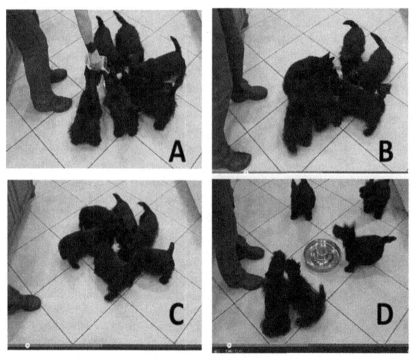

Fig. 3.9 Self-organization in puppies. (A) A group of puppies is given milk. Initially, the puppies appear distributed randomly around the saucer (A and B). As the puppies push forward to obtain the milk, they begin to move as a rotating pinwheel, in this case clockwise (C). The puppies only know to push forward toward the milk and do not communicate their intentions to each other. However, the saucer is more like a circular trough with a center without milk (D) (Perony, 2014).

Eventually, the puppies produce a pinwheel motion around the saucer in a highly coherent fashion until the milk is gone. Note that initially, the puppies went in a clockwise rotation but just as easily could have gone in a counterclockwise rotation. This is not to suggest that ethics are reducible to hungry puppies, only that the behavior of the puppies may serve in a certain way as a metaphor or analogy, a loose one to be sure.

Artificial intelligence and ethical induction

Artificial intelligence already has a profound impact on the everyday practice of medicine, which will only increase. Decision-making algorithms will have a profound impact on the roles of physicians and healthcare professionals (Case 12 in Part 1). For example, consider the situation where a physician treats a patient according to the suggestions offered by AI and there is a bad outcome. Who is responsible, the physician or AI device? If it is the AI device, how is the responsibility translated into liability and recompense to the patient? Would the liability for the AI device follow under product liability or the same medical liability of physicians and healthcare professionals?

Of interest here for ethical induction is one particular form of AI, that being machine learning. Machine learning was introduced in Chapter 1 to demonstrate that biological-plausible computational algorithms can come to behaviors suggesting inherent ethical principles and moral theories without explicitly representing the principles and theories in the neurophysiological actions that drive behavior. Machine learning AI essentially evolves to a nonrandom behavior, evidence of information, by "extracting" the nonrandomness, information, from the training data set. As the output of the machine learning operates over each particular that comprises the training data set, the output is transcendental, a type of universal notion (the importance of the transcendental notion is discussed in more detail in Chapter 2). Thus, machine learning AI induces a transcendental or universal notion from a set or particular notion; hence, it is induction. It is important to note that the induction requires *a priori* criteria (Chapter 1) embedded in the selection of training cases. Nonetheless, following organization of the training set based on *a priori* criteria from prior concepts, machine learning proceeds to extract information from the data set to generate an output that reflects the training set and predicts future outcomes from sets of data similar in a relevant way to the training set. An example, described in Chapter 1, is the machine learning involved in backing a tractor-trailer to a loading dock where no prior explicit explainable algorithm was applied and no algorithm was extractable from the computational solutions. In many ways, machine learning is analogous to grounded theory in qualitative analysis as described earlier, where a consensus evolves over discussions by participants.

Machine learning is little different from a panel of reasonable persons blinded to their interests, and competent moral judges attempting to come to some consensus that then is taken as a principle, theory, law, or precedent. Thus, machine learning can be applied to the discovery of principles, theories, laws, case precedents, and virtue where actual experiences serve as the training set to create the appropriate machine-learned principles, theories, laws, or precedents. Machine learning could also be applied to adjudication of ethical problems where the prior constructed machine-learned computational process takes the particulars of the ethical problem as inputs and the output is the adjudicated solution.

Application—Judging

Many times, the value of any ethical system is demonstrated by its application to future cases. However, as discussed earlier, establishing the consensus, which becomes the ethical principles, moral theories, laws, case precedents, and virtues, results in a loss of information, particularly the information necessary to apply the consensus to the individual subject. The loss of information may result in consequences that are subsequently considered unethical and unjust even if the consensus was applied fairly. The result is a loss of reproducibility of ethical judgments that undermine the sense of fairness and justice, just as irreproducibility in biomedical research undermines confidence in the research (Montgomery, 2019b).

As will be seen, the loss of information follows from the Second Law of Thermodynamics as applied to Information. This will be fleshed out in ethical terms momentarily. But first, a conceptual description is given. The Second Law of Thermodynamics holds that in any closed system, entropy can only increase. For example, when the cream is stirred into a cup of coffee and the two are blended, the cream and the coffee will not spontaneously separate again if there is no external energy applied, hence a closed system. The mixing of cream and coffee results in a state of increased entropy is a kind of randomness. Prior to mixing, the cream and coffee were not random as is each contained independently. Once mixed, the distribution of the cream and coffee molecules becomes random, hence an increase in entropy. Information can be considered the inverse of entropy with information being nonrandomness. For example, the letters, spaces, and punctuation marks of this text are not random and, consequently, have information extracted by reading the text. In a closed system, any step that is irreversible, such as mixing cream and coffee, increases entropy, thereby losing information irreversibly. Determining the Central Tendency and variance is an irreversible step as it is not possible to reconstruct the original distribution of observations solely from the Central Tendency and variance. This is as true for studying the effects of a medicine to reduce the risk of stroke, as described earlier, as it is for achieving an ethical consensus.

In the ethical domain, consider a group of reasonable persons blinded to their self-interest charged with reaching a consensus that is intended to be an ethical principle, moral theory, law, case precedent, or virtue. Each person may have very different operating moral theories that are oppositional. For example, one may operate from a Libertarian moral theory, while another from a Deontological moral theory and so on for Utilitarian and Egalitarian moral theories. Yet, the consensus makers come to some consensus that need not be based on any commonality but perhaps compromise out of expediency or necessity. The process of coming to a consensus counterbalances the different moral theories just as the case for the presence or absence of diabetes in the clinical study discussed earlier. The resulting consensus is a strange creature that presumably "averages out" the oppositional effects of the moral theories. It would appear that the different moral theories have been dispensed with and that the consensus is thought to reflect a judgment of the group independent of the moral theories. In other words, there has been a loss of moral information—the role of the different moral theories held by those making the consensus—resulting from the irreversible process of establishing the consensus. It is not possible based on the consensus alone to know which moral theories were held by those creating the consensus. This is an example of Ethical Incompleteness, evidenced by justice that is not provable based on ethical principles, moral theories, laws, case precedents, or virtues that were used to create the consensus, much as it is impossible to reconstruct the distribution of observations from the Central Tendency and variance alone. An example is discussed in Case 11 in Part 1.

In applying the consensus to a particular individual case by a single judge, that judge cannot both have and not have a particular moral theory, such as may be the case in the group forming the consensus; just as the individual patient cannot have and have diabetes as discussed previously. The judge has her moral theory and that is all the judge can operate from when inferring from the consensus to the individual case. Note that it is not useful to say that the judge should not use her moral theory on the basis that the moral theory is irrelevant to the consensus. The moral theories of those forming the consensus, including her's, may well be relevant even if it is not reflected in the consensus. Thus, when the judge rules in a particular case influenced by her moral theory, others may say that this is unfair, that the judge employed extrajudicial factors, specifically her moral theory. But the moral theories are not extrajudicial as they were part and parcel in the process of creating the consensus. There is not anything in the consensus that invalidates any particular moral theory.

Because of the ethical information loss in the consensus, the consensus may be an ill fit to any particular case. Thus, each judge has to bring information to the case under consideration in order to particularize the consensus to the case under consideration in order to render a judgment. In the Beauchamp and Childress system, this is called specification (Beauchamp and Childress, 2013). In this sense, the judge has little choice but to make a unique determination—in law what might be called a judge-made law or judicial activism.

It would appear preferable that the judge "suspended" her moral theory. But how would this be possible? What measure or method would be left to the judge in order to specify the consensus to the individual and particular case being judged? Perhaps if the case under consideration was exactly in a one-to-one correspondence with each term in the consensus, then the application would be relatively straightforward and no additional information needs to be brought to the case. But at best this would be a fluke. For this to happen, every case that was the substrate for creating the consensus would have to be exactly alike. Note that it would not be sufficient to say exactly alike in the relevant ways as this begs the question of that there are relevant ways as opposed to other possibilities judged irrelevant.

Alternatively, a second consensus could be created as to how the original consensus is to be interpreted. In other words, the informing moral theory can be prescribed and the judge then applies the informing moral theory, the judge's own moral theory notwithstanding. However, the second consensus may be much harder to achieve than the original consensus.

Judicial interpretation may be greater when the case is atypical of those cases used as a substrate for the ethical induction to ethical principles, moral theories, laws, case precedents, or virtues. Cases that are more atypical invite greater judicial interpretation, thus increasing the risk of judicial activism. These cases may deserve greater scrutiny and subsequent review. Borrowing from statistical concepts may be helpful. The confidence that the average of a study truly reflects the ontology or reality of the study, and thus applied prospectively with confidence, can be assessed from the variability of the observations obtained in the clinical study. Great variability gives less confidence. In quantitative analysis, the confidence in the judgment of the individual case can be proportional to its z score (statistical distance) from the consensus (held here as the mean or average). Indeed, measures can be used to determine whether the individual case under consideration is an "outlier" and, consequently, the average or consensus may not apply. In other words, the judge's application of the consensus, now codified as an ethical principle, moral theory, law, case precedent, or virtue, may be inappropriate to the case under consideration if the case under consideration is an outlier. The same concept can be applied to domains that are not quantifiable, as is much of ethics, as described earlier.

Reflection

The primary purpose of this chapter was to examine the epistemological and ontological foundations of necessary ethical inductions that generate principles, theories, laws, case precedents, and virtues that can be applied prospectively to adjudicate future ethical disputes. This process is of the same nature in ethics as it is in every other empirically derived domain of knowledge, such as science. The hope is that some of the concepts described can be applied to the development of ethical systems to make those ethical systems more robust and thus of greater certainty in the establishment of justice. Clearly, there is much work to be done, for example, how can an ethical system claim legitimacy based on the will (consensus) of those to be governed when those to be governed have not had a voice in the establishment of the consensus? While the diversity of ethical principles, moral theories, laws, case precedents, and virtues represents healthy ethics, and while perhaps accurate in their own way as to the phenomenology, such diversity does undermine some sense of rationality so critical to justice. It is not as though there are multiple ethical universes. The work to be done is not to reduce the diversity but to demonstrate how diversity arises phenomenologically and epistemically from the same ontology (Chapter 1). This chapter discussed several different approaches. With diversity maintained, each perspective can still contribute to the other perspectives, and it is not one against everyone else. Perhaps one day there will be a grand unifying ethical system.

First, however, appreciation of the complexity of the issues needs to be generated. The challenges are the complexities of human interactions when considered in their complete ecology. This was the attempt with the cases presented in Part 1. The complexity of those interactions rarely is fully considered in many texts on medical or clinical ethics.

Perhaps this chapter may serve, at least in some degree, what David Hume's work did for Immanuel Kant, who wrote:

> I freely admit that it was the remembrance of David Hume which, many years ago, first interrupted my dogmatic slumber and gave my investigations in the field of speculative philosophy a completely different direction.
>
> *Hatfield (2004)*

The effort may ultimately prove to be unsuccessful, but at the very least it is a start. To be sure, the effort may seem presumptuous to some, but that judgment can only be rendered truly after a fair hearing.

References

5-Sigma—What's That? 2012. Scientific American Observations Blog. July 17. http://blogs.scientificamerican.com/observations/2012/07/17/five-sigmawhats-that/.

Appiah, K.A., 2008. Experiments in Ethics. Harvard University Press.

Banner, D.J., Albarrran, J.W., 2009. Computer-assisted qualitative data analysis software: a review. Can. J. Cardiovasc. Nurs. 19 (3), 24–31 (Review. English, French).

Beauchamp, T.L., Childress, J.F., 2013. Principles of Biomedical Ethics, seventh ed. Oxford University Press, Oxford.

Brady, S.R., 2015. Utilizing and adapting the Delphi method for use in qualitative research. Int. J. Qual. Methods 1–6.

Gigerenzer, G., Porter, T., Swijtink, Z., 1989. The Empire of Chance: How Probability Changed Science and Everyday Life. Cambridge University Press, Cambridge.

Goodman, N., 1983. Fact, Fiction, and Forecast, fourth ed. Harvard University Press.

Gopnik, A., 2012. Scientific thinking in young children: theoretical advances, empirical research, and policy implications. Science 337 (6102), 1623–1627.

Han, F., Caporale, N., Dan, Y., 2008. Reverberation of recent visual experience in spontaneous cortical waves. Neuron 60 (2), 321–327. https://doi.org/10.1016/j.neuron.2008.08.026.

Hatfield, G., 2004. trans. and ed Immanuel Kant: Prolegomena to Any Future Metaphysics. Cambridge Texts in the History of Philosophy, p. 10. Available from: http://strangebeautiful.com/other-texts/kant-prolegomena-cambridge.pdf.

Hespos, S.J., Baillargeon, R., 2001. Reasoning about containment events in very young infants. Cognition 78 (3), 207–245.

Hespos, S.J., Ferry, A.L., Anderson, E.M., Hollenbeck, E.N., Rips, L.J., 2016. Five-month-old infants have general knowledge of how nonsolid substances behave and interact. Psychol. Sci. 27 (2), 244–256. https://doi.org/10.1177/0956797615617897.

Kelly, T., McGrath, S., 2010. Is reflective equilibrium enough? Philos. Perspect. 24, 325–359.

McLafferty, E., Farley, A.H., 2006. Analysing qualitative research data using computer software. Nurs. Times 102 (24), 34–36 (Review).

Montgomery Jr., E.B., 2019a. Medical Reasoning: The Nature and Use of Medical Knowledge. Oxford University Press, Oxford.

Montgomery Jr., E.B., 2019b. Reproducibility in Biomedical Research: Epistemic and Statistical Problems. Academic Press, New York.

Montgomery Jr., E.B., Turkstra, L.S., 2003. Evidenced based medicine: let's be reasonable. J. Med. Speech Lang. Pathol. 11, ix–xii.

Murphy, E., Dingwall, R., Greatbatch, D., Parker, S., Watson, P., 1998. Qualitative research methods in health technology assessment: a review of the literature. Health Technol. Assess. 2 (16), iii–ix. 1–274.

Osbeck, L.M., Robinson, D.N., 2005. Philosophical theories of wisdom. In: Sternberg, R.J., Jordan, J. (Eds.), A Handbook of Wisdom: Psychological Perspectives. Cambridge University Press, New York, pp. 61–83.

Palinkas, L.A., Horwitz, S.M., Green, C.A., Wisdom, J.P., Duan, N., Hoagwood, K., 2015. Purposeful sampling for qualitative data collection and analysis in mixed method implementation research. Adm. Policy Ment. Health 42 (5), 533–544. https://doi.org/10.1007/s10488-013-0528-y.

Perony, N., 2014. Puppies! Now That I've Got Your Attention, Complexity Theory. https://www.youtube.com/watch?v=0Y8-IzP01lw.

Rawls, J., 1951. Outline of a decision procedure for ethics. Philos. Rev. 60 (2), 177–197.

Rawls, J., 1974. The independence of moral theory. In: Proceedings and Addresses of the American Philosophical Association. vol. 47, pp. 5–22. Reprinted in Rawls 1999; 289, 286–302: References are to the reprinted version, cited in Kelly, T., McGrath, S., 2010. Is reflective equilibrium enough? Philos. Perspect. 24, Epistemology, 325–359.

Rawls, J., 1999. A Theory of Justice, revised ed. The Belknap Press of Harvard University Press, Cambridge, MA.

Sanborn, A.N., Mansinghka, V.K., Griffiths, T.L., 2013. Reconciling intuitive physics and Newtonian mechanics for colliding objects. Psychol. Rev. 120 (2), 411–437. https://doi.org/10.1037/a0031912.

Stigler, S.M., 2016. The Seven Pillars of Statistical Wisdom. Harvard University Press.

Sturm, T., 2017. Intuition in Kahneman and Tversky's psychology of rationality. In: Osbeck, L.M., Held, B.S. (Eds.), Rational Intuition: Philosophical Roots, Scientific Investigation. Cambridge University Press, pp. 257–286 (Chapter 12).

Weitzman, E.A., 1999. Analyzing qualitative data with computer software. Health Serv. Res. 34 (5 Pt 2), 1241–1263 (Review).

Williamson, T., Long, A.F., 2005. Qualitative data analysis using data displays. Nurse Res. 12 (3), 7–19 (Review).

The patient as parenthetical

4

To be argued...

The Belmont Report (Belmont Report: Ethical Principles and Guidelines for the Protection of Human Subjects of Research, 1978) resulted in a dramatic change in human subjects-based research in response to several scandals. It seems that the notion of autonomy was suddenly injected into the conversation and became the dominant driver of ethical behavior in human subjects-based research. Indeed, the history of human subjects-based research clearly demonstrated that obligations to beneficence and nonmaleficence, as well as appealing to the virtue of scientists, were insufficient. Thus, the surest way to enforce justice was to enforce the principle of autonomy. The principle of autonomy was proceduralized in the formality of informed consent for research participation and rapidly found adoption in medical and clinical ethics, formally or informally.

The rise of obligations to autonomy would seem to demand a patient-centric approach to the ethics of everyday medicine. But one has to ask what was the status of patient-centric ethics prior to the 1978 Belmont Report? One might think that the altruism typically attributed to physicians and healthcare professionals through the ages bespeaks patient-centric ethics. But was it ever the case that the ethics of everyday medicine was patient-centric and is it now?

A historical analysis argues that until the late 1960s and early 1970s, at least in North America, the ethics of everyday medicine was largely physician-centric. The transition that occurred was not to a patient-centric ethic but rather to a business- and government-centric ethics. This is not to say that the consequences are unjust, necessarily. However, this transition does require examination of how such business- and government-centric ethics comport the assumptions of a pluralistic modern liberal democracy that ethics (principles, theories, laws, and precedent) are derived from a consensus of those governed (for example, the citizen as a patient or potential patient) sufficient to enforce universally. Recently, particularly with the emergence of evidence-based medicine, the ethics of everyday medicine has become scientist/ academic-centric and enabled by business- and government-centric ethics.

Current status

To date, much of the discussion of medical or clinical ethics appears focused on the patient as a fulcrum by which obligations are interpreted and balanced—a patient-centric approach. Yet, this does not comport with reality, as demonstrated by the cases in Part 1. The appearance of patient-centric ethics likely resulted from reducing ethics to the dyadic, specifically the patient-physician/healthcare professional relationship.

The Ethics of Everyday Medicine. https://doi.org/10.1016/B978-0-12-822829-6.00029-1

As will be demonstrated, initially it was the physician and now business and government are the fulcrums. Perhaps it is necessary to insert citizens, present or future patients, into the discussion more transparently and rigorously.

Ethics are used here as first descriptive of how humans treat each other, the cases, such as those presented in Part 1. Inductions follow to create normative ethics in the form of codes and regulations, presumably emphasizing the best interests of patients. However, historical analyses demonstrate that patient centricity is rarely the case. From certainly reasonable perspectives, largely determined by prevailing moral theories, patient-centric ethics need not be nor should be the case exclusively. Nevertheless, the patina of patient centricity does much to deflect critical analysis of the ethics of everyday medicine. The most significant contributions of the analyses in this book are not any recommendation of a particular ethical system, as none is made. Rather, the value is in thorough vetting as an important and necessary requisite for rational change, if needed.

Patient-centric ethics

It could be argued that the ethics used in resolving the ethical status of the cases described in Part 1 were not patient-centric, depending on one's definition. Duggan et al. (2006) cited several definitions, including the patient "has to be understood as a unique human being" (Balint, 1969); "care that is closely congruent with and responsive to patients' wants, needs, and preferences" (Laine and Davidoff, 1996); and

> (1) adopting the biopsychosocial perspective (as opposed to a perspective that is narrowly biomedical); (2) understanding the patient as a person in his or her own right, not merely as a body with an illness; (3) sharing power and responsibility between the doctor and the patient; (4) building a therapeutic alliance—a relationship that is both instrumentally and intrinsically valuable; (5) understanding the doctor as a person, not merely as a skilled technician.
>
> *Duggan et al. (2006)*

The above definitions resonate with the Anti-Principlist approaches as discussed in Chapter 2.

Relative to the patient as distinct from a body with the disease, Smith distinguishes the generalized and the concrete "other" in the patient-physician/healthcare professional communication with the concrete other being a consideration of the unique individual (Smith, 1996). However, it would not be surprising that many physicians and healthcare professionals would object, saying they are practicing patient-centered ethics.

Such physicians and healthcare professionals might say they are addressing the unique concerns of the patient before them, not someone else's. To do so otherwise risks rejection by patients. Indeed, some physicians and healthcare professionals pander to the patient's wishes, for example, prescribing an antibiotic for a viral upper respiratory tract infection. The larger question is whether it is the patient's actual concerns or the physicians' and healthcare professionals' presumptions of what the patient needs. The latter may be benign paternalism (see Case 5). Of course, physicians and healthcare professionals build a unique therapeutic alliance with a particular patient, as who else prescribes treatments and follow-ups? Yet, the therapeutic alliance is often predetermined, leaving relatively few options for the patient. However, unfettered patient-centric ethics could lead to patients demanding what cannot be provided. In the setting of moderate scarcity, it would be difficult to meet every patient's desire without reducing resources for others (see Cases 6–8). Physicians and healthcare professionals are doing all they can with the resources allotted to them, or so it is said (see Case 2).

Consider a Libertarian or Contractual moral theory where the physician or healthcare professional is defined by reciprocal obligations as an exchange of goods, with the physician income representing a good received in exchange for providing good—patient care. It would be naïve to think that such considerations do not sway physicians and healthcare professionals, as any reasonable persons would seek some balance between their obligations and rewards and between their authority and responsibilities (see Case 6). Salaries have decreased when adjusted for inflation, while similarly situated nonphysician salaries have increased (Tu and Ginsburg, 2006). At the same time, physicians have increased their direct patient contact while decreasing other services, such as volunteer work. For neurologists, the average workweek consists of 55.7 h with 42.4 h spent in clinical care (Busis et al., 2017). Physicians and healthcare professional limiting their responsibilities is, in fact, a physician-centric ethic.

The issue here is not the ethics or morality of physician income. Rather, there appears little room for increasing patient contact time to accommodate addressing a wider range of psychosocial and quality of life issues. If physicians are seeing fewer patients, there likely would be income reductions and exacerbation of the severe shortage of physicians (Gudbranson et al., 2017). An increased utilization of nurse practitioners and physician assistants, as well as changing licensure requirements and practice laws, could blunt the shortage of physicians (Salsberg, 2015). However, this could change the practice of everyday medicine significantly, whose unintended consequences (see Case 3) may be difficult to predict and could prove counterproductive.

Patient-centered ethics has been touted as the standard but it is not clear how accepted it has been in practice. In a study using unannounced standardized patients presenting to internal medicine specialists, a series of case scenarios were enacted varying in biomedical complexity and contextual complexity. The former related to biomedical issues was considered generic to the disease, whereas contextual complexity related to the particulars of the specific patient, thus assessing patient-centeredness. The final evaluation and recommendation documentations were categorized as error free if the diagnosis and treatment recommendations were medically correct and reflected the patient-specific situational issues. The error-free rating in contextually complicated encounters was 22% and 9% when cases were also biomedically complex (Weiner et al., 2010). Generally, there is a paucity of studies, for example, available through PubMed that examined the physician's patient-centric actions, as most were studies of methodology and patient perspectives. Yet, normative recommendations are suspect if there is a paucity of descriptive knowledge needed to induce ecologically invalid inferences, such as consensus. In this author's opinion, the risks of unintended ethical and moral consequences of proposed ethical frameworks are high in the everyday practice of medicine, as the great majority of books and articles read rarely consider the ethical problem in the entire context or ecology of the patient embedded in an extensive and complicated healthcare delivery system. (In teaching medical and healthcare professional students, the author keeps a game of "pick up sticks," where the object is to remove a stick from a pile of sticks without moving any of the other sticks. The purpose is to demonstrate the essential difficulty of everyday medicine.)

As to treating a patient as a person in his or her own right, these physicians may be doing just that. It just so happens that what is best for the patient in front of the physician or healthcare professional is just what is right for the *l'homme moyen* (the average man [patient]) characterized by the results of evidence-based medicine randomized controlled trials. Thus, the treatment accorded to the generalized patient is just what the concrete patient needs. Indeed, to treat any person different from all the rest, as embodied in the average patient, would be contrary to good medical practice. To be sure, personalized medicine promises to sculpt treatments to the patient's idiosyncrasies but this would not be an abandonment to the current Rationalist/Allopathic Principlist approach to medicine based on the generalized patient (Montgomery, 2019a). Rather, it is possible, indeed likely, that personalized medicine actually will create stratified subgroups within the *l'homme moyen*. This is a logical necessity as one does not "invent" a new medicine for each patient—wisdom recognized by Aristotle, who wrote, "Every definition is always universal: doctors do not say what is healthy for some particular eye, but rather for every eye or else for some determinate form of eye" (Aristotle, 2001; Posterior Analytics, Chapter 14, 97b 26). To invent a new medicine for each patient would result in Solipsism of the Present Moment.

As the notion that "understanding the doctor as a person, not merely as a skilled technician" begs the question, presupposes that there is something different from a skilled technician and that difference is what makes a physician, for example, a professional. Jonathan Shapiro (2009) (08 January 2009, https://www.bmj.com/rapid-response/2011/11/02/role-doctor-professional-or-technician) responding to an editorial Understanding the role of the doctor by Godlee (2008) describe the true professional as

> well trained in the technical aspects of their trade; they will have built up enormous experience by dint of many years working at the "coal face"; they are able and keen to reflect on their current state of knowledge and remain ever curious to further hone their skills; they have the wit and the courage to apply their experience and knowledge to the subtleties of the true professional challenge; and they possess the skills to communicate with their clients in a way that is accessible and appropriate, whatever their clients' culture, knowledge or intelligence.
> *(08 January 2009, https://www.bmj.com/rapid-response/2011/11/02/*
> *role-doctor-professional-or-technician)*

Yet, it is difficult to see how these same attributes could not be ascribed to a technician. Shapiro also suggests that key to the role of the doctor is that doctors are professionals, rather than technicians. Society needs professionals when the decisions that need to be made are too complex to be consigned to a fixed pathway, algorithm, or computer programme. It's when the variables are too many or too subtle for Microsoft Vista (or even Mac OS 10.5) to reach a definite conclusion that we need professionals, whether they be doctors, lawyers, senior detectives, business CEOs, or even NHS managers.

(08 January 2009, https://www.bmj.com/rapid-response/2011/11/02/
role-doctor-professional-or-technician)

A great many physicians and healthcare professionals would likely share Shapiro's sentiments.

It seems that the real distinction between the professional and the technician does not lie in their personal attributes but rather in the situation they find themselves; if sufficiently vague, then the person is a professional. The question is at what degree of vagueness does one trip from being technical to being a professional? Further, it would seem that the current Zeitgeist would be to reduce vagueness, or at least its appearance or effect, with evidence-based medicine, algorithms, and computerized decision-making-learning artificial intelligence as examples (see Case 12). Even vagueness, perhaps due to biological and instrumental variations within the operations of physicians and healthcare professionals, can be constrained by education, licensing, and professional standards. Indeed, it is likely that patients and society would not want the quality of care provided to patients to depend on the luck of the draw of a particular expert or virtuous physician or healthcare professional. Indeed, the uniformity and quality of parts allow for the economic efficiencies of mass production. It is not clear that this is irrelevant to everyday medicine.

Perhaps the insistence of being a professional is a romantic notion to preserve some status not directly derivable from the actual practice of everyday medicine. It may be a reaction to Rationalist/Allopathic and Principlist medicine in a manner analogous to Romanticism in reaction to the Enlightenment (see Introduction). Even if aspirational, although ungrounded, it could lead to virtuous behavior that may benefit patients even if primarily for the benefit of the physician or healthcare professional. However, if that benefit is to the disadvantage of the patient, then its virtue is unclear.

Perhaps patient-centric ethics reflects a hierarchy of shareholders and stakeholders, although similar in their relations in terms of the ethical principles of beneficence, nonmaleficence, and autonomy and moral theories. If the patient in patient-centric ethics is prioritized to the apex, presumably, then it is clear that the cases of everyday medicine described in Part 1 of this text were not based solely or even primarily on patient-centric ethics. But, the question of patient-centric ethics seems so appealing when the discussion is limited to a particular patient and a particular physician or healthcare professional, a dyadic or one-on-one relationship. However, the particular patient-physician/healthcare professional does not exist in a vacuum and, indeed, the shareholders and stakeholders encompass many more interests, as demonstrated in many of the cases in Part 1. Thus, patient-centric ethics has to come to grips with competing and legitimate concerns, as discussed in each of the cases in Part 1. In that context, it is not clear that a patient-centric ethical approach, at least in its ideal form, would work. However, that is not to say that the current everyday practice of medicine could not be more patient-centric—it needs to be. Indeed, a reduction in paternalism would make remarkable progress toward more patient-centric ethics. An additional place to start would be for ethicists to become more patient-centric but in an ecologically valid sense.

Qualifying patient-centeredness

Arguing for patient-centeredness as a moral good, Duggan et al. (2006) wrote:

> On the basis of such evidence, it is fair to conclude that patient-centeredness leads to better outcomes for patients. To the extent that this is true, it follows that patient-centeredness is morally justified (perhaps even required) on consequentialist grounds. That is, unless one found an approach that led to even better outcomes, one would have no compelling reason not to be patient-centered.

However, Duggan et al. (2006) fail to appreciate the cost of such patient-centeredness, for example, in the expenditure of time and effort on the part of physicians and healthcare professionals, as demonstrated in many of the cases in Part 1.

History likely always existed with least a moderate scarcity of healthcare resources necessitating a competition of interests. To be sure, there may be wide geographic and demographic differences in the degree of scarcity, but it would seem that only the independently wealthy can escape the consequences of moderate scarcity. Dealing with a moderate scarcity of goods raises issues of justice, which then requires governance to manage, typically leading to a social contract that binds all citizens (Rawls, 1999). The social contract constrains the obligations to the ethical principles of beneficence, nonmaleficence, and autonomy by establishing which moral theory prevails. The social contract that binds all necessarily means that the ethics of everyday medicine cannot be entirely patient-centric. Note that a philosophical trivial response could be offered that the individual patient is a member of society and thus enjoys the benefits of society through any program that optimizes the entirety of everyday medicine. However, it is not clear that such a sentiment has ever been fully embraced, although certain ethicists have made somewhat similar arguments, such as feminist ethicists (see later and Chapter 2). Such an ethic would be challenged by any Kantian deontological moral theory, as an individual must be treated as an end to herself or himself, not as a means to another's ends. Rationing of access to everyday medicine clearly challenges a solely patient-centric ethic.

Although avoiding the term, rationing is de facto in most, if not all, systems of healthcare delivery, although most often implicitly. Eligibility for insurance, copays, and deductibles, preexisting conditions, investments in healthcare delivery infrastructure, formularies, underinsured policies, carve-outs in coverage, and cherry-picking marketing are just some ways of rationing healthcare. Patients with certain types of insurers, such as US Medicare versus private, may receive a different quality of care, for example, Medicare patients receive less quality of care than private insurers (Spencer et al., 2013; James et al., 2014). Interestingly, some hospitals reduce services for every patient regardless of the insurer when faced with decreasing revenue; however, patients on Medicare and Medicaid fare worse (Dranove and White, 1998). None of these situations would occur if healthcare delivery was ideal with respect to patient-centeredness. Some rationing schemes have been explicit, such as the Oregon Medicaid Experiment, where in 1990 the state of Oregon established a prioritized list of services that would be covered depending on the availability of funds (Kaplan, 1994; Perry and Hotze, 2011). Interestingly, a waiver to allow the state to proceed with the plan was initially rejected by US Health and Human Services Secretary Louis W. Sullivan based on a potential violation of the Americans with Disabilities Act of 1990 (42 USC § 12101); perhaps an example of the success of patient-centered ethics. However, it is unlikely that moderate scarcity will be eliminated and increasing efforts at rational rationing will continue to challenge a solely patient-centric ethics (Vermeulen and Krabbe, 2018).

Physician-centric ethics

The cases described in this text suggest that the ethics of the everyday practice of medicine are not patient-centric. Rather, the ethics are more likely physician-centric in both an active and a reactive manner, at least until relatively recently. It is active in the particular benefits accrued to physicians by such an ethic and reactive when the ethic performs a defense against challenges to physicians. However, a physician-centric ethic is not always incompatible with a patient-centric ethic that maximizes obligations to the patient. Indeed, a physician-centric ethic, such as virtue and care ethics, while focusing on the physician and healthcare professional, is thought to consequently provide the maximum reasonable obligation to patients (see Chapter 2). An interesting example is the physician's obligation to patient confidentiality. The adherence to such confidentiality by the physician or healthcare professional may never be recognized by the patient and not driven directly by the fear of penalties, although strong penalties exist. Rather, physicians and healthcare professionals generally believe that holding such confidences is a virtue of the physician and healthcare professional. But is it beneficence of adhering to confidentiality bestowed on the patient or on the physician and healthcare professional?

Since at least the ancient Greeks there has been a nearly absolute duty to patient privacy and confidentiality. In the Hippocratic oath it is written, "And whatsoever I shall see or hear in the course of my profession, as well as outside my profession in my intercourse with men, if it be what should not be published abroad, I will never divulge, holding such things to be holy secrets" (https://en.wikipedia.org/wiki/Hippocratic_Oath). The obligation to confidentiality would appear to be absolute given the many statements from professional organizations where breaches necessitated by legal actions or by the most extreme circumstances are the exceptions (see historical review by Higgins, 1989). The adherence to patient confidentiality is arguably the best example of what appears to be a patient-centric obligation to the ethical principle of autonomy (in the larger sense of the word to mean respect and not just self-determination). However, the question arises as to what is the motivation of such privacy and confidentiality? Is it a deontological notion where such rights are a matter of Natural Law (see Introduction), is it a virtue of the physician elicited by the patient, or is it self-serving to the medical professional that coincidentally fulfills the obligation to the ethical principle of autonomy to the patient?

In most cases, a high correlation may exist between self-interest on the part of the physician or healthcare professional and a patient-centric value; however, a correlation does not imply cause and effect. If the cause of the virtuous behavior on the part of the physician or healthcare professional is not influenced by any deontological status of the patient's rights, then there is little guarantee that any deontological right on the part of the patient will necessarily be respected in every case. This issue will be taken up later in the discussion of Thomas Percival's notion related to charity care in 1803.

The question is how widespread is the admonition to confidentiality self-serving in cases and in history? Interestingly, the aforementioned quote from the Hippocratic Oath would suggest an absolute deontological moral standard that appears to be for the primary benefit of the patient. However, as pointed out by Higgins, Hippocrates states that the physician must "say only what is absolutely necessary. For he realizes that gossip may cause criticism of his treatment" and in "Treatises of Fistula in Ano, Hemorrhoids, and Clysters," c.1370, Arderne (1910) wrote, "for if a man sees that you hold secret another's information, he will better trust you" (quoted in Higgins, 1989). Further, nearly every code of ethics was also designed to uphold the virtue of the physician class, in distinction to the virtue of the patient, as secrecy was one means of differentiating physicians from competitors. In 1803, Thomas Percival wrote, "Secrecy, and delicacy when required by peculiar circumstances, should be strictly observed. And the familiar and confidential intercourse, to which the faculty are admitted in their professional visits, should be used with discretion, and with the most scrupulous regard to fidelity and honour" (Percival, 1803, p. 30). Indeed, in the Hippocratic Oath it is written:

> To hold my *teacher* [italics added] in this art equal to my own parents; to make him partner in my livelihood; when he is in need of money to share mine with him; to consider his family as my own brothers, and to teach them this art, if they want to learn it, without fee or indenture; to impart precept, oral instruction, and all other instruction to my own sons, the sons of my teacher, and to *indentured pupils who have taken the physician's oath* [italics added], *but to nobody else* [italics added].
> *(https://en.wikipedia.org/wiki/Hippocratic_Oath)*

In defense of physician confidentiality, Engelhardt wrote "the principle of autonomy makes it morally permissible to create such special exclaves secure against such requirements or disclosure... there may be special advantages from both priests and physicians offering strict confidentiality. The capacity of priests and physicians to function in their special roles, which have social value, may be undercut by the notion that compelling State interests could force disclosure of their private communications" (Engelhardt, 1986; quoted in Higgins, 1989). Similarly, "A doctor, like a lawyer or priest, does not readily recount his professional dealings with an identifiable person; and the public's trust in the medical profession derives largely from its conviction that what transpires between doctor and patient will not be bandied about" (Higgins, 1989). The question is does the analogy to priests borrow some sense of divine right or a necessity in a Hobbesian social contract sense for physicians and healthcare professionals? At the very least, it implies a physician-centric ethos.

A striking metaphor may be in the legal context of privacy and confidentiality. Interestingly, the term "privacy" does not appear in the US Constitution, rather it appears to have "evolved" into a constitutional right (Halper, 1996). According to Halper, the first inculcation of a right to privacy was described in an article by Samuel Warren and Louis Brandeis (1890). Brandeis later became a US Supreme Court Justice. Halper and others have argued that the primary purpose for the tort of invasion of privacy was not for the interests of the individual plaintiff but for society. Enforcement of privacy was seen as maintaining the normalcy of civility. Only later in *Roe v. Wade* did the US Supreme Court establish privacy as a constitutional right. Interestingly, even then the decision was less patient-centric, the pregnant woman being the patient, and more physician-centric, that is, the abridgment of the physician's right to perform an abortion (Halper, 1996). The US Supreme Court in *Roe v Wade* [U.S. Supreme Court *Roe v. Wade*, 410 US 113 (1973) https://supreme.justia.com/cases/federal/us/410/113/] wrote:

[3](a) For the stage prior to approximately the end of the first trimester, the abortion decision and its effectuation *must be left to the medical judgment of the pregnant woman's attending physician* [italics added].

(pp. 163, 164)

(b) For the stage subsequent to approximately the end of the first trimester, the State, in promoting its interest in the health of the mother, *may, if it chooses, regulate the abortion procedure* [italics added] in ways that are reasonably related to maternal health.

(pp. 163, 164)

(c) For the stage subsequent to viability the State, in promoting its interest in the potentiality of human life, may, if it chooses, regulate, and even proscribe, abortion except where necessary, in *appropriate medical judgment* [italics added], for the preservation of the life or health of the mother.

(pp. 163–164; 164–165)

4. The State may define the term "physician" to mean only a physician currently licensed by the State, and may proscribe any abortion by a person who is not a physician as so defined.

(p. 165)

Perhaps an interesting test case of physician-centric versus patient-centric ethics is the situation of terminating an established patient-physician relationship. There are well-established mechanisms and rationales for physicians terminating their care of patients in situations where the termination was not mutually agreed upon (Mehlman, 1993). However, it is not clear that any protections for the patient are more than just "sufficient notice." These concerns are not mitigated by the fact that most patients would not want to be treated by a physician who does not want to treat them, but the situation may be very different in areas of the significant scarcity of physicians and healthcare professionals. Even in those cases, care by a recalcitrant physician or healthcare professional may be limited to emergencies. The particular scenario is where the patient cannot afford to continue paying for care. It appears a matter of common understanding that the physician or healthcare professional is not obligated to continue nonemergency care, although is expected to help facilitate the transfer of care to another physician or healthcare professional. The most common response is to direct the patient to charitable or governmental organizations.

Even the administration of charity can be physician-centric. In 1803, Thomas Percival wrote

> And the discretionary power of the physician or surgeon, in the admission of patients, could not be exerted with more justice or humanity, then in refusing to consign to lingering suffering and, and almost certain death, a numerous class of patients, inadvertently recommended as objects of these charitable institutions. "In judging of disease with regard to the propriety of the reception into hospitals," says an excellent writer, "the following general circumstances are to be considered:
>
> Whether they be capable of speedy relief; because, as it is it the intention of charity to relieve as great a number as possible, a quick change of objects is to be wished; and also because the inbred disease of hospitals will almost in inevitably creep, and some degree, upon one who continues a long time in them, but will rarely attack one whose stay is short.
>
> Whether they require in a particular manner the superintendence of skillful persons, either on account of their acute and dangerous nature, or any singularity or intricacy attending them, or erroneous opinions prevailing among the common people concerning their treatment."
>
> *Percival (1803, p. 17)*

This passage is remarkable for several statements. For example, patient-centric ethics would ask in the case of those with "lingering suffering and, and almost certain death" where are those patients to go? Perhaps the strain on the resources would render the continued hospital stay as unjust was a Utilitarian moral theory to prevail. This is suggested by the statement "Whether they require in a particular manner the superintendence of skillful persons, either on account of their acute and dangerous nature, or any singularity or intricacy attending them."

The issue here is not whether the discharge of the aforementioned patients is just or unjust but rather it demonstrates that patient-centric ethics alone does not prevail in a situation of moderate scarcity. But one notes the striking referral of patients as "objects." The question is whether such terms used to substitute for actual patients allow one to dissociate the consequences of discharging "objects" compared to discharging Jane Smith (see Case 2)? Importantly, in the everyday practice of medicine, how is using the term "objects," which may seem repugnant, different than referring to a patient as "the gallbladder in room 307?" Note that this is not to say referring to Jane Smith as "the gallbladder in room 307" necessarily obviates any patient-centeredness but the question is does it implicitly condone other than patient-centeredness? There is a movement to change from descriptions of particular patients, for example, as an epileptic or as an epileptic person to a patient with epilepsy, the concept of person-first language, but the effect of such changes on what actually happens to patients is unclear (Noble et al., 2017). However, some groups have objected to person-first language, arguing that the disease, typically chronic, defines them such as autistic rather than a person with autism. For example, in 1993, the National Federation of the Blind adopted a resolution condemning people-first language, writing:

> Be it resolved by the National Federation of the Blind in Convention assembled in the city of Dallas, Texas, this 9th day of July, 1993, that the following statement of policy be adopted: We believe that it is respectable to be blind, and although we have no particular pride in the fact of our blindness, neither do we have any shame in it. To the extent that euphemisms are used to convey any other concept or image, we deplore such use. We can make our own way in the world on equal terms with others, and we intend to do it.
>
> *Jernigan (2009)*

Indeed, more radical changes from using the term patient to something else, such as consumer or client, have been considered (Neuberger, 1999).

There may be obstacles to providing charitable care even if the physician or healthcare professional may wish to do so. For example, the author as a resident was prevented from admitting a patient with a severe migraine crisis to the hospital as a charity case. The person in charge declined, saying that they already had enough patients with migraines using the budget with the implication that any additional patients with migraine would not serve any educational purposes for the physicians in training. This is not to argue that denying admission of the patient to the hospital was just, but clearly it was not patient-centric. There may be contractual issues of the healthcare delivery system, of which physicians and healthcare professionals may be employees of, in which providing charity by waiving copays and deductibles are violations. Similarly, providing such charity may be considered an inappropriate inducement for care and violate anti-kickback laws. Further, it may be illegal for a third party to pay the insurance premiums of a patient (https://www.cms.gov/CCIIO/Resources/Fact-Sheets-and-FAQs/Downloads/third-party-qa-11-04-2013.pdf). One wonders whether these policies are sanctioned "cherry picking," if the sickest patients are least capable of paying their own insurance premiums, then perhaps they would cease being the responsibility of the insurer, private or governmental. Again, the issue is not about the justice or injustice of discouraging charitable care, it is only to demonstrate that in the real ecology of the ethics of everyday medicine, ideal patient-centric ethics may be more inspirational than aspirational.

Another manifestation of physician-centric ethics is seen in the response to the report on the extent of medical errors contributing to mortality and morbidity (Institute of Medicine, 2000). One approach is to view the delivery of care from the perspective of a business organization with the institutionalization of systems, such as in the airline industry, rather than individual approach. One might think that physicians and healthcare professionals would welcome the expertise of nonphysician and nonhealthcare professionals with critical knowledge and experience in systems organization to help ensure patient safety; at the very least, a multidisciplinary approach where each unique perspective has equal legitimacy and credence. It is not clear that such is the case. Wears and Sutcliffe (2019) argue that efforts used to improve patient safety were co-opted by physicians, which, according to Wears and Sutcliffe, may have compromised advances in patient safety.

Rise of business- and government-centric ethics

Since the late 1960s, there has been a dramatic change in insurance for healthcare, at least in North America. Canada's healthcare system prior to the 1960s was similar to that of the United States. Physicians were independent practitioners paid on a fee-for-service basis. In 1957, a serious proposal to create a national healthcare policy was opposed by businesses, insurance companies, and the Canadian Medical Association. Doctors went on strike to prevent the province of Saskatchewan from instituting a province-wide governmental insurance program in 1962 by the newly elected Co-operative Commonwealth Federation. In response to the doctors' strike, community clinics were established where physicians, for example, recruited from Great Britain, were salaried rather than paid on a fee-for-service basis (https://canadiandimension. com/articles/view/the-birth-of-medicare). The Saskatoon Agreement in 1962 ended the strike and set the model for what would become the Canadian universal health-care program (Medicare) with groups of physician representatives negotiating with the provincial government over reimbursement and policy. In many ways, Canadian Medicare is very similar to Medicaid in the United States where funding comes from both provincial and federal governments.

In the province of Ontario, generally, the provincial government negotiates a Physician Services Agreement with the Ontario Medical Association. According to the Ontario Medical Association, there has not been an agreement for 5 years (https://www.oma.org/sections/news-events/news-room/all-news-releases/state-ment-by-the-oma-regarding-negotiations-with-the-ontario-government-of-the-phy-sician-services-agreement/), during which the provincial government has reduced fees paid to Ontario physicians. Interestingly, the US government used the Medicare Sustainable Growth Rate in implementing the Balanced Budget Act of 1997 to prevent physician-related Medicare spending growth from exceeding gross domestic product growth. Each year, until the Medicare Access and CHIP Reauthorization Act of 2015, the reductions were postponed and would have resulted in a 21% reduction of Medicare fees paid to physicians. Failing mediation, the provincial government of Ontario, Canada was required to participate in arbitration proceedings. Learning of a group of specialists forming the Ontario Specialists Association, the provincial government temporarily stopped arbitration, arguing that the Ontario Medical Association did not represent physicians for the purposes of arbitration (https://www. oma.org/sections/news-events/news-room/all-news-releases/breaking-news-the-government-wants-to-stop-arbitration-the-oma-has-refused/). The provincial government pulled back some services (delisting) in order to reduce healthcare costs, such as prescription and vision care. These were re-established for persons under age 25 and over 65 and those with disabilities or other special needs.

It can be argued reasonably that the enacted Canadian policies were government-centric rather that patient-centric, at least ideally (see Case 2). To be sure, Canadian officials can make the case that such limitations on healthcare spending were to every citizen's benefit and that all patients benefit as they are citizens. However, such an argument is very problematic, as discussed in Case 2. This author is unaware of any public referendum where the public agreed to delist medication and vision care supports. It can be said that in a representative democracy, the citizens, including patients, are heard and if their representatives agree to specific policies, then so do the citizens. However, the question arises as to the veracity of such a notion of representation. Indeed, if the principles of capitalism could apply, then elections would have to be free, informed, and each person unfettered in seeking his or her interests (see discussion later).

The issues and challenges facing Canada and the United States in healthcare are strikingly similar. What is interesting is that each responded to the same challenges in a very different manner. Health insurance in the United States, except for Medicare, Medicaid, and the Veterans Health Administration, was largely effected by private companies and co-operatives, although increasingly, insuring patients in Medicare, Medicaid, and the Veterans Health Administration is being outsourced to private concerns (Nutter, 1984). Indeed, it has been argued that former US President Nixon, facing pressure for a national health insurance program, signed into law the Health Maintenance Organization Act of 1973 (Pub. L. 93-222 codified as 42 USC §300e) as a means of forestalling such efforts for a universal single-payer system and to reduce healthcare costs paid by employers. The act had the essential effect of providing funds for the development of Health Maintenance Organizations (HMOs), preempting restrictive state laws and mandating many employers to include HMOs in the options for coverage provided to patients. As essentially subsidized by the US government, HMOs were quickly able to outcompete the alternatives.

It is important to realize that the original conception of an HMO is very different than what arose after the HMO Act of 1973 and the Employee Retirement Income Security Act of 1974 (ERISA) (Paley, 1993). Indeed, the establishment of the US Marine Hospital Service in 1798 could be considered an example of an HMO. The key feature was prepaid healthcare for American seamen who essentially paid in advance for all subsequent care received. This arrangement was advanced by the War Labor Board during World War II, which allowed employers and labor unions to bargain over employee benefits, such as healthcare, in the face of wage and price freezes. It was a way for the industry to compete for the scarcity of workers.

The striking aspect of these early HMOs was that payments were made directly to providers and thus providers realized higher profits by preventing illness or worsening illness. This changed dramatically following the HMO Act of 1973 and ERISA; HMOs changed from providers to middlemen between patients and hospitals and changed from nonprofit to profit-making entities. For example, there was a remarkable and rapid shift from the public- and owner-based hospitals to investor-owned hospitals (Hoy and Gray, 1986). Further, there was a remarkable conversion from nonprofit insurers, such as various Blue Cross and Blue Shield programs, to for-profit insurers. Consequently, HMOs have diminished in terms of patient enrollment; however, the concept of managed care organizations expanded to include systems such as Preferred Provider Organizations (Gabel et al., 2001) and increasingly large healthcare provider systems that largely were their own insurer.

All share the same element of the business plan, control patient access to negotiate with providers. Many healthcare provider systems enlarged rapidly by consolidating providers in order to better negotiate with insurers. As there were large profits to be made in healthcare, it is not surprising that business-centric ethics would develop. The rationale was that market forces would align to provide innovative and better care at lower costs. It is an open question whether this succeeded or whether the change to for-profit medicine merely resulted in a transfer of wealth from healthcare providers to investors with patients as bargaining chips and government complicit (Luft, 1980; Vincenzino, 1999; Sullivan, 2001; Bundorf et al., 2004; Himmelstein and Woolhandler, 2008).

The ERISA went on to provide very strong protections from medical liability by preempting all related state laws, forcing plaintiffs to appeal to federal courts, which do not allow jury trials (that may be more sympathetic to patients), during which the patient may not be allowed to present his or her case (Pitsenberger, 2008). The frequent explanation from managed care programs was that they did not practice medicine but rather only administered contractual agreements. Thus, any care that is not stipulated as a contractual agreement need not be covered by the managed care organization. However, in a great majority of situations, uncovered care is ungiven care for which the physician, not the managed care organization, may be liable (Record, 2010). ERISA preemptions have been used to block patients' recoveries for damages consequent to denial of care if the insurer qualifies under ERISA or if the complaint is "saved" from preemption. It would appear that the window of opportunity for a patient's successful appeal for judicial relief from denied coverage is very narrow.

Another example of the increasing dominance of business- and government-centric ethics is the increasing constraints on physicians and healthcare professionals to advise and treat patients according to the patient's best interest, free from explicit or implicit coercion (Brand et al., 1998). Laws have been passed in various states to prevent physicians and healthcare professionals from inquiring whether a patient has a gun at home, although the Florida law was struck down (Court Strikes Down Florida Law Barring Doctors from Discussing Guns with Patients, NPR.org). In the early years of managed care, restrictions were often explicit but more often implicit. Even today, there are not infrequent occasions where both the patient and the physician or healthcare professional face pressures, for example, when the everyday medical care in the patient's best interest conflicts with business or government interests. An example is a referral out of network or by a specialist rather than the patient's primary care physician. A recent US Supreme Court case, *Ohio v. American Express Co.*, 585 U.S. (2018), found that American Express could prevent merchants from steering customers away from using their American Express credit card. The American Medical Association had filed an amicus brief in favor of the court holding against American Express (https://www.ama-assn.org/advocacy/

physician/us-supreme-court-case-could-limit-physician-referral-power), fearing that health insurers could similarly steer physicians away from referring to others thought to be in the patient's best interest.

Given the rise in business-centric ethics in medicine, interesting parallels are seen in those that advise others on financial matters. It is not a stretch to suggest that investment advisors or brokers share similar responsibilities and challenges as physicians and healthcare professionals, where the retail investor is in a situation similar to that of the patient. Indeed, an asymmetry of knowledge resulting in a possible asymmetry of power, and hence advantage, operates in both the investment and the medical fields (Finke, 2014). Following the global economic recession of 2008, there were concerns regarding the obligations of investment advisors and brokers to their retail investor. One issue related to the obligation to the retail investor required by a fiduciary standard or a suitability standard. Under a fiduciary obligation, the investment advisor or broker must act solely for the benefit of the investor; in contrast, a suitability standard would seem to allow other interests to influence the recommendations.

In the United States, the method of compensation (fees versus commission) and the potential punishment for exceedingly self-serving recommendations distinguish fiduciary and suitability standards, and, absent a fiduciary standard, there is little in the market to prevent or moderate self-serving recommendations (Finke, 2014). Perhaps the best means of ensuring a fiduciary obligation is to decouple compensation from the advice provided or, to have a highly effective regulatory agency. However, as investments nearly always entail some risk, there would be little inducement of a strictly fee-based advisor to recommend perhaps more risky investments that could have potentially greater returns, for example, investment in mutual funds that typically do not meet the market average compared to some directed investments. Interestingly, a major criticism against the US Patient Protection and Affordable Care Act in 2010 was a reduction in the range of insurance products available by insisting on minimum defined benefits. The effect was to reduce the number of insurance products whose policies resulted in underinsurance but at lower costs. In a sense, such products offer a greater return to the insurer but at least some modicum of insurance at a much lower cost. It could be argued that advisors recommending these insurance products are meeting a suitability standard instead of a fiduciary standard.

There is a reasoned argument that the distinction between fiduciary and suitable standards is more a continuum and that any distinction is fiction. Because it is fiction, it is not workable (Black, 2010). Relative to the everyday practice of medicine, the incursion of business- and government-centric ethics in addition to reasonable aspects of physician-centric ethics cannot hold solely to patient-centric ethics. If a solely patient-centric ethics is held as a fiduciary obligation on the part of physicians and healthcare professionals, it likely will be unworkable and have counterproductive consequences, as demonstrated repeatedly in the cases in this text. Likewise, insisting on a suitable standard of obligation may not protect patients' interests consistent with the moral theory prevailing, thus being unjust.

Interestingly, a third way has been suggested and that is professionalism, which can be sufficiently robust to cover the range of situations that fiduciary and suitable standards were hoped to produce just behavior. Black (2010) proposed four characteristics to define professional behavior in contrast to fiduciary or suitable standards. These included:

(1) Prohibition against unauthorized trading. The investment advice provider must obey the investor's instructions and cannot make decisions pertaining to the account unless the investor has authorized the investment advice provider to do so.
(2) Duty of best execution. When executing transactions on behalf of an investor, the investment advice provider must use reasonable diligence to obtain the best available price.
(3) Duty to convey accurate information. When communicating information about an investment product or strategy to an investor, the investment advice provider must exercise reasonable care to ensure that he conveys the necessary information to make an informed decision (including costs and conflicts of interest), that the information is correct, and that he conveys it accurately.
(4) Suitability obligation. When making recommendations about products and strategies (or affecting purchases if it is a discretionary account), the investment advice provider must have sufficient information about (1) the investor's financial situation, including current holdings and investment objectives and (2) the investment product or strategy he recommends, so that his recommendations are suitable for the customer.

Parallels to the obligations of physicians and healthcare professionals are robust. Item 1 is clearly analogous to the requirement that physicians and healthcare professionals obtain permission in the form of informed consent, implicitly or explicitly, before acting on the patient. Failure to do so could be considered technical battery (see Case 5). Item 2 is a little different than the requirement of physicians and healthcare professionals to practice at the standard of care. Item 3 is analogous to the notion of full informed consent. Item 4 is the requirement that the patient's particular circumstances must be understood. For example, a physician should know whether the patient is allergic or has a serious adverse reaction to a medication the physician intends to prescribe. But perhaps as importantly, what is of value to the individual and particular patient as emphasized in Phenomenological and Narrative ethics (Chapter 2). The value of Black's analyses and recommendations, when extended to the practice of everyday medicine, is a provision of an ecologically valid framework—it is more realistic. The consequence might be less risk of disappointment, thus suspicions of injustice, because of less unrealistic expectations of all parties.

Some have pointed to the incursion of business-centric ethics in the everyday practice of medicine as providing the potential advantages and benefits unique to a laissez-faire market or capitalism. In the United States, the prevailing assumption is that the "free hand" of capitalism would lead to the most optimal program of healthcare delivery. Providers and patients would come to the most optimal arrangements that were maximally mutually acceptable. However, there are several necessary presumptions or assumptions and the success of a laissez-faire system depends on the validity of the presumptions and assumptions. These include that the individual participants can act freely, are fully informed, and there is an effective choice. As described earlier, the development of managed care medicine exerted its power over physicians and healthcare professionals by capturing patients with which they bargain. To do so effectively, patients had to be bound to the managed care organizations. Further, patients are at the short end of the informational asymmetry, which is compounded by the secrecy of the insurers, affected by gag rules or invoking trade or proprietary secrets (see Case 12). Finally, it is hard for patients to fully appreciate their best interests when their health future is largely unknown, as most humans tend to be optimistic or too scared to think about it. In marketplace medicine, any move toward patient-centric ethics would have to address these distortions of the "free hand" of laissez-faire market capitalism.

One interpretation of the aforementioned situations is that the ethics of every-day medicine favor the insurer, a business, over the patient and thus is business-centric in which the government would appear complicit. Again, the issue here is not whether ERISA is just or unjust but rather to demonstrate the dominance of business-centric ethics in everyday medicine enabled by government. Indeed, any power of physician-centric ethics gives way in many cases to the business-centric ethic. Any ethical framework that narrowly draws the issues between physicians and healthcare professionals without addressing a government- and business-centric will lack ecolog-ical validity and only create more confusion.

Transactional medicine: Physician-centric versus patient-centric ethics

Increasingly heard, at least by this author, particularly in primary or family care of-fices, are questions asked quickly first in the encounter: "What do you want to get out of this visit?" and lastly "Does our plan seem reasonable to you?" These questions can be explicit but, when implicit, very common (see Case 6). It is important as "Does our plan seem reasonable to you?" is not part of informed consent if the physician or healthcare professional is imposing some sort of role for the patient for expertise the patient cannot be expected to have, reasonably. These types of questions constitute a form of transactional medicine.

In one respect, transactional medicine, as just described, harkens to the patient as a consumer where the patient expresses what is desired and the provider attempts to meet the patient's expressed wishes. A problem arises when the transactions limit the physician's or healthcare professional's ability to do more (see Case 7). An automobile mechanic who repairs a flat tire at the customer's request likely would be expected to tell the customer that the wheel is about to fall off. In the short term, the customer goes home satisfied. In the case of patients, affirming the physician's or healthcare profes-sional's question "were your expectations for today's visit met?" may have the effect of patients convincing themselves that they received justice in their care. However, it is not clear that society would not expect more.

Alternatively, asking questions such as "what do you want to get out of this visit?" as opposed to "what is bothering you?" at least allows patients a bit more leverage to assert themselves in the encounter, which is very different from the asymmetric relationship of past paternalistic physician-centric ethics. In a way, it allows the be-neficence or goods exchanged in the encounter to be defined more on the patient's terms. Perhaps, more importantly, such an explicit request in the form of the patient's answer to "what do you want to get out of this visit?" may allow more transparency in the definition of the beneficence the patient hopes to receive and the beneficence the physician or healthcare professional is prepared to offer (see Case 6).

There is a risk that questions such as "does our plan seem reasonable to you?" may imply sufficient informed consent, not only to the patient but also to the physician or healthcare professional. To be sure, the question and its affirmative answer are consent. The critical issue is that it is not *informed* consent. In order for the patient, or his or her legal representative, to give informed consent, the patient or representative must be provided with all reasonable alternative approaches defined in terms a reasonable patient would find relevant. Relevancy may be defined as information that would if known, affect the actions of a reasonable patient. The question just posed alone does not admit alternatives to the one proposed.

Scientist/academic-centric ethics (scientific paternalism)

The relationship between medical science and the everyday practice of medicine has been fraught since the ancient Greeks. Even though experimental medical science then was fairly crude by today's standards, observations and vivisections were nevertheless conducted, likely even prior to the ancient Greeks. The school of Rationalist medicine in ancient times, the precursor to modern allopathic medicine from schools that provide medical doctor degrees, was scientific in the sense of a reductionist account of disease manifestations to a fundamental set of principles. Conceptually, Galen's principle of the four humors (Arikha, 2008) is a little different than many current approaches to pharmacology. This scientific reductionism is what differentiates the Rationalist physician from the Empirics and Dogmatics (Montgomery, 2019a). Indeed, the notion of scientific medicine was a wedge allopathic physicians used to differentiate themselves from alternative medical practitioners, strikingly advocated in the American Medical Association Code of Ethics in 1847. Yet, there was little difference in the care provided even until the early 1900s.

The battle over scientific medicine was clearly evident in the Flexner report on medical education, which produced a profound change in medical education and thus in the eventual practice of medicine. Indeed, the report called for the education of medical students in a laboratory by medical scientists (as well as a salaried rather than fee-for-service physician payment scheme). This was vigorously opposed by the leading physicians of the time, such as William Osler (Montgomery, 2019a). The compromise that lasted until the 2000s was Solomon-like—the medical student was "cut in half," with the first 2 years given over to medical scientists and the last 2 years given over to clinicians. In this author's opinion, the introduction of case- or problem-based learning in the medical school curriculum did little to change the fundamental conceptual approaches, although it is a significant change in pedagogy.

A dramatic change occurred with the championing of evidence-based medicine in the 1980s, with emphasis placed on randomized controlled trials, essentially a specific type of experimentation. Prior to that, other forms of evidence, such as case reports, case series, and expert opinion, were considered legitimate forms of medical practice. However, fundamentally evidence-based medicine is not science in the sense of generating new knowledge, just as deductive logic cannot produce new knowledge other than a recognition of tautologies. Evidence-based medicine is an epistemic statistical tool that only informs about the internal consistency of the observations or data from which inferences are drawn (Montgomery, 2019a). Evidence-based medicine makes no scientific claims. Any claim of causality requires substantial inferential leaps, exposing a high epistemic risk (Montgomery, 2019a, 2019b). For example, if administration of a ground up kitchen sink generates a statistically significant improvement in disease A without adverse effects compared to patients with disease A who received a placebo or alternative treatments, then the logic of current evidence-based medicine would compel physicians and healthcare professionals to administer a ground up kitchen sink to all patients with disease A. One could object, saying there is no causal explication why a ground up kitchen sink would affect the pathoetiology or pathophysiology of disease A. However, this is an argument outside of evidence-based medicine.

One could argue that there must be some reason why a ground up kitchen sink improved patients with disease A, as otherwise the results of randomized clinical trials of ground up kitchen sink would have to be a fluke. But this is not a fair argument (arguing *ad ignorantiam*—arguing from what is unknown). Further, the history of medicine is replete with many examples of successful therapeutics in the absence of any understanding of the causal mechanisms (Montgomery, 2019a).

The fundamental basis of present evidence-based medicine is purely epistemological, making no comment on what is real (ontological) and based solely on the internal consistency of the observations or data. For example, there is nothing in the statistics used in evidence-based medicine that would prevent or hold invalid determination of the average (mean) weight of all adult humans and duckbill platypuses in Australia. It is not clear what meaning such an average weight would have but nonetheless, such a determination would be entirely consistent with statistics. So, what is there to prevent statisticians from creating averages (purported to have some relevance to reality) that serve little purpose or convey little meaningful knowledge other than some prior knowledge that is independent of the randomized controlled trial?

In the 1800s, Claude Bernard, a preeminent physiologist, facetiously suggested as a measure of the European male, averaging samples of urine from every train station in Europe (Stigler, 2016, p. 35). The famous early statistician Francis Galton would reply "No statistician dreams of combining objects of the same generic group that do not cluster towards a common center; no more should we attempt to compose generic portraits out of heterogeneous elements, for if we do so the result is monstrous and meaninglessness" (Galton, 1883; quoted in Stigler, 2016, p. 37). But that statement begs the question of presuming a common center prior to the application of statistics. Such presumptions risk tautology, a self-fulfilling prophecy, yielding no new knowledge but rather a misunderstanding.

The fundamental assumption held by many statisticians and scientists is that statistics of the common center (Central Tendency) somehow reflects reality, better than observations or data [see The Meaningless of the Mean (Montgomery, 2019a) and Chapter 3]. In other words, the Central Tendency has an ontological status (reality) rather than a purely epistemic status (a measurement) supervening over the actual observations or data. Indeed, any variance between the Central Tendency and actual observations or data is attributed to noise, biological variability, or instrumental error, a problematic attribution that risks patient care.

It could be that patients and potential patients surrender their authority to a panel of scientists, by common will. But what of future patients and potential patients, such as those not born? When they are born, will the issue have to be revisited and, if so, when and how often? Alternatively, the citizens can engage in a type of Ulysses pact with their elected government representatives who can then delegate such authority to a panel of scientists to which the patient—as like Ulysses who instructed his crew to disregard subsequent commands (see Case 5)—has little recourse. Indeed, this is the basis for governments creating regulatory agencies, such as the Food and Drug Administration (FDA), and issuing laws and regulations stipulating what medications, biologics, or devices are permitted. However, in the United States, such approval for the latter does not guarantee access for those patients covered by insurance.

An example of the discrepancy where governmental regulatory agencies and business diverge from physicians and patients is in the use of FDA-approved medicines, biologics, and devices but for indications (diseases and disorders) not specified by the FDA. For example, the FDA can approve treatment B for disease A; however, it is up to the discretion of the physician whether to use treatment B for disease C with certain provisos. The following is from the FDA website:

> Good medical practice and the best interests of the patient require that physicians use legally available drugs, biologics and devices according to their best knowledge and judgement. If physicians use a product for an indication, not in the approved labeling, they have the responsibility to be well informed about the product, to base its use on firm scientific rationale and on sound medical evidence, and to maintain records of the product's use and effects.
> *(More extensive discussion can be found at https://www.fda. gov/regulatory-information/search-fda-guidance-documents/ label-and-investigational-use-marketed-drugs-biologics-and-medical-devices)*

In the United States, some insurers often do not cover off-label uses, which generally translates into the patient not receiving treatment even when the insurer is petitioned consistent with the position of the FDA just described. Note that the insurer need not provide any justification for disallowing the off-label use (see Case 12).

The introduction of evidence-based medicine for some insurers has become another reason to not provide coverage to access FDA-approved treatments. Merely the absence of randomized controlled trials is sufficient to deny proven efficacy; consequently, patients are denied coverage. Logically and to paraphrase Carl Sagan, the astronomer, "the absence of evidence [demonstrating benefit] is not evidence of absence [benefit]." In some cases, academic physicians appear to abet denial of care not demonstrated effective or safe with randomized control trials. It is not clear that a reasonable patient presented with alternative forms of evidence would acquiesce to the denial of care (Montgomery and Turkstra, 2003). Such cases demonstrate scientific/academic-centric ethics.

Evidence-based medicine has become the mantra of academic medical centers, particularly those involved in clinical research, the physician/healthcare professional scientist. One merely needs to review position statements by various professional organizations on accepted therapies to see the difference that evidence-based medicine has made. Yet, evidence-based medicine has had difficulty finding traction in nonacademic medical centers, particularly those in rural, presumably not academic, areas (Tracy et al., 2003; Hisham et al., 2016).

Consumer-centric ethics

It may be important to distinguish patient-centric from consumer-centric ethics, of which transactional medicine described previously may be an example. To be sure, consumers generally are patients but that does not mean that consumer-centric ethics is synonymous with patient-centric ethics. They differ in that a consumer may be a patient but a patient may not be a consumer. For example, insurers may court consumers but not all patients. In the United States, many insurers undertook active and passive measures to exclude certain patients from their consumers. For example, an exclusion for preexisting conditions or differential and higher premiums discouraged patients with chronic diseases. Even a cursory review of the advertisements by insurers picture the healthy. This author has never seen an advertisement to attract subscribes to an insurance company that featured a debilitated and seriously sick patient. Closure of healthcare facilities in areas associated with poverty, which often means higher rates of chronic diseases, may be another example.

Many have argued that structuring healthcare delivery based on consumers allows the "invisible hand of the market" of capitalism to operate. Competition among insurers and healthcare delivery organizations would be to the best interest of patients, allowing innovation and focusing on delivering value to the patient. However, it is not at all clear that the status of the insurance and healthcare provider systems is such that necessary prerequisites for the invisible hand to operate fairly are in place.

For consumerism to work, the patient must be an informed free agent and the marketplace of insurers and healthcare providers must be sufficiently plentiful to encourage competition. Inevitably, this requires a degree of "excess capacity." Moderate scarcity would not allow the patient to act freely. To be sure, excess capacity frequently allows a timelier delivery of services (see Case 2), but there is a cost to the excess capacity. It is not at all clear that any country could afford sufficient excess capacity to allow the patient full freedom in the patient's choice. Thus, some scarcity must be imposed and means are necessary to enforce the scarcity. The critical issue is that one should not conflate consumer-centric ethics with patient-centric ethics. This is not to say that either is more just.

Romanticism and misplaced virtue

Humans appear to be hardwired to be empathetic and some appear to be "more" wired than others (see Chapter 1). Patients seeking help likely increase the empathetic reward response in physicians and healthcare professionals, to which they may be predisposed as evidenced by their choice of professions. Further, for some, physicians and health-care professionals may be more affected when they perceive injustice, particularly if the patient appears to be aware of the injustice (Fondacaro et al., 2005). Particularly of concern here is countertransference and its effects on notions of patient-centric ethics.

Countertransference generally refers to emotional responses engendered in a physician or healthcare professional by a patient, particularly a patient that is emotive (Marshall and Smith, 1995; Rentmeester and George, 2009; Consoli and Consoli, 2016). The emotions aroused in the physician or healthcare professional can be of positive or negative valence (emotional value), or some combination, which can be productive or counterproductive. Most authors have commented on the counterproductive effects of negative valence countertransference (see Case 10). Few have commented on the counterproductive effects of positive valence countertransference. Interestingly, Thomas Percival commented on the issue in 1803, writing:

> It is, I trust, an ill-founded opinion, that compassion is not the virtue of a surgeon. This branch of the profession has been charged with hardness of heart: And some of its members have formally justified the stigma, by ridiculing all softness of manners; by assuming the contrary deportment; and by studiously banishing from their minds that sympathy, which they falsely suppose would be unsuitable to their character, and unfavorable to the practical exercise of their art. But different sentiments now prevail. *And a distinction should ever be made between true compassion, and that unmanly pity which enfeebles the mind; which shrinks from the sight of woe; which inspires timidity; and deprives him, who is under its influence, of all capacity to give relief. Genuine compassion arouses the attention of the soul; gives energy to all its power; suggest expedients in danger; incites to vigorous action in difficulty; and strengthens the hand to execute, with promptitude, the purposes of the head. The pity which you should repress is a turbulent emotion* [italics added]. The commiseration which you should cultivate is a calm principal. It is benevolence itself directed forcibly to a specific object. And the frequency of such object diminishes not, but augments its energy: For it produces a tone or constitution of mind, constantly in unison with suffering; and prepared, on every call, to afford the full measure of relief.
>
> *Percival (1803, pp. 126–127)*

The possibility offers that positive valence countertransference could lead to the subjugation of the patient's autonomy, thereby preventing truly patient-centric ethics. The quote from Thomas Percival demonstrates the difference between positive valence countertransference that is productive from that which may be counterproductive.

Some writers have offered the notion of double risk—the risk of the disease and the risk of seeing the physician or healthcare professional. The latter risk is the possible adverse outcome consequent to seeing the physician. But is there such a risk and, if so, what is its nature? Churchill et al. (2016, p. 118) offer as evidence the extent patients go to actually avoid seeing a physician, such as consulting family, friends, and the internet first. But is this just a matter of convenience rather than fear? Countering the inference of Churchill and colleagues, physicians continue to be the most trusted source of information (Hesse et al., 2010). To be sure, some patients suffer from iatrophobia, which is the fear of seeing a physician. Yet, this is estimated to be only 3% of the population (https://health.usnews.com/health-news/patient-advice/articles/2014/07/01/how-to-overcome-extreme-fear-of-doctors), although it is hard to find evidence to substantiate this estimation. It would seem important to distinguish iatraphobia from social issues such as the reticence of African-Americans to seek medical help based on a long history of disclination and unfair treatment (Washington, 2006). Further, iatraphobia needs to be distinguished from the potential bad news of a serious or threating diagnosis. To equate this reasonable fear with a fear of physicians is to confuse the fear in the message with the messenger.

Asymmetries in the patient-physician/healthcare relationship

Some ethics analyses find the patient-physician/healthcare professional relationship is asymmetric, which likely is true. But to understand the effects of the asymmetry, one has to critically understand the nature of the asymmetry. Too often, perhaps, the asymmetry is characterized in terms of power without any qualification of the term. On its own, the word "power" can have a significant rhetorical impact quite beyond its actual physical effect (see Narrative ethics in Chapter 2). Left vague, the asymmetry of power is then couched in sociological connotations of the powerful and the powerless, the shareholder versus the stakeholder, respectively. The powerless then are forced to subjugate themselves. Churchill et al. (2016, p. 119) write "The risk entailed in supplication, the surrender involved in accepting the authority of the clinician to greatly influence or even direct some aspects of our lives and to interact with our bodies in certain unusual and powerful ways, is at times a truly terrifying element of engaging with clinicians." It is difficult to see this as not intended to elicit a sense of compassion that risks falling over into pity, which only reinforces the notion of an asymmetry of power.

To be sure, the structure of the traditional patient-physician/healthcare professional may contribute to an asymmetry of psychological power; it is difficult to be assertive if one is in a flimsy examining gown before a fully uniformed physician or healthcare professional and provided little opportunities in the typical flow of discussion. Psychological asymmetries could result from any number of factors, including race, ethnicity, gender, and age; some perhaps coincidental, others perhaps not. Insurers, private or governmental, may create asymmetries that typically draft physicians and healthcare professionals into being complicit (see Case 2).

Ultimately, the patient or the patient's representative has the absolute right to reject recommendations and therapies, and, in this aspect, the patient is the one in power, at least in the negative sense. Perhaps what is required is the routine and consistent application to what would be analogous to the Miranda warning where everyone questioned by the police is first informed that they have a right to an attorney. In this case, every suggestion by the physician or healthcare professional should include the stipulation that the patient has the right to refuse any recommendation, such would be necessary for informed consent in its fullest connotation (see Case 5). However, the situation is different when the patient desires a treatment that the physician or healthcare professional declines to offer. But even at that, if fulfillment of the patient's request is consistent with the standard of care, then arguably the patient is in a more powerful position.

Any asymmetry fundamentally relates to information that the patient seeks and the physician or healthcare professional is expected to provide (Cases 5 and 6). All else follows from the information conveyed. However, there are a great many circumstances where there is an asymmetry of information. For example, a judge often depends on the information provided by expert witnesses but it does not follow that the judge is less powerful than the expert witnesses. The judge is the finder of facts, not the expert witness, and the judge determines what follows from the findings of fact. It is the patient who finds the facts as relating to the patient's own sense of self, it is the patient who determines what follows. The judge commands information from the expert witness, as does the patient when the physician or healthcare professional accepts a relationship with the patient. Further, the judge expects the experts, perhaps through questioning by the attorneys, to make their case to the judge by conforming to what the judge can expect, not what the expert witnesses wish to convey. In the same way, the physician or healthcare professional should be expected to make the case to the patient. Anything less is paternalism. Such paternalism may not be unjust but that would require an ethical system that minimizes patient autonomy and, as every citizen is likely to become a patient, such minimization of patient autonomy would likely be rejected.

It may well be the case, and often, that the patient or the patient's representative invites an asymmetry of power. Franz Ingelfinger, a physician and esteemed editor of the *New England Journal of Medicine*, the preeminent medical journal, wrote of his own experience when diagnosed with cancer. He wrote:

I received from physician friends throughout the country a barrage of well-intentioned but contradictory advice. As a result, not only I, but my wife, my son, and daughter-in-law (both doctors), and other family members became increasingly confused and emotionally distraught. Finally…one wise physician friend said, "What you need is a doctor." He was telling me to forget the information I already had and the information I was receiving from many quarters, and to *seek instead a person who would…tell me what to do, who would in a paternalistic manner assume responsibility for my care* [italics added]. When that excellent advice was followed, my family and I sensed immediate and immense relief.

Ingelfinger (1980)

Not infrequently, a patient will ask of the physician or healthcare professional what the physician or healthcare professional would do if they were the patient. The patient may say "whatever you [the physician or healthcare professional] think is best." It can be argued that the physician or healthcare professional should avoid this temptation (see Ulysses pact in Introduction). While perhaps innocent in a particular case, it risks establishing a precedent and a habit of procedure that can be counterproductive in the wider consideration and application. Further, one should expect that a well-trained physician or healthcare professional should be able to help the patient or patient's representative in such a way to help the patient find an answer in his or her own terms in respecting the obligation of the ethical principle of autonomy, much in the manner that an expert witness helps the judge as described previously.

Why do some ethicists appear to play on the fears of patients by adding an unnecessary fear, such as fear of a physician? There is a saying that those intent on being white knights often have to invent their own dragons. But creating dragons so as to allow one to be a white knight would not strike most as virtuous. Certainly, it makes the patient parenthetical to ethics.

Feminist bioethics

Feminist medical ethics provides a valuable perspective by which to view the ethical condition of many patients. However, as a caveat, a single, precise, and succinct definition of feminism as it relates to feminist bioethics is difficult with many variations (Wolf, 1996, pp. 3–43). This should not be surprising given the fluidity of concepts of sexual orientation and gender beyond biological sex. Further, at least gender varies on a continuum from poles that can be described as masculine or feminine. Thus, it is not surprising to see differences in positions taken by feminist ethicists.

Feminist ethics proceeds from several central themes. One theme has been the effects of the oppression of women based on sex and gender. As such, feminism can be related to the patient as parenthetical, particularly if the patient is of the female gender. But the lessons of feminist ethics go far beyond gender. Certainly, the oppression of women has been both extremely ubiquitous and, at times, subtle throughout history and contemporaneously. The subtlety then requires fine and sharp dissecting tools to begin to appreciate the full ramifications. Such tools have frequently been provided by feminist ethicists and make important contributions beyond gender studies (Tomlinson, 2012).

Another theme has been framed as attention to individualism versus the community, for example, the community of those of the predominantly female gender (see Chapter 2). The culture of individualism is attributed to male dominance, resulting in a characterization as "…atomistic and self-serving individuals, it strips away relationships that are morally central. This not only is impoverished, but may be harmful, because it encourages disregard for those bonds" (Wolf, 1996, p. 16). Wolf goes on to write:

> Moreover, there are some harms that only achieve the status of moral wrong when one considers groups. An individual woman, for example, may have no right to participate in an experimental drug protocol. But if that protocol or all protocols systematically exclude a group of people on account of their gender, that may well be an ethical wrong.
>
> *Wolf (1996, p. 17)*

The issue here is the apparent distinction between the individual's rights and that of the group in a manner where the group's rights appear to be transcendent or at least dominant. In that case, the individual's right would become subordinate. To be sure, the individual can surrender her rights to facilitate the group's claim of right, but in most pluralistic modern liberal democracies, one cannot give up his or her individual basic or fundamental rights. Subordination of an individual's right to that of a group would appear to be inconsistent with patient-centric ethics in everyday medicine. Most reasonable persons would likely find such surrendering of an individual's rights unacceptable, if only that it risks setting a precedent affecting others. Thus, ultimately, patient-centric ethics is necessarily atomistic. In the case of a woman's right to participate in an experimental drug protocol, an Egalitarian moral theory would hold that everyone, including the woman, would have a right to participate, although the obligation to the individual's right would be qualified based on other relevant and legitimate interests. For example, the logistics of having a specific person participate may require an expenditure that endangers the interest of others to participate and the potential benefits intended for society. Stating that a particular woman does not have the right confuses obligation to a principle with the principle itself. Distinguishing the obligation from the right allows workable solutions.

There is a possible pernicious presumption that group or gender rights are superordinate of those of an individual unless the individual grants such a relation. In the absence of consent, the consensus, perhaps achieved by Reflective Equilibrium (Chapter 3), would find the situation unjust. A motivation to allow such superordinate relations to a group may be that the individual cannot appreciate the advantages of the relationship without doing so and that their de facto coercion was necessitated to ensure the success of the group. However, how is this not paternalism and a violation of the obligation to autonomy? However, one cannot let a few people drive on the wrong side of the road, thus endangering many others. This is an example where the obligation to the principle of autonomy is not sacrosanct but is balanced by obligations to nonmaleficence to others and justice is achieved. Yet, the principles of autonomy and nonmaleficence are not jettisoned. This situation also demonstrates that justice is paramount over the obligation to the ethical principles (see Chapter 1).

Toward a more patient-centric ethics of everyday medicine

The aforementioned discussions demonstrate that any workable notion of hegemonic patient-centric ethics or ethos in the everyday practice of medicine risks being counterproductive even if possible. It is not a question of whether patients deserve beneficence, autonomy, and nonmaleficence but rather what is the obligation to these ethical principles to patients in the wider context of society, government, and business? The answer depends on the prevailing moral theory. For example, a Libertarian or Utilitarian moral theory necessarily holds that ethics that there are not patient-centric but are just. The Libertarian moral theory, particularly in its Contractualist form, holds that the degree or a patient-centric ethic is negotiated. The Utilitarian moral theory holds a group- or society-centric ethics. An Equalitarian moral theory is mute on patient-centric ethics, provided that all patients are treated the same. A Deontological moral theory may champion patient-centric ethics but this is problematic in a society where an adjudication of disputes often is on a contractual basis, limiting any Deontological moral theory. Perhaps new moral theories can be induced that would readily accommodate the complexities described.

External regulation to establish patient centricity

If it were deemed morally better to increase the patient-centricity of the ethics of everyday medicine, how would that happen? One might look to a patient's bill of rights, for example, the European Cancer Patient's Bill of Rights, which stipulated that:

> (1) The systematic and rigorous sharing of best practice between and across European cancer healthcare systems and (2) the active promotion of Research and Innovation focused on improving outcomes; (3) Improving access to new and established cancer care by sharing best practice in the development, approval, procurement and reimbursement of cancer diagnostic tests and treatments. Work with other organisations to bring into being a Europe based centre that will (1) systematically identify, evaluate and validate and disseminate best practice in cancer management for the different countries and regions and (2) promote Research and Innovation and its translation to maximise its impact to improve outcomes.
>
> *From abstract of Højgaard et al. (2017)*

The rights stipulated would seem uncontroversial as they are inherent in everyday practice. Who would argue otherwise? The primary means used to achieve these goals are educating other physicians and healthcare professionals and demonstrating cost-effectiveness (National Cancer Policy Forum, Board on Health Care Services, Institute of Medicine, 2013; Lawler et al., 2014).

At first glance, it would appear that a patient's bill of rights enforces a patient-centric ethic. The question is what is the nature of the rights? Are rights such as the "right to Life, Liberty, and the pursuit of Happiness" in the Preamble to the US Declaration of Independence an aspiration that always remains to be aspired as those rights do not have the force of law (Smith, 2002)? For example, many states in the United States have independent external review boards where a patient can object to an action by an insurer. However, unless the insurer is "saved" from the preemption of ERISA, it is not likely that any adverse determination by any reviewer will find redress in a court of law.

Flood and Epps (2002) make a distinction between rights *in* treatment and rights *to* treatment. Rights *in* treatment refer to the actual conduct of treatment and involve issues of informed consent, confidentiality, and provision of the standard of care. These rights seem relatively understood, if not respected, and there generally are means to address conflicts. Rights *to* treatment are very different and relate to access to care where the issues of standard of care, informed consent, and confidentiality are not issues. In the United States, rights *to* treatment typically involve coverage for the cost of care within the contractual terms of the insurance policy, and thus access, and in Canada, timely access to standard care is limited by governmental policy, whether by statute or practice.

Both the United States and Canada generally have access to external and independent reviews and administrative tribunals review insurer decisions. In Canada, administrative tribunals appear to have legal force and patients still have access to the courts, unlike in the United States (Flood and Epps, 2002). In the United States, it is clear that making significant legal options for patients to enforce whatever rights they may claim depend greatly on mitigating the effects of the ERISA preemptions. There have been several attempts to change the preemptions but all have failed (Benjamin, 2001; Mayers, 2003). However, courts still may reflect government-centric ethics. Flood and Epps (2002), describing such a case, wrote:

> *Cameron* [the legal case] concerned the Nova Scotia government's decision not to fund a male infertility treatment. The Nova Scotia Court of Appeal ruled that the government had discriminated against the infertile but that this decision was saved by Section 1 of the Charter (a reasonable limit prescribed by law and demonstrably justified in a free and democratic society) because of the high cost of fertility treatments. Leave to appeal to the Supreme Court was denied.

Unless every disease is prevented, note that cure is not applicable as it entails costs, or there is no scarcity of care, whether by nature or purpose, then patient-centric ethics will always have to compete with other interests, such as business- and government-centric ethics.

This author remembers an episode of the television show "The 21st Century" hosted by Walter Cronkite in 1967 in which he asked scientists to predict the future (The Futurists, May 28, 1967). With regard to the future of medicine, the scientists all agreed that the future would be shaped by societies' will to provide what medical science already has achieved. It would appear that moderate scarcity, which, according to John Rawls, drives systems of justice (see Introduction), is not due to a scarcity of medical technology. Rather, the scarcity lies in what society is willing to make available, which is why a bill of rights, in the sense of rights *in* treatment, while important, is not the limiting factor in a patient's care. It is in the right *to* care where the greatest challenge to dominant patient-centric ethics lies.

It is the moderate scarcity perhaps created by societies' will to enforce a right *to* treatment that will constrain patient-centric ethics. However, an important question to answer early in the process is what is the scope of "patient" in patient-centric ethics? For example, in a pluralistic modern liberal democracy, can society be satisfied when 90% of the citizens have their civil rights respected? If not 90%, is 99% or 80% required? If patient-centric ethics requires every patient to have rights *in* and *to* treatment, there may be diminishing returns in terms of increasing cost of resources for incrementally increasing the number of patients protected. At some point, a break-even point may be exceeded. Alternatively, setting a threshold or quota would appear arbitrary and violate Egalitarian and Deontological moral theories, assuming that adhering to Egalitarian or Deontological moral theories is important (see the case of the Oregon Medicaid Experiment of 1990 described earlier).

Targeting a threshold or quota invites an actuarial approach where a summary measure, such as a mean or average, is the control variable. In other words, success may be claimed if the average compliance, over multiple jurisdictions, when an individual's right *in* and *to* treatment is reached. However, substituting the average, such as *l'homme moyen* (average man), as the target invites a logical fallacy when used to assess whether the system is sufficient to protect an individual patient. The logical fallacy is the Fallacy of Four Terms as seen in Argument 4.1 here:

Argument 4.1

Major premise: *patients* (bridging term) *had their rights in and to treatment respected* (major term)
Minor premise: *John Doe* (minor term) *is a patient* (bridging term)
Conclusion: *John Doe* (minor term) *had his rights in and to treatment respected* (major term)

The bridging term, *patients*, in the major premise may not be the same as the bridging term, *patient*, in the minor premise, then there are four terms, a major, a minor, and two bridging terms. The argument is the Fallacy of Four Terms and no confidence in the conclusion can be taken from the argument. *Patients* in the bridging term in the major premise may be a sample of patients actually surveyed, which could be a different population than the population that contains the *patient*, John Doe, in the bridging term of the minor premise. It just may be that the population surveyed to establish *patients* in the major premise all had their rights *in* and *to* treatment respected, while the population, of which John Doe is a member, did not have their rights *in* and *to* treatment respected. Inferences drawn from sampling one population cannot be transferred to the other population (Montgomery, 2019b).

The problem is worsened when *patient* in the major premise is some summary measure, such as a measure of central tendency, necessary when there is variation, or an abstraction, necessary when the constructed measure is complex. However, healthcare delivery is often assessed and managed on some composite score, typically related to the most pressing healthcare problems faced by a group or society or those problems whose "return" on the resource investment is maximal (see discussion of the Oregon Medicaid Experiment discussed earlier). However, for patients with other problems, the approach would not appear to ensure their patient-centric ethical care, but rather that of the group or society.

Most policies and procedures used to assure patient-centric ethics in the entirety of everyday medicine likely rely on summary measures as a means to enforce patient-centric ethics may find the Fallacy of Four Terms inescapable. Yet, the alternative is to establish some means where administration of the policies and procedures assures every patient their rights *in* and *to* treatment, but this is challenged by business- and government-centric ethics (see Case 12). Indeed, the conundrum of wanting to attend to the patient as a "concrete" (particular and individual) rather than generalized (such as a summary measure or principle) "other" lies at the heart of many versions of feminism (Smith, 1996), virtue, and care ethics (see Chapter 2). Perhaps by aligning patient-, business-, and government-centric ethics, as Sir William Osler said, one might obtain the promise of Bernard, "pax sine crimine, pax sine turbine, pax sine rixa" (peace without crime, peace without turmoil, peace without quarrel). However, experience, such as those described in the cases in Part 1 of this text, suggests that such an alignment has not yet been achieved.

Internal regulation to establish patient centricity

An alternative to imposing direct external assertion of patient's rights is allowing Adam Smith's "Invisible Hand" of the free market to regulate ethical provision of everyday medicine (Smith, 1759, 1776). Indeed, this is often raised in opposition to governmental regulations of healthcare, for example, what is known in the United States as Obamacare (Patient Protection and Affordable Care Act of 2010). The presumption is that patients, as free agents in the consumption of healthcare services, would evolve to some relationship with healthcare providers that was mutually optimal based on the exchange of goods, monies for services that did not require external influences. Indeed, external influences, such as governmental regulation, were seen as "tipping the scales" to impair the efficient mechanisms of the invisible hand of the free market.

What are the requirements of a laissez-faire system, such as the healthcare system, that allow the "invisible hand" to operate so as to bring justice to everyday medicine? First, the system must be understood at the fundamental level at which the system operates, that is, the individual consumer and provider. In everyday medicine, the individual physician or healthcare professional deals directly with the individual patient. Second, the most optimal system evolves as an equilibrium between opposing or offsetting forces. The provider wants to maximize return and the consumer wants to minimize cost. Wherever the equilibrium settles is what is justice, as the obligations to the autonomy of the provider and the consumer are fully respected. Both the provider and the consumer can be assured that their beneficence and nonmaleficence have been optimized by fulfilling the obligation to autonomy on both parties. How could either reasonably argue for more?

For such an equilibrium to be reached, there must be real dynamics, such as the ability of either the provider or the consumer to end their relationship and to seek relationships with other providers or consumers. In other words, both parties must have the ability to "vote with their feet." Further, to optimize justice in achieving equilibrium, both parties must be capable, in the fullest understanding of the term, to enter into the "contract" that is established by the equilibrium. The question is whether a reasonable healthcare system can comport with the necessary requirements for a laissez-faire system affecting the "invisible hand" work (Le Grand, 2009).

The first requirement that the fundamental level of interactions is directly between the patient (customer) and the physician or healthcare professional (provider) is how the Canadian healthcare system operates, at least in most cases. For most healthcare services, the Canadian system is a single-payer and thus there is no competitive advantage to pit one physician or healthcare provider against another in a competitive bid to drive down cost. In the United States, the large majority of physicians are employees of healthcare provider systems. Thus, the patient "contracts" with the healthcare provider system, not with the individual physician or healthcare professional, even though the actual provision of care is affected by the physician or the healthcare professional. Thus, "negotiation" between the patient and the physician or healthcare professional is curtailed by the healthcare system that is external to the patient–physician/healthcare professional direct interactions. As to the second requirement, any equilibrium that is obtained between the patient and the physician or healthcare professional is constrained by the contractual relationship between the patient and the healthcare provider system. Case 2 in Part 1 demonstrates the problems that can arise where an administrator deciding healthcare policy is too far removed from the actual consequences of those decisions on the individual patient.

The patient could "vote with his or her feet" and the physician knowing this may be more consumer-oriented, but this can only happen so long as the patient remains in the specific healthcare provider system. If there is no scarcity of the needed type of physician or healthcare professional, then the autonomy of the patient is preserved. However, for a patient to truly "vote with his or her feet," the patient would have to be free to leave the particular system. Many healthcare systems can only accept a transfer during limited annual enrollment periods except under limited circumstances. Further, significant penalties accrue if there is any break in healthcare coverage. It is not as though the patient could leave one system, wait a while, and then be allowed to join a different healthcare provider system.

In a single-payer system such as in Canada, patients are technically free to choose whichever physician or healthcare professional they wish. However, significant scarcity makes finding a new physician willing to accept a new patient very difficult. The scarcity of physicians then is an external force tipping the scales in any equilibrium reached between the individual patient and the individual physician or healthcare professional. This author has seen, in his opinion, the extraordinary reluctance of patients to seek a new physician in patients clearly dissatisfied with their care, and frequently with good reason. Often it is a psychological reticence but it is clearly exacerbated by the scarcity of physicians.

A power equilibrium shifting between a patient and a physician or healthcare professional based on scarcity need not be unjust and therefore unethical. To shift the dynamic in the opposite direction where patients can migrate freely to different physicians and healthcare professionals would require a relative surplus of physicians, healthcare professionals, and health resources. Note that the notion of scarcity versus shortage has to be considered locally. What is the distance a patient can be reasonably expected to commute to seek specific healthcare? Thus, while there may be global shortages of physicians, healthcare professionals, and resources, there may be areas, such as affluent urban areas, where there is a relative surplus of physicians, healthcare professionals, and resources. In areas where there is a real surplus of physicians, healthcare professionals, and resources, it becomes a buyer's market and patients are better able to "vote with their feet."

This relative surplus typically implies an excess capacity that has to be paid one way or another. Reimbursements and other inducements must be sufficient to keep a physician or healthcare professional in an area of relative surplus. The question is whether this is acceptable to society to enable patient consumerism.

Ethics and ethos

Ethos is taken here as the "character of the times." Understandably, ethos is complex and often indistinct. Yet, often there is a discernible tone. The tone is set by shared values, a sense of comradery, common purpose, similar expectations, and others. It allows people to have a sense of their position in the larger scheme. However, no particular ethos is assured to be permanent and changes in ethos can be jarring.

It is no coincidence that physician-centric ethics would coexist with a "Captain of the Ship" mentality (see Cases 1 and 12). In the structure involving physicians and healthcare professionals, there is asymmetry and hierarchy in which physicians are at the apex (see Case 1). Thus, it would not be a stretch to conclude that physicians in that era had a sense of entitlement, and likely still do. Contrast to that ethos with commercialism where the patient becomes a consumer and the physician becomes a means to another end, ultimately to the financial benefit of the insurers and healthcare delivery systems. Physicians become a cog in the machine and from a commercial perspective, this is necessary. The physician as a cog in the machine is replaceable. Means by which physicians can establish their unique and individual value are obviated by employment contracts and marketing that feature the healthcare system or insurer, not the physicians.

Physicians become replaceable parts (see Case 4). This is most obvious in many academic medical centers where physicians are no longer eligible for tenure, with its unique protections conveyed upon the tenured professor of medicine or surgery. Tenure became important in the Glided Age where professors advocating for child work and welfare laws were threatened by industrialists on the boards of directors of academic centers. Indeed, the original notion of tenure was not a "job for life" but rather an expectation that the tenured professor of medicine or surgery could continue to safely advocate for the care of patients and others. Now, most physicians in academic centers are "clinical professors," subject to annual reappointments that could be withheld for any reason. The position of the nontenure community physician is far more precarious.

Today, many physicians are abandoning patient care or retiring early. Many discourage their children from entering the medical profession. The discouragement is palpable and documented in books such as Sandeep Jauhar's "Doctored: The Disillusionment of an American Physician" (Jauhar, 2014), in a more cynical vein in Samuel Shem's "The House of God" (Shem, 1979), or in a lament in "God's Hotel: A Doctor, a Hospital, and a Pilgrimage to the Heart of Medicine" by Victoria Sweet (2012). To be sure, the work of physicians is long, hard, and demanding of the physician and all those around the physician. The sacrifices are many. But they always have been. What has changed? Perhaps the change from physician-centric ethics to business- and government-centric ethics has been jarring in many ways. Perhaps physicians have lost a sense of themselves. Perhaps the ethos that physicians have created and manifested in their physician-centric ethic is no longer relevant in the commercial or governmental ethos reflected in the business- and government-centric ethics, and perhaps many physicians and healthcare professionals are ill-prepared to deal with the change. Tragic as it might be, it is hard to see that a patient should feel sorry for the physician, although many may. It is an open question whether patients are better served by physician-centric ethics or business- and government-centric ethics.

Perhaps physicians, healthcare professionals, businesses, and governments could be better served in total by patient-centric ethics and the ethos it generates. In that case, the patient is the deciding factor in the context of all patients and future patients, essentially all of society, that decides and the roles of physicians, healthcare professionals, businesses, and governments are constructed correspondingly. It is almost hard to visualize what an ethos based on patient-centric ethics would look like. Perhaps, as of today, what such an ethos and ethics might be as a "homework assignment."

The best of all possible worlds

A central theme of this book is that the ecology in which the ethics of everyday medicine operates is very complex. Hopefully, that point has been amply demonstrated. Despite the very long history of medicine, even from prehistory in the traditions that passed from generation to generation, it is not at all clear that justice is being served even today. It would seem that the problem of justice in medicine is intractable and that efforts to create justice are fatiguing to the point of despondency. That despondency is similar to religion with a juxtaposition between an omnipotent and loving God and a world of suffering, death, and evil. How could such a God allow the world that humans inhabit? To be sure, some theologians offer some notion of original sin that caused humans to be cast out from a truly perfect world, yet such retribution would not seem compatible with a loving god. Perhaps the god's love is limited but then how could the god be omnipotent?

In response to the theological conundrum, Gottfried Leibniz offered the notion of the "best of all possible worlds" (Leibniz, 1710). The argument offered is that God is perfect (omnipotent); therefore, the world created is perfect by definition. It just so happens that the "perfect" world contains injustice, suffering, and death. It would be a nonsensical question to ask whether there is a world more perfect. If it were possible to have a more perfect world, God would have created it and humans would inhabit that world. Because humans do not inhibit that world, then it is not possible to have a more perfect world. The question is nonsensical, as would be the question "what is outside the universe?"

It might be tempting to say that the current ethics of everyday medicine is the "best of all worlds." Being able to believe that would relieve the nagging of trying to do better and some sort of peace could be obtained. But how is one to know whether there is no other ethical system that is better, if not globally then at least locally? There is no *deus ex machina* (god in the machine) often used as a literary device to resolve an irresolvable problem, the "corner that the author painted herself into." In a pluralistic modern liberal democracy, there is no appeal to a *deus ex machina*. The citizens are left to their own devices. And that takes effort. There are tools and methods to help, which are the focus of Chapters 3 and 5.

Societal level

There are multiple levels at which the effort to develop ethical systems that help ensure justice operates. There are levels of the aggregate: the community, state, nation, and humankind. To be sure, the fundamental unit of operation is between the individual patient and the individual physician or healthcare professional; however, this relationship is embedded repeatedly in a complex ecology. And yet, in the complex ethical ecology of a pluralistic modern liberal democracy, ultimately ethics is determined by the consensus of those affected, which includes the patient, potential patients (thus, all citizens), physicians, healthcare professionals, and those charged with ensuring the just operations of the ethical system.

There are enormous challenges in developing a societal consensus and, in many ways, the success of modern medicine is a big culprit. As discussed earlier, vaccinations, public health, and antibiotics have reduced or obliterated any communal sense of disease by minimizing the dependence of one's own health on the health of others. An example is AIDS, as discussed in Chapter 1. Egalitarian efforts to argue that one's own health is dependent on the health of others, as, for example, one's own civil rights are as safe to the degree that others have their rights protected, does not seem effective. Indeed, in a survey regarding healthcare costs, the conclusion was that everyone usually thinks their health insurance is reasonable while everybody else's is too expensive. In many ways, modern healthcare insurance in the United States has exacerbated the problem by stratifying risk and off-loading those with the greatest risk. Exclusions of healthcare insurance based on preexisting disease filtered out those with chronic diseases that risk increasing complications and thus the need for healthcare. It is interesting that despite the controversies over Obamacare, elimination of the restrictions based on preexisting conditions seems to have a more secure basis in future healthcare systems in the United States. On a broader scale, there is the issue of environmental justice, particularly environmental racism, where some groups of people, such as the poor, are saddled with exposure to environmental dangers and health risks disproportionately to others. There are many more examples of leveraging risk stratification that are beyond the scope of this effort.

The central theme of Chapter 1 is that the fundamental process or primordial substrate of ethics lies in the affective (emotive) neurophysiological response from which declarative and deliberative ethical principles, moral theories, laws, ethical precedents, and virtues are post hoc, and initially derivative. This position is little different from the views of many philosophers and ethicists throughout human history, particularly in the form of ethical virtue (Chapter 2) and now intellectual virtue (Chapter 5). The extension from those discussions is that any societal change requires a change in the societal consensus. Fundamentally, this requires the establishment of an emotive (emotional) societal consensus that hopefully can lead to a principled societal consensus that can be operationalized into laws, policies, procedures, and systematic precedents.

How to establish an alternative emotive (emotional) consensus? The answers to such a question are far beyond the understanding and experiences of this author, who has a Bachelor's of Science degree in biochemistry, then an MD, followed by a career in basic neuroscience and subspecialty neurology left little room for the humanities. Yet, even to this author, humanities seem the best means of establishing an emotive (emotional) consensus. The personal experiences of patients, relevant to all citizens, as they are potential patients, discovered through phenomenological, narrative, care, virtue, and feminist ethics can be important starting points.

Optimism is buoyed by the impact of the work of the humanities to change society. The examples are numerous and include (1) "Oliver Twist" by Charles Dickens, (2) "Uncle Tom's Cabin" by Harriet Beecher Stowe, (3) "The Jungle" by Upton Sinclair, (4) "All Quiet on the Western Front" by Erich Maria Remarque, (5) "1984" by George Orwell, (6) "To Kill a Mocking Bird" by Harper Lee, and (7) "Silent Spring" by Rachel Carson, among others. A remarkable feature of these works is that they resonate with affective (emotive) mechanisms already present within the reader (Chapter 1). Having engaged the reader, the experiences received from the works continue to shape and differentiate the original affective (emotive) mechanisms. Further, this process occurs in an expanding number of citizens evolving into a collective response that can drive change. A deep philosophical analysis of art, as the product of the humanities, to affect human history can be found in Heidegger (2008).

Individual level

As suggested, the change is first affected in the individual, but through the use of mass media, the individual responses become a collective response. The question becomes how to cultivate the individual response. Perhaps surprisingly, this is a very simple issue as most reasonable persons already have a built-in ability to sense justice and, most importantly, injustice. The problem appears to be one of changing the personal discomfort of observing injustice in others to some action. Lisa Dodson, studying terrible economic inequalities and injustices, interviewed persons who responded in personal ways (Dodson, 2009). Frequently, the motivating factor for these persons to resist economic inequalities and injustices was the resisters' recognition of their role in perpetuating the injustice. Even though they may see themselves as an intermediary, not personally responsible, they did not create the system that led to the injustice, that they were just a "cog" in the machine that generates and perpetuates injustice, nonetheless, they acted.

Even though just a "cog" in the injustice machine, they played a role in the operations of the machine that enabled them some power or leverage. For example, it might be a supervisor adding (padding) the number of hours recorded to have been worked by another person in order to provide some modicum of economic justice to that person. It might be a physician who was less than truthful on an insurance form in order to obtain care for her patient. For the physician, the process has a long tradition in medicine, known as "gaming" the system. For example, with the implementation of reimbursements based on diagnostic codes in the United States known as the Diagnosis Related Group (DRG), physicians quickly learned of ways to game the system. At least initially, giving the patient a diagnosis of seizure disorder rather than epilepsy allowed the physician to keep the patient in the hospital longer, even if the diagnosis of epilepsy would have been the most accurate. Another practice involved adding comorbidities and complications to increase the complexity in the DBG code, thereby increasing reimbursement so as to allow the physician or healthcare professional to provide more charitable care.

Some physicians have waived copays but this practice has been challenged by insurers as insurers have argued that these charitable acts actually were an attempt to induce inappropriate use of medical services by reducing the barrier to those services. Similarly, some insurers in the past have contractual agreements that limit the charitable care that can be provided by a healthcare provider system, presumably to force those patients into the insurance system so as to recover potentially lost premiums. Some physicians have "down coded," for example, charging a "nursing visit" instead of a comprehensive medical evaluation. However, coding has increasingly been taken out of the hands of physicians and now is done by professional coders responsible to the administrator of the healthcare system where there is an incentive to "up code."

Each of these attempts to game the system often provides the actor with a sense of virtue, and such virtues lead to a sense of a good, in the nonmaterial sense of good, or blessed life leading to eudaimonia in Aristotle's term, that conveys happiness. However, in an important sense, such actions may be counterproductive. The effects are local as they are secretive. One can appreciate the necessity of secrecy as some of these efforts can be considered fraud or theft, which could carry criminal and civil penalties or job dismissal at the very least. The secrecy of these acts means that there can be no collective response necessary to change the system at the societal level. When the individual patient benefits, there is a reduction in the emotional discomfort on the part of the actor, which may dissipate the very energy needed to create the collective and drive change. It was the wide readership of the atrocities in Dickens's "Oliver Twist" and other works that helped Dickens encourage and thus congeal public outrage at the Poor Laws of 1934, even if that outrage did not originate with Dickens's work (Davis, 2012). One could almost hear a modern Oliver Twist patient asking "please, sir, I want some more"; in Oliver's case, it was thin gruel and in the patient's case, a bit more of the physician's or healthcare professional's time, understanding, and concern.

The remarkable success of modern medicine has changed the individual's experience with the disease. The risk of mortality has shifted greatly from the young to the old (Preston, 1976). Thus, concern for mortality goes from the relatively young and middle-aged to the elderly where efforts other than empathy and comfort seem less important. It would seem that concern, particularly that sufficient for change, is stratified and that for the majority of shareholders, such as citizens, the disease is someone else's concern and, similarly, the responsibility for it. Perhaps the ideal situation would be a lottery where the disease is distributed by chance rather than by age or socioeconomic circumstance. Perhaps it may be critical for the healthy to recognize their culpability for the care of the unhealthy. But this may require a superhuman, thus making the current ethical systems the "best of all possible worlds." It is for society to decide whether the effort to create a better world is worth the effort.

References

Arderne, J., 1910. In: Power, D. (Ed.), Treatises of Fistula in Ano, Hemorrhoids and Clysters. Paul, Trench, and Trübner, and H. Frowde, Oxford University Press, London.

Arikha, N., 2008. Passions and Tempers: A History of the Humours. Paperback Harper Perennial.

Aristotle, 2001. Posterior analytics. In: McKeon, R. (Ed.), The Basic Works of Aristotle. The Modern Library Classics, New York. (Chapter 14, 97b 26).

Balint, E., 1969. The possibilities of patient-centred medicine. J. R. Coll. Gen. Pract. 17, 269–276.

Belmont Report: Ethical Principles and Guidelines for the Protection of Human Subjects of Research, National Commission for the Protection of Human Subjects of Biomedical and Behavioral Research, Department of Health, Education and Welfare. 1978. United States Government Printing Office, Washington, DC.

Benjamin, G.C., 2001. Patient bill of rights 2001. Physician Exec. 27 (3), 75–77.

Black, B., 2010. How to Improve Retail Investor Protection After the Dodd-Frank Wall Street Reform and Consumer Protection Act. Faculty Articles and Other Publications. (Paper 184) http://scholarship.law.uc.edu/fac_pubs/184.

Brand, G.S., Munoz, G.M., Nichols, M.G., Okata, M.U., Pitt, J.B., Seager, S., 1998. The two faces of gag provisions: patients and physicians in a bind. Yale Law Policy Rev. 17, 249–280.

Bundorf, M.K., Schulman, K.A., Stafford, J.A., Gaskin, D., Jollis, J.G., Escarce, J.J., 2004. Impact of managed care on the treatment, costs, and outcomes of fee-for-service Medicare patients with acute myocardial infarction. Health Serv. Res. 39 (1), 131–152.

Busis, N.A., Shanafelt, T.D., Keran, C.M., Levin, K.H., Schwarz, H.B., Molano, J.R., Vidic, T.R., Kass, J.S., Miyasaki, J.M., Sloan, J.A., Cascino, T.L., 2017. Burnout, career satisfaction, and well-being among US neurologists in 2016. Neurology 88 (8), 797–808.

Churchill, L.R., Fanning, J.B., Schenck, D., 2016. What Patients Teach: The Everyday Ethics of Health Care. Oxford University Press, p. 118.

Consoli, S.G., Consoli, S.M., 2016. Countertransference in dermatology. Acta Derm. Venereol. 96 (217), 18–21.

Davis, M., 2012. Did Charles Dickens really save poor children and clean up the slums? BBC News Mag. (7 February). Available from: https://www.bbc.com/news/magazine-16907648.

Dodson, L., 2009. The Moral Underground: How Ordinary Americans Subvert an Unfair Economy. The New Press.

Dranove, D., White, W., 1998. Medicaid-dependent hospitals and their patients: how have they fared? Health Serv. Res. 33, 163–185.

Duggan, P.S., Geller, G., Cooper, L.A., Beach, M.C., 2006. The moral nature of patient-centeredness: is it "just the right thing to do"? Patient Educ. Couns. 62 (2), 271–276. Abstracted from: Mead, N., Bower, P., 2000. Patient-centredness: a conceptual framework and review of the empirical literature. Soc. Sci. Med. 51, 1087–1110.

Engelhardt, H.T., 1986. Foundation of Bioethics. Oxford University Press, Oxford, pp. 297–301.

Finke, M.S., 2014. Fiduciary Standard and Financial Advice: Findings From Academic Literature. Available from: https://ssrn.com/abstract=2419727. https://doi.org/10.2139/ssrn.2419727.

Flood, C.M., Epps, T., 2002. A Patients' Bill of Rights: A Cure for Canadians' Concerns About Health Care? Available from: http://irpp.org/wp-content/uploads/assets/research/health-and-public-policy/new-research-article-3/pmvol3no12.pdf.

Fondacaro, M., Frogner, B., Moos, R., 2005. Justice in health care decision-making: patients' appraisals of health care providers and health plan representatives. Soc. Justice Res. 18 (1), 63–81.

Gabel, J., Levitt, L., Pickreign, J., Whitmore, H., Holve, E., Rowland, D., Dhont, K., Hawkins, S., 2001. Job-based health insurance in 2001: inflation hits double digits, managed care retreats. Health Aff. (Millwood) 20 (5), 180–186.

Galton, F., 1883. Inquiries Into Human Faculty and its Development. MacMillan, London350.

Godlee, F., 2008. Understanding the role of the doctor. BMJ 337, a3035. https://doi.org/10.1136/bmj.a3035. 19097980.

Gudbranson, E., Glickman, A., Emanuel, E.J., 2017. Reassessing the data on whether a physician shortage exists. JAMA 317 (19), 1945–1946.

Halper, T., 1996. Privacy and autonomy: from Warren and Brandeis to Roe and Cruzan. J. Med. Philos. 21 (2), 121–135.

Heidegger, M., 2008. Basic writings. In: Farrell, D.K. (Ed.), On the Origin of the Work of Art. 1st Harper Perennial Modern Thought ed. Harper Collins, New York, pp. 143–212.

Hesse, B.W., Moser, R.P., Rutten, L.J., 2010. Surveys of physicians and electronic health information. N. Engl. J. Med. 362, 859–860.

Higgins, G.L., 1989. The history of confidentiality in medicine: the physician-patient relationship. Can. Fam. Phys. 35, 921–926.

Himmelstein, D.U., Woolhandler, S., 2008. Privatization in a publicly funded health care system: the U.S. experience. Int. J. Health Serv. 38 (3), 407–419.

Hisham, R., Liew, S.M., Ng, C.J., Mohd Nor, K., Osman, I.F., Ho, G.J., Hamzah, N., Glasziou, P., 2016. Rural doctors' views on and experiences with evidence-based medicine: the FrEEDoM qualitative study. PLoS One 11 (3), e0152649.

Højgaard, L., Löwenberg, B., Selby, P., et al., 2017. The European Cancer Patient's Bill of Rights, update and implementation 2016. ESMO Open 1 (6), e000127.

Hoy, E.W., Gray, B.H., 1986. Trends in the growth of the major investor-owned hospital companies. In: For-Profit Enterprise in Health Care, Bradford H. Gray for the Committee on Implications of For-Profit Enterprise in Health Care and Institute of Medicine Staff. National Academies Press, pp. 250–259.

Ingelfinger, F.J., 1980. Arrogance. N. Engl. J. Med. 303 (26), 1507–1511.

Institute of Medicine, 2000. To Err Is Human: Building a Safer Health System. National Academy Press, Washington, DC.

James, M.L., Grau-Sepulveda, M.V., Olson, D.M., Smith, E.E., Hernandez, A.F., Peterson, E.D., Schwamm, L.H., Bhatt, D.L., Fonarow, G.C., 2014. Insurance status and outcome after intracerebral hemorrhage: findings from get with the guidelines-stroke. J. Stroke Cerebrovasc. Dis. 23 (2), 283–292.

Jauhar, S., 2014. Doctored: The Disillusionment of an American Physician. Farrar, Straus and Giroux, New York.

Jernigan, K., 2009. The pitfalls of political correctness: euphemisms excoriated. Braille Monit. 52 (3).

Kaplan, R.M., 1994. Value judgment in the Oregon Medicaid experiment. Med. Care 32 (10), 975–988.

Laine, C., Davidoff, F., 1996. Patient-centered medicine. A professional evolution. JAMA 275, 152–156.

Lawler, M., Le Chevalier, T., Murphy Jr., M.J., et al., 2014. A catalyst for change: the European cancer patient's Bill of Rights. Oncologist 19 (3), 217–224.

Le Grand, J., 2009. Choice and competition in publicly funded health care. Health Econ. Policy Law 4 (Pt 4), 479–488.

Leibniz, G., 1710. Essais de Théodicée Sur la bonté de Dieu, la liberté de l'homme et l'origine du Mal [Essays of Theodicy on the Goodness of God, the Freedom of Man and the Origin of Evil]. https://www.gutenberg.org/files/17147/17147-h/17147-h.htm.

Luft, H.S., 1980. Trends in medical care costs. Do HMOs lower the rate of growth? Med. Care 18 (1), 1–16.

Marshall, A.A., Smith, R.C., 1995. Physicians' emotional reactions to patients: recognizing and managing countertransference. Am. J. Gastroenterol. 90 (1), 4–8. (Review).

Mayers, U.W., 2003. The Bipartisan Patient Protection Act: greater liability on managed care plans. Ann. Health Law 12 (2), 341–365. (Table of Contents).

Mehlman, M.J., 1993. The patient-physician relationship in an era of scarce resources: is there a duty to treat? Conn. Law Rev. 25, 349–391.

Montgomery Jr., E.B., 2019a. Medical Reasoning: The Nature and Use of Medical Knowledge. Oxford University Press, Oxford.

Montgomery Jr., E.B., 2019b. Reproducibility in Biomedical Research: Epistemic and Statistical Problems. Academic Press, New York.

Montgomery Jr., E.B., Turkstra, L.S., 2003. Evidenced based medicine: let's be reasonable. J. Med. Speech Lang. Pathol. 11, ix–xii.

National Cancer Policy Forum, Board on Health Care Services, Institute of Medicine, 2013. Delivering Affordable Cancer Care in the 21st Century: Workshop Summary. National Academies Press, Washington, DC.

Neuberger, J., 1999. Let's do away with "patients" and Raymond Tallis, Commentary: leave well alone. In: Do we need a new word for patients? BMJ 318, 1756–1758.

Noble, A.J., Robinson, A., Snape, D., Marson, A.G., 2017. 'Epileptic', 'epileptic person' or 'person with epilepsy'? Bringing quantitative and qualitative evidence on the views of UK patients and carers to the terminology debate. Epilepsy Behav. 67, 20–27.

Nutter, D.O., 1984. Access to care and the evolution of corporate, for-profit medicine. N. Engl. J. Med. 311 (14), 917–919.

Paley, W.D., 1993. Overview of the HMO movement. Psychiatr. Q. 64 (1), 5–12.

Percival, T., 1803. Medical Ethics: A Code of Institutes and Percepts Adapted to the Professional Conduct of Physicians and Surgeons. Printed by S. Russell, for J. Johnson, Saint Paul's Church Yard, and R. Bickerstaff, Strand, London, p. 30.

Perry, P.A., Hotze, T., 2011. Oregon's experiment with prioritizing public health care services. Virtual Mentor. 13 (4), 241–247.

Pitsenberger, W., 2008. "SEZ WHO?": state constitutional concerns with external review laws and resulting conundrum posed by Rush Prudential HMO V. Moran. Connecticut Insurance Law J. 15 (1), 85–118.

Preston, S.H., 1976. Mortality Patterns in National Populations: With Special Reference to Recorded Causes of Death. Academic Press, New York.

Rawls, J., 1999. A Theory of Justice, revised ed. The Belknap Press of Harvard University Press, Cambridge, MA.

Record, K.L., 2010. Wielding the wand without facing the music: allowing utilization review physicians to Trump doctors' orders, but protecting them from legal risk ordinarily attached to the medical degree. Duke Law J. 59 (5), 955–1000.

Rentmeester, C.A., George, C., 2009. Legalism, countertransference, and clinical moral perception. Am. J. Bioeth. 9 (10), 20–28.

Salsberg, E.S., 2015. Is the physician shortage real? Implications for the recommendations of the Institute of Medicine Committee on the governance and financing of graduate medical education. Acad. Med. 90 (9), 1210–1214.

Shapiro, J., 2009. Available from: https://www.bmj.com/rapid-response/2011/11/02/role-doctor-professional-or-technician.

Shem, S., 1979. The House of God. Penguin Random House, New York.

Smith, A., 1759. The theory of moral sentiments. https://www.ibiblio.org/ml/libri/s/SmithA_MoralSentiments_p.pdf.

Smith, A., 1776. An inquiry into the nature and causes of the wealth of nations [the wealth of nations].

Smith, J.F., 1996. Communicative ethics in medicine: the physician-patient relationship. In: Wolf, S.M. (Ed.), Feminism & Bioethics: Beyond Reproduction. Oxford University Press, Oxford, pp. 184–215.

Smith, M., 2002. Patient's Bill of Rights: a comparative overview (PRB 01-31E). In: Law and Government Division. Available from: http://publications.gc.ca/Collection-R/LoPBdP/BP/prb0131-e.htm.

Spencer, C.S., Gaskin, D.J., Roberts, E.T., 2013. The quality of care delivered to patients within the same hospital varies by insurance type. Health Aff. (Millwood) 32 (10), 1731–1739.

Stigler, S.M., 2016. The Seven Pillars of Statistical Wisdom. Harvard University Press, p. 37.

Sullivan, K., 2001. On the "efficiency" of managed care plans. Int. J. Health Serv. 31 (1), 55–65. (Review).

Sweet, V., 2012. God's Hotel: A Doctor, a Hospital, and a Pilgrimage to the Heart of Medicine. Riverhead, New York.

Tomlinson, T., 2012. Methods in Medical Ethics: Critical Perspectives. Oxford University Press, Oxford. (Chapter 6).

Tracy, C.S., Dantas, G.C., Upshur, R.E., 2003. Evidence-based medicine in primary care: qualitative study of family physicians. BMC Fam. Pract. 4, 6.

Tu, H.T., Ginsburg, P.B., 2006. Losing ground: physician income, 1995–2003. Track Rep. (15)1–8.

Vermeulen, K.M., Krabbe, P.F.M., 2018. Value judgment of health interventions from different perspectives: arguments and criteria. Cost Eff. Resour. Alloc. 16, 16.

Vincenzino, J.V., 1999. Medical care costs: trends and outlook. Stat. Bull. Metrop. Insur. Co. 80 (4), 28–33.

Warren, S.D., Brandeis, L.D., 1890. The right to privacy. Harv. Law Rev. 4 (5), 193–220.

Washington, H.A., 2006. Medical Apartheid: The Dark History of Medical Experimentation on Black Americans From Colonial Times to the Present. Harlem Moon/Doubleday Broadway Publishing Group, New York.

Wears, R., Sutcliffe, K., 2019. Still Not Safe: Patient Safety and the Middle-Managing of American Medicine. Oxford University Press, Oxford.

Weiner, S.J., Schwartz, A., Weaver, F., Goldberg, J., Yudkowsky, R., Sharma, G., Binns-Calvey, A., Preyss, B., Schapira, M.M., Persell, S.D., Jacobs, E., Abrams, R.I., 2010. Contextual errors and failures in individualizing patient care: a multicenter study. Ann. Intern. Med. 153 (2), 69–75.

Wolf, S.M. (Ed.), 1996. Feminism & Bioethics: Beyond Reproduction. Oxford University Press, Oxford.

A workable framework

5

To be argued...

The ethics of everyday medicine in its full complexity has yet to be sufficiently explored. The cases presented in Part 1 attest to the state of ignorance resulting from not addressing the ethics in a sufficiently robust manner to match the complexities of the ecology of these ethical problems. Consequently, no definitive guiding algorithm can be offered that directly leads to answers. Instead, a method of exploration is needed to fully flesh out the issues and construct the questions. More effective and robust ethical inductions are needed to develop principles to guide future algorithms (Chapter 3). These algorithms can aid in finding practical and effective solutions for physicians and healthcare professionals in the everyday practice of medicine, guide those who establish healthcare laws and policies, and assist those who adjudicate ethical conflicts.

The unknown need not generate paralysis. Even when there are no sufficient answers and perhaps not even well-formed questions, nonetheless there are habits and modes of thought that can be applied. Indeed, many of these habits and modes of thought can be considered a form of intellectual virtue. Analogous to the hope that virtue in ethics leads to justice, intellectual virtue may lead to new and more certain knowledge. This chapter explores what might constitute intellectual virtue in the construction of ethics and morality to complement the more formal explorations of ethical induction of Chapter 3.

For the patient, justice cannot wait

One of the greatest differences between the scientist and the physician or healthcare professional is that the patient demands answers now even if the knowledge base for answering the demand is incomplete and, worse, inconsistent. A scientist whose study of 50 rodents fails to unambiguously answer a scientific question can study 100 more rodents and postpone any claim to an answer. The patient frequently does not have the luxury of postponing and must "call the question"; the physician or healthcare professional must respond. Even choosing not to respond is a response in the negative. It is a little different in the ethics of everyday medicine. A patient in the need of justice cannot wait for a convening of ethicists to consider the problem by gathering relevant experiences to form the substrate for ethical induction, elaborate a consensus, and then adjudicate disputes to achieve justice. The problem is the lack of ethical systems composed of ethical principles, moral theories, laws, case precedents, and virtues sufficient to address the complexities of everyday medicine as demonstrated in the cases presented in Part 1.

The Ethics of Everyday Medicine. https://doi.org/10.1016/B978-0-12-822829-6.00030-8

Patients and all those who are involved in the care of the patient must employ some current ethical system even while simultaneously constructing a better system. The metaphor of Otto Neurath, philosopher, is apt:

> We are like sailors who on the open sea must reconstruct their ship but are never able to start afresh from the bottom. Where a beam is taken away a new one must at once be put there, and for this, the rest of the ship is used as support. In this way, by using the old beams and driftwood the ship can be shaped entirely anew, but only by gradual reconstruction.
>
> *Neurath (1973)*

As has been discussed in Chapter 3, ethical decision-making is not deductive as in logical deduction but does trade on the certainty afforded by deduction, the latter being the Logic of Certainty. Rather, one must use the Logic of Discovery that necessarily employs the Fallacy of Confirming the Consequence and the Fallacy of Four Terms, the injudicious use of which results in irreproducibility (Montgomery, 2019b).

The Fallacy of Four Terms has been addressed in some detail in Chapter 3. The focus here is on the Fallacy of Confirming the Consequence, which is of the form *if a implies b is true and b is true, therefore a is true*. This is invalid, which means that *a* may be true or may be false but the argument cannot ascertain either. It may be that *b* is true for any number of reasons other than *a*. The Scientific Method, in its most common application, is the Fallacy of Confirming the Consequence where *hypothesis a* stands in for *a* and *prediction b* stands for *b* creating the form *if hypothesis a implies prediction b is true and prediction b is true, therefore the hypothesis is true*. Again, the prediction may be true because of any number of hypotheses. Translating to ethics, the form may be *if the principle of autonomy is determining then the physician should act in this manner, the physician is acting in this manner, therefore the principle of autonomy is determining*. Clearly, there may be another reason why the physician acts as she/he does other than out of a knowledge of or obligation to the ethical principle of autonomy. The danger is that an ethicist making the argument would conclude that the obligation to autonomy is the determinant factor and apply this claim to other situations and circumstances, risking results that would be deemed unjust. This would result in an irreproducibility of justice in cases thought relevant. The situation described is an example of the Inverse Problem. If there are multiple causes of the exact same outcome, then there is no way to know which cause is operating by only observing the outcome.

The centrality of hypothesis

As can be readily appreciated, the Scientific Method, as typically used for any empirically based system of knowledge, including the ethics of everyday medicine, is only as good as the hypothesis (discussed in detail in Montgomery, 2019a, 2019b). Yet, even the best hypothesis does not obviate the Inverse Problem. What is needed is a set of hypotheses. This is clearly evident in the everyday practice of medicine in the form of the differential diagnosis. The critical nature of fully informed differential diagnosis is demonstrated by the consequence of the failure to establish such a differential diagnosis, as demonstrated in Cases 6 and 9 in Part 1.

The only way to rescue the Fallacy of Confirming the Consequence is to change the logical form from *if a implies b is true and b is true, therefore a is true* to *if and only if a implies b is true and b is true, therefore a is true*. Note that this construction obviates the Inverse Problem. However, how is it practical to construct the logical form *if and only if hypothesis a implies prediction b is true and prediction b is true, therefore hypothesis a is true*? The problem is that the Fallacy of Induction prevents complete confidence in the term *if and only if hypothesis a implies prediction b*. For example, the induction *all ravens seen have been black, therefore all ravens are black* cannot exclude the possibility that somewhere unbeknownst to the observer a non-black raven exists that disproves the induction. One is left with resorting to a "differential diagnosis" of all reasonable possibilities, recognizing that what is reasonable is problematic (see Chapter 3). For example, in the case of ethics, are those persons, charged with establishing the consensus by induction, reasonable persons blinded to their self-interest among other qualifications and sufficiently broadminded and experienced? How would one define a reasonable person and ensure that she is blinded to her self-interest (Chapter 3). What is sufficient experience? Indeed, the ethical differential diagnosis of ethical hypotheses likely will evolve as Neurath's boat as ethical knowledge is reconstructed upon the sea of experiences. The task and challenge are to ensure a sufficient number of good hypotheses, despite the human tendency to "jump to conclusions." As Leo Tolstoy wrote:

> The totality of causes of phenomena is inaccessible to the human mind. But the
> need to seek causes has been put into the soul of man. And the human mind, without
> grasping in their countlessness and complexity the conditions and phenomena,
> of which each separately may appear as a cause, takes hold of the first, most
> comprehensible approximation and says: here is the cause.
>
> *Tolstoy (1869)*

The irreproducibility of biomedical research and medical diagnostic errors suggests that scientists and clinicians have not considered all reasonable alternatives in either the experimental design or inference from the results, particularly if the prediction is found.

The surer way to shape knowledge is to use the valid, modus tollens, a form of the Scientific Method, although this is rarely if ever used. Modus tollens is of the *form if a implies b is true and b is false, therefore a is false*. The statement *a* cannot be anything else but false. The form in ethics would be *if the principle of autonomy is determining, then the physician should* not *act in this manner, the physician is acting in this manner, therefore the principle of autonomy is not determining*. One could use the modus tollens form for all reasonable elements of the ethical differential diagnosis. However, the Fallacy of Induction makes this problematic for the reasons described earlier. Nonetheless, the use of the modus tollens form can start to reduce the number of elements in the differential diagnosis.

Hypotheses are necessary as *a priori* principles to organize experiences as a substrate for ethical induction to subsequent principles, theories, laws, and precedents as described in Chapters 2 and 3. However, seldom is that any concern as to how hypotheses are generated (Montgomery, 2019a). Generally, when considered, hypotheses are thought to originate spontaneously, by happenstance, or generally as the result of some psychological predisposition. However, more likely hypotheses arise from metaphors (discussed in detail in Montgomery, 2019a). Psychological constructs can be the basis for a metaphor that generates an ethical hypothesis but these are problematic as discussed in Chapter 1. Also, similar judgments or actions in other fields such as law or business ethics can serve as metaphors by which to generate ethical hypotheses to be brought to bear on the ethics of everyday medicine. For example, the physician's or healthcare professional's obligation to beneficence to the patient can conflict with contractual obligations to insurers. The analogy of the obligations of investment advisors to their clients can be a useful metaphor, as described in Chapters 2 and 4. Similarly, the asymmetry in power, in terms of knowledge, between the patient and the physician or healthcare professional can be compared to a judge who must consider expert testimony. In this case, the patient would be analogous to the judge while the physician or healthcare professional would be analogous to the expert witness in a legal case, also discussed in Chapter 4.

Intellectual virtues

Given the incompleteness of any understanding of the ethics of everyday medicine, as there are no precise and accurate algorithms to solving ethical problems that arise in the complex ecology of everyday medicine, the question becomes how to proceed. In fact, there are few good questions, let alone answers, and perhaps the first effort is to generate good questions that will be productive of good answers. The hope is that in responding to the cases presented in Part 1, questions will start to form. However, as Louis Pasteur said, "In the fields of observation chance favours only the prepared mind" (Lecture, University of Lille, December 7, 1854, Pearce, 1912).

The question becomes how to prepare the mind. One aspect is to cultivate habits and modes of thinking that facilitate insights. One potential is to have a set of intellectual virtues analogous to the ethical virtues (see Chapter 2). The hope that an ethically virtuous physician or healthcare professional naturally will behave ethically to provide justice to patients, intellectually virtuous persons involved in developing ethical systems will be able to create productive ethical systems that reliably lead to justice for the patient. Indeed, while many thinkers attribute the first declarative notion of intellectual virtue (and related epistemological virtue) to the 1980s, ethical virtues, such as those proposed by Aristotle, were the inspiration (Zagzebski and DePaul, 2003).

Intellectual virtues, like ethical virtues, are "excellences" in Aristotle's term, meaning competencies that result in the good or truth. Importantly, arriving at the good or truth has a reward for the person doing so, and in acting intellectually virtuous, there is an intrinsic reward (Sherman and White, 2003). The intellectual virtues may be analogous to moral virtue. Indeed, virtue ethicists have argued that the primary focus on virtuous actions is on the actor—such as physician, healthcare professional, ethicist, and healthcare policy maker or adjudicator—that benefits the physician or healthcare professional. Any benefit conveyed to others, such as patients, is a byproduct; it certainly is not patient-centric (see Chapter 4). Whether this is the most efficient route to justice for patients is an open question (see Introduction and Chapter 4).

Some have argued that intellectual virtues are derivative of the motivation to knowledge, reflecting an innate desire for a cognitive contact with reality (Zagzebski, 1996). While that may be true, it does not explain the particular intellectual virtues commonly held, such as open-mindedness, carefulness, imagination, and discipline. The argument to be made here is that there are specific intellectual virtues of epistemic necessity.

Intellectual virtues, as discussed here, are epistemic devices, methods used to gain knowledge, such as ethics and morals that may or may not link to an ontology, what is reality such as inherent ethical and moral rights (Hookway, 2003). At least since the 1950s, epistemology has been preoccupied with knowledge as a justified true belief (leaving relatively untouched the origins of belief much in the same manner that the origins of hypotheses have been neglected, at least according to this author). The focus on intellectual virtue, initially as a means of justification of belief, departs from traditional epistemology (Sherman and White, 2003). One approach holds belief arrived at through the use of intellectual virtues, in itself, justifies ontological claims (termed the Reliabilist approach). In a sense, this approach would appear to convey some ontological status on intellectual virtues. This would seem a bit problematic because it would appear to discount consideration of the belief content, both its premises and its propositions that make up the belief argument. Yet, in the absence of any other means of justifying belief, this approach would be reasonable. Historically, the Reliabilist approach seems to be the default position with the defeat of Logical Empiricism (where all knowledge is derivable from self-evident observations and rules of inference) and the rise of postmodernism. The alternative approach holds that intellectual virtues are important guides in acquiring beliefs that likely will be justifiable, termed the Responsibilist approach. This approach would be an epistemological device that does not necessarily convey any ontological status on intellectual virtues.

Interestingly, modern science discounts both the Reliabilist and the Responsibilist approaches, while ethical science, such as the development of ethical systems, would embrace one or both approaches. The presumptions underlying the response of modern science would appear to contrast sharply with those of the ethical scientist and likely account for some of the animosity between the two knowledge domains. Indeed, some scientists objected to the American Medical Association's Code of Ethics in 1847 (American Medical Association, 1848), believing scientific advancement would obviate the need for any code of ethics. Modern science (dating to the founding of the Royal Society in 1660) appears to presume that observations (data) speak for themselves and all that was needed was the appropriate experimental device to generate the observations. This presumption likely was the bases for Thomas Hobbes' criticism of the Royal Society, writing "not everyone that brings from beyond seas a new gin [machine], or other jaunty device, is, therefore, a philosopher" (Hobbes, 1839). Unsurprisingly, many ethicists lacking the experimental devices (although there are ethical experiments) might feel estranged from modern science. However, data do not speak for themselves (Montgomery, 2019a, 2019b) and thus modern science and ethics are in the same position, as is any empirically based knowledge domain.

The perspective taken here is that of the Responsibilist, the intellectual virtues help one in determining content but they are not content. In the same way, understanding the logic of statistics and its proper use enables confidence in claims made; logic and statistics does not determine the claims (see Chapter 3). Perhaps new and novel here, intellectual virtues are not analogous to the ethical virtues, where, in the latter, comporting with the ethical values is justified by helping to achieve a good life (Aristotle's term *eudaimonia*). Rather, the intellectual virtues reviewed here are related to the fundamental epistemic problems described earlier, such as the Inverse Problem and the necessary use of the logical Fallacy of Confirming the Consequence and Fallacy of Four Terms.

Open-mindedness

Open-mindedness often is held as openness to multiple possibilities, such as different ethical systems. However, the notion of open-mindedness is complex, admitting of multiple connotations, if not different definitions. Problems arise when people use the term without explicitly defining their use, while the shared ability to use the term falsely conveys the impression that each understands each other. The philosophical tool helpful in this case is to "unpack" the notion of open-mindedness to determine the actual definitions or connotations meant.

One dimension of open-mindedness is either passive, a willingness to entertain different views, or active. If passive, one need not be open-minded until challenged to consider different views. This open-mindedness is an elicited condition, not an ongoing or proactive one. Another view is that open-mindedness is active, a responsibility to constantly consider different views, indeed a responsibility to seek out those different views. For example, Aristotle began many of his treaties by first reviewing the philosophical views of his processors and continually referring back to them as he advanced his own views. Such an active responsibility is rarely met in much of recent writing, particularly in biomedical research (the area of this author's expertise), and perhaps science in general. Ethics, as an empirically based knowledge domain (see Chapter 3), likewise benefits from the virtue of proactive open-mindedness necessitated by a pluralistic modern liberal democracy.

Charles Bazerman analyzed scientific reports in the *Philosophical Transactions of the Royal Society* from 1665 to 1800 (Bazerman, 1997, pp. 169–186). He described four stages. From 1665 to 1700, scientific papers were uncontested reports of events. From 1700 to 1760, discussions centered over results were added. In the third period, 1760–80, more theoretical aspects were addressed in papers that "explored the meaning of unusual events through discovery accounts" (Bazerman, 1997, p. 184). From approximately 1790 to 1800, papers reported claims for which the experiments were evidence. Interestingly, the claims focused on the experimenter and not "recognizing the communal project of constructing a world of claims.... Although the individual scientist has an interest in convincing readers of a particular set of claims, he does not yet explicitly acknowledge the exact placement of the claims in the larger framework of claims representing the shared knowledge of the discipline" (Bazerman, 1997, p. 184). It would seem that many scientists became more concerned with their own answers rather than anyone else's, suggesting that the scientist's personal achievement is more important than advancing science.

As described earlier, scientists publishing in the Royal Transactions between 1790 and 1800 focused on the experimenter (Bazerman, 1997, p. 184). It is interesting to speculate whether the trends noted between 1790 and 1800 persist. Interestingly, in a metaanalysis of randomized controlled clinical trials, less than 25% of the publications cited prior publications on the same subject (Robinson and Goodman, 2011). A vernacular has arisen reflecting widespread concerns about a failure to cite others, particularly competitors (Grant, 2009). These terms include "citation amnesia," "citation negligence," and "disregarding syndrome." In a study of citations in publications over a wide range of intellectual disciplines, including the physical and behavioral sciences and humanities, citations could be divided into the research front (as in at the front lines of a battle), typically by their recent publication, and foundational, typically referring to earlier publications (Hargens, 2000). The validity of the classification was supported by how papers in these two categories were used. Focusing on the physical sciences, specifically celestial masers (astronomy), chiral separations (chemistry), and physics (theoretical nuclear physics), research from recent papers was cited the most. Hargens suggested that perhaps one reason for the paucity of foundational papers is the presumption by the physical scientists of a common and shared understanding of the foundations of their science. Hargens described the notion of obliteration by incorporation (essentially citation amnesia), attributed to Garfield (1977) and Merton (1967), suggesting that physical sciences evolve in an incremental manner in which older knowledge just becomes part of the operating presumptions. The history of science, particularly as described by Thomas Kuhn (1962), suggests otherwise.

Why should open-mindedness be considered a virtue? Perhaps a person enjoyed some beneficence from another to which open-mindedness was attributed. Perhaps open-mindedness is a feature of a person held in esteem, thus one wishing to emulate. Perhaps the esteemed person is so because of fame or fortune. In science, someone is esteemed because of a recognized history of success—however, success is defined. More likely, open-mindedness is invoked when there is a failure, with the lack of open-mindedness offered as an explanation of failure. The lack of open-mindedness may be a very important cause of the crisis of irreproducibility in biomedical research (Montgomery, 2019b). In ethics, it is possible that a lack of open-mindedness resulted in the Principlists and the Anti-Principlists positions seen as irreconcilable and incommensurate. Chapter 2 argues against that position.

Open-mindedness seems to be based on a fundamental skepticism taken as a virtue. For example, if one had perfect knowledge, such as a consistent and complete explication of some set of phenomena, there would be no need for open-mindedness. Indeed, insisting on open-mindedness by valuing alternatives beyond the already perfect claim could only mean that the alternatives are not perfect. Insistence on open-mindedness would be counterproductive. Thus, open-mindedness cannot be an end in itself. It must be a means to some other end. As a means, it is an epistemic device. It is clear that there is no perfect empirical knowledge and, for that reason, open-mindedness is still a necessity but not an inevitability.

Open-mindedness is a necessity because of the Inverse Problem and susceptibility to the Fallacy of Confirming the Consequence. When considering an experiment, either physical or thought (Gedanken experiment), both possible in ethics, postulating only a single hypothesis as the cause of a prediction could have other causes is an example of the Fallacies of Confirming the Consequence and Limited Alternatives. The experiment will appear true; however, it is invalid because the experiment is a Fallacy of Confirming the Consequence and another cause was responsible. The fallacy is of the form *if hypothesis **a** implies prediction **b** is true, prediction **b** is found true, therefore hypothesis **a** is true.* The only way to make the fallacy valid is to change the form to *if and only if hypothesis **a** implies prediction **b** is true, prediction **b** is found true, therefore hypothesis **a** is true.* For example, if the principle of beneficence implies the physician acts thusly, the physician does act thusly, therefore the principle of beneficence is true as the cause of the physician's action. However, the physician could have been acting to avoid maleficence to himself. If the only possible cause of the physician's actions was fulfillment of the obligation to beneficence, then the physician's actions would be proof that the obligation to beneficence is the controlling factor.

The only way to assert *if and only if hypothesis **a** implies **b** as true* requires that all other hypotheses are considered and found not to actually predict **b**, which requires open-mindedness. Alternatively, the logical argument could be rephrased as *if (hypothesis **a** or **c** or **d** or ...) implies prediction **b** is true, prediction **b** is found true, and hypotheses **c**, **d**, ... are false, therefore hypothesis **a** is true.* Again, appropriate open-mindedness would acknowledge the possibility of hypotheses **c**, **d**, ... as being true so that they could be proven false subsequently.

There may be many reasons why a person may not be open-minded. There may be a lack of awareness of the necessity to be open-minded. There may be the heartfelt conviction that one's own position is the best. Indeed, assuming the person to be relatively sophisticated in the issues, the person may have reasons to value his own position over those of others and therefore see no need to address the other positions. While understandable, this is dangerous; it does not deny the "theoretical" importance of open-mindedness but thwarts any practical positive consequences derived from open-mindedness.

To have the kind of proactive positive open-mindedness would imply the need for humility and modesty, typically considered among the ethical virtues. However, the difficulty arises when considering humility and modest as nouns, suggesting an independent, perhaps ontological, status—there exists such a thing as humility and modesty that a person can have or hold. The problem of humility and modesty as nouns extends to humility and modesty to an adjective that is, modifying the character, again a noun. In both cases, such use results in self-undermining, can a person claim being, as a noun or adjective, to be humble and modest without demonstrating pride (considered here the opposite of humble) or immodesty? It is important to note that just because one can speak or write the sentence "the person is humble and modest" does not mean there is any such thing as humble and modest as nouns or adjectives any more than the statement "there are unicorns" means there are such things as unicorns, as a noun. One saying that "justice was done" or "beneficence was conveyed" does not mean that justice and beneficence have an ontological status as a noun (see Chapter 1).

Perhaps it would be more productive to consider terms such as "open-mindedness," "humility," "modesty," "justice," "beneficence," and others as adverbs, a description of actions taken. Indeed, very young infants can (re-)act to situations and circumstances to which "justice" as a noun or an adjective can be attributed. Note that an action need not be acted out to qualify. Indeed, thoughts regarding actions that are not executed often activate the same regions of the brain that are activated during the actual conduct of the action (Chapter 1). In a very real sense, terms such as "open-mindedness," "humility," "modesty," "justice," "beneficence," and others are gerunds, verbs modified to have the status of nouns. "Run," as a verb, becomes "running," as a noun or adjective, as many other verbs to which "ing" is attached. If one is not careful, the focus could shift to the "ing" rather than the verb and give the sense of a whole class of "objects," the gerunds that have some independent status, perhaps ontological, from the verb from which it was derived.

Reconsidering open-mindedness as an adverb that qualities or characterizes actions may have the advantage of requiring specific actions. For example, considering justice as a verb or an adverb impels just actions. Similarly, open-mindedness, as an intellectual virtue, is an adverb that impels actions to compensate for the problems imposed by the Inverse Problem and the Fallacy of Confirming the Consequences. Thus, open-mindedness is not a noun or adjective, rather it is an adverb that qualifies an action that requires inclusion and consideration of a range of possible actions. Too often, the first action that comes to mind is taken as the action that should be taken. As the action may be an unconscious one, it would be a type of gut reaction. When considering ethical problems, the admonition to "check your gut at the door" is good advice.

This author taught a course of the ethics of everyday medicine to undergraduate students. The author naively thought the students would not have any prior positions on the various topics, that the students would be a *tabula rasa*, a blank slate onto which the author could write. Every student had an opinion on the question that was the basis for discussion. In many ways, the opinions held by the students might be called "folk ethics" analogous to "folk psychology." As discussed in Chapter 1, the students' position could not help being driven by an affective (emotional) response to the situation and circumstances of the question, often a *prima facie (on the face of it)* or unsophisticated view of the case. For example, when asked whether a physician should lie to an insurer in order to obtain financial coverage for a patient's treatment the physician suspected the insurer would not cover, the consensus initially was that it is never good to lie. When pressed further to consider the student's mother as the patient, many students said it was permissible to lie to the insurer, if the insurer was a bad insurance company. One can readily see how the conversation would continue, for example, what is a good or bad insurer? The admonition is to "check your gut at the door" when considering an ethical problem. But this does not mean not having an affective (emotional) response but rather check its influence.

Studies on the sense of fairness, taken as the analog of justice (Rawls, 1999), suggest those with a low sense of empathy may also have an impaired sense of fairness that could be considered an example of an impaired sense of justice. For example, children with high functioning autism who score low on empathy are also more willing to accept an unfair offer than other children (Wang et al., 2019). The complexity of these psychological constructs regarding autism makes attribution of an impaired sense of justice to low empathy, implying a reduced or impaired emotional reactivity, very problematic. Nevertheless, findings suggest that this is a possible causal relationship and suggest that a normal affective (emotional) response is important to a sense of justice.

As discussed in Chapter 1, the innate sense of justice may be poorly differentiated in social situations of increasing complexity. Thus, the innate affective response to injustice undergoes modifications and increases in complexity with experience and social learning, whether implicit or explicit. The admonition "check your gut at the door" invites more deliberation that can lead to behaviors more attuned to the complexity of the ethical situation. In other words, avoid the quick emotional response that would otherwise prevent an orderly deliberative consideration of the issues, particularly other causes or formulations that lead to the same ethical problem (the Inverse Problem). Thus, checking one's gut at the door would be an action to which open-mindedness can be attributed.

Carefulness

In considering all the possible hypotheses for a given prediction in order to prevent falling into the Fallacy of Confirming the Consequence, the Fallacy of Induction argues that one cannot know whether all the possible hypotheses have been considered, as discussed previously. One is left with the requirement of considering all *reasonable* possible hypotheses but this only exchanges the difficulty from identifying all possible hypotheses to the difficulty of defining what is reasonable [a similar problem confronts the selection of reasonable persons for consensus making, leading to ethical principles, moral theories, laws, and precedents (see Chapter 3)].

One possible tool that could be used to determine reasonableness is to consider the relationship between hypotheses and predictions as a metaphor—an example of a metaphor is *love is like a rose*. In this case, love is the target domain and borrows its meaning from a rose, which is the source domain (Kövecses, 2010). As a rose is beautiful, pleasant, alluring, and with thorns, so too may be love. The metaphor works as there are aspects in common between love and a rose such that they resonate and thus reinforce. Note that the metaphor *love is like an old shoe* would be more difficult to appreciate, unless the old shoe implies a sense of comfort after the shoe has been worn in, perhaps not unlike love. The argument made is that the empirical investigation can be considered a metaphor where the *hypothesis is like the prediction*, the hypothesis gains certainty, of a kind, from the prediction demonstrated to be true.

In any empiric investigation, biomedical or ethical, there must be a particular relationship between the hypothesis and the prediction. The relationship cannot be one of identity (exactly the same) as otherwise, the logical device (even though a fallacy) used *if hypothesis a implies prediction b is true and prediction b is true, therefore hypothesis a is true* would be the same as *if hypothesis a implies hypothesis a is true and hypothesis a is true, then hypothesis a is true*. The result is a tautology that does not provide any new information. Indeed, in the degree that there is a "distance" between the hypothesis and the prediction, there is a similar potential for new knowledge. However, too great a distance between the hypothesis and the prediction increases the risk of irreproducibility, either in the experiment itself or in the subsequent applications of the hypothesis to other predictions thought reasonably relevant to the prediction in the case tested.

One measure of "distance," hence reasonableness, is the Epistemic Risk in assuming that the hypothesis under consideration is falsely considered reasonable. Carelessness can contribute to Epistemic Risk. Epistemic Risk consists of Epistemic Distance, differences in degree between the hypothesis and the prediction, and Epistemic Degrees of Freedom, qualitative differences (Montgomery, 2019a, 2019b). Consider the creation of an ethical consensus as the hypothesis resulting from induction that is to be applied to another case separate and independent from the cases used as the substrate for the ethical induction. The consensus, now hypothesis, would be expected to predict whether the judgment of the separate and independent case would be held independently as just or unjust. From the discussions in Chapter 3, the greater similarity between the cases used for the consensus and the subsequent separate and independent case would suggest greater agreement between the substrate cases and the separate independent case. The result would be a (1) greater reproducibility of outcomes when the consensus is applied to many subsequent separate and independent cases and (2) greater confidence in justice being served.

Epistemic Distance relates to the distance of the consensus and the prediction on the same dimension. For example, an ethical consensus derived from adults and now applied to children might constitute one dimension. In this sense, the "distance" between the adult and a 6-year-old is quite different from the "distance" between the adult and a 16-year-old. There would be less Epistemic Risk applying a consensus principle, theory, law, or precedent derived from the adult to the 16-year-old (see Case 13) than applying the consensus to a 6-year-old. It would seem careless to apply the consensus derived from adults to a child aged 6 but perhaps less careless to apply the consensus to a child aged 16.

Epistemic Degrees of Freedom relate to the number of dimensions on which the consensus principle, theory, law, or precedent can differ from the subsequent separate and independent case. How would the different dimensions be combined into a single metric called reasonableness that allows a dichotomous decision into yes, reasonable or no, not reasonable? For example, consider a consensus created by a group of male-gendered persons that subsequently is applied to a separate and independent case of a female-gendered person. As pointed out by feminist ethics, there are a great many dimensions on which male and female-gendered persons differ, so how could one quantitate and then reduce to a single dimension the differences between male and female-gendered persons (see Chapter 2)? Thus, applying the consensus derived from cases involving male-gendered persons to cases involving female-gendered persons would be of high Epistemic Degrees of Freedom, and thus high Epistemic Risk, that could be viewed as excessive carelessness. The notion of Epistemic Risk is more fundamental. As argued in Chapter 2, there may be less Epistemic Distance and Degrees of Freedom between an explanation of just behaviors based on neurophysiology compared to an explanation based on psychology.

The aforementioned discussion on the Fallacy of Confirming the Consequence can be applied to use of the Fallacy of Four Terms in ethics as discussed in detail in Chapter 3. Importantly, the Fallacy of Four Terms is fundamental to any induction, ethical and otherwise. From that chapter, the following argument was considered:

Argument 5.1

Major premise: *A sample of citizens* (bridging term) *holds principle, theory, law, or precedent A* (major term)
Minor premise: *The citizen/patient B* (minor term) *is the sample of citizens* (bridging term) in the major premise
Conclusion: *The citizen/patient B* (minor term) (should) *holds principle, theory, law, or precedent A* (major term)

The major premise is obtained by observations of a sample of citizens from which a consensus, induction, was made, that being *principle, theory, law, or precedent A is held*. The validity of this argument (syllogism) depends on the truth of the minor premise, which would only be true if the sample of citizens (the bridging term) in the major premise, that was the substrate for induction, is the same as the sample containing the citizen/patient (bridging term) in the minor premise. If the sample of citizens in the major premise was all male-gendered while the citizen/patient in the minor premise was female-gendered, there is considerable Epistemic Distance and Degrees of Freedom and thus considerable Epistemic Risk. There would be a high risk for the Fallacy of Four Terms.

Carefulness as an intellectual virtue would require consideration of Epistemic Risk. In both the cases of ethical induction involving the Fallacy of Four Terms and the quasi-deduction involving the Fallacy of Confirming the Consequence, the ontological and epistemological similarity between the hypothesis and the prediction or a term, such as the bridging term in syllogism, invites misunderstanding. But it is important to note that in the discovery of new ethical principles, moral theories, laws, case precedents, and virtues, the bridging term cannot be identical as otherwise, the argument is a tautology and no new ethical knowledge is gained. For example, the bridging term in the major premise is a sample of citizens whose actions are considered and judged just. As David Hume would note, this consensus is retrospective. The bridging term in the minor premise, which allows the conclusion, is the present-day sample, specifically the individual citizen/patient, and is applied prospectively and thus is subject to Hume's Problem of Induction. Hume wrote, "instances of which we have no experience [the future by prospection], *must necessarily resemble* [italics added] those, of which we have [the past by retrospection]" (Hume, 1739).

One way to reduce the Epistemic Risk is to reduce the Epistemic Degrees of Freedom. In the example given here where a consensus was created over the observations of male-gendered participants, the application of the consensus to the female gender is a source of Epistemic Risk due to the many degrees of freedom involved in relating the male and female genders. The Epistemic Degrees of Freedom could be reduced by creating the consensus by female-gendered reasonable persons blinded to their self-interest over the observations of female-gendered participants with the intent of applying the consensus to female-gendered persons. This example is very consistent with the Principle of Causational Synonymy that dates back to the Ancient Greeks.

The Principle of Causational Synonymy holds that whatever is "in" the cause of an effect is also "in" the effect. For example, the cause of a person's hand stirring water lies in the fact that the electrons that make up the surface of the hand repel the electrons on the surface of the atoms in the water. This principle was used in Chapter 1 to demonstrate why psychological theories, particularly as they relate to ethics, cannot be considered causal to ethical behavior. The cause of behaviors is the change in the electrical potentials across the cell surface membrane of nerve and muscle cells. The presumed mechanisms underlying a psychological theory are highly unlikely to be explicated at the level of electrical activity in the nervous and muscle systems. Often, attempts to involve psychological mechanisms to explain ethical behavior result in conundrums and paradoxes as discussed in Chapter 1. One could argue that it is intellectual carelessness to assert a psychological mechanism as a necessary or sufficient explication of ethical behavior. If the only alternative is psychological explication, then a caveat is in order before the psychological explication is given, not as an afterthought, as an appreciation of the caveat would be blunted by the intuitive appeal of the psychological explication. This is not to say that all ethical behaviors in the future will be understood explicitly in the changes in electrical potentials across the cell surface of neurons and muscles. Nevertheless, considerable improvements in the understanding of ethical behaviors can be achieved by understanding, in a general way, those operations in Chaotic and Complex nervous systems as discussed in Chapter 1. In this regard, the understanding is through the Process Metaphor as discussed in Chapter 2.

Another aspect of carefulness is the recognition and subsequent avoidance of a quasifact, which is an inference or opinion drawn from observations (facts or data) that has the epistemic status of facts. For example, a set of ethical behaviors are observed from which the inference is made that they demonstrate principle A even if no single observation corresponds exactly to principle A. The result is an abstraction that may or may not be representative of the ontology of the observations (see Chapter 3). Now, principle A is taken as a fact when it is used to defeat alternative inferences, even if the alternative inferences are equally consistent with the observations. One merely has to read almost any scientific report to find that the inferences of others are used to build an argument rather than the actual observations, data, originally used to generate the inference. In the absence of any critical discussion of the validity of the inferences by contrasting with a reasonable alternative, the inferences are falsely taken as true. This is particularly the case if the inferences have an intuitive appeal quite independent of the observations (Johnson-Laird, 2008). One would think that such argumentation is at a high risk of being in error, as evidenced by its subsequent failure, assuming they ever were tested.

Imagination

As will be discussed, imagination is not the same thing as open-mindedness. The latter indicates the willingness to consider alternatives. Imagination has to do with creating alternatives. The conundrum of imagination is well expressed by Lewis Carroll in "Through the Looking-Glass and What Alice Found There" (1871):

> Alice laughed. "There's no use trying," she said: "one *can't* believe impossible things."
>
> "I daresay you haven't had much practice," said the Queen. "When I was your age, I always did it for half-an hour a day. Why, sometimes I've believed as many as six impossible things before breakfast."

Imagination is critical to the generation of testable hypotheses, which can become new knowledge and, importantly, alternatives to current hypotheses. It is the imaginative hypothesis relative to other hypotheses and relative to the prediction that offers the greater potential for the reward of new knowledge.

There are a great many obstacles to imagination such that a person's lack of imagination does not mean they are incapable of imagination. Paraphrasing the famous physiologist Claude Bernard in the 1800s, "We are more often fooled by things we think we know than things we do not." This is particularly true when what one thinks they know is very intuitive (Johnson-Laird, 2008). This is abetted by the use of quasifacts as described earlier. This issue of obstacles to the imagination is deep seated, as evidence provided by Thomas Kuhn in "The Structure of Scientific Revolutions" (Kuhn, 1962). Kuhn posited that science advances by paradigm shifts in a revolutionary rather than evolutionary manner (also see Feyerabend, 1975).

Previous paradigms are rejected in the course of new paradigms establishing themselves. Central to this thesis is Kuhn's notion of incommensurability as it relates to the enterprise of gaining new knowledge in any empirically based knowledge domain, although Kuhn focused on physics. Kuhn wrote, "the proponents of different paradigms practice their trades in different worlds" (Kuhn, 1962). The proponents of one paradigm just cannot understand the proponents of alternative paradigms, leading to irreconcilable differences that are resolved in the competition for resources by sociological rather than scientific means.

In some ways, the contention between Principlist and Anti-Principles approaches (paradigms) may seem irreconcilable, suggesting incommensurability among the proponents of each paradigm relative to the proponents of the alternative paradigm. Yet, there is one universe, not two—one inhabited by the Principlists and the other by the Anti-Principlists. Incommensurate paradigms more likely reflect epistemology rather than ontology. Epistemology, as it is human-made, can change. The antagonism between the two approaches is more likely the result of the forced epistemic response to the fundamental ontological conundrum of the potential infinity of ethical behaviors from which to extract a set of principles, theories, laws, case precedents, and virtues. Indeed, the two approaches are opposite sides of the same epistemic coin (Chapter 2). Advances in the ethics of everyday medicine are more likely with an iterative use of both approaches rather than discounting one or the other altogether.

Discipline

Different from carefulness, discipline requires consistency of action, even if there are mental reservations. Often, discipline is established by force of will. Mental reservations prevent discipline reducing to perseverance, as will be discussed later. The value of discipline as an intellectual virtue lies in the admonition "one learns more from error than chaos." In important ways, discipline does not prevent error. Indeed, error, as a source of productive irreproducibility, is more valuable than being correct (Montgomery, 2019b). Chaos is likely to lead to error but in the manner of unproductive irreproducibility from which no conclusion with any degree of certainty can be obtained.

Recognizing and avoiding the use of quasifact, described previously, may require discipline. Forcing one's self to consider the works of others to avoid the problems of citation amnesia often requires discipline. Often, the use of specific procedures can be a way to help enforce discipline, as will be discussed later.

Perseverance

Some epistemologists have argued for perseverance as an intellectual virtue, seeing it analogous to the ethical virtue of courage. However, in some respects, perseverance could appear hostile to open-mindedness. In an intellectual context, it would seem that persons should persevere in their position or claim, contrary to competing claims. If one assumes that there is no clear advantage of the persevering person position or claim over the alternatives, then such perseverance would seem contrary to open-mindedness. Interestingly, Thomas Kuhn, who revolutionized the understanding of scientific change, argued for such perseverance, recognizing that there are a great many obstacles for any new position or knowledge claim to eventually succeed (Kuhn, 1962). In other words, Kuhn suggested that the scientist with a new and emerging knowledge claim should persist and essentially ignore competing claims. Interestingly, Galileo congratulated Copernicus for ignoring the considerable contravening observations and persist with his sun-centered solar system (Feyerabend, 1975). However, if open-mindedness is considered a virtue, it would seem that perseverance would be a vice rather than a virtue.

Christopher Hookway pointed out that when applied to an individual person, an intellectual virtue may actually be a vice but when applied to a group or organization, that same notion is a virtue (Hookway, 2003). Perhaps perseverance is an example. It just may be that it is extraordinarily difficult for an individual to champion alternatives, particularly in the setting of severe scarcity in resources to support research and scholarly work generating competition that risks injustice. This is particularly true in the era of the professional scientist/scholar whose livelihood depends on the successful competition for grant funding. Yet, if open-mindedness is critical to successful research and scholarship, it would be difficult to leave it to the tender mercies of the individual in such fraught conditions.

Perhaps an alternative is to allow individuals to persevere in their particular position or knowledge claim, recognizing that the individuals may not give alternatives a fair chance and, at the same time, have institutions or organizations support a plurality of positions or knowledge claims. Guaranteeing a plurality of positions and knowledge claims is perhaps most appreciated in government, particularly those parliamentary governments with institutionalized loyal opposition, such as seen in the United Kingdom, Canada, and others (Helms, 2004). While it may be too much to ask for an individual to be completely open-minded, it may be possible to have an institution or organization that is open-minded by the calculated effort to be inclusive of all reasonable alternatives. Unfortunately, history has suggested that such an ideal is rarely realized (Kuhn, 1962). Note that not only do the victors write the history, they also right the rules. The implications for ethical induction are discussed in detail in "The experts, judges, and officials as those who make the consensus" section in Chapter 3.

Embrace the iconoclast

An iconoclast is a person who critiques accepted theories with an eye to replacing those theories. The term originates from icon, which is an image, originally of a religious entity, and clast, which means to break. An iconoclast is not a curmudgeon or contrarian, although an iconoclast is often equated with a curmudgeon or contrarian in a pejorative manner and dismissed. The latter do not distinguish between accepted fact, which the iconoclast usually accepts, and theory. However, is it important to note that quasifact, as described later, is fair game, indeed preferably so, for the iconoclast.

There is a very good reason to encourage the iconoclast, which is the Pessimistic Induction, particularly as it relates to empirically based knowledge. The Pessimistic Induction holds that every theory in the past has been proven wrong and thus every theory in the future (including the present) will be proven wrong. This likely means that the iconoclast has a reasonable chance at changing knowledge in the present and the future.

As a corollary to the Pessimistic Induction, one looks to current theory to find its weaknesses and self-deceptions as a starting point. Perhaps the archetypical iconoclast was Socrates, who was informed by the Oracle of Delphi that he, Socrates, was the smartest person in the world:

> Everyone here, I think, knows Chaerephon, he said, he has been a friend of mine since we were boys together; and he is a friend of many of you too. So you know the eager impetuous fellow he [was]. Well, one day he went to Delphi, and there he had the impudence to put this question—do not jeer, gentlemen, at what I am going to say—he asked, "Is anyone wiser than Socrates?" And the Pythian priestess answered, "No one."

> Well, I was fully aware that I knew absolutely nothing. So what could the god mean? for gods cannot tell lies. For some time I was frankly puzzled to get at his meaning; but at last I embarked on my quest. I went to a man with a high reputation for wisdom—I would rather not mention his name; he was one of the politicians—and after some talk together it began to dawn on me that, wise as everyone thought him and wise as he thought himself, he was not really wise at all. I tried to point this out to him, but then he turned nasty, and so did others who were listening.

> So I went away, but with this reflection that anyhow I was wiser than this man; for, though in all probability neither of us knows anything, he thought he did when he did not, *whereas I neither knew anything nor imagined I did* [italics added].
> *(Plato, Apology 21a-d, https://www.roangelo.net/logwitt/delphi.html)*

As a note of caution, Socrates was made to drink hemlock and die. But it was not for showing self-congratulating "wise" men to be fools, but for corrupting the youth, an excuse that is readily transparent, and impiety. Socrates chose to drink the hemlock rather than escape, as the latter would only be used to tarnish his reputation. Socrates and his iconoclast nature survived through the millennia and continues to inspire those fortunate to study his words. These days, iconoclasts find it difficult to get grant funding, papers published, or speaking invitations.

Procedure to actualize the intellectual virtues

Often it is helpful to create and use checklists. However, the presumption of sufficiency creates risks. Rather, a checklist should be considered the minimum and used primarily to ensure that important considerations are not overlooked. Further, the checklist often helps reduce the analysis of the ethical problem to a procedure that can help reinforce the intellectual virtue of discipline consistent with virtue ethics. It may be useful to use rhetorical devices in the checklist to engage the user and to facilitate remembering to use the checklist items. With those considerations in mind, the following checklist is offered.

Check your gut at the door

As one's prior ethical and moral philosophy likely will have an impact, it is important that the potential impact be recognized and checked.

Who's got skin in the game?

As seen in a number of cases in Part 1, the number, range, and diversity of potential shareholders and stakeholders are quite large, often far larger than most other considered analyses (see Introduction). Consideration of only a few, most often two, in the traditional dyadic approach to ethics likely leads to the logical Fallacy of Confirming the Consequence and the Fallacy of Four Terms as described earlier. In order to identify all reasonable shareholders and stakeholders, it is helpful to have loose and flexible definitions of "reasonableness" in the initial identification procedure, as not doing so risks premature elimination of shareholders or stakeholders. Later, the list can be culled to those reasonable stakeholders and shareholders.

Who is doing what to whom?

The answer to that question helps in identifying the dynamics involved in the ethical problem, who are the stakeholders and who are the shareholders. Note that any one person or entity can be both a stakeholder and a shareholder. Indeed, in a pluralistic modern liberal democracy, nearly all those involved are both. The next step is to characterize the relationship between every pair of shareholders and stakeholders based on some set of ethical principles, moral theories, laws, and precedents. For the purpose of this book, the analysis will largely center on the obligation to the ethical principles of beneficence, autonomy, and nonmaleficence in the context of Libertarian, Utilitarian, Deontological, and Equalitarian moral theories. Yet, it is important to note that Anti-Principlist ethics plays an important role in the checklist, providing items such as "what is the narrative?" and "what's at stake?" Further, the question requires an understanding of the context and ecology in which the ethical problem takes place. Feminist ethics can be highly effective in this effort, particularly but not limited to when issues across genders are involved.

What's at stake?

This question relates to what is at stake or what value is at risk for all those involved. At first glance, what is at stake would seem quite straightforward, the curing of disease, reducing pain, and restoring abilities to the patient, as well as reducing costs (considered in the widest term) to the shareholders. However, this would be an extremely and unrealistically narrow view. The impact of disease and disorders has an effect on virtually every aspect of a person's life—it affects the notion of self-authenticity as discussed in phenomenological ethics in Chapter 2. To be sure, many physicians and healthcare professionals take a very narrow view of the beneficence that they are expected to provide (see Cases in Part 1, particularly Cases 6 and 10, and Chapter 4). However, it is not clear that a sufficient number of patients and potential patients share that expectation sufficient to set those expectations as policy or procedure for all patients, now and in the future.

What is the narrative?

One of the first efforts involved in checking one's gut is to fully and as clearly as possible describe the ethical problem. To the shareholders and stakeholders, the ethical problem can be considered a narrative, each is telling a story. Importantly, that story often originates well before the ethical problem at hand, and stories have a trajectory that makes considerations of the future important to the ethical problem at hand. Most often that story is from the perspective of the shareholder or stakeholder. This is not to say that the particular shareholder or stakeholder does not appreciate the narrative of others but that requires a facility referred to as the Theory of Other Minds, which may be varied and inconsistent at best. It is here that the system of narrative ethics plays a major role. Narrative ethics utilizes methods and models from literary criticism (Jones, 1999). Modifying from Jones, aspects of literary criticism include (1) who is the narrator, (2) is the narrator reliable, (3) from which angle of vision does the narrator tell the story, (4) what has been left out of the narrative, (5) whose voice is not being heard and why, (6) what kind of language and images does the narrator use, and (7) what effect does that kind of language have in creating patterns of meaning that emerge from the text? As Jones points out, literary criticism analysis then enables the participants to recast the ethical problem into a more logical or structured form, lending itself to quasi-deductive analysis based on ethical principles and moral theories (note that the term quasi-deductive is used because ethical logical analysis does not lend itself to true valid forms of logic such as syllogistic or propositional logic as discussed in Chapter 1). Assiduous application of such literary criticism analysis may help mitigate the effects of prior ethical and moral presumptions.

Pick a tool to decide

Once the nature of the ethical problem has been assessed as described earlier, an initial attempt to adjudicate the problem is undertaken that is lawful. This may be only the first step in a series of negotiations to arrive at a final determination. An impasse may require subsequent ethical or legal professional assistance. Using the Principlist approach, any conflicting obligations to the ethical principles due to each participant require some adjudication in terms of the prevailing or informing moral theory. The moral theories include Libertarian (Contractual), Utilitarian, Deontological, and Egalitarian, but do not exclude other moral theories or even the development of new or novel theories. The latter may be particularly important as the expanded territory of the ethics of everyday medicine is explored. This approach is demonstrated in many of the cases presented in Part 1.

An Anti-Principlist approach would argue against the use of ethical principles or moral theories, although still constrained by law. In this case, an attempt is made to determine which precedent is most similar or relevant to the ethical problem at hand. Matching to the relevant case precedent requires the precise characterization of the ethical problem in the manner described previously. The similarities can be assessed by means of Epistemic Risk. The relevance of the precedent may be estimated by consideration of the Epistemic Distance and Epistemic Degrees of Freedom between the proposed case precedent and the ethical problem at hand. It may be that one has to look afar for a precedent, for example, policies regarding fiducial responsibilities of investment advisors with respect to conflict of interest relative to conflict of interests that confront the physician or healthcare professional between obligations to the patient and to the insurers. This is discussed in Chapter 4.

It may be that there are no precedents, particularly given the rapid changes in healthcare, and precedents must be created. In doing so, the general approaches afforded by the intellectual virtues described earlier, coupled with the specific issues discussed in Chapter 3, may be helpful.

Calling the question

Any adjudication of the ethical problem must be lawful. However, as the medical ethicist Arthur Schafe, was quoted as saying, "Our law is a blunt instrument, totally unsuitable for cases like this" regarding the conviction of Robert Latimer in the murder of his severely disabled 12-year-old daughter (https://archive.macleans.ca/article/1994/11/28/a-blunt-instrument). Often, the law is no less blunt in the ethics of everyday medicine given the complex ethical ecology in which everyday medicine is practiced. An example would be telishment, where presumably an innocent suffers in order to make the system work. When a charitable hospital is responsible for injuring a patient, it is the physician who is held liable even if the circumstances were beyond the physician's control rather than risk the charitable hospital (Chapter 1). It is possible that the necessary resolution of a particular ethical problem would set a difficult precedent. Kant's categorical imperative holds "Act only according to that maxim whereby you can, at the same time, will that it should become a universal law" (Kant, 1785). The categorical imperative may be a very high bar, and its applications to every instance of everyday medicine may be very problematic. There may be circumstances where a formal "calling the question" can lead to unintended adverse consequences. Yet, justice demands that the question be called in one form or another.

At the same time, "settling out of court" can become a cover for injustice. As US Supreme Court Justice Louis Brandies wrote, "Sunlight is said to be the best of disinfectants; electric light the most efficient policeman" (Brandeis, 1914). Resolving this conundrum in the context of the ethics of everyday medicine may require research and scholarly work. Importantly, it is not clear that there is any answer today that all reasonable persons blinded to their self-interest would consent to, now or in the future.

References

American Medical Association, 1848. Code of Ethics of the American Medical Association. Philadelphia, T.K. and P.G. Collins, Printers.

Bazerman, C., 1997. Reporting the experiment: the changing account of scientific doings of the Philosophical Transactions of the Royal Society, 1665-1800. In: Harris, R.A. (Ed.), Landmark Essays on Rhetoric of Science: Case Studies. vol. 11. Hermagoras Press, pp. 169–186.

Brandeis, L.D., 1914. Other People's Money and How the Bankers Use It. Frederick A. Stokes, New York.

Feyerabend, P., 1975. Against Method: Outline of an Anarchistic Theory of Knowledge. New Left Books, London.

Garfield, E., 1977. The 'obliteration phenomenon' in science—and the advantage of being obliterated! In: Garfield, E. (Ed.), Essays of an Information Scientist. vol. 2. Institute for Scientific Information, Philadelphia, PA, pp. 396–398.

Grant, B., 2009. The Scientist. (June 25). Available at: https://www.the-scientist.com/the-nutshell/citation-amnesia-the-results-44065.

Hargens, L.L., 2000. Using the literature: reference networks, reference contexts, and the social structure of scholarship. Am. Sociol. Rev. 65 (6), 846–865.

Helms, L., 2004. Five ways of institutionalizing political opposition: lessons from the advanced democracies. Gov. Oppos. 39 (1), 22–54.

Hobbes, T., 1839. English Works of Thomas Hobbes of Malmesbury, eighteenth ed. vol. iv Molesworth, p. 82.

Hookway, C., 2003. How to be a virtue epistemologist. In: DePaul, M., Zagzebski, L. (Eds.), Introduction to Intellectual Virtue: Perspectives From Ethics and Epistemology. Oxford University Press, Oxford. (ProQuest Ebook Central). Available at: https://ebookcentral.proquest.com/lib/wisc/detail.action?docID=422848.

Hume, D., 1739. Treatise of Human Nature. Available at: http://www.gutenberg.org/files/4705/4705-h/4705-h.htm.

Johnson-Laird, P., 2008. How We Reason. Oxford University Press, Oxford.

Jones, A.H., 1999. Narrative based medicine: narrative in medical ethics. Br. Med. J. 318 (7178), 253–256 (Review).

Kant, I., 1785. Fundamental principles of the metaphysics of morals. In: Wood, A.W. (Ed.), Basic Writings of Kant. The Modern Library, New York.

Kövecses, Z., 2010. Metaphor: A Practical Introduction, second ed. Oxford University Press, Oxford.

Kuhn, T., 1962. The Structure of Scientific Revolutions. University of Chicago Press, Chicago.

Merton, R.K., 1967. On Theoretical Sociology. Free Press, New York.

Montgomery Jr., E.B., 2019a. Medical Reasoning: The Nature and Use of Medical Knowledge. Oxford University Press, Oxford.

Montgomery Jr., E.B., 2019b. Reproducibility in Biomedical Research: Epistemic and Statistical Problems. Academic Press, New York.

Neurath, O., 1973. Anti-spengler. In: Neurath, M., Cohen, R.S. (Eds.), Otto Neurath: Empiricism and Sociology. D. Reidel Publishing Company, Dordrecht-Holland/Boston-USA, pp. 158–214.

Pearce, R.M., 1912. Chance and the Prepared Mind. Science Vol. 35 (912), 941–956.

Rawls, J., 1999. A Theory of Justice, revised ed. The Belknap Press of Harvard University Press, Cambridge, MA.

Robinson, K.A., Goodman, S.N., 2011. A systematic examination of the citation of prior research in reports of randomized, controlled trials. Ann. Intern. Med. 154 (1), 50–55. https://doi.org/10.7326/0003-4819-154-1-201101040-00007.

Sherman, N., White, H., 2003. Intellectual virtue: emotions, luck and the ancients. In: DePaul, M., Zagzebski, L. (Eds.), Introduction to Intellectual Virtue: Perspectives From Ethics and Epistemology. Oxford University Press, Oxford. (ProQuest Ebook Central). Available at: https://ebookcentral.proquest.com/lib/wisc/detail.action?docID=422848.

Tolstoy, L., 1869. War and Peace, vol. 4, Part 3, Section 1 (R. Pevear, L. Volokhonsky, Trans.). Penguin Random House, New York.

Wang, Y., Xiao, Y., Li, Y., Chu, K., Feng, M., Li, C., Qiu, N., Weng, J., Ke, X., 2019. Exploring the relationship between fairness and 'brain types' in children with high-functioning autism spectrum disorder. Prog. Neuro-Psychopharmacol. Biol. Psychiatry 88, 151–158.

Zagzebski, L., 1996. Virtues of the Mind: An Inquiry Into the Nature of Virtue and the Ethical Foundations of Knowledge. Cambridge University Press.

Zagzebski, L., DePaul, M., 2003. Introduction. In: DePaul, M., Zagzebski, L. (Eds.), Introduction to Intellectual Virtue: Perspectives From Ethics and Epistemology. Oxford University Press, Oxford. (ProQuest Ebook Central). Available at: https://ebookcentral.proquest.com/lib/wisc/detail.action?docID=422848.

Postscript: COVID-19 pandemic and Black Lives Matter

This book was well into the publication process when the COVID-19 pandemic struck and the social unrest related to Black Lives Matter erupted into social consciousness and onto the streets. How these events will impact the ethics of everyday medicine is anything but clear. Perhaps taking the relationship of the patient with the physician or healthcare professional as strictly dyadic may seem as a way to isolate the relationship from the larger social struggles outside the clinic or away from the bedside.

The central theme of this book is that the ethics of everyday medicine has never been just at the dyadic microlevel justice in the clinic and at the bedside, such as between the patient and the physician/healthcare professional, and inevitably involves the macrolevel, the actions taken by or on behalf of society. Further, the patient, family member, caregivers, physicians, and healthcare professionals do not exist in a social vacuum and bring to the clinic and bedside their own narratives that are shaped by all their experiences. It is through the lens of the individual's narrative that the encounter in the clinic or at the bedside is shaped.

COVID-19

It seems inevitable that patients, physicians, and healthcare professionals will have to deal with the COVID-19 pandemic. In many ways, the COVID-19 pandemic has laid bare and exacerbated the scarcity of resources, which constrains how justice is to be effected in healthcare at the micro- and macrolevel. The issue is whether the scarcity remains manageable or whether it becomes so severe that the only recourse is that it is everyone for themselves. We have seen this in the hoarding of commodities, medical and otherwise, as well as in the "hoarding" of medical care where elective treatments are postponed, in many cases perhaps indefinitely, as hospital beds are held in reserve for anticipated COVID-19 patients. All of this places serious ethical stresses on the notion of elective. Yet, this really is triage but it seems no one wants to say "triage" with all that the word portends. The scarcity of resources demands triage even if no one says the name or by its other name, rationing.

The stress of the COVID-19 has infected nearly every aspect of medicine. It has become a source of great emotional stress on physicians and healthcare professionals doing their upmost to mitigate the scarcity, often at their own personal risk and those of their families. For many physicians and healthcare professionals, there have been great financial losses and, for the most part, no plans for restitution by societies' governing proxies. Yet, from the viewpoint of this author, the overwhelming numbers of physicians and healthcare professionals persevere, explore, and exploit every opportunity to mitigate those losses and barriers to patients.

There is another scarcity that is relatively unrecognized but whose effect continues to challenge the ethics of everyday medicine. That scarcity is personal freedoms and liberties in a modern pluralistic liberal democracy. This issue takes form in sacrifices, accepted willingly or imposed. It is a form of maleficence, harm, but that is not necessarily pejorative. But often such harm is necessary, and perhaps our surest guide is Thomas Aquinas' Principle of Double Effect that has been addressed in this book. How we navigate the maleficence of self-isolation, social distancing, mask wearing, hand washing, and scarcity of resources, including jobs, as these infringements affect nearly every aspect of life, is telling of who we are and what we expect of ourselves and others. The ethics of everyday medicine does not escape.

The infectivity of COVID-19 essentially places a dangerous weapon in the hands of nearly every individual. Indeed, COVID-19 becomes a means of asymmetric warfare where the most marginalized individual merely has to walk out the door of her home without a mask and does not maintain social distance, even if she has fever, cough, and fatigue. This is a challenge and a reversal in healthcare in the United States since the 1960s where the risk and consequent burden of disease could be segregated to those who were relatively disenfranchised. Consequently, much rises and falls with the effectiveness of leadership and the willingness and justification of imposing constraints on individual freedoms. While the ethical principles can demarcate the lines of ethical concerns, descriptions of the dynamics, what is needed is a normative moral theory that translates into right or wrong, legal or illegal, and moral or immoral. The greater question is what will be the supervening moral theory? Nowhere is this more evident in the politicization of the pandemic in a post-truth era.

It takes little imagination to anticipate some of the questions and concerns that will confront physicians and healthcare professionals. There are a great many precedents. Consider the following:

- A parent approaches a pediatrician for a medical excuse from having his/her child vaccinated against COVID-19 just as he/she did for vaccinations against rubella (measles).
- A state that attempts to prevent physicians from inquiring about firearms in the household could similarly prohibit discussion of social distancing, mask wearing, and handwashing.
- How does the physician or healthcare professional act when the patient suffers from COVID-19 while ignoring the physician's advice? It was not so long ago that HIV/AIDS was held by some as justifiable punishment for deviant lifestyles.
- Will the physician or healthcare professional listen with professional sympathy and respect to a patient who says it is his right not to wear a face mask? How will the physician or healthcare professional attempt to educate the patient? Will the physician or healthcare worker bar those not wearing a face mask from the clinic?
- Many countries have lists of reportable diseases. Will physicians or healthcare professionals report a patient they suspect as having a COVID-19 infection?
- What if COVID-19 is never fully eradicated? How do physicians and healthcare professionals learn "to live with it?" What happens to the elderly or those at high risk? Is the answer perpetual self-isolation?

Black Lives Matter

The Black Lives Matter movement has brought attention to racial injustice and, importantly, the very long history of racial injustice. The movement is widespread, which means that the issue will find its way into the ethics of everyday medicine. It already has, at least as the precursors to today's movement were long embedded in the ethics of everyday medicine, now they will be more acute. Consider the disproportionate burden of COVID-19 on Black, Latino and poor populations.

The role of the physician, healthcare professional, administrator, and legislator, indeed all those involved in healthcare, at the micro- and macrolevel justice in healthcare is difficult, as has been discussed frequently throughout this book. Historically, policies, procedures, laws, cases, and precedents have not been patient-centric. Even if physicians and healthcare professionals can bracket their opinions regarding macrolevel justices or injustices, it is highly unlikely that the patient, caregiver, family member, or friend can or will. This means that patients, caregivers, family members, and friends will bring their feelings at the macrolevel into the clinic and bedside as these feelings are very much in their microlevel ecology. Physicians and healthcare professionals are likely to do so as well if uneducated or untrained in how to bracket their opinions, which presumes a recognition of the need to do so.

As Caucasian males continue to dominate among physicians and healthcare professionals, particularly in leadership, and indeed for very long periods of time, it is reasonable to raise the question of historical, systemic, and systematic racism and sexism raised, for example, by the Black Lives Matter movement. As historical injustices continue beyond the life span of any individual person, who then perpetuates them and how? Invariably there will be accusations of white male privilege; how could there not? It would be naïve for the Caucasian male physician or healthcare professional to think that non-Caucasian or nonmale patients would not harbor suspicions of white male privilege that could influence the patient and physician/healthcare professional relationship. Similarly, it would be naïve to think that such patients would not have similar suspicions of the actual healthcare provided. History is replete with examples, such as the Tuskegee syphilis study and the lack of non-Caucasian and nonmale representation in randomized control trials that increasingly dictate everyday medical care. More problematic is the fact that many physicians and healthcare professionals would be unaware of such biases.

Society has long had difficulty in relating macrolevel justice or injustice to the microlevel. Balancing equality, that does not differentiate the standing among persons, with equity that requires cognizance and accommodation of differences, for example, to persons with disabilities, has always been difficult. The issue of equity also applies to those with social and economic disadvantages not due to personal individual actions or inactions but rather imposed by a history of inequality. For Caucasian males, physician, healthcare professional, or otherwise, it may be hard to relate their personal sense of responsibilities to the societal injustices inherited from the long history of racial and gender inequality. What is owed by the privileged person who has benefited through acts of omission to rectify unjust acts of commission by others? Likely, those who are the opposite of privileged would expect some redress through equity if equality is ever to be reached. Yet, a fundamental premise of modern liberal democracies is that no individual can be deprived of life, liberty, or property without due process. This is where macrolevel injustice must find redress at the microlevel. The clinic and bedside are no exceptions.

Ethical analyses, such as by examining the obligations to principles, can demarcate the lines of contention, as some supervening moral theory is required to adjudicate a solution. Libertarian and Utilitarian moral theories do not necessitate corrections of historic macrolevel justices that redirect resources to equity even at the risk of inequality at the microlevel. Deontological moral theories become problematic as unenforceable. Egalitarian theories have difficulty in relating equity to equality. It is not at all clear whether we have the appropriate moral theory.

At the everyday level of medicine, the patient and physician/healthcare professional relationships can become a minefield for what are called microaggressions. For example, the physician or healthcare professional might say to the obese patient "you should lose weight." The obese patient's feelings of indignation and anger likely would be held just by most physicians and healthcare professionals. The problem is that many forms of colloquial speech have lost their connection to racism, sexism, and other forms of discrimination and injustice from which they originated. But it is likely that those who have suffered from discrimination and injustice have not forgotten. For example, while a Caucasian physician may not think anything of the phrase "being sold down the river," the patient of African descent sees this as a reference to slavery. Similarly, "extracting his pound of flesh" can be seen by Jews as an anti-Semitic microaggression related to the deceitful denial of justice to Shylock in Shakespeare's "Merchant of Venice," even if non-Jews did not mean it as microaggression and were unaware.

The problem becomes where to draw the line. Do we remove terms such as "master bedroom," "slave cylinder" in mechanical engineering, or "bullet points" in presentations? Yet, one cannot dismiss the maleficence of such phrases even if unintentional. At the same time, most reasonable persons blinded to their own self-interests would hold the unintentional maleficence usually as accidents and not crimes. But then how are such accidents prevented? Any governing moral theory will need to find a sufficient consensus among reasonable persons blinded to their own self-interests among those to be governed to impose societal actions on all.

"What we can do, we will do"

In many ways, the current situation and circumstances of the pandemic and social unrest over racial and gender injustice are like Pandora's box. The box was opened and many demons were released. But a moment's hesitation by Pandora in closing the box allowed hope to escape. At least there is hope in the sense that in many ways, medicine has been in these binds before and hopefully we can learn from them. Another hope is that in the confines of the clinic and bedside, physicians and healthcare professionals can be resilient in the face of the pandemic and in the reckoning for racial injustice even as the wider society continues to struggle and those struggles continue to spill over into the clinic and bedside. Perhaps physicians and healthcare professionals can teach, even if only by being examples, resilience to their patients. Perhaps it may only be by finding ways to help by redress while not finding blame in the individual microlevel. The hope may be only that "what we can do, we will do."

Index

Note: Page numbers followed by *f* indicate figures and *t* indicate tables.

A

Absolutism, 129–130
Abstraction, nature of, 364–366
Academic-centric ethics, 456–459
Acceptable contracts, 205, 226–227
Acquired immunity deficiency syndrome (AIDS), 322–323
Act of commission, 103–104
Act of omission, 103–104
Adult-like moral motivations, 274
Adverse effects
  deep brain stimulation, 172
  error and, 205–207
  estimation of, 176–177
  long-term, 122, 139
  short-term, 123
Advocacy, 46–47
Aequanimitas, 165
Affordable Care Act, 71, 322
Age of license, 269–270
Age of majority, 269–270
Altruism, 312–313
American Academy of Pediatrics, 271
American Medical Association (AMA) Code of Ethics of 1847, 107–111, 242, 290, 456
American Medical Association's Principle of Medical Ethics of 1912, 48
American Psychiatric Association, 220–221
American Speech and Hearing Association (ASHA), 46–47, 49
Amygdala neuronal activities, 312
Angiogram, 145
Anglo-American common law, 267–268
Anthropomorphism, 57
Anti-Parkinson's disease medications, 105–106
Anti-Principlism, 336–337, 347, 356–359
  approach, 103, 355, 440, 497
  epistemic components of, 348
  Ethical phenomenalism, 373–374
  ethical theories, 371–373

ethics, 226–227, 299
  problem for, 347–348
Antiretroviral therapy, 128–129
Aphasia, 252–253
Aphorism, 165
Appeal to virtue, 163
A Priori problem of induction, 351–356, 397–398
Aquinas' Principles of Double Effect, 105–107, 182–183, 207–208, 319, 325–326
Aquinas, Thomas, 69
Aristotle, 229, 402, 442
  ethics, 289
  logos, 133–134
  potentiality and actuality, 264
Arizona Supreme Court, 126–127
Artificial intelligence (AI), 247, 299. *See also* Machine learning artificial intelligence (AI)
Assembly line, 114
Assent (dissent), 271–274
Atmospheric turbulence, 428
Audiology and Speech-Language Pathology Act, 49
Authenticity, 378
Autonomy, 283–284, 367
  coronary artery disease, 149–151
  ethical principle, 196–198
  nature of, 262–265
  Parkinson's disease, 123–125
  patient-centric ethics, 264
  positive, 284
  as respect, 265–266
  stoic, 284–285
  types of, 284
Availability bias, 131–132

B

Balanced Budget Act of 1997, 450–451
Bandwagon effect, 131–132
Basal ganglia disorder, 200

Battery, *vs.* negligence, 126–127
Bazerman, Charles, 484–485
Belmont Report, 263, 276, 439
Beneficence, 276
    ethical principle, 148–149, 193–195
    maleficence *vs.*, 66–69
    obligation to, 44–46, 147, 162–163,
        166–168, 238–239, 249, 275
Bernard, Claude, 65, 457, 492
Best interests
    children, 266
    justification, 268
    standard of, 275–276
Bifurcations (transitions), 426–427, 427*f*
Binary ASCII encoding, 369
Biological-plausible computational
        algorithms, 432
Biomedical research, 389–390, 393, 424
Blue Cross and Blue Shield programs,
        451–452
Borderline personality disorders, 223
Botulinum toxin, 193
Bowie, Norman E., 88–89
Brain, mirror neurons in, 311–312
Brandeis, Louis, 447, 498
Brooklyn Medical Society, 111
Brownian motion, 414, 415*f*
Bundling, 162–163
Business-centric ethics, 176, 450–455
Business/government-centric ethics, 176,
        450–455

C

Calling the question, 497–498
Canadian Medical Protective Association, 81,
        89, 126, 128, 161, 269
Capitation, 177
Captain of the Ship theory, 40–41, 43, 299
Carbidopa, Parkinson's disease, 101–102
Care ethics, 383
Carefulness, 488–491
Carroll, Lewis, 492
Case mix complexity, 181
Case precedents, 389
    continuous *vs.* dichotomous character,
        418–419
    interpretation of, 419–420
    nonequilibrium steady-state experiences,
        430

Caucasian children, 315
Causality, 254, 396
Centrality of hypothesis, 480–482
Central Tendency, 391–392, 395, 426–427
Cervical dystonia, 193
    context of physician's training, 199
    diagnosis, 194
    disallowance, 194
    misdiagnosis, 202
    prevalence rate, 194
Chaos, 370–371, 427–428
Chaotic system, 329–336, 428
Chattel, 266
Cherry-picking, 93, 449
Children
    assent (dissent), 271–274
    best interest, 266
    decision-making capacity, 262–263
    notion of, 266
    potentiality and actuality, 264
Chronic pain, treatment, 167–168
Churchland, 339
Citation amnesia, 485
Citation negligence, 485
Civil Code, 269
Cleary v. Hansen, 106
Cobbs v. Grant and Wilkinson v. Vesey,
        127–128
Cognitive control, 306–307
Coleridge, Samuel Taylor, 401–402
College of Audiologists and Speech
        Language Pathologists of Ontario,
        252
College of Physicians and Surgeons of Nova
        Scotia, 90–91
College of Physicians and Surgeons of Ontario,
        92
Commerce Clause of US Constitution, 322
Committee on Bioethics, 271
Community immunity. *See* Herd immunity
Complex ethical ecology, 42–44
Complexity system, 329–336, 370–371, 428
Computer circuitry, 304
Concerned physician, 293
Conflict of interest, 238–239
Consensus approach, 129–130
Consent, 271–273
Conservation, 133
Conservative ethics, 418

Constant contraction, 199–200
Consultant. *See also* Family physician
   letter to family physician, 85–87
   responsibility to patient, society, and
      family physician, 97–98
Consultation procedure, 91
Consulting neurologist, 103–105
Consumer-centric ethics, 459
Contracts, 157, 181
   acceptable, 205, 226–227
   constant, 199–200
   elements of, 174–175, 181, 239
   fundamental principles, 239–240
   Hobbesian social contract, 446–447
   with insurance organization, 175
   with physicians and healthcare
      professionals, 177
   requirement, 194, 267–268
   social, 444, 446–447
   tonic, 199–200
   valid, 125–126
Contractual agreement, 87
Contractualist moral theory
   balancing obligations, 109
   beneficence, 194
   educational systems failure, 201–203
   fear of obligation, 187
   medical record, 162, 166
   patient-centric ethics, 441
   transparency, 181
Contractual obligation, 44, 247, 249, 276
Conversion disorder, 197, 219
Copernicus, 493
Core assessment program for intracerebral
      transplantations (CAPIT), 242–243
Core assessment program for surgical
      interventional therapies in Parkinson's
      disease (CAPSIT-PD), 242
Coronary artery angiogram, 145, 147
Coronary artery disease
   abnormality of, 145–147
   beneficence, 148–149
   ethical responsibilities of system, 151–152
   false-positive angiogram, 147–148
   lapses in medical reasoning, 152–156
   medical facts of matter, 145–147
   nonmaleficence, 148–149
   obligation to autonomy, 149–151
   patient characteristics, 146*t*

Corporal punishment, 288
Correlational analysis, 420
Counterbalancing, 425
Countertransference, 226–227, 460
Cream skimming, 93
Cronkite, Walter, 466
Curbside consultations, 90–91

D

DBS. *See* Deep brain stimulation (DBS)
Decision-making capacity, 262–263
Deductive approach, 136–137, 200–201,
      285–286
Deductive syllogistic logic, 403–409
Deep brain stimulation (DBS), 139, 270
   adverse effects, estimation of, 176–177
   benefit-to-risk ratio, 172
   candidacy criteria, 241–243
   cost of providing, 174–175
   ethical analysis, 238–241
   fear of obligation, 186–188
   justice, effects on, 177–182
   medical facts of the matter, 171–172,
      237–238, 261–262
   motor functions, 172
   neurologist is not expert on, 243–244
   physician-centric estimation of risk, 176
   primum non nocere, 182–186
   scientist-centric ethics, 241–243
   surgery, 173
   transparency, unintentional lack of,
      177–182
   "you are not bad enough", 172–174
Demonic possession, 217–218
Demonstration, 365–366, 365*f*
Deontological moral theory
   complex ethical ecology, 43–44
   consultant's responsibility, 97
   contract, 109
   educational systems failure, 202–203
   effects on justice, 180–181
   family physician, 95
   justice, 156
   Kant's notion, 211–212, 263
   medical error, 205
   medical record, 163, 166
   micro- and macrolevel justices, 211–212, 239
   minimum beneficence, 241
   moderate scarcity, 99

Deontological moral theory *(Continued)*
  obligations, 54–56, 150
  patient-centric ethics, 464–465
  psychological disorders treatment, 222–223
  reproducibility as justice, 362
  right to refuse any treatment, 140
  unalienable right, 271
  video fluoroscopic evaluation of
    swallowing, 70, 73
Deontological sense, 289
Descartes dualism, 230–232
Deus ex machina, 324
Diabetes mellitus type I, 228–229, 232
Diagnosis-related group (DRG) method,
    178–180
Diagnostic and Statistical Manual (DSM-5),
    220–221
Diagnostic errors, 152–153, 156
Dictated but not read report, 159–160
Discipline, 493
Discovery *vs.* judgment, 398–399
Disregarding syndrome, 485
Dissent (assent), 271–274
Dissociative disorders, 219
Dissuasion, 131–132
Dodson, Lisa, 473
Dogmatic school of Medicine, 242
Donne, John, 165
Dopamine, 101, 121–123
Double-blinded study, 122
Drug holidays, 105–106
Dualism, 230–232, 324
Duran, Betty, 282
Dyadic approach, 232–233
Dyskinesias, 237
Dysphagia, 39
  examination, 40
  systematic reviews, 39–40
Dyspnea, 145–147, 146*t*
Dystonia, 261

E

Educational costs, for medical student, 203
Educational systems failure, 199–204
Egalitarian moral theory, 313, 321
  aphasia, 252–253
  contract, 109
  educational systems failure, 202–203
  effects on justice, 180–181

ethical analysis, 240–241
family physician, 95
justice, 157
medical record, 163, 166
micro- and macrolevel justices, 211–212,
    239
moderate scarcity, 99
obligations, 54–56, 150, 249
Parkinson's disease, 241
patient as suspect, 226–227
patient-centricity, 466
psychological disorders treatment,
    222–223
reproducibility as justice, 362
right to refuse any treatment, 140
video fluoroscopic evaluation of
    swallowing, 73–74
Electronic medical record, 159–161
Emancipation of minor, 267–268
Emergency Medical Technicians (EMTs),
    293
Emergency room care, 293
Empiric/Irregular school, 349
Empirics, 253–254
Empiric school of medicine, 290
Employee Retirement Income Security Act of
    1974 (ERISA), 77, 81, 155–156, 451
Enlightened Self-Interest, 322–323
Entropy, 430
Enumeration, 396
Epistemic Degrees of Freedom, 367, 489
Epistemic Distance, 489
Epistemic privilege, 196
Epistemic risk, 138, 366–367, 489
Ergodicity, 417–418
Ethical analysis
  deep brain stimulation, 238–241
  Egalitarian moral theory, 240–241
Ethical consensus
  criteria for moral judges, 423
  reasonable person, 423–424
Ethical dissonance, 280–284
Ethical ecology, 274–276
Ethical experiences, 297
Ethical imperative, 280–281
Ethical incompleteness, 241, 433–434
Ethical induction, 435
  and artificial intelligence, 432–433
  empirically derived inductions, 393

epistemological issues, 395–397
measures of robustness, 393
Meno's paradox, 397–398
moral and ethical objective truth, 390–392
necessity of, 393–394
ontological issues, 394–395
A Priori problem, 397–398
reflective equilibrium, 393
Ethical issues, insurance coverage, 248–250
Ethical principles, 66, 69–74, 83–84,
    319–321, 394–395
continuous *vs.* dichotomous character,
    418–419
degree of obligation, 420
implications for, 324–329
nonequilibrium steady-state experiences,
    430
resolve conflicting obligations to, 54–56
Ethical property dualists, 368
Ethical resources, 275
Ethical system, 389
Anti-Principlist ethical system, 226–227
Chaos and complexity, 370–371
diversity variety, 348–351
epistemic risk, 366–367
first principle, 356–357
indigenous persons, 289–290
Principlist approach, 348
transcendent principles, 351–356
variation variety, 348–351
Ethical virtues, 319
Ethics, 440
an exercise in chaos and complexity,
    329–336
care, 383
of clinicians, 78–80
conscious deliberative declarative
    analyses, 302–303
and consciousness, 367–370
and ethos, 469–471
experimental psychological studies, 301
feminist, 381–382
feminist bioethics, 463–464
from gut, 40–41
of healthcare systems, 75–77
implication for, 327–329, 334
institutions, committees, and agencies to
    ensure, 53–54
narrative, 378–381

neurophysiological explication of, 297
patient-centric, 440–444
phenomenological, 373–385
physician-centric, 445–450
problems, 334
psychological mechanisms, 304
romanticism and misplaced virtue, 459–461
scientist/academic-centric, 456–459
virtue, 383–385
Ethos, 132–135
European Society of Cardiology Education
    Committee, 151–152
Evidence-based medicine, 251–254, 456–457
"Explorations of Justice", 317

F
Face-vase illusion, 332–333, 333*f*
Facial recognition, 315
Factitious disorder, 224–227
Fairness, 298
emotive response, 298–299
ethical experiences, 298
moral theories, 299
principlism, 298–299
reproducibility, 298
sense of, 298
Fait accompli, 151
Fallacy of Affirming a Disjunctive, 149, 198
Fallacy of Confirming the Consequence, 130,
    149, 198, 480
Fallacy of Four Terms, 409–410, 480
Fallacy of Limited Alternatives, 50, 153, 198
Fallacy of Pseudotransitivity, 136–137, 137*f*
False-negative angiogram, 145–147
False-positive angiogram, 145–147
False-positive mammogram, 148
Family Health Team, 98–99
Family physician, 159
consultant responsibility to, 97–98
duty of care for patient, 88–91
duty of care of society, 91–97
medical facts of matter, 85–87
moderate scarcity of expert care, 98–99
prescription, writing, 85–87
stakeholders and shareholders, 87–88
Fatigue, 145–147
Fear of obligation, 186–188
Fee-for-service primary care physicians,
    92–93

Fee-for-service reimbursement, 177
Feminism, 381
Feminist bioethics, 463–464
Feminist ethics, 381–382, 495–496
Fiberoptic endoscopic examination of
        swallowing (FEES), 40
Firearm Owners' Privacy Act, 45
Flexner Report, 164
Folk ethics, 487
Folk psychology, 487
Foot, Philippa, 183
Framing effects, 132
Free hand of laissez-faire market capitalism,
        454–455
Frugality, 174
Functional neurological disorder, 218, 226–227
    differential diagnosis, 218–219, 224–225
    motor system and, 228
    systematic review of, 223

G

Gage, Phineas, 324–325
Galenic medicine, 254
Galileo, 493
Gambler's Fallacy, 198
Gaming, 473
General Data Protection Regulation, 255
Generalizations, 350
Gestalt psychology, 374–375, 376f
Good medical practice, 241–242
Good Samaritan laws, 106
Government-centric ethics, 176, 450–455
Gowers, William, 226
Grounded theory, 413–416
Group differences, 274
Gut reaction, 41, 65–66

H

Happiness, 133
Hard paternalism, 130–131
Harm Principle, 276
Healthcare delivery, 177
    balkanization of, 116
    complexity of, 56–57
Healthcare insurance, 116
Healthcare professionals, 111–114
    contracts with, 177
    education and training, 232–233
    indigenous patient, implications for, 290

machine learning artificial intelligence,
        251–252, 255, 258–260
Health Insurance Portability and
        Accountability Act of 1996 (HIPPA),
        161
Health Maintenance Organization Act of
        1973, 155–156, 451
Health Maintenance Organizations (HMOs),
        451
Health problem, 154–155
Hellenic schools of medicine, 254
Herd immunity, 263, 293
Heuristic devices, 131–132
Hidden curriculum, 184
Higgins, 446
Hippocrates, 446
Histopathological correlations, 220–221
Hobbesian social contract, 446–447
Hobbes, Thomas, 71, 401–402, 483–484
Holmes, Oliver Wendell, 240
Homeostasis, 430
Homework assignment, 470–471
Homogenization of patients, 114–115
Hookway, Christopher, 493
Hospital-based errors, 204
Human Development Index survey, 280
Human nervous system, 329
Human Tissue Act 2004, 265
Hume, David, 490
Hume's Problem of Induction, 405
Hypothesis testing, 413
Hypothetico-deductive approach, 153, 155,
        200, 209

I

ICD-10-CM/PCS MS-DRG v34.0
        Definitions Manual, 181
Iconoclast, 494–495
Identity Principle, 403–404
Idiopathic cervical dystonia, 193
Idiopathic Parkinson's disease, 243
Idiosyncrasy, 167
Illness anxiety disorder, 219
Illness scripts, 155
Imagination, 492
Immaterial mind, 228–232
Impact bias, 131–132
Inborn mechanisms, 307
Indigenous persons, 281

communication and presumptions,
    problems, 282
elasticity of time, 287
ethical dissonance, 280–284
ethical system, 289–290
ethics as moral-ought/prudential-ought,
    287–289
implications for physicians and healthcare
    professionals, 290
mental illnesses, 281
nonindigenous physician interaction with, 283
noninterference, 285–286
no privileged knowledge, 286–287
physician's perspective evolution, 280–281
sharing of knowledge, 286–287
stoicism, 284–285
Influenza, 322–323
Information-theoretic incompleteness, 241
Informed consent, 125–126, 237, 456
    elements, 125–126
    informing in, 127–129
    patient's voluntary admission, 250
Ingelfinger, Franz, 462
Inherent biases, 132
Inherent logic, 400
Inherent mechanisms, 316
Injustice, 300, 314–317
Inscrutability of reference, 282
Insel, Thomas, 221–222
Instinct, 134
Institute of Medicine, 204–207
Institution, obligations and, 107–111
Instrumental (epistemological) sense, 297
Insurance coverage, suicide risk and, 247–248
    denied, 247
    ethical issues, 248–250
    machine learning artificial intelligence
        (see Machine learning artificial
        intelligence (AI))
    net effect, 247–248
    obligation of nonmaleficence, 249
Insurers, 174–175
Integration of care, 56–57
Intellectual virtues, 482–495
    argument 5.1, 490–491
    carefulness, 488–491
    discipline, 493
    iconoclast, 494–495
    imagination, 492

open-mindedness, 484–488
perseverance, 493–494
procedure to actualize the, 495–498
Interchangeability, 114–117
International Classification of Diseases-10
    (ICD-10), 221
Intrinsic neurophysiological reward
    mechanisms, 310
Inverse Problem, 153–155, 154f
Invisible hand of the market, 459

J

James, William, 196–197
Jauhar, Sandeep, 470
Jeopardy, 251
Judge-made law, 434
Judicial activism, 434–435
Justice, 69–74, 156–158, 396
    affective neurophysiological
        mechanisms, 300
    complex ecology, 317
    effects on, 177–182
    emotions, 300
    emotive response, 297
    ethical mechanisms, 297
    from ethical principles, 317–318
    fundamental test of, 320–321
    imposition of, 298
    in law, 318
    neurophysiological basis for, 302
    neurophysiological representations of,
        305–311
    operationalization of, 322
    physician/healthcare professional,
        317–318
    psychological-neuroanatomical concepts,
        302
    reproducibility, 359–364
    sense of, 297–298
    subconscious mechanisms, 304–305
    telishment, 319
    types of, 313
    virtues, 319
Justice Stewart's approach, 298–299
Justifiable paternalism, 268

K

Kantian moral theory, 43–44, 70, 99
Kant, Immanuel, 112, 129, 133, 436

King, Martin Luther, Jr., 65
Kuhn, Thomas, 485, 492–493

L

Lack of transparency, 177–182
Laissez-faire capitalism, 166, 285
Lapses, in medical reasoning, 152–156
Laws, 69–74, 389
  continuous *vs.* dichotomous character,
    418–419
  Good Samaritan laws, 106
  nonequilibrium steady-state experiences,
    430
  physical, 391
  Poor Laws in England, 177
Levodopa
  Parkinson's disease, 101–102, 121–123,
    135–136, 279–280
  side effects, 237
Liberal ethics, 418
Libertarian moral theory, 321, 441
  aphasia, 252–253
  beneficence, 194
  consultant's responsibility, 97
  contract, 109
  educational systems failure, 201–203
  effects on justice, 180–181
  ethical analysis, 239–241
  ethical ecology, 43–44
  family physician, 96
  fundamental principles of contracts, 249
  justice, 157
  medical error, 205
  medical record, 162–163, 166–167
  microlevel justice, 211–212
  obligations, 54–55, 150
  patient as suspect, 226–227
  reproducibility as justice, 362
  video fluoroscopic evaluation of
    swallowing, 70, 73, 76, 78
Literary criticism, 496–497
Locke, John, 267
Logical fallacy, 198
Logic of Certainty, 390, 393–394, 396, 400,
    480
Logic of Discovery, 390, 393–394, 396, 400,
    480
Logics, 394

centrality of certainty, 401–402
inherent probability and statistical
    thinking, 400–401
and philosophy, 399–409
Logistic regression, 420–421
Logos, 131, 133–134
Loomis v. Wisconsin, 255
Lord President of the Council, 227
Lorenz attractor, 428, 429*f*
Lottery, 157
Lymphomas, 180

M

Machine learning artificial intelligence (AI),
    248, 327, 421, 432
  architecture, 329
  evidence-based medicine and, 253–254
  relationship to physicians and healthcare
    professionals, 251, 255
  software as a medical device
    physician and healthcare professional
      responsibility, 258–260
    US Food and Drug Administration,
      255–258
  speech-language pathologists, 253–254
  structure of, 326
Macrolevel justice, 208–209, 239
Magnetic resonance imaging (MRI), 73–74,
    261–262
Maleficence, 66–69, 218–219
Malingering, 224–227
Mammograms, 148
Managed care, 177
Manic-depressive illness, 220
Manufacturing mentality, 114
Material risk, 128
Mature minors
  concepts, 267
  doctrine, 271
  issue of, 270
  legal limitations to, 270
  notion of, 270
McClelland model, 357
McNaughton, George, 111
Medical consent, 269–270
Medical disorders, psychological *vs.* real,
    220–222
Medical educational costs, 203

Medical educators, 209–211
Medical errors, 204–209
Medical liability, 240
Medical license, 115
Medical model, 217–218
Medical necessity
  Emergency Medical Technicians, 293
  insurer's criteria, 247–248, 250
  vaccinations under, 293
Medical reasoning
  failure of, 198–199
  lapses in, 152–156
Medical record, 159, 161
  electronic, 159–161
  patient-generated requests amendments,
    159–160
  patients' access issue, 162
  subsequent copy, 160–161
Medical science, 167
Medical technology, 166–167
Medical tourism, 141–142, 175
Medicare, 92–94, 141–142
Medicare Access and CHIP Reauthorization
    Act of 2015, 450–451
Memorial Sloan-Kettering Cancer Center, 251
Memory loss, 283
Meno's paradox, 397–398
Mental disorder, 222
Mental illness, 217–218, 222
  diagnosis of, 232
  indigenous persons, 281
  nonphysical, 228–232
Metaphor, 135–139
Methodic school, 254
Microlevel justice, 208–209, 211–212, 239
Miniature humanoid, 225–226
Mini-miniature humanoid, 225–226
Mixed-methods research, 416–417
Mixis, 356
Moderate scarcity, 73, 98–99
Modern ethics, 349
Modern liberal democracies, 322
Modus ponens in deductive logic, 403–404
Monoamine oxidase type B (MAO-B)
    inhibitors, 86
Moral and ethical objective truth, 390–392
Moral culpability, 337–338
Moral imperative, 280–281
Moral insight, 423

Morality, 311–314, 322–323
Moral-ought, 287
Moral theories, 211, 321, 389, 394–395, 434
  continuous *vs.* dichotomous character,
    418–419
  degree of obligation, 420
  Deontological (*see* Deontological moral
    theory)
  Egalitarian (*see* Egalitarian moral theory)
  implications for, 324–329
  individual, 241
  Libertarian (*see* Libertarian moral theory)
  from macrolevel justice, 319–321
  micro- and macrolevel justices, 239
  nonequilibrium steady-state
    experiences, 430
  state-of-being relationships, 404
  supervening, 178, 186, 201–202,
    205, 239
  Utilitarian (*see* Utilitarian moral theory)
  video fluoroscopic evaluation of
    swallowing, 69–74, 83–84
Morbidity and mortality conferences, 113
Mortal things, 137–138
Motor assessments, 237–238
Movement disorders, 98, 101, 105, 112–113,
    194
Muscogee Nation, 288–289
Mutual intent, 76–77

N

Narrative ethics, 264, 378–381, 496–497
National Academies of Science, Engineering,
    and Medicine, 204
National Institute for Mental Health (NIMH),
    221
Natural ontology, 287–288
Necker cube illusion, 332, 332*f*
Negative Predictive Value, 64
Negative valence countertransference, 460
Negligence, technical battery *vs.*, 126–127
Neural dynamics, 324–329
Neural network computer algorithm,
    309–310, 310*f*, 326, 357
Neurath, Otto, 480
Neurocognitive testing, 171
Neurological dysfunction, 218
Neurometabolic imaging, 221, 225, 311
Neuropharmacology, conflation of, 339–341

Neurophysiology, 337–338
  concepts, 325
  genetics with, 339–341
  mechanisms, 225–226
  of morality, 311–314
  psychology *vs.*, 303–305
Neurosurgery, 139
"Neutral-smelling dry-goods store", 335
NIMH's Research Domain Criteria (RDoC)
      project, 221
Noninterference, 285–286
Nonlinear dynamics, 428
Nonlinearity, 329–330
Nonmaleficence, 69, 195–196, 276, 325–326
  coronary artery disease, 148–149
  obligation to, 162, 167–168, 238–239,
      249, 275
Nonphysical mental illness, 228–232
Novum Organum, 349–350
Nuclear option, 97
Nurse practitioner, 253, 294

O

Oberlander, J., 82–83
Objective ethical and moral truths, 390–392
Objective truth, 391
Obligation, 103–105
  to beneficence, 44–46, 147, 162–163,
      166–168, 238–239, 249, 275
  and institution, 107–111
  to nonmaleficence, 162, 167–168,
      238–239, 249, 275
"Off-label" use, 241–242
Ohio v. American Express Co., 452–453
Omission bias, 183–184, 186
Ontological sense, 297
Ontology, of nature, 287–288
Open-mindedness, 484–488
Operational thought, 273
Operative concept, 54
Oregon Experiment, 75–76
Oregon Healthcare program, 95
Oregon Health Plan, 74, 82–83
Oregon Medicaid Experiment, 444–445
Organic brain syndrome, 281
Osler, William, 78, 163–165
Out-of-network system, 116, 141–142
OxyContin, 168
Oxytocin

behaviors, 339
  chemical structure, 340–341, 340*f*

P

Pain empathy, 309
Pain management programs, 168
Palliative care, 166
Parental paternalism, 266–269
Parkinson's disease, 64, 279
  autonomy, 123–125
  informed consent, 125–126
  initial treatment of, 153
  levodopa, 121–123, 135–136, 237, 279–280
  medical facts, 101–103
  metaphor, 135–139
  obligations, 103–105
  paternalism (*see* Paternalism)
  patient's mind, changing, 132–135
  patients right to refuse any treatment,
      139–141
  patients with, 39
  Rating Scale, 121–122
  reasonable, 129–130
  shareholders, 103–105
  stakeholders, 103–105
  suasion, 131–132
  symptoms, 121
  technical battery *vs.* negligence, 126–127
"Participant reactive attitudes", 337
Paternalism, 123–124
  by disinformation, 130–131, 139
  hard, 130–131
  justification of, 268
  by lack of alternatives, 141–142
  parental limitation, 266–269
  soft, 130–131, 174
Pathoetiological mechanisms, 220–221
Pathology, 152–156, 154*f*
Pathophysiological mechanisms, 220–221
Pathos, 132–135
Patient-centeredness, 444–445
Patient-centric ethics, 248, 286–287, 290,
      440–444
  autonomy, 264
  of everyday medicine, 464–469
Patient-physician/healthcare professional
  asymmetries in, 461–463
  communication, 440
Patient-physician relationship, 90, 461–463

Patient Protection and Affordable Care Act (ACA), 96, 116, 141–142, 468
Patient's mind, changing, 132–135
Patients, workable framework, 479–480
Pattern recognition, 199–200
Pennsylvania Medical Society, 53
Pennsylvania Superior Court Boyer v. Smith, 126–127
Percival and the American Medical Association Code of Ethics of 1847, 104–105
Percival, Thomas, 110, 112–113, 446, 448, 460
Perseverance, 493–494
Personal health records, 159
Personality disorders, 223
Personal moral dilemmas, 306
Person-first language, 448–449
Persuasion, 131–132
Pessimistic Induction, 494
Phenomenological ethics, 373–385
Philosophers, 402
Phronesis, 402
Physical laws, 391
Physician-centric estimation of risk, 176
Physician-centric ethics, 286, 445–450, 455–456
Physicians, 111–114
  concerned, 293
  contracts with, 177
  education and training, 232–233
  extenders, 98–99
  indigenous patient, implications for, 290
  as judges, 58
  liability, 78
  machine learning artificial intelligence, 251–252, 255, 258–260
  office staff, 294
  perspective evolution, 280–281
Physicians News Digest, 126–127
Piaget's balance beam problem, 357
Pick up sticks game, 441–442
Pluralistic modern liberal democracy, 208, 252–253, 298, 320, 347
  acceptable contract in, 205
  consciousness, 367–368
  feminist ethics, 381–382
  justice in, 240
  patient-centricity, 466
  psychological disorders, 222

Pneumonia, 39
Poll taxes, 124
Poor Laws in England, 177
Positive Predictive Value, 64
Positive valence countertransference, 460
Potentiality and actuality, 156–157, 264–265
Practical exercise, 133
Prescription, writing, 85–87
Primary care physicians, 197–198
Primum non nocere, 182–186
Principle of Causational Synonymy, 301, 491
Principle of Double Effect, 69, 105–107, 182–183, 207–208, 319, 325–326
Principle of Harmony, 288
Principle of the Excluded Middle, 372, 406, 418–419
Principle of Transitivity, 403–404
Principlism, 129–130, 275, 336–337, 348, 356–359
  application of, 355
  reproducibility, 359–364
Principlist approach, 42, 87–88, 103
Principlist ethical theory, 66, 372
Principlist formalism, 319
Probability syllogism, 395–396
Problem-based learning (PBL), 209–211
Product liability, 259
Prudential-ought, 287
Psychiatric disorder, 195–196, 217–218, 220–221
Psychogenic disabilities, 225
Psychogenic disorder, 193, 218, 223–224, 226, 228–229, 232
Psychogenicity, 195–196, 199
Psychogenic nonepileptiform seizures, 195–196
Psychological constructs, 305, 482
Psychological disorder, 195–196, 220–222
  challenge for, 220–221
  equitable and inequitable decisions, 225
  ethical obligations, 222
  neurometabolic imaging and, 225
  treatment, 222–224
Psychological problems, 209–210
Psychology, *vs.* neurophysiology, 303–305

Q

Qualitative Brownian motion effect, 415
Qualitative research for ethics, 410–417
Quine, Willard Van Orman, 282

R

Racial development, 314–315
Rationalist/Allopathic medicine, 154*f*,
    165–167, 220–221, 290
Rationalist/Allopathic physician, 166–167,
    220–221, 252–254, 290
Rationalist/Allopathic Principlist
    approach, 442
Rationalist/Allopathic/Regular school, 349
Rationalist/Allopathic system of medicine,
    114–115
Rationalist/Allopathic tradition, 153
Rawls, John, 398, 423, 425–426, 429–430,
    466
Rawls' veil of ignorance, 211–213
Reaction times (RTs), 306–307
Reasonableness, 495
Received knowledge, 298
Reciprocity, 288
Reductionism, 114–115, 365–366
Reductionist medicine, 166–168
Referral, 91
Reflective equilibrium, 129–130, 173, 393,
    425, 428
Regulated Health Professions Act
    (RHPA), 252
Reimbursement
    fee-for-service, 177
    of hospitals, 178
    for surgery, 162–163, 167
Reliabilist approach, 483
Reproducibility, 298, 350–351
    biomedical research, 361–362
    justice, 359–364
    one-to-one correspondence, 370
    in science and medicine, 361
Revulsion, 265
Rhetoric, 131
Right to healthcare, 156–157
Right to purchase insurance for healthcare,
    156–157
Right to refuse any treatment, 139–141
Romanticism and misplaced virtue,
    459–461
Ropinirole, 121–122
Rubin vase illusion. *See* Face-vase illusion
Russell, Bertrand, 394

S

Sagan, Carl, 458
Salgo v. Stanford, 127–128
Saskatoon Agreement in 1962, 92, 450
Schafe, Arthur, 497–498
Schizophrenia, 220
Scientific communication, 211
Scientific medicine, 153
Scientific Method, 480–481
Scientific paternalism, 456–459
Scientism, 153, 211
Scientist-centric ethics, 241–243, 456–459
Second Law of Thermodynamics, 430
Selective serotonin reuptake inhibitor, 193
Selegiline, 86–88
Self-agency, 196
Self-determination, 262–264
Self-interest, 266
Self-organization, 332, 431–432, 431*f*
Self-serving sense, 165
Sense of fatalism, 281
Sense of self, 196–197, 252–254
Sensitivity
    AI/ML SaMD, 257
    child to assent or dissent, 274
Settling out of court, 58–60
Shades of gray, 273–274
Shareholders, 87–88, 103–105
Shem, Samuel, 470
Slippery slope, 124
Smith, Adam, 468
Smith, Jane, 448–449
Social context, 198
Social contract, 444, 446–447
Social hazards, 324
Society, 87
Socrates, 137, 404–405, 409, 494
Soft paternalism, 130–131, 174
Software as a medical device (SaMD),
    255–260
Solipsism, 129–130, 220–221
Somatic symptom disorder, 219
Specificity
    AI/ML SaMD, 257
    child to assent or dissent, 274
Speech-language pathologists (SLPs), 39–40,
    42, 252–253

Spine surgery, surgeon refuses, 294
Stakeholders, 65–66, 87–88, 103–105
Standards of care, 202
Stare decisis, 68
Statistics, 390, 394
Stoicism, 284–285
Stream of consciousness, 196–197, 378–379
Suasion, 131–132
Suboptimal cognitive acts (SCAs), 152–153, 204
Substantia nigra pars compacta, 101, 121
Suicide risk, 247–249
Summa Theologica (1485), 69
Sunk cost effect, 131–132
Sunthesis, 356
Supervening moral theory, 55, 178, 186, 201–202, 205, 239
Supreme Court of Canada, 128
Syllogism, 137–139
Syllogistic deduction, 404
Systems errors, 152, 177–178

T

Technical battery, *vs.* negligence, 126–127
Teleological evolutionary explication, 305
Teleological speculation, 315
Telephone game phenomenon, 415
Telishment, 299, 319
Texas Medical Board, 53–54
The Lancet, 226
"The Organization of Behaviour", 304
A Theory of Justice, 65
Theory of Other Minds, 496–497
Therapeutic privilege, 125
Thomson, Judith, 183
Thrombolytic therapy, 188
Tissue plasminogen activator (TPA), 130–131, 188
Tolstoy, Leo, 148–149, 481
Tonic contraction, 199–200
Torque, physical notion of, 359
Traditional science, 389
Transactional medicine, 455–456
Transcendent principles, 351–356
Transparency, unintentional lack of, 177–182
"Treatise on Human Nature", 134

Two-Eyed Seeing concept, 280–283, 286–287, 290

U

Ulysses pact, 226–227
Unfettered patient-centric ethics, 440–441
United Nations Convention on the Rights of the Child, 267, 274
United States and Canadian medical schools survey, 184, 185*t*
United States v. Rutherford, 442 U.S. 544, 140–141
Unpredictability, 330
Urinary tract infection, 101
US Court of Appeals for the Second Circuit, 240
US Food and Drug Administration (FDA), 140–141, 255–258, 261
US Health Insurance Portability and Accountability Act of 1996, 125
US Health Maintenance Organization Act of 1973, 77
US Institute of Medicine, 113, 204
US Joint Commission, 168
US National Institutes of Health (NIH), 243
US Patient Protection and Affordable Care Act in 2010, 453
US Supreme Court Justice Potter Stewart, 356
Utilitarian moral theory, 321
  aphasia, 252–253
  autonomy, 263, 265
  best interest standard, 275–276
  cervical dystonia, 194–195
  consultant's responsibility, 97
  effects on justice, 180–181
  justice, 157
  medical error, 205
  medical record, 163, 166–167
  micro- and macrolevel justices, 239
  microlevel justice, 211–212
  obligations, 54–56
  patient as suspect, 227
  psychological disorders treatment, 223
  reproducibility as justice, 362
  suicide risk, 249
  unalienable right, 271
  video fluoroscopic evaluation of swallowing, 70

V

Valid contract, 125–126
Vices virtue, 202
Video fluoroscopic evaluation of swallowing
(VFES)
   clinical examination, 64–65
   clinicians, ethics of, 78–80
   cost *vs.* benefit, 66–69
   gut reaction, 65–66
   healthcare systems, ethics of, 75–77
   limiting obligation to beneficence, 81–82
   medical facts of matter, 63
   Oregon Health Plan, 82–83
   policing, 80–81
   virtues, 69–74
Virtues, 111–114
   appeal to, 163
   and care ethics, 445
   continuous *vs.* dichotomous character,
418–419
   ethical principle, 165
   ethics, 157–158, 176–177, 187, 207, 289,
383–385
   implications for, 324–329
   induction from microlevel justice, 319–321
   nonequilibrium steady-state experiences,
430
   self-serving sense of, 165
   stoicism, 284–285
   transactional ethic, 289
   video fluoroscopic evaluation of
swallowing, 69–74

Visual Analog Scale, 283
Visual illusions, 307, 308*f*

W

"War on Cancer" metaphor, 136
Warren, Samuel, 447
Wearing-off effects, 237
Welfare, 133
WellPoint, 251
Whewell, William, 401–402
Wiggle room, 117
Wilson's disease, 200–201
Wisdom of Solomon, 47–48
Workable framework
   calling the question, 497–498
   centrality of hypothesis, 480–482
   intellectual virtues, 482–495
      argument 5.1, 490–491
      carefulness, 488–491
      discipline, 493
      iconoclast, 494–495
      imagination, 492
      open-mindedness, 484–488
      perseverance, 493–494
      procedure to actualize the, 495–498
   patients, 479–480

X

X^2 test, 408

Z

Zero-sum proposition, 275

Printed in the United States
By Bookmasters